Free Radicals in Biology

VOLUME VI

Contributors

Alberto Boveris
Enrique Cadenas
Ercole L. Cavalieri
Britton Chance
David G. Cornwell
Richard G. Cutler
Roberto Docampo
Robert W. Egan
Paul H. Gale
B. Kalyanaraman
R. J. Kulmacz

W. E. M. Lands
Lawrence J. Marnett
P. J. Marshall
Silvia N. J. Moreno
Nobuhiro Morisaki
Peter J. O'Brien
John E. Repine
Eleanor G. Rogan
K. Sivarajah
Robert M. Tate

Free Radicals in Biology

Volume VI

Edited by

William A. Pryor

Thomas and David Boyd Professor in the
Departments of Chemistry and Biochemistry
Louisiana State University
Baton Rouge, Louisiana

1984

 Academic Press, Inc.

(Harcourt Brace Jovanovich, Publishers)
ORLANDO   SAN DIEGO   SAN FRANCISCO   NEW YORK   LONDON
TORONTO   MONTREAL   SYDNEY   TOKYO   SÃO PAULO

ACADEMIC PRESS, INC.
Orlando, Florida 32887

United Kingdom Edition published by
ACADEMIC PRESS, INC. (LONDON) LTD.
24/28 Oval Road, London NW1 7DX

Main entry under title:

Free radicals in biology.

    Includes bibliographical references and indexes.
    I. Radicals (Chemistry)--Collected works.
2. Biological chemistry--Collected works. I. Pryor,
William A.
QP527.F73   574.1'9282   75-13080
ISBN 0-12-566506-7 (v.6)

PRINTED IN THE UNITED STATES OF AMERICA

84 85 86 87   9 8 7 6 5 4 3 2 1

Contents

CHAPTER 1   **Prostaglandin Endoperoxide Synthase-
             Catalyzed Oxidation Reactions**
             Paul H. Gale and Robert W. Egan

CHAPTER 2   **Lipid Peroxide Actions in the Regulation
             of Prostaglandin Biosynthesis**
             W. E. M. Lands, R. J. Kulmacz, and P. J. Marshall

CHAPTER 3   Hydroperoxide-Dependent Oxidations during Prostaglandin Biosynthesis
Lawrence J. Marnett

CHAPTER 4   Fatty Acid Paradoxes in the Control of Cell Proliferation: Prostaglandins, Lipid Peroxides, and Cooxidation Reactions
David G. Cornwell and Nobuhiro Morisaki

CHAPTER 8   **Free-Radical Intermediates in the Antiparasitic Action of Drugs and Phagocytic Cells**
Roberto Docampo and Silvia N. J. Moreno

CHAPTER 9   **Multiple Mechanisms of Metabolic Activation of Aromatic Amine Carcinogens**
Peter J. O'Brien

Contributors

Numbers in parentheses indicate the pages on which the authors' contributions begin.

Alberto Boveris (211), Departamento de Química Biológica, Facultad de Farmacia y Bioquímica, 1113 Buenos Aires, Argentina

Enrique Cadenas (211), Institut für Physiologische Chemie I, Universität Düsseldorf, D-4000 Düsseldorf 1, Federal Republic of Germany

Ercole L. Cavalieri (323), Eppley Institute for Research in Cancer and Allied Diseases, University of Nebraska Medical Center, Omaha, Nebraska 68105

Britton Chance (211), Johnson Research Foundation, School of Medicine, University of Pennsylvania, Philadelphia, Pennsylvania 19104

David G. Cornwell (95), Department of Physiological Chemistry, The Ohio State University, Columbus, Ohio 43210

Richard G. Cutler (371), Gerontology Research Center, National Institute on Aging, Baltimore City Hospitals, Baltimore, Maryland 21224

Roberto Docampo (243), Laboratory of Molecular Biophysics, National Institute of Environmental Health Sciences, Research Triangle Park, North Carolina 27709

Robert W. Egan (1), Merck Institute for Therapeutic Research, Rahway, New Jersey 07065

Paul H. Gale (1), Merck Institute for Therapeutic Research, Rahway, New Jersey 07065

B. Kalyanaraman (149), National Biomedical ESR Center, Department of Radiology, Medical College of Wisconsin, Milwaukee, Wisconsin 53226

R. J. Kulmacz (39), Department of Biological Chemistry, University of Illinois at Chicago, Chicago, Illinois 60612

W. E. M. Lands (39), Department of Biological Chemistry, University of Illinois at Chicago, Chicago, Illinois 60612

Lawrence J. Marnett (63), Department of Chemistry, Wayne State University, Detroit, Michigan 48202

P. J. Marshall (39), Department of Biological Chemistry, University of Illinois at Chicago, Chicago, Illinois 60612

Silvia N. J. Moreno (243), Laboratory of Molecular Biophysics, National Institute of Environmental Health Sciences, Research Triangle Park, North Carolina 27709

Nobuhiro Morisaki[1] (95), Department of Physiological Chemistry, The Ohio State University, Columbus, Ohio 43210

Peter J. O'Brien (289), Department of Biochemistry, Memorial University of Newfoundland, St. John's, Newfoundland A1B 3X9, Canada

John E. Repine (199), Webb-Waring Lung Institute and the Departments of Medicine and Pediatrics, University of Colorado Health Sciences Center, Denver, Colorado 80262

Eleanor G. Rogan (323), Eppley Institute for Research in Cancer and Allied Diseases, University of Nebraska Medical Center, Omaha, Nebraska 68105

[1]Present address: Second Department of Internal Medicine, Chiba University, Chiba, Japan.

K. Sivarajah[2] (149), Laboratory of Pulmonary Function and Toxicology, National Institute of Environmental Health Sciences, Research Triangle Park, North Carolina 27709

Robert M. Tate[3] (199), Webb-Waring Lung Institute and the Departments of Medicine and Pediatrics, University of Colorado Health Sciences Center, Denver, Colorado 80262

[2]Present address: Division of Environmental Health, Pennsylvania Department of Health, Harrisburg, Pennsylvania 17120.
[3]Present address: Pulmonary Division, Department of Medicine, Denver General Hospital, Denver, Colorado 80204.

General Preface

This multivolume treatise had its genesis in April, 1970, when a number of chemists and biologists interested in free radical biology met in Atlantic City at the President's Symposium of the American Society for Experimental Pathology [*Federation Proceedings* **32,** 1859–1908 (1973)]. In a discussion following the meeting, the speakers all agreed that no adequate textbook or monograph existed in the fascinating and diverse field of free radical biology. This lack is felt both by workers in the field who wish up-to-date reviews and by biologists and physicians who are not working in the field but who wish to learn of recent developments.

The areas included under the general rubric of free radical biology are so varied that no single author could possibly have expertise in all of them. For example, relevant topics include the organic and physical-organic chemistry of free radicals; the various reactions of oxygen, including oxygen toxicity, acute respiratory distress syndrome, autooxidation, reactions of the superoxide radical, and reactions of singlet oxygen; the chemistry of antioxidants, including vitamin E; the chemistry of polyunsaturated fatty acids and their role in membrane chemistry and physics; photochemistry, photobiology, and radiation biology; oxidases, hydroxylating enzymes, and detoxification systems; radical involvement in ischemia and reprofusion injury; heat shock; light-induced damage to the eye; electron-spin resonance (ESR) studies of enzymes and substrates, spin-label studies, and ESR studies of tissue samples; the toxicity of chlorinated hydrocarbons; the chemistry and biochemistry of smog; the chemistry and toxicity of cigarette smoke and other organic combustion products; emphysema; prostaglandin chemistry and biology; one-electron oxidation of polycyclic aromatic hydrocarbons and other compounds to carcinogens; tumor promotion; quinone drugs and electron-transfer reactions; and finally, the role of free radicals in aging.

In view of the need for an up-to-date review of free radical biology and the enormous diversity of the areas involved, the participants in the 1970 FASEB meeting agreed that a series of monographs was needed. It has been my pleasure and privilege to serve as editor of these volumes.

I have asked the authors involved in these series to write both for novices and for specialists. I wanted chapters that would not only serve as a first place to look for an introduction to a field, but also as up-to-date reviews for experts. This has proved to be a difficult task. In some cases the subject matter could easily be presented at an elementary level; in others, however, the very nature of the material dictated a more detailed and advanced review. I hope, nonetheless, that most of the chapters in these volumes are at a level that allows them to serve both as a brief introduction to each area and also as an up-to-date survey of each topic.

It seems particularly appropriate that the first of these volumes was published on the two-hundredth anniversary of the discovery of oxygen by Joseph Priestley. Certainly the necessity of organisms tolerating oxygen in their energy-producing systems gives rise to many of the problems and interesting topics in this field. Had glycolysis, or some similar anaerobic process, never been replaced with respiration, organisms would not have had to learn to protect themselves against the oxidative threat that oxygen presents. Also, oxygen appears to be particularly susceptible to one- as well as two-electron reactions, and thus is responsible for producing some of the one-electron intermediates found in the cell.

I hope that these volumes, which bring together many of the diverse subjects in free radical biology, will make these topics accessible to chemists, biologists, and physicians. I also hope that the reader will agree that this is a fascinating, sometimes controversial, and important field.

William A. Pryor

Preface to Volume VI

This volume of *Free Radicals in Biology* differs from the previous volumes in several ways. First, readers will notice considerably more emphasis on biology in this volume and a clinical orientation in some of the chapters. Dramatic and quite breathtaking progress has been made in this field in recent years, with new developments in prostaglandins, superoxide and superoxide dismutases, and in the study of the oxidative reactions of xenobiotics tumbling over one another. This rich profusion of exciting advances in the current literature has led to new theories that are reflected in many of these chapters. Another departure in this volume is the grouping of chapters in a single area of free radical biology: Six of the chapters treat the free radical chemistry of the arachidonic acid cascade and the biochemistry of the prostaglandins, leukotrienes, and other products from arachidonic acid. The enormous biological significance of the products of the arachidonic acid cascade has become apparent in the last decade. However, the impact of the arachidonic acid cascade on free radical processes in the cell and, contrariwise, the impact of free radical processes on the arachidonic acid cascade (such as feedback control of the PGI/TXB ratio) has not been reviewed in a single volume until now. This sixth volume of *Free Radicals in Biology*, therefore, brings together a number of experts in different aspects of arachidonic acid chemistry to explore the important implications of free radical biology in this field.

Chapter 1, by Paul Gale and Robert Egan, introduces readers to the chemistry of the eicosanoids and gives the structures of all of the prostaglandin and leukotriene compounds that are referred to in this volume. This list of names, acronyms, and structures serves as a key to all of the chapters on arachidonic acid. Gale and Egan, who have made important contributions to this field, go on to treat the PGG-catalyzed oxidation of organic compounds, developing the theme that the redox chemistry of the PG synthase enzymes has profound effects on xenobiotic metabolism.

Chapter 2, by W. Lands, R. Kulmacz, and P. Marshall, stresses the role of lipid hydroperoxides in controlling prostaglandin biosynthesis. At submicromolar concentrations, lipid hydroperoxides activate PGH synthase, and this raises peroxide concentrations to levels at which peroxidases become effective in destroying the peroxides. At higher than micromolar levels, hydroperoxides inactivate PG synthesis by destroying the enzymes involved in the biosynthetic reactions.

Lawrence Marnett discovered in the 1970s that xenobiotics are oxidized during PGH biosynthesis, and in Chapter 3 he discusses the implications of this area, stressing developments in his own laboratory. In Chapter 4, David Cornwell and Nobuhiro Morisaki discuss what they call "fatty acid paradoxes," the contradictory effects that fatty acids at different concentration levels often have on cell proliferation, tumorigenesis, and metastasis. Intervention strategies for the control of cell proliferation can produce either positive or negative signals and can lead to unanticipated results in biological systems.

In Chapter 5, B. Kalyanaraman and K. Sivarajah review the study of the arachidonic acid cascade process and the oxidation of xenobiotics using electron spin resonance, a field that has developed more than most readers are aware. There is now an enormous body of ESR-related data that bear on mechanisms in the arachidonic acid cascade.

In Chapter 6, Robert Tate and John Repine discuss the causes of lung injury in conditions such as hyperoxia. Their hypothesis is that the excessive accumulation of phagocytes in lung tissue is the cause of the damage, and the possible role of alveolar macrophages and neutrophils in disorders such as pulmonary oxygen toxicity and the adult respiratory distress syndrome is reviewed.

In Chapter 7, Enrique Cadenas, Alberto Boveris, and Britton Chance review the origin of low-level chemiluminescence in cells. The systems reviewed include model systems and both *in vitro* and *in vivo* exposed organs. The data are analyzed to determine the mechanisms for light emission and the excited species that are involved.

Roberto Docampo and Silvia Moreno review oxy-radical involvement in parasitic diseases in Chapter 8. Parasitic diseases are among the most widespread of human diseases, currently affecting about three billion persons per year. In recent years it has been apparent that many of the antiparasitic compounds produce free radicals; the sensitivity of parasites to these radicals often depends on a deficiency in enzymes that moderate against oxy-radical damage in the parasite.

In Chapter 9, Peter O'Brien reviews mechanisms for activation of aromatic amine carcinogens. Multiple mechanisms of activation, both ionic and free radical-mediated, are proposed, depending on the target organ.

Ercole Cavalieri and Eleanor Rogan review in Chapter 10 one- and two-electron mechanisms for activation of polynuclear hydrocarbon (PAH) carcinogens. This is a controversial and somewhat clouded field, and these authors outline both types of mechanisms and attempt to put the one-electron pathway in perspective. The one-electron pathway, which produces a radical cation as the first intermediate, applies to PAH with low ionization potentials (below 7.35 eV) and is catalyzed by P-450 and peroxidases (including prostaglandin synthase). The two-electron oxidation is catalyzed by similar enzyme systems and often produces arene oxides and vicinal diol-epoxides as the ultimate carcinogens.

In Chapter 11, Richard Cutler presents a hypothesis to rationalize the effects of radicals on the life span of mammals. He proposes that aging results from toxic by-products of metabolism, and longevity is determined by the ability of an organism to deal with these products. The levels of antioxidants, peroxides, and enzymes (particularly guanylate cyclase and cyclooxygenase) in various species are correlated with life span data.

The variety of these chapters and the clear importance of many of the topics make this an exciting volume. I hope you enjoy your voyage through it.

William A. Pryor

List of Abbreviations

AA	Arachidonic acid
AAF	2-Acetylaminofluorene
ADP	Adenosine diphosphate
AIA	2-Alkyl-2-isopropylacetamide
ANFT	2-Amino-4-(5-nitro-2-furyl)arylamide
AP	Aminopyrine
ARDS	Adult respiratory distress syndrome
B[a]A	Benz[a]anthracene
B[a]P	Benzo[a]pyrene
B[a]P-7,8-diol	7,8-Dihydroxy-7,8-dihydrobenzo[a]pyrene
BHA	Butylated hydroxyanisole
BHT	Butylated hydroxytoluene
BW 755C	3-Amino-1-[m-(trifluoromethyl)phenyl]-2-pyrazoline
Cyclic AMP (cAMP)	Cyclic adenosine monophosphate
Cyclic GMP	Cyclic guanosine monophosphate
DAA	Diaminoanisole
DAB	N,N-Dimethyl-4-aminoazobenzene
DB[ah]A	Dibenz[ah]anthracene
DBB	Dibenzoylbenzene
DB[ah]P	Dibenzo[ah]pyrene
DB[ai]P	Dibenzo[ai]pyrene
DDC	Diethyl dithiocarbamate
DDEP	3,5-Bis(ethoxycarbonyl)-4-ethyl-2,6-dimethyl-1,4-dihydropyridine
DEAE-cellulose	Diethylaminoethylcellulose
DES	Diethylstilbestrol
DIES	Dienestrol
anti-Diol-epoxide	7β,8α-Dihydroxy-9α,10α-epoxy-7,8,9,10-tetrahydrobenzo[a]pyrene
syn-Diol-epoxide	7β,8α-Dihydroxy-9β,10β-epoxy-7,8,9,10-tetrahydrobenzo[a]pyrene

DMBA	7,12-Dimethylbenz[a]anthracene
DMPO	5,5-Dimethyl-1-pyrroline N-oxide
DMSO	Dimethyl sulfoxide
DPBF	Diphenylisobenzofuran
DPPH	2,2-Diphenyl-1-picrylhydrazyl
ED_{50}	50% maximally inhibitory dose
EFA	Essential fatty acid(s)
ESR	Electron spin resonance
ETYA	Eicosatetraynoic acid
$Eu(hfc)_3$	Tris[3-(heptafluoropropylhydroxymethylene)-d-camphorato]europium (III)
FANFT	N-[4-(5-nitro-2-furyl)-2-thiazolyl]formamide
FBS	Fetal bovine serum
FFA	Free fatty acid(s)
G	Gauss
GAC	Guanylate cyclase–arachidonate–cyclooxygenase
GPX	Glutathione peroxidase
GSH	Glutathione
$H_2B[a]P$	7,8-Dihydroxy-7,8-dihydrobenzo[a]pyrene
HCHO	Formaldehyde
HETE	Hydroxyeicosatetraenoic acid
15-HPE_1	15-Hydroperoxyprostaglandin E_1
HPETE	Hydroperoxyeicosatetraenoic acid
HPLC	High-pressure liquid chromatography
HRP	Horseradish peroxidase
IC_{50}	50% maximally inhibitory dose
IP	Ionization potential(s)
K_I	Inhibition constant
K_M	Michaelis–Menten constant
LES	Life span energy potential
LPS	Life span potential
LT	Leukotriene
MAB	Methyl-4-aminoazobenzene
MAM	methylazoxymethanol
MC	3-methylcholanthrene
MDA	Malondialdehyde (or malonaldehyde)
MK-477	2-Aminomethyl-4-tert-butyl-6-iodophenol
MNP	2-methyl-2-nitrosopropane
MP	Methylphenazine
MPO	Myeloperoxidase
MPS	Methyl phenyl sulfide
MPSO	Methyl phenyl sulfoxide

NA	Naphthylamine
NMR	Nuclear magnetic resonance
N-OH-AF	*N*-Hydroxyaminofluorene
$[O_X]$ $([O_X]•)$	Oxidizing species
PAH	Polycyclic aromatic hydrocarbons
PB	Phenylbutazone
PBN	α-Phenyl-*tert*-butylnitrone
PES[a]	Prostaglandin endoperoxide synthase
PG	Prostaglandin
PGHS[a]	Prostaglandin H synthase
PMA[b]	Phorbol myristate acetate
4-POBN	α-(4-pyridyl 1-oxide)-*N-tert*-butylnitrone
RBL-1	Rat basophilic leukemia cells
ROOH	Lipid hydroperoxide
RSV	Ram seminal vesicle
RSVM	Ram seminal vesicle microsomes
SMR	Specific metabolic rate
SOD	Superoxide dismutase
TBA	Thiobarbituric acid
TBARM	Thiobarbituric acid-reacting material
TMB	Tetramethylbenzidine
TMH	Tetramethylhydrazine
TPA[b]	12-*O*-Tetradecanoylphorbol 13-acetate
TPH	Tetraphenylhydrazine
TXA_2	Thromboxane A_2
TXB_2	Thromboxane B_2
U	Unit
18:2 (n-6)	Linoleic acid; example of the numbering system used for fatty acids; numbers separated by a colon indicate number of carbon atoms and number of double bonds, respectively; position of first double bond from methyl terminus of acyl chain given in parentheses

[a]These two are synonymous, different descriptions of the same enzyme systems.
[b]These are the systematic and common nomenclature names for the same molecule.

Contents of Other Volumes

CHAPTER **1**

Prostaglandin Endoperoxide Synthase-Catalyzed Oxidation Reactions

Paul H. Gale and Robert W. Egan

Merck Institute for Therapeutic Research
Rahway, New Jersey

I. INTRODUCTION

Prostaglandins, thromboxanes, and leukotrienes constitute a family of compounds derived from polyunsaturated eicosanoic acids such as arachidonic acid. These compounds are collectively referred to as eicosanoids. Many are biologically active, and the free-radical mechanisms involved in their syntheses can lead to the cooxidation of xenobiotics. Because this volume has a number of chapters on eicosanoids, this chapter includes an overview of their structures, their biosynthetic pathways, and their pathophysiological relevance. After the introduction, we critically review the irreversible oxidative deactivation of certain enzymes in the arachidonic acid cascade and the organic oxidation reactions catalyzed by the hydroperoxidase aspects of these enzymes. Both of these phenomena are related to the control of eicosanoid biosynthesis and are the result of oxidative species formed enzymatically by prostaglandin endoperoxide synthase and other hydroperoxidases.

A. Nomenclature and Structures

The field of prostaglandin research began with the independent discoveries by von Euler [1] and Goldblatt [2] that substances from human seminal plasma and sheep vesicular glands can lower blood pressure and contract smooth muscle. Von Euler also found a small amount of the active material in prostate gland tissue and named it prostaglandin [3]. Since the isolation and identification of a natural prostaglandin in 1962 [4], there has been an explosion of activity in all areas of this research. By the use of radioactive substrate, it was discovered in 1964 that arachidonic acid is the precursor of prostaglandin E_2 (PGE_2) synthesis [5, 6] (see Table I for structures). This knowledge led to the proposal that biosynthesis proceeded through endoperoxide intermediates [7–9], which were isolated, characterized, and named PGG_2 and PGH_2 in 1973 [10–12].

In addition to their role as intermediates in the synthesis of the classical prostaglandins PGE_2, PGD_2, and $PGF_{2\alpha}$, these endoperoxides can be metabolized to two other biologically important substances. It was reported in 1975 that brief incubations of washed human platelets with thrombin produced an intermediate named thromboxane A_2 (TXA_2, Table I) because of its oxetane ring structure and extreme potency in aggregating platelets [13, 14]. The oxetane ring of this very unstable compound is rapidly hydrolyzed to a pyranyl, forming TXB_2, a more stable product that had been isolated and identified from similar incubation mixtures during the preceding year [15].

Prostacyclin (PGI_2) is the most recently identified member of the family of compounds derived from PGH_2. Discovered in 1976 during a study of arachidonic acid metabolism in isolated pig aorta microsomes [16], its major biological properties are in opposition to those of TXA_2. The structure of PGI_2 (Table I) was determined through a collaboration between the Wellcome and Upjohn Laboratories [17] and is closely related to Δ^7-6(9)-oxy-$PGF_{1\alpha}$ discovered earlier by Pace-Asciak [18]. Like TXA_2, PGI_2 is chemically unstable and degrades rapidly in aqueous media to 6-keto-$PGF_{1\alpha}$. The general structure of each of these prostaglandins is a central ring system with two side chains. The carboxyl chain contains a 5,6-cis-double bond and the ω-chain has a 13,14-trans-double bond

TABLE I   Structures of Eicosanoids

Name	Abbreviation	Structure
Arachidonic acid (5,8,11,14-eicosatetraenoic acid)	AA	
Prostaglandin G_2	PGG_2	
Prostaglandin H_2	PGH_2	
Prostaglandin E_2	PGE_2	
Prostaglandin D_2	PGD_2	
Prostaglandin $F_{2\alpha}$	$PGF_{2\alpha}$	
Thromboxane A_2	TXA_2	
Thromboxane B_2	TXB_2	
Prostaglandin I_2	PGI_2	

(continued)

TABLE I (*Continued*)

Name	Abbreviation	Structure
6-Ketoprostaglandin $F_{1\alpha}$	6-Keto-PGF$_{1\alpha}$	
5-Hydroperoxyeicosatetraenoic acid	5-HPETE	
5-Hydroxyeicosatetraenoic acid	5-HETE	
Leukotriene A_4	LTA$_4$	
Leukotriene B_4	LTB$_4$	
Leukotriene C_4	LTC$_4$	
Leukotriene D_4	LTD$_4$	
Leukotriene E_4	LTE$_4$	

and a hydroxyl group at C-15. Clearly, the biological specificity of these compounds resides in the central ring systems, which are joined stereospecifically to these chains by chiral centers at C-8 and C-12.

Along with the discovery of TXA_2, it was established that platelets contain a lipoxygenase, the first demonstration of such an enzyme in mammalian systems [15]. It has subsequently been shown that slow-reacting substance of anaphylaxis, (SRS-A), an important mediator of asthma [19], is synthesized from arachidonic acid by enzymes other than endoperoxide synthase [20]. In 1979 these discoveries culminated in the delineation of the chemical structures of slow-reacting substance of anaphylaxis as conjugates of arachidonic acid and peptides. These were named leukotrienes because they contain a conjugated triene and are synthesized by leukocytes [21, 22].

Structures of this group of compounds are also shown in Table I. The initial intermediate in the synthesis of leukotrienes is 5-hydroperoxyeicosatetraenoic acid (5-HPETE), a hydroperoxide with conjugated double bonds at C-6 and C-8. Both 5-HPETE and its reduction product, 5-hydroxyeicosatetraenoic acid (5-HETE), are S-enantiomers at the chiral center with absorbance maxima at 235 nm. Leukotriene A_4 (LTA_4), another unstable intermediate in this pathway, contains a (5S,6R)-epoxide and a 7E,9E,11Z,14Z-double-bond arrangement. Physiologically active leukotrienes are of two structure classes: LTB_4, with (5S,12R)-hydroxyl groups, and sulfidopeptide leukotrienes (LTC_4, LTD_4, and LTE_4), with a sulfide linkage to a peptide at the 6R-position of arachidonic acid. The double bonds in sulfidopeptide leukotrienes are arranged as in LTA_4, whereas LTB_4 is 6Z,8E,10E,14Z. Sulfidopeptide leukotrienes and LTA_4 absorb strongly at 280 nm ($\epsilon = 37,000$), whereas LTB_4 absorbs maximally at 270 nm.

B. Biosynthetic Pathways

The *arachidonic acid cascade* describes collectively the many reactions by which this fatty acid is enzymatically transformed into a multitude of physiologically active products (Fig. 1). These biosynthetic sequences, many of which involve several steps, can be classified according to the type of enzyme catalyzing the primary oxygen addition reaction. One category, initiated by prostaglandin endoperoxide synthase (EC 1.14.99.1), leads through bisdioxygenation to the endoperoxides PGG_2 and PGH_2, which are isomerized to prostaglandins and related products. The other class comprises a group of lipoxygenases that add a molecule of oxygen to various positions on the arachidonic acid molecule to form hydroperoxides. Like the prostaglandin endoperoxides, these unstable molecules can be transformed into biologically active substances such as leukotrienes. Although a number of other polyunsaturated fatty acids can be used by these dioxygenase enzymes, arachidonic acid is of most significance because it is, by far, the most abundant natural substrate.

Arachidonic acid occurs primarily as a constituent of cell membrane phos-

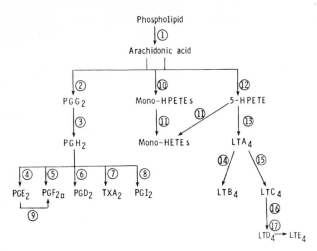

Fig. 1.   Enzymology of arachidonic acid metabolism. Structures and abbreviations for the compounds are listed in Table I. The enzymes catalyzing these transformations are as follows: (1) phospholipase A$_2$ or phospholipase C and diacylglycerol lipase, (2) endoperoxide synthase-oxygenase, (3) endoperoxide synthase-hydroperoxidase, (4) endoperoxide-PGE$_2$ isomerase, (5) PGH$_2$ reductase or nonenzymatic, (6) endoperoxide-PGD$_2$ isomerase, (7) thromboxane synthase, (8) PGI$_2$ synthase, (9) 9-ketoreductase, (10) 8-, 9-, 11-, 12-, or 15-lipoxygenase, (11) hydroperoxidases, (12) 5-lipoxygenase, (13) LTA$_4$ synthase-dehydrase, (14) epoxide hydrolase, (15) glutathione transferase, (16) γ-glutamyl transpeptidase, and (17) aminopeptidase.

pholipids and must be released before it can participate as the free acid in subsequent oxygenation reactions. Because arachidonic acid is bound to the central carbon of the glycerol in phospholipids, a phospholipase A$_2$ can catalyze its release. These enzymes are controlled by peptide regulators [*23, 24*], the synthesis of which is inhibited by glucocorticoids. Alternatively, a phospholipase C can specifically catalyze the formation of diacylglycerol from phosphatidylinositol containing stearic and arachidonic acids associated with the first and second carbons, respectively [*25*]. This diacylglycerol can be subsequently cleaved to release arachidonic acid either directly by a lipase [*26, 27*] or indirectly by a kinase-catalyzed rephosphorylation to phosphatidic acid, which activates a latent phospholipase A$_2$ [*28*].

Prostaglandin endoperoxide synthase (Fig. 1), also called cyclooxygenase, incorporates a molecule of dissolved oxygen into arachidonic acid to form an endoperoxide bridge between C-9 and C-11. Subsequent stereospecific rearrangement leads to a bicyclic structure with a bond between C-8 and C-12. A second molecule of oxygen is then added, generating a hydroperoxide at C-15 (PGG$_2$) [*29, 30*]. The peroxidase component of the same bifunctional enzyme [*31*] reduces this hydroperoxide to PGH$_2$, a key intermediate, which is the substrate for enzymes that form biologically active prostaglandins [*10, 11*].

Glutathione-dependent isomerases convert PGH$_2$ to PGE$_2$ [*32*] and PGD$_2$ [*33,*

34]. Although these autocoids are relatively resistant to nonenzymatic degradation, they can be enzymatically metabolized to less active substances by de-
hydrogenases and ketoreductases [*35*]. Prostaglandin $F_{2\alpha}$ can be formed either
by direct reduction of PGH_2 or by 9-ketoreductase-catalyzed PGE_2 metabolism
[*36*]. Catalyzed by thromboxane and PGI_2 synthases, PGH_2 can be isomerized to
TXA_2 [*13*] and PGI_2 [*16*], respectively. However, as opposed to PGE_2, PGD_2,
and $PGF_{2\alpha}$, these compounds are rapidly and nonenzymatically deactivated by
hydration to TXB_2 and 6-keto-$PGF_{1\alpha}$. Finally, PGH_2 can be cleaved either
enzymatically by thromboxane synthase [*37*] or chemically into malondialdehyde and 12-hydroxy-5,8,10-heptadecatrienoic acid. Not every cell is
competent to synthesize each prostaglandin, the distribution of products being
established by the complement of enzymes in a particular cell.

Lipoxygenases, which catalyze the incorporation of only a single molecule of
oxygen into arachidonic acid to form a hydroperoxide, are widely distributed in
mammalian tissues. These lipoxygenases and hydroperoxides are named according to the carbon atom to which the oxygen is added, giving a series of mono-
HPETEs from 5-, 8-, 9-, 11-, 12-, and 15-lipoxygenases. Hydroperoxidases
rapidly reduce these hydroperoxides to the corresponding alcohols (mono-
HETEs). Although further transformations of these mono-HETEs are known, the
biological significance of most of these compounds has not been established
[*38*]. In contrast, 5-HPETE, which is formed by the 5-lipoxygenase, is the
precursor of leukotrienes [*39*]. Catalyzed by a dehydrase, 5-HPETE is metabolized to LTA_4, an unstable 5,6-epoxide, which undergoes enzymatic transformations by an epoxide hydrolase to LTB_4 or by a glutathione transferase to LTC_4
[*21, 22, 40*]. In certain cells, LTC_4 is metabolized to LTD_4 by a γ-glutamyl
transpeptidase [*41*] or subsequently to LTE_4 by an aminopeptidase [*42*].

Although these lipoxygenase and endoperoxide synthase pathways ultimately
convert arachidonic acid to unique molecules, these differences arise primarily
from variations in the enzymology after the addition of oxygen. Despite adding
at different carbons on the arachidonic acid backbone, each enzyme incorporates
molecular oxygen by the same mechanism [*29, 43*] (Fig. 2). In each case the

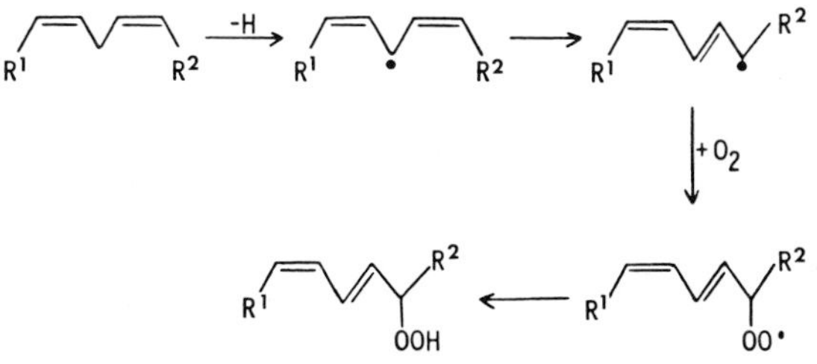

Fig. 2. Mechanism of lipoxygenase reactions. The R^1 and R^2 are the remaining hydrocarbon
chains of the fatty acid.

primary step is abstraction of a hydrogen atom from one of the doubly allylic carbons at the 7, 10, or 13 position [44]. Either adjacent double bond can then shift into conjugation and transfer the electron density by two carbons, providing six possible sites of molecular oxygen addition. A molecule of oxygen adds at the electron-rich site, followed by reduction to the hydroperoxide. After hydrogen atom abstraction, this reaction can proceed spontaneously at a diffusion-controlled rate. However, these lipoxygenases stereospecifically abstract the hydrogen and retain the fatty acid radical throughout the reaction to form (S)-mono-HETEs [44].

C. Physiological Actions

Although the actions of prostaglandins and leukotrienes in normal physiology and in pathophysiology have been studied in detail, their extensive and frequently antagonistic actions preclude the identification of a specific role for these agents. Although far from complete, Table II lists some of their important actions and emphasizes the significant effects of these eicosanoids on smooth muscle tone, cyclic nucleotide levels, platelet aggregation, and cell migration. The mutually antagonistic actions of sets of prostaglandins such as PGE_2 and $PGF_{2\alpha}$ [45] and, more recently, PGI_2 and TXA_2 [46] make control of biosynthesis at a level subsequent to endoperoxide synthase extremely important. Most inhibitors such as nonsteroidal antiinflammatory agents block this primary enzyme and thereby depress all prostaglandins equally [47–49]. An exception is aspirin, which inhibits irreversibly by acetylation [50] with a preference for platelets, where protein resynthesis cannot occur and TXA_2 is the major product. Platelets therefore, cannot form prostaglandin products for the remainder of their circulating lifetime, and TXA_2 synthesis is inhibited in preference to PGI_2 synthesis [51]. Inhibitors of thromboxane synthase are also available [52, 53], and hydroperoxides inhibit PGI_2 synthase in vitro [54, 55]. Like endoperoxide synthase inhibitors, these more selective agents will help define the physiological role of eicosanoids.

Because both prostaglandins and leukotrienes are rapidly metabolized, their effects are expressed locally, as exemplified by the homeostatic control exerted by TXA_2 and PGI_2. Thromboxane A_2, a potent aggregating agent, is formed by platelets, whereas PGI_2, an equally effective inhibitor of platelet aggregation, is formed by the endothelium [56]. Hence, a break in the endothelial layer results in a local decrease in PGI_2 concentration and increased TXA_2-induced platelet aggregation to form a clot and seal the lesion. Even the more stable prostaglandins do not circulate to any degree because they are rapidly metabolized in the lung and kidney [57–59], rendering local or even intracellular effects most relevant.

TABLE II   Physiological Actions of Arachidonic Acid Metabolites

Eicosanoid	Effect	References
PGE_2	Vasodilation	62
	Raise cAMP levels	63
	Decrease gastric acid secretion	64–66
	Pain sensitization	67
PGI_2	Relax smooth muscle	68
	Vasodilation	69
	Inhibit platelet aggregation	46,69
	Raise cAMP levels	70,71
	Decrease gastric acid secretion	66,91
	Renin release	72
TXA_2	Contract smooth muscle	46,73
	Cause platelet aggregation	74
	Bronchoconstriction	75
PGD_2	Inhibit platelet aggregation	76,77
	Raise cAMP	78
LTC_4–LTD_4	Contract smooth muscle	39,79
	Constrict peripheral airways	80,81
	Leakage in microcirculation	82
	Decrease cAMP levels	83
LTB_4	Neutrophil and eosinophil chemotaxis	84,85
	Leakage in microcirculation	82,86
	Raise cAMP levels	87
	Cause neutrophil aggregation	88
12-HETE–12-HPETE	Neutrophil chemotaxis	89
	Stimulate insulin secretion	90

A host of tissues are influenced by eicosanoids as a result of their capacity to mediate vascular tone and alter cAMP concentrations. These effects are noted on the reproductive, cardiovascular, respiratory, renal, gastrointestinal, endocrine, and skin as well as the hematological systems. As a result of these actions, imbalances in prostaglandin or leukotriene levels have been implicated in a variety of disorders including inflammation, pain, ulcers, atherosclerosis, renal failure, asthma, psoriasis, and gout. In addition, $PGF_{2\alpha}$ is used as a regulator of the reproductive cycle, and PGI_2 is being tested as a replacement for heparin in hemodialysis [60]. In most of these disorders, contractile and aggregatory prostaglandins would be considered causative, whereas those that relax smooth muscle would be palliative. Total inhibition of prostaglandin biosynthesis may therefore not be beneficial, depending on the relative importance of PGI_2 or TXA_2. In contrast, inhibition of the synthesis of sulfidopeptide leukotrienes or antagonism of their effects should be generally beneficial, especially in asthma, where sulfidopeptide leukotrienes play a significant role in the human disease [61].

II. HYDROPEROXIDASE ACTIVITY OF ENDOPEROXIDE SYNTHASE

A. Hydroperoxide Metabolism

Metabolism of the hydroperoxide PGG_2, the natural substrate for endoperoxide synthase, follows the general reaction scheme for this enzyme,

$$ROOH \rightarrow ROH + [O_x]$$

and represents the enzymatic reduction of a lipid hydroperoxide, ROOH, to its corresponding alcohol, ROH [10, 11], with the concurrent formation of oxidizing equivalents, $[O_x]$ [92, 93]. Shown in the upper panels of Fig. 3 are the results of experiments in which microsomal endoperoxide synthase was incubated with PGG_2, 15-hydroperoxy-PGE_1 (15-HPE_1), or 15-HPETE to yield their respective alcohols, PGH_2, PGE_1, or 15-HETE [94]. These reactions were monitored by incubating microsomes with ^{14}C-labeled substrate for about 30 sec, then extracting the products and remaining substrate into chilled ether, chromatographing in an appropriate solvent system, and observing the pattern of radioactivity either with a radiochromatogram scanner or by scintillation counting. The identities of the products were established by comparing their chromatographic mobilities with authentic standards.

Panel A (Fig. 3) shows the metabolism of PGG_2 to PGH_2. Although more PGH_2 would have been formed had more enzyme been used, conditions were chosen to demonstrate the dramatic stimulation of hydroperoxidase by 500 μM

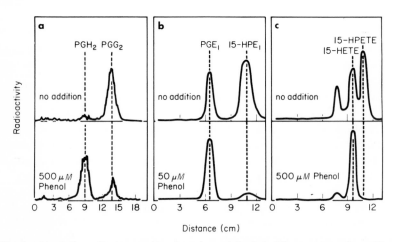

Fig. 3. Metabolism of hydroperoxides. In each panel the dashed lines represent the position of chromatographic standards detected by mass assay. The upper panels indicate reactions with no additive, whereas the lower traces depict the same reactions in the presence of phenol.

phenol from <10 to 75% conversion (lower panel A). Reduction of 15-HPE$_1$ exclusively to PGE$_1$ is shown in panels B, with the stimulatory effect of 50 μM phenol illustrated in the lower trace. Metabolism of 15-HPETE formed primarily 15-HETE plus a more polar unknown material (panel C). The second product was not a peroxide and, because no 15-HETE metabolism was detected, it must have been formed enzymatically from 15-HPETE. In the presence of 500 μM phenol, 15-HPETE was metabolized almost exclusively to 15-HETE.

On the basis of these and other observations [31, 94, 95] it is apparent that this hydroperoxidase can utilize a variety of hydroperoxides but not endoperoxides. The specificity of the enzyme has been studied by comparing the relative rates of metabolism of PGG$_1$, PGG$_2$, 15-HPE$_1$, and 15-HPETE [55] using N-demethylation of aminopyrine, an oxidizable enamine [96, 97], to monitor the rate of oxidant release during enzymatic reduction of the hydroperoxides. Absorbance of the blue cation-radical was measured at 580 nm. Using the time to reach maximum absorbance as an indicator of aminopyrine oxidation, the relative rates are PGG$_1$ > PGG$_2$ > 15-HPE$_1$ > 15-HPETE. Incubations were performed at 4°C, at which temperature the enzyme was still active, because reaction rates were too rapid to monitor conveniently at 25°C.

In a similar evaluation of substrate specificity Yamamoto *et al.* used purified endoperoxide synthase and monitored reactions as the increase in absorbance at 436 nm resulting from oxidation of guaiacol to 3,3'-dimethoxydipheno-4,4'-quinone during reduction of the hydroperoxides [31]. Again, PGG$_1$ was reduced most rapidly and extensively, followed by 15-HPE$_1$ and 15-HPETE. Hydrogen peroxide, *p*-menthane hydroperoxide, cumene hydroperoxide, and *tert*-butyl hydroperoxide were less suitable substrates, having higher K_m values. A similar preference of this hydroperoxidase for lipophilic hydroperoxides compared with hydrogen peroxide was discovered using intensities of the ESR signal generated during incubation of substrates with microsomal enzyme [94].

B. Cosubstrates

As shown in Fig. 3 phenol increases the extent of the hydroperoxidase reaction with all three hydroperoxide substrates. Stimulation results from the reducing potential of phenol, rendering it a cosubstrate for this reaction, in which hydroperoxide (ROOH) reduction is accompanied by incorporation of $[O_x]$ into cosubstrate (A):

$$ROOH + A(\text{reduced}) \rightarrow ROH + A(\text{oxidized})$$

This is the generalized expression for hydroperoxidase reactions catalyzed by endoperoxide synthase, horseradish peroxidase [98], pea seed peroxidase [99], thyroid peroxidase [100], and others [101].

Many organic and inorganic reducing agents serve as cosubstrates for the hydroperoxidase activity of endoperoxide synthase (Table III) [94]. In these experiments the extent of hydroperoxide reduction was measured using the incubation procedures described in Fig. 3. Percent change is the extent of hydroperoxide reduction in the presence of the cosubstrate compared with a control reaction in the absence of additive. Those compounds listed above the dashed line stimulate the reduction of each hydroperoxide, by as much as 600% in the case of phenol with PGG_2.

Several of the cosubstrates such as aminopyrine [96] and methional [102] are known to scavenge oxidizing radicals, and the reaction of methional with hydroxyl radical has been studied in detail [102, 102a]. Although many of these compounds proceed by radical mechanisms giving characteristic intermediates [103, 104], tryptophan catalytically converts the oxidizing equivalents into luminescence [105], and sulindac sulfide undergoes a nonradical mechanism to a sulfoxide (see Section IV,A,1). Nevertheless, each compound has the capacity to be oxidized. In addition to these organic molecules, inorganic reducing agents such as NaI, which can be oxidized to iodine, also serve as cosubstrates [106]. Although many other cosubstrates have been studied [31] and some are described elsewhere in this volume (Lands *et al.,* Chapter 2; Marnett, Chapter 3; and Kalyanaraman and Sivarajah, Chapter 5), this list highlights the plethora of compounds that can be oxidized by endoperoxide synthase-hydroperoxidase.

TABLE III   Stimulation of Hydroperoxide Metabolism by Cosubstrates

Cosubstrate	Cosubstrate concentration (μM)	Percent change[a]		
		PGG_2	15-HPE$_1$	15-HPETE
Phenol	500	+610	+275	+290
Aminopyrine	1000	+570	+307	+150
Diethyl dithiocarbamate	200	+480	+274	+188
Promethazine	100	+525	+264	+179
Sulindac sulfide	100	+525	+257	+178
Lipoic acid	100	+465	+228	+306
Methional	200	+345	+111	+39
Tryptophan	500	+130	+57	+100
Anisole	2000	−10	−5	−27
Salicyclic acid	2000	+10	−5	+9
Glutathione (red)	500	ND[b]	−5	−18
Methionine	500	−20	−7	0
Sulindac	2000	+10	−8	+6
Indomethacin	500	0	−31	−17

[a]Plus indicates stimulation and minus indicates inhibition of the reaction relative to control incubations containing no cosubstrate.

[b]ND, Not determined.

The lack of specificity for reducing cosubstrate is not unexpected, because most peroxidases utilize a variety of reducing agents [98]. The diversity of oxidation pathways indicates that the mechanism is determined primarily by the chemistry of the cosubstrate rather than by the enzyme itself.

In contrast, those compounds listed below the dashed line in Table III either have no significant effect on the hydroperoxidase or are mild inhibitors. Like the active compounds, each hydroperoxide substrate is influenced similarly. Although most of these inactive agents are not antioxidants, reduced glutathione is; the failure of the hydroproxidase to utilize this substance is unexpected and imparts some selectivity to the oxidant generated by the enzyme–hydroperoxide complex. The consistent action of cosubstrates with all three hydroperoxides indicates that a single, nonspecific enzyme rather than a series of specific enzymes is responsible for these reactions, an observation that has been confirmed using purified enzyme [31, 103].

C. Endogenous Radical Formation

Because lipoxygenases employ a free-radical mechanism for adding molecular oxygen to arachidonic acid [29], ESR spectroscopy has been used to detect a radical intermediate [92, 107]. An ESR signal is detected upon mixing ram seminal vesicle microsomes with arachidonic acid and then freezing in liquid nitrogen. The origin of this signal has been identified as the hydroperoxidase, using a variety of cosubstrates and inhibitors with either arachidonic acid or hydroperoxide as substrate [92]. The signal formed by incubating 100 μM 15-HPE$_1$ with microsomes for 30 sec is shown in Fig. 4. The mixture was freeze-quenched at the peak of activity and maintained frozen for observation in the spectrometer, where a broad signal with no discernible hyperfine structure was observed. Neither enzyme alone, substrate alone, nor denatured enzyme and substrate elicit a signal. The upper trace shows that PGE$_1$, the reduction product of the reaction, gives no signal and the lower trace demonstrates the effect of 500 μM phenol. Depression of the signal by phenol is concentration dependent and occurs as the hydroperoxidase reaction is stimulated (Table III).

Comparable results are obtained with other hydroperoxides, as shown in Table IV [94]. Substrates and concentrations are listed in the first and second columns, with or without phenol in the third. Finally, the peak-to-peak signal intensities normalized to arachidonic acid at 100 μM without phenol are listed in the fourth column. Arachidonic acid (actually the PGG$_2$ from its metabolism) gives a signal that is depressed by phenol, whereas PGH$_2$, the first nonhydroperoxide in this cascade, forms no significant signal even at 500 μM. As demonstrated in Fig. 4 for 15-HPE$_1$, 15-HPETE also forms an ESR signal, which is depressed by phenol; 15-HETE, its reduction product, does not. The primary product of platelet lipoxygenase, 12-HPETE, also generates a phenol-dependent signal that is

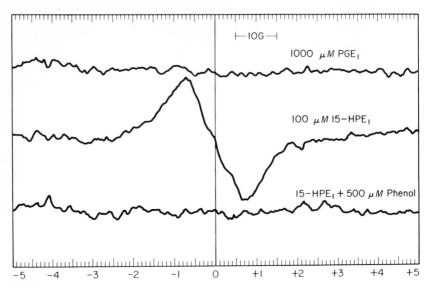

Fig. 4. Electron paramagnetic resonance signal from endoperoxide synthase-hydroperoxidase and 15-HPE$_1$. After the reaction the signal was recorded on a Varian E-109E spectrometer at $-185°C$ over a 200-G scan range in the $g = 2$ region, with a modulation amplitude of 3.2 G, a modulation frequency of 100 KHz, and a microwave power of 64 mW.

94% as intense as that of arachidonic acid. As expected, 12-HETE is inactive. Likewise, hydrogen peroxide is a substrate for this enzyme and gives a phenol-dependent ESR signal about as intense as that of 15-HPE$_1$. Ascaridole, a non-prostaglandin endoperoxide, elicits no response. Therefore, those substrates for the hydroperoxidase generate phenol-sensitive ESR signals.

By a determination of their capacity to depress this ESR signal, various amino acids were examined as possible sites of attack on enzymes by hydroperoxidase-generated oxidants [103]. The signal formed with microsomal enzyme and 15-HPETE and the effects of 1 mM concentrations of amino acids are shown in Table V. There is no significant effect of methionine, histidine, cysteine, cystine, or tyrosine, wheras tryptophan depresses the signal 51%, designating this as the most susceptible site in the peptide portion of enzymes. In addition to phenol and tryptophan, all other cofactors that stimulate the peroxidase depress this signal, whereas those materials without significant stimulatory capacity do not.

In biological systems, radicals and transition metals are the most likely sources of ESR signals. The hydroperoxidase-dependent signal has a linewidth of about 25 G, a spin ½ line shape, and a g-factor of 2.0056. Power saturation of the signal occurs below 1 mW, characteristic of a radical. This signal is observed at $-185°C$ but not at ambient temperature. Because it is generated by microsomal

TABLE IV   Radical Formation by Hydroperoxidase

Substrate	Substrate concentration (μM)	Phenol concentration (μM)	ESR signal intensity[a]
AA[b]	100	0	100
	100	500	7
PGH$_2$	500	0	10
15-HPE$_1$	100	0	63
	100	500	2
PGE$_1$	1000	0	3
15-HPETE	100	0	84
	100	500	8
15-HETE	1000	0	5
12-HPETE	100	0	94
	100	500	5
12-HETE	500	0	8
H$_2$O$_2$	100	0	61
	100	500	8
Ascaridole	1000	0	6

[a]All values normalized to 100 for arachidonic acid at 100 μM without phenol.
[b]AA, Arachidonic acid.

and solubilized preparations but not after DEAE-cellulose chromatography or with purified enzyme, the unpaired electron is not on the enzyme itself. Clearly, this radical must reside on endogenous materials isolated along with the microsomes rather than on the primary oxidant, further evidence that intact cells contain natural antioxidants that can be oxidized by endoperoxide synthase. Although ascorbate is present in the microsomes, this is not the origin of the ESR signal, because its intensity decreases with cooling. The natural cofactor has not been identified, although uric acid is a cosubstrate for the hydroperoxidase [108].

TABLE V   Effects of Amino Acids on the ESR Signal

Amino acid	Concentration (mM)	Percent change in ESR intensity[a]
Methionine	1	+14
Histidine	1	−5
Cysteine	1	−9
Cystine	1	+13
Tyrosine	1	−10
Tryptophan	1	−51

[a]Plus indicates an increase and minus a decrease.

D. Differential Modulation of Oxygenase and Hydroperoxidase

Although both oxygenase and hydroperoxidase activities of endoperoxide synthase reside on the same enzyme and probably involve the same heme center [109], they can be altered independently. For example, indomethacin competitively inhibits oxygenase activity with an IC_{50} of 0.2 μM, yet has no significant effect on hydroperoxidase at up to 500 μM (Table III). Most other nonsteroidal antiinflammatory agents, including aspirin, which acetylates an essential serine residue [110], selectively depress the oxygenase. Hydroperoxidase activity can therefore be expressed even while oxygenase inhibitors are bound reversibly or irreversibly in the vicinity of the active site.

A related phenomenon is exemplified by sulindac sulfide, sulfinpyrazone, and N-phenylanthranilic acids, which stimulate the hydroperoxidase while inhibiting the oxygenase. Sulindac sulfide inhibits oxygenase activity with an IC_{50} of 0.1 $\mu M;$ higher concentrations stimulate the hydroperoxidase (Table III) while continuing to inhibit the oxygenase. Hydroperoxidase stimulation arises from sulfide oxidation, whereas oxygenase inhibition results from the overall molecular structure, which is similar to other acidic endoperoxide synthase inhibitors. Although binding is essential for oxygenase inhibition and sulfide oxygenation [103], the binding that renders sulindac sulfide a cyclooxygenase inhibitor has no bearing on hydroperoxidase stimulation, because indomethacin does not alter cooxygenation (see Table VIII) [111].

Phenylbutazone is also a cosubstrate for this enzyme [112]. An analog of phenylbutazone, sulfinpyrazone sulfide, contains a methyl phenyl sulfide moiety and thereby possesses the same potential for oxidation as sulindac sulfide. As shown in Table VI this agent also inhibits oxygenase while stimulating hydroperoxidase reactions. However, as opposed to sulindac sulfoxide, sulfinpyrazone sulfoxide also stimulates the hydroperoxidase, because phenylbutazone can be oxidized [112].

N-Phenylanthranilic acids manifest the same dual actions as these sulfides. At 200 μM, flufenamic acid increases the extent of hydroperoxidase reaction about 300% when 15-HPETE is used as substrate [R. W. Egan, unpublished data]. In contrast, with arachidonic acid as substrate, these compounds inhibit the oxy-

TABLE VI   Effects of Sulfinpyrazones on Endoperoxide Synthase

Compound	Oxygenase at 1 μM (%)[a]	Hydroperoxidase at 200 μM (%)[a]
Sulfinpyrazone	−9	+80
Sulfinpyrazone sulfide	−61	+148

[a]Plus indicates stimulation and minus inhibition.

genase [113, 114]. Hence, this entirely separate class of compounds has the same effects as sulindac and sulfinpyrazone sulfides on endoperoxide synthase. However, fenamates differ from the other inhibitors in that their inhibitor potency is increased by the presence of phenol and other antioxidants [114]. These observations suggest that phenol either unmasks a binding site for N-phenylanthranilic acids that is distinct from that of other nonsteroidal antiinflammatory drugs or that both phenol and the fenamate are antioxidants and are collectively at a sufficiently high concentration to cause inhibition. Although a large number of oxygenase inhibitors have been discovered, there are no effective inhibitors of the hydroperoxidase.

III. DEACTIVATION OF ENZYMES BY HYDROPEROXIDASES

In addition to attacking cosubstrates, as described in Section II,B, oxidizing equivalents, $[O_x]$, released during the reduction of hydroperoxides to alcohols by endoperoxide synthase are capable of reacting with enzymes. As a consequence, enzymes such as endoperoxide synthase itself [30, 92] and PGI_2 synthase [55] can be irreversibly deactivated. The 5-lipoxygenase from rat basophilic leukemia (RBL-1) cells is also modulated by hydroperoxidases [115]. Other redox systems such as cytochrome P-450 [116], galactose oxidase [117], and tyrosinase [118] undergo substrate-dependent autoinactivation, suggesting that this mechanism may extend beyond prostaglandin synthesis.

A. Endoperoxide Synthase-Oxygenase

The kinetics of arachidonic acid oxygenation by endoperoxide synthase are shown in Fig. 5 as an oxygen monitor trace with solution oxygen tension plotted against time [92]. As oxygen is incorporated into arachidonic acid by prostaglandin cyclooxygenase, solution oxygen tension decreases, and this is manifested as a downward-sloping line with the slope reflecting the reaction rate. Two moles of oxygen are consumed per mole of arachidonic acid metabolized, and 1 μM indomethacin blocks the reaction, as expected. The initial rate is about 600 nmol oxygen per milligram protein each minute. However, oxygen uptake ceases before all the arachidonic acid or all the oxygen is used. Although a second addition of arachidonic acid does not cause increased oxygen consumption, further addition of fresh enzyme does induce renewed oxygen uptake, which precludes substrate depletion. Hence, the cyclooxygenase is naturally autodeactivating, and this deactivation is irreversible. Because oxygenation of 11,14-eicosadienoic acid causes deactivation, this phenomenon is also not due to product inhibition.

Moreover, the cyclooxygenase is deactivated during a 30-sec preincubation

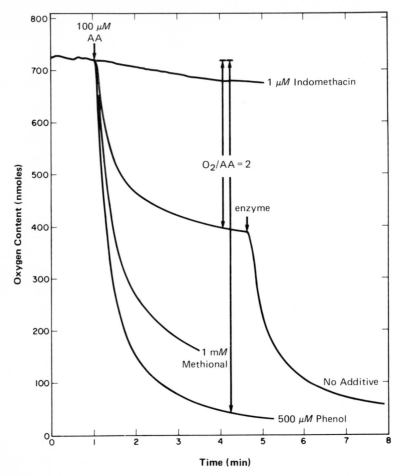

Fig. 5. Oxygenation of arachidonic acid by endoperoxide synthase. Ram vesicular gland microsomal enzyme was mixed for 30 sec with buffer containing either no additive, phenol, or methional before arachidonic acid addition. Indomethacin was preincubated with the enzyme for 2 min.

with cumene or linoleic acid hydroperoxides [*119*] or by a 2-min preincubation with 15-HPE$_1$, 15-HPETE, or PGG$_2$ [*92, 103*]. Although preincubation with as little as 3.6 μM PGG$_2$, 2.5 μM 15-HPETE, or 15 μM 15-HPE$_1$ dramatically diminishes subsequent oxygenation of arachidonic acid (Fig. 6), in the absence of preincubation these hydroperoxides do not significantly affect either the initial or the secondary oxygenase reactions. Likewise, the metabolic products 40 μM PGH$_2$, 2.5 μM 15-HETE, and 1000 μM PGE$_2$ have only minimal effects on the enzyme. However, 100 μM 15-HETE inhibits the primary reaction 50% and the secondary reaction 41% by a nonredox mechanism. At these hydroperoxide concentrations, the presence of phenol during preincubation protects very effectively against cyclooxygenase deactivation. For example, when preincubated with the enzyme for 2 min, 3.3 μM 15-HPETE inhibits the primary reaction 89%. The presence of 500 μM phenol during the 2-min preincubation reduces the deactivation to 7%. In contrast, it has no protective effect when added just

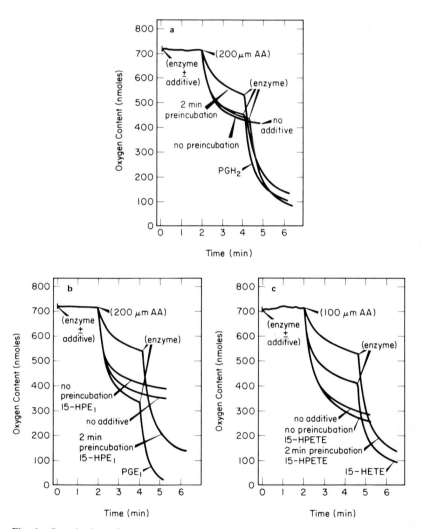

Fig. 6.   Deactivation of oxygenase activity by hydroperoxides. (a) PGG_2–PGH_2, (b) 15–HPE_1–PGE_1, (c) 15-HPETE–15-HETE. In the control reaction with no additive, microsomal enzyme was added to buffer 2 min before arachidonic acid, giving an initial rate of about 600 nmol O_2 per milligram protein each minute.

before arachidonic acid, indicating that the protection is the result of phenol being present during the preincubation period. In addition, 500 μM phenol or 1 mM methional roughly doubles both the initial rate and the total reaction before deactivation (Fig. 5). Although the absolute quantity of reaction increases in these instances, the molar ratio of oxygen consumed to arachidonic acid reacted remains 2.

These experiments indicate that the cyclooxygenase is inactivated by being present during the reductive breakdown of hydroperoxide, but not by the hydroperoxide itself or by the organic metabolites. The deactivation most likely occurs as a consequence of something transient being released during the reduction of the hydroperoxide, the stage at which oxidizing equivalents, $[O_x]$, are generated. Because this inactivation occurs with several hydroperoxides in the absence of oxygenase reaction and is prevented if phenol is present during the preincubation, a peroxidase-dependent species is clearly implicated.

B. Endoperoxide Synthase-Hydroperoxidase

The kinetics of peroxidase reactions utilizing radioactive PGG_2, 15-HPE$_1$, and 15-HPETE as substrates were determined [94], and the percentage of substrate remaining was plotted against time. To simplify the interpretation, PGG_2 and 15-HPETE reactions were performed in the presence of 500 μM phenol. Under those conditions PGH_2 was the sole product of PGG_2 reduction, and only 15-HETE resulted from 15-HPETE (Fig. 3). Following the initial rapid metabolism of each substrate during the first 15 to 30 sec, further reaction was insignificant for an additional 150 sec. Although a second addition of substrate at 60 sec resulted in no further conversion, the addition of fresh enzyme at that point initiated a second burst of rapid metabolism with each of the three substrates, followed by a plateau of inactivity. In each instance the initial and secondary aliquots of protein were identical, although less enzyme was added with PGG_2, leading to less metabolism.

Therefore, the initial cessation of reaction is due neither to product inhibition nor substrate depletion. Coupled with the observations that neither standing for 20 min nor adding further phenolic cosubstrate could reinstate activity, these results led to the conclusion that enzyme deactivation is irreversible. Deactivation of the hydroperoxidase occurs more rapidly than that of the cyclooxygenase, perhaps reflecting a closer proximity to the site of oxidant formation.

C. Prostaglandin I$_2$ Synthase

Although both the oxygenase and hydroperoxidase activities of endoperoxide synthase are oxidatively deactivated during reaction with hydroperoxide, this might be expected because they reside on the enzyme responsible for generating the oxidant. However, deactivation by this process of an independent enzyme, PGI_2 synthase, the catalyst for the nonredox reaction that isomerizes PGH_2 to PGI_2, has also been observed. Inhibition of porcine aorta PGI_2 synthase by several hydroperoxides with IC_{50} values as low as 1 to 2 μM was discovered by Salmon et al. [54], and more detailed studies revealed that inhibition of the enzyme in ram seminal vesicle microsomes requires preincubation with hydroperoxide [55]. Ham et al. found that 1 mg of this preparation, which also contains endoperoxide synthase, converts 179 pmol PGH_2 to about 50 pmol PGI_2 in 1 min. When preincubated with the enzyme for 1 min before the addition of PGH_2, PGG_2, PGG_1, 15-HPE_1, 12-HPETE, or 15-HPETE inhibits the reaction, whereas failure to preincubate greatly reduces their potency. The organic reduction products have no significant effect. For example, 34 μM PGG_1 inhibits the enzyme 6% when added just before PGH_2 and 87% when preincubated, whereas PGH_1 inhibits <10% with or without pretreatment. As found previously with endoperoxide synthase, hydroperoxidase cosubstrates present during the preincubation can ameliorate the inhibition. With PGG_1 as inhibitor, 2.5 μM 2-aminomethyl-4-tert-butyl-6-iodophenol (MK-447) decreases the inhibition from 87 to 6%. However, the effects of PGG_2, the most potent inhibitor, are unaffected by MK-447, suggesting that stronger reducing agents would be required to protect against this natural substrate.

In contrast, Moncada and Vane demonstrated that thromboxane synthase from platelets is not inhibited by hydroperoxides [119a]. In guinea pig lung microsomes it was confirmed by Ham et al. that the inactivity of PGG_1 and PGG_2 on thromboxane synthase is independent of preincubation [55]. The ratio of TXA_2 to PGI_2 is as important as the absolute amounts in determining their physiological effects. Whereas TXA_2 is an aggregator of platelets [74] and a contractile agent for smooth muscle [46], PGI_2 prevents or reverses platelet aggregation [69] and relaxes smooth muscle [68]. In concert with hydroperoxidases, lipid hydroperoxides generated by endoperoxide synthase, other lipoxygenases, or nonenzymatic lipid peroxidation can alter the ratio of TXA_2 to PGI_2 in favor of TXA_2. As a result, disorders associated with high lipid peroxidation or constriction of the vasculature such as atherosclerosis and hypertension can be exacerbated [120].

D. 5-Lipoxygenase

In addition to the enzymes involved in prostaglandin biosynthesis, other mammalian lipoxygenases form hydroperoxides that can regulate oxygenase activity while being metabolized to alcohols. Of particular interest is the 5-lipoxygenase, because it catalyzes the primary step in leukotriene synthesis. In order to inhibit these pathways selectively, it is relevant to know whether this enzyme responds to the same type of redox regulation as endoperoxide synthase. Whether hydroperoxidase activity is associated directly with 5-lipoxygenase or is an independent enzyme in these cells is not yet known. As opposed to prostaglandin synthesis, which requires the reduction of PGG_2 to PGH_2, leukotriene formation competes with hydroperoxide reduction, because dehydration to LTA_4 utilizes 5-HPETE as substrate. In addition to deactivation by hydroperoxidases, most lipoxygenases require low concentrations of endogenous hydroperoxides for activity [121–123] (also see Lands et al., Chapter 2, this volume). These two conflicting influences contribute to the in vitro regulation of lipoxygenases, activation occurring at low hydroperoxide concentrations and deactivation at higher levels.

Redox regulation of the 5-lipoxygenase from RBL-1 cells has been examined using a variety of antioxidants and hydroperoxides to alter arachidonic acid

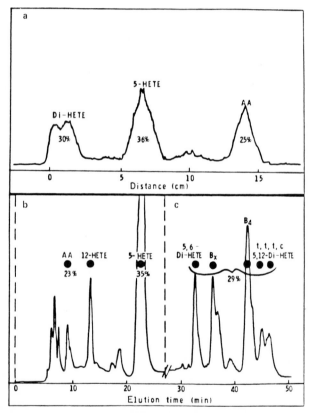

Fig. 7. Arachidonic acid metabolism by RBL-1 supernatant fraction. (a) Paper chromatography in 50:50:0.1 ether–hexane–acetic acid. (b) HPLC in 5:95:0.1 ethanol–hexane–acetic acid; absorbance, 235 nm; 0.1 AUF (absorbance units full scale) at 0.5 ml/min. (c) HPLC in 5:95:0.1 ethanol–hexane–acetic acid; absorbance, 270 nm; 0.01 AUF at 2 ml/min.

oxygenation [115]. This enzyme has been studied in the supernatant fraction of a 100,000 g centrifugation of sonified cells. Activity was monitored as the conversion of [^{14}C]arachidonic acid to the products shown in Fig. 7. The upper trace represents a paper chromatographic analysis of a standard reaction mixture recorded by a radiochromatogram scanner. The major product (36%) was 5-HETE, with 30% di-HETEs unresolved near the origin. These results are confirmed by the HPLC trace shown in Fig. 7b, where the di-HETE region is resolved. The filled circles indicate positions of arachidonic acid, 5-HETE, and LTB$_4$ standards. Below the circles are the percentages of the components based on radioactivity, which agree closely with the percentages established by paper chromatography. As expected, this supernatant fraction contains 5- and 12-lipoxygenases, hydroperoxidases, a dehydrase, and an epoxide hydrolase.

The extent of this metabolism can be modulated by a variety of redox agents (Table VII). For example, Lands et al. demonstrated with soybean lipoxygenase that depletion of hydroperoxides from the incubation mixture using glutathione peroxidase and reduced glutathione depresses enzyme activity [124]. This phenomenon also occurs with 5-lipoxygenase, in which case pure glutathione peroxidase in the presence of glutathione decreases 5-HETE formation in a concentration-dependent fashion. In accord with this concept chemical antioxidants such as nordihydroguaiaretic acid and 3-amino-1-[m-(trifluoromethyl)phenyl]-2-pyrazoline also inhibit 5-lipoxygenase. In contrast, phenol, which modulates other lipoxygenases, has no effect on 5-lipoxygenase at up to 1 mM. Hence, there is more specificity to chemical inhibitors than would be expected solely on the basis of their antioxidant capacity.

Because 5-HPETE is metabolized to 5-HETE, the supernatant fraction also possesses significant hydroperoxidase activity, a peroxidase that also reduces H$_2$O$_2$ and 15-HPE$_1$. Indeed, incubation with H$_2$O$_2$ or 15-HPE$_1$ deactivates the 5-lipoxygenase. This effect is not observed when hydroperoxide and arachidonic

TABLE VII   Redox Inhibition of 5-Lipoxygenase

Additive[a]	Concentration	Inhibition (%)
GPX–GSH	0.02 unit, 200 μM	2
	0.05 unit, 200 μM	66
	0.1 unit, 200 μM	97
NDGA	1.7 μM	68
BW 755C	4.6 μM	55
Phenol	1 mM	0
H$_2$O$_2$	100 μM	33
	200 μM	87
15-HPE$_1$	60 μM	100

[a]Abbreviations: GPX, glutathione peroxidase; GSH, glutathione; NDGA, nordihydroguaiaretic acid; BW 755C, 3-amino-1-(m-trifluoromethyl)phenyl-2-pyrazoline.

acid are added simultaneously, characteristic of the peroxidase-dependent inactivation described for other cell-free lipoxygenases. Redox regulation of 5-lipoxygenase by high and low levels of hydroperoxides is therefore sufficiently similar to that for endoperoxide synthase that selective inhibition on that basis would be difficult.

E. Heme Involvement in Endoperoxide Synthase

Initial studies on the purification of endoperoxide synthase resulted in the isolation of a protein with <0.1 heme group per molecule [125]. However, it was discovered that this preparation was the apoenzyme, which could be reconstituted and activated with heme from a variety of sources including hemoglobin [126, 127]. Although protoporphyrin complexes with other transition metals such as manganese and cobalt restore some oxygenase activity to the apoenzyme, it is significantly below the level with heme, and the hydroperoxidase activity is even less functional [126]. Reports on the heme content of reconstituted enzyme indicate either one [109] or two [128] heme groups per subunit. The most compelling evidence shows that there is only one per subunit and that it is involved in both oxygenase and hydroperoxidase activity [109].

The reconstituted enzyme has an absorption maximum in the Soret region at about 412 nm [109, 128], the intensity of which is dependent on the concentration of heme. The addition of linoleic acid hydroperoxide to the reconstituted enzyme causes a decrease in this absorption [128]. On this basis, oxidation of the active site or actual displacement of heme from the enzyme was suggested as the mechanism of substrate-dependent autoinactivation. However, we demonstrated that the absorption at 412 nm persists following heat denaturation (Fig. 8b).

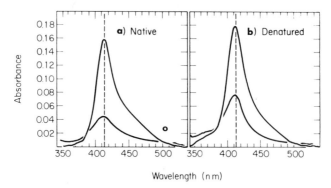

Fig. 8.   Endoperoxide Synthase Hemin Absorption. Reactions were conducted at ambient temperature in a Beckman model 35 dual-beam spectrophotometer. Additions of 8 μ*m* hemin and 40 μ*M* 15-HPETE were made simultaneously to sample and reference cuvettes. The adjustment for free hemin brought the absorbance at 350 nm to zero.

Hence, despite altering the peptide portion of the molecule sufficiently to inactivate the enzyme, heat treatment does not dislodge the heme [*103*]. The heat-inactivated enzyme that possesses no significant oxygenase or hydroperoxidase activity also has a decreased 412-nm absorption in response to 15-HPETE, indicating that the change is based not on enzyme activity, but on a chemical reaction between heme and hydroperoxide.

Because hemin itself absorbs with a maximum at 385 nm, a comparable concentration of free hemin must be placed in sample and reference cells when endoperoxide synthase is measured at 412 nm. Furthermore, hemin is a weak hydroperoxidase that is capable of reducing the hydroperoxide in the reference cell, rendering these experiments technically difficult. The hemin absorption at 385 nm is very sensitive to hydroperoxides such that 40 μM 15-HPE$_1$ decreases the absorbance of 8 μM hemin 65%, whereas 5% of the 15-HPE$_1$ is metabolized to PGE$_1$. The Soret region absorbance of endoperoxide synthase at 412 nm, therefore, seems to be an accurate monitor of the enzyme–heme complex. However, the role of this complex in the deactivation by hydroperoxides must be examined in more detail.

Although it is not normally a cosubstrate for endoperoxide synthase, it has been reported that, in the presence of flufenamic acid, the haloenzyme reconstituted in phospholipid vesicles can metabolize NADH [*129*]. The extent of this metabolism is doubled by cytochrome b_5 and cytochrome b_5 reductase. The requirement for flufenamic acid suggests that it is critical to the process. Because this compound is itself a cosubstrate for the hydroperoxidase [R. W. Egan, unpublished data], one explanation for this observation is that endoperoxide synthase oxidizes flufenamic acid, which in turn chemically oxidizes NADH, as opposed to a direct interaction between NADH and endoperoxide synthase. The increased oxidation by cytochrome b_5 and its reductase may reflect the normal NADH oxidation catalyzed by this redox couple.

IV. COSUBSTRATE OXIDATION

The disposition of the hydroperoxide during these reactions having been characterized, the fates of the low molecular weight cosubstrates were investigated. These radical-scavenging reducing agents should be metabolized by hydroperoxidase-dependent reactions, a large number of which are described elsewhere in this volume (Lands *et al.*, Chapter 2; Marnett, Chapter 3). Many of these studies have been conducted with purified endoperoxide synthase to avoid interference from other drug-metabolizing systems in microsomes. On the basis of the effects of inhibitors, it seems that even those reactions using microsomes do not involve cytochrome *P*-450, the most likely alternative to endoperoxide synthase [*103*].

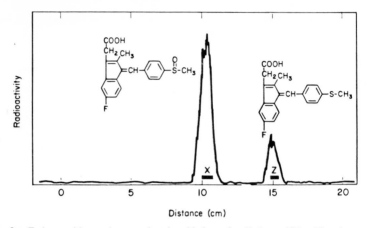

Fig. 9.   Endoperoxide synthase-catalyzed oxidation of sulindac sulfide. The chromatogram shows the pattern of radioactive products in the ether extract from a 60-sec incubation of 100 μM [^{3}H]sulindac sulfide with 0.56 mg of microsomal protein and 150 μM 15-HPE$_1$. The bar labeled x indicates the location of nonradioactive authentic sulindac (structure to the left of the large peak), whereas that labeled z designates sulindac sulfide (structure over the small peak).

A. Sulfides

1. Sulindac Sulfide

As described in Table III sulindac sulfide is a potent stimulator of the hydroperoxidase activity of endoperoxide synthase, whereas its oxidation product, the sulfoxide (sulindac), is not. Shown in Fig. 9, along with the structures, is an example of hydroperoxidase-dependent metabolism of sulindac sulfide to sulindac [*94*]. Although the extent of reaction varies with protein concentration, the exclusive product with 15-HPE$_1$ as oxidizing substrate is the sulfoxide. The metabolism is protein dependent, destroyed by heat denaturation of the enzyme, and elicted by PGG$_2$ and 15-HPETE as well as 15-HPE$_1$. Furthermore, at low levels of enzyme and high sulindac sulfide concentrations this is an equimolar reaction [*94*]. The stoichiometry was established by tandem incubations, monitoring [^{3}H]sulindac sulfide oxidation in the presence of nonradioactive 15-HPE$_1$, and 15-[1-^{14}C]HPE$_1$ reduction in the presence of nonradioactive sulindac sulfide.

The mechanism of oxygen transfer to sulfide has been examined by mass spectral determination of the distribution of ^{18}O in the products, with [^{18}O]15-HPE$_2$ as reactant [*111*]. Labeled 15-HPE$_2$ was prepared by the endoperoxide synthase-catalyzed reaction of arachidonic acid with ^{18}O$_2$. Because of the dioxygenase reaction mechanism, the isotopic composition of the resulting 15-HPE$_2$ and PGE$_2$ would be identical, and the PGE$_2$ was used to establish this parameter because it produces a more easily interpreted mass spectrum. The molecular ion region of the mass spectra for the tetratrimethylsilyl derivatives of both enzymatically prepared and authentic PGE_2 were the basis for comparison. The molecular ion for authentic silylated PGE_2 is 640, the largest ion in this region. Because the labeled PGE_2 could contain either zero, one, two, or three ^{18}O atoms, the ratios of m/e 640, 642, 644, and 646 were compared to establish the isotopic distribution, which shows that 78% of the molecules contained ^{18}O in the hydroxyl at C-15. Consequently, 78% of the 15-HPE_2 would possess ^{18}O atoms in the hydroperoxide, the functional group of interest in these studies.

The [^{18}O]15-HPE_2 was used as substrate for the oxidation of sulindac sulfide to sulindac by microsomal enzyme. The PGE_2 resulting from this reaction contains 76% ^{18}O at C-15 compared with 78% in the starting 15-HPE_2, confirming that the hydroperoxide moiety is not exchanging in some unexpected fashion. Partial mass spectra obtained for authentic sulindac indicated an intense ion at m/e 356 (the molecular ion), an ion of low intensity at m/e 358, and no signal at m/e 354. Consequently, the mass spectrum of sulindac with an ^{18}O atom in the sulfoxide would exhibit an intense ion at m/e 358, a low-intensity ion at m/e 360, and no signal at m/e 356. Analysis of the isotopic composition of the reaction product indicates that 78% of this sulindac contains ^{18}O. Hence, the oxygen atom transferred to sulindac sulfide arises exclusively from [^{18}O]15-HPE_2. Virtually identical results are obtained using a purified preparation of endoperoxide synthase [128, 130] in the presence of hemoglobin.

Because sulindac sulfide also inhibits the oxygenase activity of endoperoxide synthase (see Section II,D), it seemed possible that the oxygen atom was transferred while the sulfide was bound to this site. Although the two catalytic activities are functionally exclusive, binding to an oxygenase site could place the thiomethyl group in correct juxtaposition for oxidation. However, indomethacin, an oxygenase inhibitor [49] that has no significant effect on the peroxidase [94], does not alter the stoichiometry of the sulindac sulfide oxidation by 15-HPE_1 (Table VIII). The capacity of indomethacin to prevent binding of sulindac sulfide

TABLE VIII   Effect of Indomethacin on Sulindac Sulfide Oxidation

Sulindac sulfide (μM)	Indomethacin (μM)	Product formed (nmol/ml)		Ratio[a]
		PGE_1	Sulindac	
0	0	42	—	—
100	0	100	70	0.70
100	0	4.5[b]	0[b]	—
0	100	40	—	—
100	100	94	64	0.68

[a]The molar ratio of sulindac to PGE_1 formed.
[b]Enzyme heated at 85°C for 15 min.

assumes that both affect the oxygenase at the same site, which is likely because they inhibit competitively.

2. Methyl Phenyl Sulfide

In contrast to sulindac sulfide, which inhibits the oxygenase activity of endoperoxide synthase at <1 μM, methyl phenyl sulfide (MPS) stimulates the microsomal enzyme at up to 500 μM, eliminating the concern that binding to the oxygenase site could alter the normal kinetics and mechanism of sulfide oxidation [103]. Characteristic of oxidizable cosubstrates [131], low levels of MPS stimulate the oxygenase, whereas higher concentrations cause inhibition (Table IX), probably by depleting the level of endogenous hydroperoxide below that required to initiate catalysis [131a]. The more pronounced effect of MPS on microsomal compared with purified endoperoxide synthase may result from a larger endogenous antioxidant concentration in the former.

Using similar procedures and the same [^{18}O]15-HPE$_2$ described for sulindac sulfide, the mechanism of MPS oxidation to methyl phenyl sulfoxide (MPSO) has been studied with purified endoperoxide synthase. The PGE$_2$ resulting from the reaction contained 74% ^{18}O at C-15 compared with 78% in the starting material. Spectra of the oxidation product, MPSO, were obtained following gas chromatography without chemical treatment. The molecular ion of MPSO at m/e 140 was predominant in this region, with no 138 ion and a peak only 5% as intense at 142. Hence, the predominant ion of [^{18}O]MPSO would be 142. The product of the reaction with [^{18}O]15-HPE$_2$ and MPS had a large 142 and a small 140 ion. By comparing peak heights, it was determined that 86% of the oxygen in the sulfoxide is ^{18}O. Therefore, the oxygen atoms in this sulfoxide are also transferred directly from the hydroperoxide.

The characteristics of MPS binding to the enzyme during oxidation have been

TABLE IX   Methyl Phenyl Sulfide Effects on Endoperoxide Synthase

		Extent of reaction	
Tissue	MPS (μM)	Oxygenase (nmol O_2/mg protein)	Hydroperoxidase (nmol PGE$_2$/mg protein)
Microsomes	0	170	20
	500	236	93
	5,000	52	290
	10,000	19	280
Purified	0	3,533	1,120
	500	8,938	2,920
	5,000	6,253	7,160
	10,000	4,028	6,520

investigated by determining the stoichiometry and chiral character of the reaction. The stoichiometry was determined using purified enzyme with the procedures described for sulindac sulfide (Section IV,A,1). Over a fivefold concentration range from 100 to 500 μM MPS, the reaction remains equimolar, indicating closely coupled oxidation and reduction [103]. Furthermore, the MPSO prepared from the reaction of MPS and 15-HPE$_1$ with microsomal enzyme is chiral, in contrast to chemically oxidized MPS, which is a racemic mixture [103]. However, compared with a literature standard ($[\alpha]_D^{25}$ of $-127.5°$ [132]), the optical rotation of this material is low ($[\alpha]_D^{25}$ of $-14.5°$). This observation was confirmed by analysis of the NMR spectrum of enzymatically prepared MPSO in the presence of the chiral shift reagent tris[3-(heptafluoropropylhydroxymethylene)-d-camphorato]europium(III) [Eu(hfc)$_3$].

Resonances at 10.4, 7.9, and 5.3 ppm were the ortho, meta and para, and methyl protons of MPSO, respectively. These assignments [133] gave the expected 2 : 3 : 3 ratio of areas and were further confirmed by studies at varying Eu(hfc)$_3$ levels. The 0.126 ppm splitting of the methyl resonance is due to nonequal complex formation of the two MPSO diastereomers, with the (S)-($-$)-MPSO complex at higher field. In agreement with the polarimetry data, the average ratio of enantiomers in the NMR spectrum is 44:56. The low chirality of the enzymatically formed MPSO indicates that the sulfide is not held rigidly within its binding site. As a result, electrophilic attack on the sulfur could involve either lone pair with almost equal frequency. Such a lack of stereospecificity is also noted with serine hydroxymethylase [134] and certain bacterial sulfide oxidations [135, 136]. With other sulfides or different experimental conditions, the chirality of the reaction might be increased. However, it is clear that there is an optical preference of the hydroperoxidase, further indicating that the sulfide is bound when oxidized.

The molecular transformations occurring between the hydroperoxide on 15-HPE$_2$ and the sulfur on sulindac sulfide or MPS are summarized in Fig. 10.

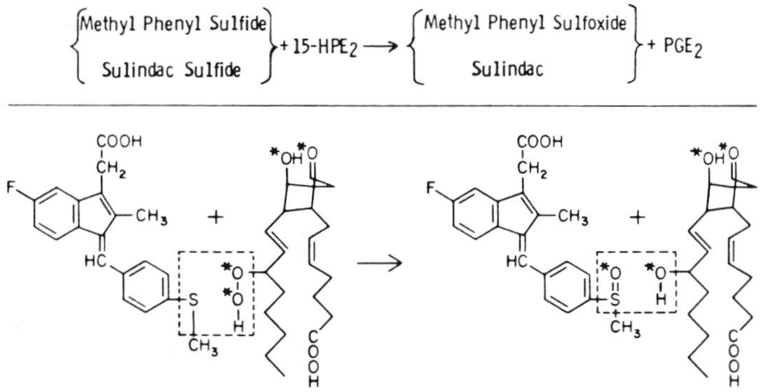

Fig. 10. Mechanism of sulfide oxidation by endoperoxide synthase.

There is a stoichiometric equimolar reaction between sulfide and 15-HPE$_2$. Shown below the equation are the structures of the reactants and products with the relevant functional groups placed in proximity and enclosed within dashed boxes with asterisks designating ^{18}O atoms. The oxygen atom is transferred to the sulfide, and after the reaction all the atoms are accounted for. This precludes indirect oxidizing species such as molecular or singlet oxygen as contributors to this cooxygenation.

B. Free-Radical Reactions

Endoperoxide synthase-catalyzed cosubstrate oxidations involving radical rather than two-electron reactions are reviewed elsewhere in this volume (Marnett, Chapter 3; Kalyanaraman and Sivarajah, Chapter 5). For example, the same oxidative N-demethylation of aminopyrine that is characteristic of this compound with other chemical and enzymatic oxidation systems has been demonstrated for this enzyme [137]. We have also detected the characteristic ESR signals of aminopyrine and dimethylphenylenediamine oxidations formed by microsomal endoperoxide synthase [103]. These radicals are transient, coincide with blue color formation for the former and pink for the latter, and are identical in duration and ESR hyperfine structure to those obtained by chemical oxidation with Fenton's reagent or bromine water. Although enzymatic sulfide oxidation follows a nonradical mechanism, it is the same route as chemical oxidation in that instance also. Hence, cosubstrates that increase hydroperoxide reduction are metabolized by a mechanism dictated more by their chemical structure than by the enzyme.

Because cooxidation by the hydroperoxidase occurs by both radical and non-radical mechanisms, we studied competition between substrates from these two classes for the oxidant generated by endoperoxide synthase [103]. As described by Marnett et al. [138], phenylbutazone oxygenation utilizes dissolved molecular oxygen, and the stoichiometry can be as high as several thousand phenylbutazone molecules oxidized for each hydroperoxide reduced, indicating a chain reaction. With the knowledge that the sulfide is bound when oxidized, it seemed possible to distinguish the location of the primary oxidation of these molecules by studying competition with MPS. Although it is clear that chain propagation proceeds in solution, it seemed possible that the initiation step could involve enzyme-bound phenylbutazone. The hydroperoxidase reaction was investigated at as much as a sevenfold molar excess of phenylbutazone over MPS. As shown in the third column of Table X both MPS and phenylbutazone stimulate the extent of 15-HPE$_2$ metabolism. In combination they stimulate more than either one individually. Were phenylbutazone competing with MPS for the oxidizing equivalents released by the hydroperoxidase, it would decrease the oxidation of MPS to MPSO. As shown in the last column the extent of MPS conver-

TABLE X   Effect of Phenylbutazone on Methyl Phenyl Sulfide Oxidation[a]

MPS (μM)	Phenylbutazone (μM)	Product formed	
		PGE$_2$ (nmol/ml)	MPSO (nmol/ml)
0	0	42	—
150	0	69	66
0	500	91	—
0	1000	96	—
150	500	106	71
150	1000	125	87

[a]Reactions measuring 15-HPE$_2$ reduction utilized 15-[1-^{14}C]HPE$_2$ and nonradioactive MPS, whereas those measuring MPS oxidation used [$methyl$-^{14}C]MPS and nonradioactive 15-HPE$_1$. Purified enzyme was 150 μg/ml, hemin was 10 μM, and 15-HPE$_2$ was 160 μM.

sion increases 32% at 1000 μM phenylbutazone. Hence, the primary locus of phenylbutazone oxidation is different from that of MPS, possibly in solution via an enzyme-generated oxidant. Nevertheless, radical-forming substrates such as phenylbutazone protect endoperoxide synthase from self-destruction, indicating that the enzyme inactivation can result from a species present in solution. Such an oxidant possesses many characteristics akin to those of the hydroxyl radical.

C. Reaction Mechanisms

The active site of endoperoxide synthase may resemble those of other peroxidases, the most completely studied of which is cytochrome c peroxidase (Fig. 11). As depicted by Poulos and Kraut [139], in that enzyme a histidine residue and a water molecule are coordinated to the Fe(III)–protoporphyrin. It is evisioned that the aquo ligand is displaced by the hydroperoxide as the iron is increased in valence. Other amino acid side chains such as an indole from tryptophan, an imidazole from histidine, and a guanidinium from arginine stabilize the transition state as the oxygen is transferred to the iron via heterolytic cleavage of the peroxide bond. The alcohol dissociates, leaving an oxidized active site involving Fe(IV)–protoporphyrin and an indole radical. In the case of cytochrome c peroxidase, the enzyme is specific for cytochrome c as reducing substrate, which these authors suggest may be retained with its heme parallel to the heme in the peroxidase by aspartate–lysine hydrogen bonding [140]. As opposed to cytochrome c peroxidase, which remains oxidized until cytochrome c is available, the endoperoxide synthase-hydroperoxidase can release a reactive oxidant, [O$_x$], into solution to initiate a radical chain. In contrast to either electron transfer or homolytic cleavage to a radical, oxidation of MPS to MPSO involves formation of a sulfur–oxygen bond. Hence, the various mechanisms by

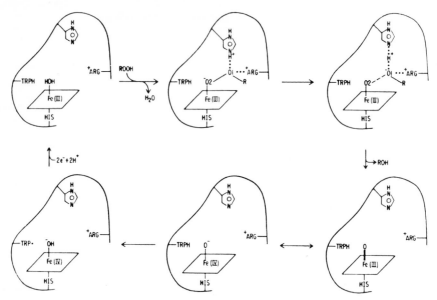

Fig. 11.   Mechanism of cytochrome c peroxidase activation.

which oxidizing equivalents are transferred to the oxidizable substrate vary among enzymes and among substrates for a given enzyme.

Putting these two possibilities in a diagrammatic sequence (Fig. 12), endoperoxide synthase-hydroperoxidase contains heme with iron probably in a valence state of 3 [E(III)]. This may or may not have sulfide cosubstrate bound. It is oxidized in a formal sense to Fe(V) upon reduction of the hydroperoxide [E(V)]. From this stage there are two routes for discharge of this oxidizing potential: via bond formation as with the sulfides to give E(III) and sulfoxide or via release of an oxidant, $[O_x]$, into solution. This radical can then initiate a chain reaction by hydrogen atom abstraction from substrate A. In the nonradical case the oxygen atom in the product arises from the hydroperoxide, and an equimolar reaction with enzyme-bound cosubstrate ensues. In fact, with sulfides the substrate may be bound during hydroperoxide reduction. In the radical case virtually any stoichiometry could be observed, with the oxygen in the product coming from dissolved molecular oxygen.

Because an oxygen atom is transferred directly from MPS to MPSO, the sulfur must be in proximity to the iron complex. Nevertheless, it is not bound in the inner sphere of the iron because one site is occupied by the hydroperoxide and the sixth coordination position is on the other site of the planar protoporphyrin group. Hence, sulfide binding occurs on the same side of the heme as hydroperoxide binding and must permit adequate motion to account for the low chiral character of the product.

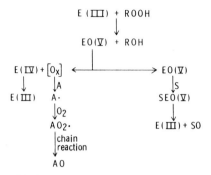

Fig. 12. Mechanisms of hydroperoxidase-catalyzed oxidations. E(III), Hydroperoxidase with Fe(III)–heme; EO(V), hydroperoxidase with bound oxygen atom and Fe(V)–heme; E(IV), hydroperoxidase with Fe(IV)–heme; SEO(V), EO(V) with S bound to the active site; $[O_x]$, oxidizing species released into solution; A, substrate oxidized in single-electron steps; A·, radical form of A; $AO_2·$, peroxyl radical of A; AO, oxygen atom adduct of A; S, substrate oxidized in two-electron steps; SO, oxygen atom adduct of S.

V. SUMMARY

In mammalian physiology, endoperoxide synthase and a variety of lipoxygenases catalyze the formation of fatty acid hydroperoxides, which are subsequently metabolized to alcohols by hydroperoxidases. During hydroperoxide reduction, a broad spectrum of organic materials can be oxidized. In fact, the oxidant formed by endoperoxide synthase, an enzyme that catalyzes both oxygenase and peroxidase reactions, can attack and irreversibly deactivate the enzyme itself. Because this autoinactivation occurs with several hydroperoxides in the absence of oxygenase reaction and is prevented by antioxidants, a hydroperoxidase-dependent species must be involved. Even though the two activities of endoperoxide synthase are functionally exclusive, they are associated with a common heme that absorbs at 412 nm. Although this absorption decreases when exposed to hydroperoxide, it does not reflect the functional state of the enzyme. In addition to endoperoxide synthase, oxidative deactivation occurs with PGI_2 synthase, 5-lipoxygenase, and other redox enzymes beyond the realm of arachidonic acid metabolism.

Oxidation of low molecular weight organic molecules such as sulfides and amines by two-electron or radical mechanisms is dictated more by innate chemical reactivity than by the enzyme. Some reagents such as sulfides are metabolized stoichiometrically while associated with the enzyme, whereas other oxygenations are propagated in solution by chain mechanisms. Although these reactions should proceed by similar mechanisms *in vivo*, cells with high levels of natural antioxidants should be more refractory to attack on enzymes, and their principal response could be enzyme inhibition due to decreased endogenous hydroperoxide levels. The primary function of these enzymes is synthesis of eicosanoids; however, their redox chemistry may also have profound effects on xenobiotic metabolism.

ACKNOWLEDGMENT

The authors wish to acknowledge Ms. D. Dell'Aquila and Ms. P. Levan for their assistance in preparing the manuscript.

REFERENCES

1. U. S. von Euler, *Naunyn-Schmiedebergs Arch. Exp. Pathol. Pharmakol.* **175,** 78 (1934).
2. M. W. Goldblatt, *J. Physiol. (London)* **84,** 208 (1935).
3. U. S. von Euler, *Klin. Wochenschr.* **14,** 1182 (1935).
4. S. Bergström, R. Ryhaye, B. Samuelsson, and J. Sjövall, *Acta Chem. Scand.* **16,** 501 (1962).
5. S. Bergström, H. Danielsson, and B. Samuelsson, *Biochim. Biophys. Acta* **90,** 207 (1964).
6. D. A. van Dorp, R. K. Beerthuis, D. H. Nugteren, and H. von Heman, *Biochim. Biophys. Acta* **90,** 204 (1964).
7. B. Samuelsson, *J. Am. Chem. Soc.* **87,** 3011 (1965).
8. D. H. Nugteren, R. K. Beerthuis, and D. A. van Dorp, *Recl. Trav. Chim. Pays-Bas* **85,** 405 (1966).
9. B. Samuelsson, E. Granström, K. Green, and M. Hamberg, *Ann. N.Y. Acad. Sci.* **180,** 138 (1971).
10. D. H. Nugteren and E. Hazelhof, *Biochim. Biophys. Acta* **326,** 448 (1973).
11. M. Hamberg and B. Samuelsson, *Proc. Natl. Acad. Sci. U.S.A.* **70,** 899 (1973).
12. M. Hamberg, J. Svensson, T. Wakabayashi, and B. Samuelsson, *Proc. Natl. Acad. Sci. U.S.A.* **71,** 345 (1974).
13. M. Hamberg, J. Svensson, and B. Samuelsson, *Proc. Natl. Acad. Sci. U.S.A.* **72,** 2994 (1975).
14. M. Hamberg, J. Svensson, and B. Samuelsson, *Adv. Prostaglandin Thromboxane Res.* **1,** 19 (1976).
15. M. Hamberg and B. Samuelsson, *Proc. Natl. Acad. Sci. U.S.A.* **71,** 3400 (1974).
16. S. Moncada, R. Gryglewski, S. Bunting, and J. R. Vane, *Nature (London)* **263,** 663 (1976).
17. R. A. Johnson, D. R. Morton, J. H. Kinner, R. R. Gorman, J. C. McGuire, F. F. Sun, N. Whittaker, S. Bunting, J. Salmon, S. Moncada, and J. R. Vane, *Prostaglandins* **12,** 915 (1976).
18. C. Pace-Asciak and L. S. Wolfe, *Biochemistry* **10,** 3657 (1971).
19. W. E. Brocklehurst, *J. Physiol. (London)* **151,** 416 (1960).
20. B. A. Jakschik, S. Falkenhein, and C. W. Parker, *Proc. Natl. Acad. Sci. U.S.A.* **74,** 4577 (1977).
21. R. C. Murphy, S. Hammarström, and B. Samuelsson, *Proc. Natl. Acad. Sci. U.S.A.* **76,** 4275 (1979).
22. S. Hammarström, R. C. Murphy, B. Samuelsson, D. A. Clark, C. Mioskowski, and E. J. Corey, *Biochem. Biophys. Res. Commun.* **91,** 1266 (1979).
23. R. J. Flower and G. J. Blackwell, *Nature (London)* **278,** 456 (1979).
24. F. Hirata, B. A. Corcoran, K. Venkatasubramanian, E. Schiffmann, and J. Axelrod, *Proc. Natl. Acad. Sci. U.S.A.* **76,** 2640 (1979).
25. S. Rittenhouse-Simmons, *J. Clin. Invest.* **63,** 580 (1979).
26. R. L. Bell, D. A. Kennerly, N. Stanford, and P. W. Majerus, *Proc. Natl. Acad. Sci. U.S.A.* **76,** 3238 (1979).

27. S. M. Prescott and P. W. Majerus, *J. Biol. Chem.* **258**, 764 (1983).
28. M. M. Billah, E. G. Lapetina, and P. Cuatrecasas, *J. Biol. Chem.* **256**, 5399 (1981).
29. B. Samuelsson, E. Granström, K. Green, M. Hamberg, and S. Hammarström, *Annu. Rev. Biochem.* **44**, 669 (1975).
30. W. L. Smith and W. E. M. Lands, *Biochemistry* **11**, 3276 (1972).
31. S. Ohki, N. Ogino, S. Yamamoto, and O. Hayaishi, *J. Biol. Chem.* **254**, 829 (1979).
32. N. Ogino, T. Miyamoto, S. Yamamoto, and O. Hayaishi, *J. Biol. Chem.* **252**, 890 (1977).
33. E. Christ-Hazelhof and D. H. Nugteren, *Biochim. Biophys. Acta* **572**, 43 (1979).
34. T. Shimizu, S. Yamamoto, and O. Hayaishi, *J. Biol. Chem.* **254**, 5222 (1979).
35. H.-H. Tai and D. Guey-Bin Chang, *in* "Methods in Enzymology" 86, p. 142. (W. E. M. Lands and W. L. Smith, eds.), Vol. Academic Press, New York, 1982.
36. S.-C. Lee and L. Levine, *J. Biol. Chem.* **249**, 1369 (1974).
37. U. Diczfalusy, P. Falardeau, and S. Hammarström, *FEBS lett.* **84**, 271 (1977).
38. P. Borgeat, B. Fruteau de Laclos, and J. Maclouf, *Biochem. Pharmacol.* **32**, 381 (1983).
39. B. Samuelsson, *Science* **220**, 568 (1983).
40. P. Borgeat and B. Samuelsson, *Proc. Natl. Acad. Sci. U.S.A.* **76**, 3213 (1979).
41. B. Samuelsson and S. Hammarström, *Prostaglandins* **19**, 645 (1980).
42. K. Bernstrom and S. Hammarström, *J. Biol. Chem.* **256**, 9579 (1981).
43. A. L. Tappel, P. D. Boyer, and W. D. Lundberg, *J. Biol. Chem.* **199**, 267 (1952).
44. M. Hamberg and B. Samuelsson, *J. Biol. Chem.* **242**, 5336 (1967).
45. F. A. Kuehl, Jr., V. J. Cirillo, E. A. Ham, and J. L Humes, *Adv. Biosci.* **9**, 155 (1973).
46. S. Moncada and J. R. Vane, *Pharmacol. Rev.* **30**, 293 (1979).
47. J. R. Vane, *Nature (London) New Biol.* **231**, 232 (1971).
48. J. B. Smith and A. L. Willis, *Nature (London), New Biol.* **231**, 235 (1971).
49. T. Y. Shen, *in* "Anti-Inflammatory Drugs" (J. R. Vane and S. H. Ferreira, eds.), p. 305. Springer-Verlag, Berlin and New York, 1979.
50. G. J. Roth, N. Stanford, and P. W. Majerus, *Proc. Natl. Acad. Sci. U.S.A.* **72**, 3073 (1975).
51. C. Patrono, G. Ciabattoni, E. Pinca, F. Pugliese, G. Castrucci, A. De Salvo, M. A. Satta, and B. A. Peskar, *Thromb. Res.* **17**, 317 (1980).
52. H.-H. Tai, C. L. Tai, and N. Lee, *Arch. Biochem. Biophys.* **203**, 758 (1980).
53. J. B. Smith and W. Jubiz, *Prostaglandins* **22**, 353 (1981).
54. J. A. Salmon, D. R. Smith, R. J. Flower, S. Moncada, and J. R. Vane, *Biochim. Biophys. Acta* **523**, 250 (1978).
55. E. A. Ham, R. W. Egan, D. D. Soderman, P. H. Gale, and F. A. Kuehl, Jr., *J. Biol. Chem.* **254**, 2191 (1979).
56. R. Korbut and S. Moncada, *Thromb. Res.* **13**, 489 (1978).
57. P. Y.-K. Wong, J. C. McGiff, L. Cagen, K. U. Malik, and F. F. Sun, *J. Biol. Chem.* **254**, 12 (1979).
58. E. Anggard, K. Gréen, and B. Samuelsson, *J. Biol. Chem.* **240**, 1932 (1965).
59. E. Anggard and B. Samuelsson, *Ark. Kemi* **25**, 293 (1966).
60. H. F. Woods, G. Ash, M. J. Weston, S. Bunting, S. Moncada, and J. R. Vane, *Lancet* **2**, 1075 (1978).
61. J. W. Weiss, J. M. Drazen, N. Coles, E. R. McFadden, Jr., P. F. Weller, E. J. Corey, R. A. Lewis, and K. F. Austen, *Science* **216**, 196 (1982).
62. C. V. Wedmore and T. J. Williams, *Nature (London)* **289**, 646 (1981).
63. F. A. Kuehl, Jr., J. L. Humes, J. Tarnoff, V. J. Cirillo, and E. A. Ham, *Adv. Cyclic Nucleotide Res.* **1**, 493 (1972).
64. A. Bennett, I. F. Stamford, and W. G. Unger, *J. Physiol. (London)* **229**, 349 (1973).
65. A. Robert, J. E. Nezamis, C. Lancaster, and A. J. Hanchar, *Gastroenterology* **77**, 433 (1979).
66. B. J. R. Whittle, *Brain Res. Bull.* **5**, Suppl. 1, 7 (1980).
67. S. H. Ferreira, *Nature (London), New Biol.* **240**, 200 (1972).

68. G. J. Dusting, S. Moncada, and J. R. Vane, *Prostaglandins* **13,** 3 (1977).
69. S. Moncada, R. J. Gryglewski, S. Bunting, and J. R. Vane, *Nature (London)* **263,** 663 (1976).
70. R. R. Gorman, S. Bunting, and O. V. Miller, *Prostaglandins* **13,** 377 (1977).
71. J. E. Tateson, S. Moncada, and J. R. Vane, *Prostaglandins* **13,** 389 (1977).
72. R. Franco-Saenz, S. Suzuki, and S. Y. Tan, *Prostaglandins* **20,** 1131 (1980).
73. P. J. Piper and J. R. Vane, *Nature (London)* **223,** 29 (1969).
74. B. B. Vargaftig and P. Zirinis, *Nature (London), New Biol.* **244,** 114 (1973).
75. P. Needleman, S. Moncada, S. Bunting, J. R. Vane, M. Hamberg, and B. Samuelsson, *Nature (London)* **261,** 558 (1976).
76. J. B. Smith, M. J. Silver, C. M. Ingerman, and J. J. Kocsis, *Thromb. Res.* **5,** 291 (1974).
77. E. E. Nishizawa, W. L. Miller, R. R. Gorman, G. L. Bundy, J. Svensson, and M. Hamberg, *Prostaglandins* **9,** 109 (1975).
78. D. C. B. Mills and D. E. MacFarlane, *Thromb. Res.* **5,** 401 (1974).
79. C. H. Kellaway and E. R. Trethewie, *Q. J. Exp. Physiol. Cogn. Med. Sci.* **30,** 121 (1940).
80. G. Smedagard, P. Hedqvist, S.-E. Dahlen, B. Revenäs, S. Hammarström, and B. Samuelsson, *Nature (London)* **295,** 327 (1982).
81. J. W. Weiss, J. M. Drazen, N. Coles, E. R. McFadden, Jr., P. F. Weller, E. J. Corey, R. A. Lewis, and K. F. Austen, *Science* **216,** 196 (1982).
82. S.-E. Dahlen, J. Bjork, P. Hedqvist, K.-E. Arfors, S. Hammarström, J.-A., Lindgren, and B. Samuelsson, *Proc. Natl. Acad. Sci. U.S.A.* **78,** 3887. (1981).
83. R. G. G. Andersson, L. E. Gustafsson, S. E. Hedman, P. Hedqvist, and B. Samuelsson, *Acta Physiol. Scand.* **116,** 97 (1982).
84. M. J. H. Smith, A. W. Ford-Hutchinson, and M. A. Bray, *J. Pharm. Pharmacol.* **32,** 517 (1980).
85. E. J. Goetzl and W. C. Pickett, *J. Immunol.* **125,** 1789 (1980).
86. M. A. Bray, F. M. Cunningham, A. W. Ford-Hutchinson, and M. J. H. Smith, *Br. J. Pharmacol.* **72,** 483 (1981).
87. H.-E. Claesson, *FEBS Lett.* **139,** 305 (1982).
88. A. W. Ford-Hutchinson, M. A. Bray, M. V. Doig, M. E. Shipley, and M. J. H. Smith, *Nature (London)***286,** 264, (1980).
89. E. J. Goetzl, J. M. Woods, and R. R. Gorman, *J. Clin. Invest.* **59,** 179 (1977).
90. S. A. Metz, W. Y. Fujimoto, and R. P. Robertson, *Endocrinology* **111,** 2141 (1982).
91. I. H. M. Main and B. J. R. Whittle, *Br. J. Pharmacol.* **53,** 217 (1975).
92. R. W. Egan, J. Paxton, and F. A. Kuehl, Jr., *J. Biol. Chem.* **251,** 7329 (1976).
93. F. A. Kuehl, Jr., J. L. Humes, M. L. Torchiana, E. A. Ham, and R. W. Egan, *Adv. Inflammation Res.* **1,** 419 (1979).
94. R. W. Egan, P. H. Gale, and F. A. Kuehl, Jr., *J. Biol. Chem.* **254,** 3295 (1979).
95. L. J. Marnett, P. Wlodawer, and B. Samuelsson, *J. Biol. Chem.* **250,** 8510 (1975).
96. B. W. Griffin, *FEBS Lett.* **74,** 139 (1977).
97. H. Sayo and M. Masui, *J. Chem. Soc., Perkin Trans. 2* p. 1640 (1973).
98. B. C. Saunders, *Inorg. Biochem.* **2,** 988 (1973).
99. A. Ishimaru and I. Yamazaki, *J. Biol. Chem.* **252,** 6118 (1977).
100. M. L. Coval and A. Taurog, *J. Biol. Chem.* **242,** 5510 (1967).
101. H. B. Dunford and J. S. Stillman, *Coord. Chem. Rev.* **19,** 187 (1976).
102. C. Beauchamp and I. Fridovich, *J. Biol. Chem.* **245,** 4641 (1970).
102a. W. A. Pryor and R. H. Tang, *Biochem. Biophys. Res. Commun.* **81,** 498 (1978).
103. R. W. Egan, P. H. Gale, E. M. Baptista, K. L. Kennicott, W. J. A. Vanden Heuvel, R. W. Walker, P. E. Fagerness, and F. A. Kuehl, Jr., *J. Biol. Chem.* **256,** 7352 (1981).
104. J. M. Lasker, K. Sivarajah, R. P. Mason, B. Kalyanaraman, M. B. Abou-Donia, and T. E. Eling, *J. Biol. Chem.* **256,** 7764 (1981).

105. T. Yoshimoto, S. Yamamoto, K. Sugioka, M. Nakano, C. Takyu, A. Yamagishi, and H. Inaba, *J. Biol. Chem.* **255**, 10199 (1980).
106. R. W. Egan, P. H. Gale, G. C. Beveridge, G. B. Phillips, and L. J. Marnett, *Prostaglandins* **16**, 861 (1978).
107. D. H. Nugteren, R. K. Beerthuis, and D. A. van Dorp, *Recl. Trav. Chim. Pays-Bas* **85**, 405 (1966).
108. N. Ogino, S. Yamamoto, O. Hayaishi, and T. Tokuyama, *Biochem. Biophys. Res. Commun.* **87**, 184 (1979).
109. G. J. Roth, E. T. Machuga, and P. Strittmatter, *J. Biol. Chem.* **256**, 10018 (1981).
110. G. J. Roth, C. J. Siok, and J. Ozols, *J. Biol. Chem.* **255**, 1301 (1980).
111. R. W. Egan, P. H. Gale, W. J. A. Vanden Heuvel, E. M. Baptista, and F. A. Kuehl, Jr., *J. Biol. Chem.* **255** 323 (1980).
112. L. J. Marnett, P. Wlodawer, and B. Samuelsson, *J. Biol. Chem.* **250**, 8510 (1975).
113. E. A. Ham, V. J. Cirillo, M. Zanetti, T. Y. Shen, and F. A. Kuehl, Jr., *in* "Prostaglandins in Cellular Biology" (P. W. Ramwell and B. B. Pharriss, eds.), p. 345. Plenum, New York, 1972.
114. R. W. Egan, J. L. Humes, and F. A. Kuehl, Jr., *Biochemistry* **17**, 2230 (1978).
115. R. W. Egan, A. N. Tischler, E. M. Baptista, E. A. Ham, D. D. Soderman, and P. H. Gale, *Adv. Prostaglandin, Thromboxane, Leukotriene Res.* **11**, 151 (1983).
116. G. D. Nordblom, R. E. White, and M. J. Coon, *Arch. Biochem. Biophys.* **175**, 524 (1976).
117. G. R. Dyrkacz, R. D. Libby, and G. A. Hamilton, *J. Am. Chem. Soc.* **98**, 626 (1976).
118. K. Lerch, *Proc. Natl. Acad. Sci. U.S.A.* **75**, 3635 (1978).
119. P. J. O'Brien and A. Rahimtula, *Biochem. Biophys. Res. Commun.* **70**, 832 (1976).
119a. S. Moncada and J. R. Vane, *in* "Biochemical Aspects of Prostaglandins and Thromboxanes" (N. Kharasch and J. Fried, eds.), p. 155. Academic Press, New York, 1977.
120. C. R. Pace-Asciak, *Prostaglandins* **13**, 811 (1977).
121. M. E. Hemler, G. Graff, and W. E. M. Lands, *Biochem. Biophys. Res. Commun.* **85**, 1325 (1978).
122. W. L. Smith and W. E. M. Lands, *J. Biol. Chem.* **247**, 1038 (1972).
123. M. O. Funk, S. H.-S. Kim, and A. W. Alteneder, *Biochem. Biophys. Res. Commun.* **98**, 922 (1981).
124. W. L. Smith and W. E. M. Lands, *Ann. N.Y. Acad. Sci.* **180**, 107 (1971).
125. M. Hemler, W. E. M. Lands, and W. L. Smith, *J. Biol. Chem.* **251**, 5575 (1976).
126. N. Ogino, S. Ohki, S. Yamamoto, and O. Hayaishi, *J. Biol. Chem.* **253**, 5061 (1978).
127. R. Ueno, T. Shimizu, K. Kondo, and O. Hayaishi, *J. Biol. Chem.* **257**, 5584 (1982).
128. F. J. Van der Ouderaa, M. Buytenhek, F. J. Slikkerveer, and D. A. van Dorp, *Biochim. Biophys. Acta* **572**, 29 (1979).
129. P. Strittmatter, E. T. Machuga, and G. J. Roth, *J. Biol. Chem.* **257**, 11883 (1982).
130. F. J. Van der Ouderaa, M. Buytenhek, D. H. Nugteren, and D. A. van Dorp, *Biochim. Biophys. Acta* **487**, 315 (1977).
131. R. W. Egan, P. H. Gale, G. C. Beveridge, L. J. Marnett, and F. A. Kuehl, Jr., *Adv. Prostaglandin Thromboxane Res.* **6**, 153 (1980).
131a. M. E. Hemler, H. W. Cook, and W. E. M. Lands, *Arch. Biochem. Biophys.* **193**, 340 (1979).
132. F. Wudl and T. B. K. Lee, *J. Chem. Soc., Chem. Commun.* p. 61 (1972).
133. H. Nozaki, K. Oshima, and Y. Yamamoto, *Bull. Chem. Soc. Jpn.* **45**, 3495 (1972).
134. C. M. Tatum, Jr., P. A. Benkovic, S. J. Benkovic, R. Potts, E. Schleicher, and H. G. Floss, *Biochemistry* **16**, 1093 (1977).
135. C. E. Holmlund, K. J. Sax, B. E. Nielsen, R. E. Hartman, R. H. Evans, Jr., and R. H. Blank, *J. Org. Chem.* **27**, 1468 (1962).
136. R. M. Dodson, N. Newman, and H. M. Tsuchiya, *J. Org. Chem.* **27**, 2707 (1962).

137. K. Sivarajah, J. M. Lasker, T. E. Eling, and M. B. Abou-Donia, *Mol. Pharmacol.* **21,** 133 (1982).
138. L. J. Marnett, M. J. Bienkowski, W. R. Pagels, and G. A. Reed, *Adv. Prostaglandin Thromboxane Res.* **6,** 149 (1980).
139. T. L. Poulos and J. Kraut, *J. Biol. Chem.* **255,** 8199 (1980).
140. T. L. Poulos and J. Kraut, *J. Biol. Chem.* **255,** 10322 (1980).

CHAPTER **2**

Lipid Peroxide Actions in the Regulation of Prostaglandin Biosynthesis

W. E. M. Lands, R. J. Kulmacz, and P. J. Marshall

Department of Biological Chemistry
The University of Illinois at Chicago
Chicago, Illinois

I. INTRODUCTION

The biological occurrence of lipid peroxidation and other free-radical reactions *in vivo* has been recognized for many years. Many studies have emphasized the damaging effect of these reactive agents on cell membranes and examined the cellular mechanisms that can limit the extent of peroxidation and free-radical injury. However, more recent studies have suggested that lipid peroxidation is also related to the normal function of cells. In particular, studies of prostaglandin biosynthesis have demonstrated that the first compounds produced by the action of cyclooxygenase and lipoxygenase on arachidonic acid are unstable peroxides that can be converted to potent physiological mediators. Furthermore, there is accumulating evidence that the reactions that form these peroxide intermediates

FREE RADICALS IN BIOLOGY, VOL. VI

proceed through stereospecific, controlled free-radical mechanisms initiated by lipid hydroperoxides. Thus, the generation of hydroperoxide activator immediately after stimulation of cells may be an important regulatory mechanism in normal cell physiology. In this chapter we review the evidence that a hydroperoxide activator has a necessary role in prostaglandin biosynthesis, and we consider the consequences of this mechanism in controlling biological events.

II. THE HYDROPEROXIDE REQUIREMENT OF PROSTAGLANDIN H SYNTHASE

Prostaglandin H synthase (PGHS) is a microsomal glycoprotein with a molecular weight of about 70,000. It has been purified to homogeneity from bovine [1] and ovine seminal vesicles [2, 3] and from human platelets [4]. The same polypeptide catalyzes two distinct reactions: the bisdioxygenation of arachidonic acid to form PGG_2 (cyclooxygenase activity) and the reduction of PGG_2 to PGH_2 (peroxidase activity) [5, 6]. Heme is a required cofactor for both activities [1, 4, 7].

The chemical rearrangements that occur during the biosynthesis of PGG_2 have been extensively studied and are the subject of several reviews [8–10]. The 13-L hydrogen is abstracted from arachidonic acid in the rate-determining step, with a considerable kinetic isotope effect for the abstraction of 13-L tritium [11]. In addition, two molecules of oxygen are added to the fatty acid, accompanied by two cyclizations to yield the endoperoxide and cyclopentane structures of PGG_2 [12]. A variety of observations indicate the involvement of free radicals in the enzymatic conversion of arachidonate to PGG_2. The cyclooxygenase reaction is affected by many compounds known to be free-radical inhibitors, and it can be mimicked by model free-radical reactions [for review, see 10]. An ESR signal attributed to a free radical has been reported to occur when microsomal preparations of PGHS are incubated with arachidonate [13, 14]. The significance of this signal to an understanding of the cyclooxygenase mechanism is diminished somewhat by the fact that it is not observed with the pure protein [14] and, when present, it seems to be associated primarily with the peroxidase activity of PGHS [15].

Two other heme-containing dioxygenases, tryptophan dioxygenase and indoleamine dioxygenase, which are thought to activate molecular oxygen in the course of catalysis, are sensitive to inhibition by the oxygen analog carbon monoxide [16–18]. The failure of carbon monoxide to inhibit the cyclooxygenase [6, 19, 20] has led to the suggestion that the biosynthetic reaction involves enzymatic activation of the fatty acid substrate rather than activation of molecular oxygen [10, 20]. Such activation of an organic substrate has been proposed for the mechanism of another dioxygenase, protocatechuate 2,4-dioxygenase [21].

An important aspect of the PGHS-cyclooxygenase activity is that it exhibits autoaccelerative reaction kinetics. This can be clearly seen when the oxygen consumption is monitored during the reaction of the synthase with arachidonic acid at pH 9.5 [20]. The time taken for the cyclooxygenase to reach maximal velocity is decreased by added lipid hydroperoxides in general [22] and PGG_2 in particular [22, 23]. Although H_2O_2 at 0.1 mM could stimulate the reaction, the long-chain lipid hydroperoxides are effective at much lower concentrations [22, 24]. Thus, experimenters must use care in removing lipid peroxides from the substrate, arachidonic acid, because only 0.01% contamination of 100 μM arachidonic acid will abolish lags. The important role of peroxides in the mechanism of PGHS-cyclooxygenase is further demonstrated by the dramatic effect of added glutathione peroxidase on the cyclooxygenase activity [25, 26]. This peroxide scavenger can inhibit cyclooxygenase activity even when added after the reaction with arachidonate is in progress [20, 27]. Thus, the hydroperoxides that act as activators of cyclooxygenase are continuously required for its activity. The requirement for activator peroxide has important implications for *in vivo* regulation of the enzyme activity, because small increases in the level of cellular peroxides would be expected to result in dramatic increases in cyclooxygenase activity [28]. Activation of the cyclooxygenase reaction by peroxide does not seem to involve PGHS-peroxidase activity becasue the PGHS reconstituted with manganoprotoporphyrin IX has no peroxidase activity itself [29], but it exhibits a cyclooxygenase activity that is sensitive to inhibition by glutathione peroxidase [R. J. Kulmacz, unpublished results]. Thus, the two catalytic events can proceed independently of each other, and the naturally occurring heme-containing enzyme can both form and destroy lipid hydroperoxide.

The paradox of the capacity of the PGHS-cyclooxygenase activity, which requires peroxide as activator, to function in the presence of PGHS-peroxidase activity, which consumes peroxide, can be resolved by the fact that the concentration of peroxide required to activate the cyclooxygenase [24, 30] is about two orders of magnitude lower than the K_m of PGHS-peroxidase [6, 30]. Thus, the cyclooxygenase can be activated by peroxide levels that are too low to be efficiently removed by the PGHS-peroxidase activity. However, when additional peroxidase activity is present in sufficient quantity (such as the cellular glutathione peroxidase might provide *in vivo*) the PGG_2 produced by the cyclooxygenase can be removed rapidly enough to keep the accumulated level of peroxide below that needed to activate the cyclooxygenase fully. It is the balance between the generation and removal of hydroperoxides in a tissue that determines the ambient level of peroxide activator and the proportion of cyclooxygenase that is catalytically active. Cyclooxygenase activity tends to be manifested when lipid hydroperoxide concentrations range between 0.01 and 1 μM, the lower concentration that begins to activate the enzyme [30] and the upper level that starts to cause irreversible inactivation of the synthase [14].

The mechanism proposed for PGHS is depicted in Fig. 1; the cyclooxygenase catalytic cycle is on the right of the figure, and that of the peroxidase is on the left [20, 31, 32]. In this proposal the resting heme-containing holoenzyme binds activator hydroperoxide (ROOH) in step 1. Activation of the cyclooxygenase occurs in step 2 with the formation of enzyme-bound hydroperoxyl radical. Substrate fatty acid (FH) is then bound and converted to the F· radical by transfer of the 13-L hydrogen to the enzyme-bound hydroperoxyl radical. The F· radical subsequently reacts with 2 mol of oxygen to produce the PGG_2 radical (step 5), which then regenerates the enzyme-bound hydroperoxyl radical and diffuses from the enzyme (step 6), completing the cycle. Radical intermediates (ROO· and F·) can be quenched by radical chain transfer reagents (AH in Fig. 1), which break the catalytic cycle. This would presumably be the mechanism of the inhibition of cyclooxygenase that occurs with antioxidants under some conditions (see Section VI).

The failure of the n-3 type of fatty acids to be effectively converted to prostanoids [33] appears to be due to a lipid hydroperoxide requirement that is much higher than that for the n-6 type of acids [24, 34]. This phenomenon thus permits the n-3 acids to be effective competitive inhibitors of prostaglandin biosynthesis from the n-6 acids *in vivo*, where cellular peroxide concentrations are low. Although trienoic prostaglandins can be synthesized *in vitro* from the eicosapentaenoic acid, 20:5 (n-3) [34, 35], there is no clear evidence for their biosynthesis *in vivo*. Thus, the decreased thrombotic tendencies of Eskimos [36] and other

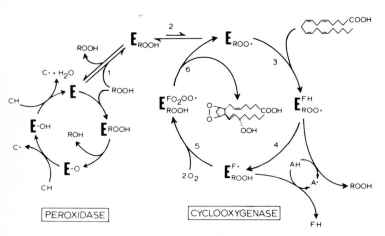

Fig. 1.   Proposed free-radical mechanism for PGHS. Abbreviations: ROOH, lipid hydroperoxide; FH, arachidonic acid; AH, free-radical-trapping agent; CH, peroxidase cosubstrate.

humans ingesting appreciable amounts of n-3 fatty acids [37] is probably a reflection of the high peroxide requirement for reaction of the cyclooxygenase with these acids. In addition, this situation emphasizes the essential role of lipid peroxides in prostaglandin biosynthesis and reaffirms the expected low value for steady-state concentrations of lipid peroxides *in vivo*.

III. EVIDENCE FOR A PHYSIOLOGICAL ROLE OF PEROXIDES DURING PROSTAGLANDIN SYNTHESIS

The stimulation of prostaglandin synthesis by hydroperoxides or by agents that affect hydroperoxide metabolism have been reported by a number of investigators. Morse *et al.* [38] found that synchronous spawning in mollusks can be stimulated by exogenous H_2O_2 (5 mM) and inhibited by catalase or aspirin. A cell-free preparation of PGHS from the mollusks was stimulated proportionally by H_2O_2 up to 0.3 mM, with higher concentrations being progressively inhibitory. Furthermore, this H_2O_2-induced stimulation was also blocked by either catalase or aspirin. Stewert *et al.* [39], using smooth muscle strips prepared from lung, showed that H_2O_2 causes contraction, which is maximal with 1 mM. This contraction is most likely mediated by an increase in prostaglandin synthesis because the contraction is blocked by indomethacin or meclofenamate. Schaefer *et al.* [40, 41] noted a close relationship between the stimulation of prostaglandin biosynthesis and the production of H_2O_2 by monoamine oxidase in the particulate fraction of rat brain homogenates. This group has clearly demonstrated that the enzymatic decomposition of catecholamine analogs by monoamine oxidase produces H_2O_2, which in turn stimulates $PGF_{2\alpha}$ synthesis. This stimulation is blocked by an inhibitor of monoamine oxidase, pargyline (0.2 mM), by catalase, or by indomethacin (0.2 mM) [42]. When H_2O_2 is continuously generated (94 μM/20 min) by glucose oxidase during the incubation period, there is a marked stimulation of $PGF_{2\alpha}$ formation. This effect of glucose oxidase is abolished by catalase or indomethacin, but not by pargyline. The increase in $PGF_{2\alpha}$ synthesis is not due to an increase in the mobilization of endogenous arachidonate from phospholipids because $PGF_{2\alpha}$ synthesis is enhanced by H_2O_2 even in the presence of exogenous arachidonic acid (5–50 μg/ml). Therefore, Schaefer *et al.* concluded that the H_2O_2 directly stimulates PGHS.

Polgar and Taylor [43] reported that ascorbic acid (5–15 mM) stimulates prostaglandin synthesis 10- to 20-fold in lung fibrobalsts, and this stimulation is blocked by catalase (220 units/ml). Added H_2O_2 (0.88 mM) also increases the synthesis of prostaglandin E_2 (PGE_2) (6-fold), 6-keto-$PGF_1\alpha$ (10-fold), thromboxane B_2 (8-fold), and either catalase or very high concentrations of H_2O_2 (4.4 mM) can prevent this increase in synthesis. Ascorbic acid (15 mM) also increases the release of arachidonate 3-fold, and this increase is likewise blocked by catalase. The increase in the release of engoenous arachidonate accounts for only a fraction of the increase in prostaglandin synthesis and does not account totally for the 10- to 20-fold increase observed. However, as indicated by Polgar and Taylor [44], the actions of ascorbic acid may not be identical for all cell types. In smooth muscle cells ascorbic acid activates release but inhibits synthesis, whereas in endothelial cells it has no effect on release but inhibits prostaglandin synthesis. These results may be a reflection of the varied capacity of individual cell types to handle hydroperoxides and their varied sensitivity to destructive actions of high levels of H_2O_2 on PGHS.

Baud et al. [45] were interested in the effects that oxygen radicals generated by extracellular xanthine oxidase might have on prostaglandin synthesis in isolated glomeruli. Their experimental model was designed to mimic the conditions observed during the accumulation of phagocytic cells in glomerular capillaries. When a mixture of xanthine and xanthine oxidase was added, prostaglandin synthesis was clearly enhanced. Increasing concentrations of catalase progressively inhibited this stimulation of PGE_2 and $PGF_{2\alpha}$ synthesis, whereas the radical scavengers mannitol (0.1 and 10 mM) and chlorpromazine (0.25 mM) had no effect. Hydrogen peroxide at concentrations of 1 to 100 μM stimulated PGE_2 and $PGF_{2\alpha}$ synthesis, and this action was blocked by mepacrine (0.1 mM), a putative phospholipase inhibitor [46]. The release of preincorporated [^{14}C]arachidonate in isolated glomeruli was stimulated in the presence of xanthine oxidase. These results led Baud et al. [45] to conclude that H_2O_2 increases phospholipase activity and therefore enhances prostaglandin synthesis through an increase in the availability of arachidonate. The failure of chlorpromazine to block prostaglandin synthesis detracts from the hypothesis because the concentration used can inhibit phospholipase action [47]. Perhaps phospholipase action is not the primary rate-limiting reaction, but indequate amounts of activating peroxide restrain prostaglandin biosynthesis.

A direct activation of PGHS by lipid hydroperoxide may be reflected in the results of Adcock et al. [48]. This group noted that perfusion of resting or antigen-stimulated guinea pig lung with 15-hydroperoxyarachidonic acid (3–60 nmol/ml/min) or 13-hydroperoxylinoleic acid (60 nmol/ml/min) causes a marked enhancement of the release of rabbit aorta-contracting substance (thromboxane A_2). The fatty acid hydroperoxides also increase (3- to 15-fold) the antigen-stimulated release of the lipoxygenase pathway product, slow-reacting substance of anaphylaxis. Mickel and Horbar [49] found that platelets respond with a slow aggregation to hydroperoxy-free arachidonate, whereas a rapid response results from the exposure to hydroperoxy arachidonate. The concentration of exogenous hydroperoxyarachidonic acid was approximately 5 mM in this work. Even though the authors could partly recover the unreacted peroxide, they contended that the peroxide did not directly cause aggregation. This concentration, however, is well in excess of that necessary to activate PGHS. Diamant *et al.* [50] reported a correlation between lipid peroxidation and prostaglandin levels with gestational age of placental tissue and suggested that lipid peroxidation is connected to the initiation of labor by way of increased prostaglandin synthesis.

Indirect evidence for a hydroperoxide activator comes from the action of transition metals that promote hydroperoxide formation and from the agents that prevent the formation of or eliminate hydroperoxides. Vargaftig *et al.* [51] observed that copper ions and aminothiols potentiate arachidonic acid-induced platelet aggregation and the release of pharmacologically active substances. They speculated that free radicals and/or H_2O_2 generated during the oxidation of aminothiols in the presence of metal ions might be required for peroxidation of arachidonic acid with the resultant generation of active substances other than prostaglandins. In 1975 Vargaftig *et al.* [52] confirmed that Cu^{2+} and and Zn^{2+} (1–5 mM) potentiate the platelet aggregation induced by subthreshold amounts of arachidonic acid (25–50 μM). Certain metal chelators and catalase block this arachidonic acid-induced aggregation, and catalase also prevents the formation of rabbit aorta-contracting substance (thromboxane A_2). The effects of metal-chelating agents or catalase on prostaglandin synthesis appear to be specific because metal-chelating agents or catalase do not affect adhesion induced by ADP (1–5 μM). These investigators concluded that H_2O_2 or an equivalent chemical species (possibly a lipid hydroperoxide) is generated by the valence change of the metal. The hydroperoxide interacts with arachidonic acid to activate prostaglandin synthesis, and this process is inhibited by catalase or glutathione peroxidase, which degrades the hydroperoxide, or by indomethacin, which directly inhibits prostaglandin synthesis. Catalase was used also by Panganamala *et al.* [53] and Rahimtula and O'Brien [54] to prevent prostaglandin synthesis in seminal vesicle microsomes. Panganamala *et al.* [53] were interested in the roles of peroxides and free radicals in prostaglandin synthesis and platelet aggregation. The addition of catalase to bovine seminal vesicles blocked the conversion of arachidonic acid (50 μM) PGE$_2$ or PGF$_2$ by greater than 75%. Panganamala *et al.* [53] also demonstrated that the radical scavengers aminotriazole (1 mM) and chlorpromazine (0.24 mM) inhibit prostaglandin synthesis in microsomes prepared from bovine seminal vesicles and also prevent platelet aggregation. They speculated that an active intermediate is required for the oxygenation of arachidonic acid, and this intermediate if formed by the enzymatic breakdown of H_2O_2. Inhibitors of active oxygen formation, such as cyanide, azide, and 3-amino-1,2,4-triazole, and scavengers of active oxygen species, such as α-tocopherol, dimethyl sulfoxide, methional, and nitroblue tetrazolium, have been reported to block prostaglandin formation by quenching the generation of hydroperoxides [52, 53, 55, 56].

Lipid peroxides play a role in both activating and inactivating PGHS. As the level of lipid peroxide increases above that needed for the activation of PGHS, increased inactivation or inhibition occurs. This inactivation is separate from the self-inactivation that occurs during cyclooxygenase catalysis [57]. The self-inactivation occurs despite the protective action of glutathione peroxidase even when added glutathione peroxidase maintains the peroxide levels at submicromolar concentration. Thus, it is not associated with an accumulation of undesirably high levels of cellular peroxides. Egan et al. [58] suggested that peroxidase action on the hydroperoxide releases oxidizing equivalents, which deactivate PGHS. They noted that the peroxidase cosubstrates phenol and methional promote PGH_2 formation at the expense of PGG_2 and increase the initial rate and the extent of reaction before deactivation, although they failed to emphasize that the self-catalyzed inactivation is not prevented. The destructive action of micromolar amounts of peroxides was shown by Seigel et al. [59], who demonstrated that added 12-hydroperoxyeicosatetraenoic acid (12-HPETE) inhibits PGHS activity in intact platelets with a half-maximal concentration of 15 μM, whereas in homogenates of platelets the half-maximal concentration is 3 μM. The sensitivity of PGHS to peroxides was particularly evident in attempts to purify the enzyme. High yields were possible only after recognizing and accommodating that sensitivity [60]. The destruction of PGHS by hydroperoxide was quantitatively assessed by Hemler and Lands [20]. They compared the rate of inactivation due to added hydroperoxide with the rate of self-catalyzed inactivation during catalysis. Destruction of PGHS activity was hastened by the addition of 15-hydroperoxyarachidonic acid. Inactivation with 7.3 μM hydroperoxide occurred at a rate (4.2 min^{-1}) comparable to the rate of self-inactivation (5.1 min^{-1}). The addition of phenol (0.67 mM), a cosubstrate for the peroxidase activity of PGHS, reduced the rate of destruction by 7.3 μM peroxide 60-fold, from 4.2 to 0.07 min^{-1}, but offered essentially no protection against the self-catalyzed inactivation. It thus appears that the PGHS is inactivated by micromolar levels of hydroperoxide but not appreciably inactivated by submicromolar amounts of lipid hydroperoxide.

Fujimoto and Fujita [61] reported that lipid peroxidation induced by 0.4 mM Fe^{2+} and 1 mM vitamin C causes a rapid decrease in PGE synthesis in kidney slices. The level of thiobarbituric acid-reactive material produced under the incubation conditions was greater than 40 μM when the level of prostaglandin synthesis was decreased by 70%. The data do not permit a distinction to be made between the cyclooxygenase step and the 9-keto isomerase step as the site of inhibition.

Prostaglandin I synthase (PGIS) is especially sensitive to the deleterious action of peroxides. Moncada et al. [62] observed that preincubation with micromolar amounts of 15-HPETE (IC_{50} = 1.2 μM) inhibits PGIS activity in blood vessel microsomes. Ham *et al.* [*63*] investigated the destructive action on PGIS of oxidants generated by PGHS activity. Preincubation of ram seminal vesicles microsomes with 15-hydroperoxyarachidonic acid (74 μM) for 1 min inhibits PGI synthesis by 70%, whereas the addition of phenol or 2-aminomethyl-4-*tert*-butyl-6-iodophenol (0.1 mM), substrates for PGHS-peroxidase activity, prevents this inhibition. In contrast, inhibition resulting from PGG_2 is neither time dependent nor blocked by 2-aminomethyl-4-*tert*-butyl-6-iodophenol, suggesting that PGG_2 directly inhibits the PGIS. In contrast, thromboxane A_2 synthase is completely resistant to the action of the presumed oxidants released by the hydroperoxides. Beetens *et al.* [*64*] confirmed the results of Ham *et al.* and observed that vitamin C (0.5 mM) increases PGI synthesis (as measured by 6-keto-$PGF_{1\alpha}$) from arachidonic acid by 85%. When PGH_2 is used as the substrate, vitamin C (0.5 mM) increases PGI_2 synthesis by only 12%. The authors concluded that vitamin C neutralizes the oxidative species generated during PGHS activity and thus protects PGIS. The observation that PGIS is inhibited by micromolar amounts of hydroperoxide, whereas relatively higher levels seem to be required to inhibit PGHS has major significance in human health. Prostacyclin, which is synthesized by vascular endothelial cells, antagonizes platelet aggregation and also prevents the proliferation of smooth muscle cells. A hydroperoxide concentration that impairs PGIS activity but activates PGHS would lead to the unopposed synthesis and action of thromboxane A_2. Protection against accumulating undesirably high levels of hydroperoxides is provided physiologically by cellular peroxidases. In platelets the peroxidase activity that convert 12-HPETE to 12-hydroxyeicosatetiaenoic acid (12-HETE) appears to be half-maximally saturated by 10 μM of exogenous 12-HPETE [*59*]. The K_m for lipid hydroperoxides with glutathione peroxidase has a similar magnitude [*65*]. Therefore, PGHS susceptibility to inactivation would be a function of the level of cellular peroxidase activity relative to the intensity of the hydroperoxide challenge.

IV. FORMATION OF HYDROPEROXIDES

The mechanism(s) by which hydroperoxides are formed after appropriate cell stimulation remains to be elucidated. Accurate measurements of endogenous hydroperoxide levels are difficult to perform because experimental handling of the tissue can lead to uncontrolled increases in peroxide levels and any detected peroxide may be artifactual rather than physiological. The level of peroxide generally estimated to be present in normal cells is low. The potential for forming peroxides can be inferred, nevertheless, from the presence of enzymes restricting their accumulation (e.g., catalase [*66*] and glutathione peroxidase [*65*] and from indirect evidence of their increase in some pathological conditions [67–69]. The exhalation of hydrocarbons such as ethane and pentane has been offered as evidence that lipid peroxidation does occur *in vivo* [70–72]. Measurements employing the thiobarbituric acid assay have been used to monitor lipid peroxidation and to indicate changes in peroxidation following insult or trauma to the tissue [67, 69]. Several excellent reviews have been written concerning the mechanism of lipid peroxide formation [68, 69, 73, 74] and the controls restricting peroxide accumulation [66, 75, 76]. In this section we highlight features of these processes with emphasis on their role in the generation of a hydroperoxide activator of PGHS.

Hydroperoxides can result directly from an enzymatic (lipoxygenase) process or indirectly as a consequence of oxidative free-radical attack on unsaturated fatty acids. The latter event may be important in initiating changes because all fatty acid oxygenases studied in detail require a lipid peroxide activator (cyclooxygenase [25, 57], soybean lipoxygenase [77, 78], and reticulocyte lipoxygenase [79]). Thus, these enzymes can catalytically amplify the presence of a small amount of any activating peroxide by forming larger amounts of their specific hydroperoxide products. The initiating lipid peroxide molecules are generally regarded to be formed from "activated" oxygen species intermediate between O_2 and HOH. Heme-containing enzymes, in general, and NADPH-cytochrome c reductase, in particular, have been implicated in the production of "activated" oxygen, which can cause lipid peroxidation [80]. The production of radicals, which can cause peroxidation, occurs especially during the activity of oxidative flavoprotein enzymes. NADPH oxidase, which plays an important role during the oxidative burst of phagocytic cells, transfers an electron to molecular oxygen to yield the superoxide anion [Eq. (1)] [81, 82].

$$2 O_2 + NADPH \rightarrow 2 O_2^- + NADP^+ + H^+ \tag{1}$$

Superoxide dismutase (SOD) converts superoxide anion to H_2O_2

$$2 O_2^- + 2 H^+ \xrightarrow{\text{SOD}} H_2O_2 + O_2 \tag{2}$$

[Eq. (2)] [83]. Hydrogen peroxide itself can serve as a PGHS activator at concentrations of greater than 20 μM [22], or it can initiate the formation of reactive species such as hydroxyl radicals and singlet oxygen, which have been proposed to be rate limiting in the process of radical-mediated lipid peroxidation [84–87]. Hydroxyl radicals that may be formed in cellular systems are thought to result from a metal-catalyzed Haber–Weiss reaction [88, 89], which can also produce

$$O_2^- + H_2O_2 \xrightarrow{(Fe^{3+})} OH^- + OH\cdot + O_2 \quad (\text{or } {}^1O_2) \tag{3}$$

singlet oxygen [Eq. (3)] [*90*]. Singlet oxygen, by far the most reactive and elusive of the oxygen species [*91, 92*], has also been implicated as a product in the reaction between H_2O_2 and hypohalite, which is formed by the myeloperoxidase (MPO)-catalyzed reaction between H_2O_2 and a halide [Eqs. (4) and (5)] [*93, 94*].

$$H_2O_2 + Cl^- \xrightarrow{\text{MPO}} OCl^- + H_2O \qquad (4)$$

$$OCL^- + H_2O_2 \rightarrow H_2O + Cl^- + {}^1O_2 \qquad (5)$$

The participation of these excited states of oxygen in lipid peroxidation has been inferred by studying the effect on microsome or liposome systems of agents that are known to scavenge a particular form of oxygen. The production of hydroxyl radicals from the interaction of superoxide and H_2O_2 has been proposed to be responsible for membrane lipid peroxidation *in vitro* [*85*], and this peroxidation can be diminished by the addition of SOD, catalase, and scavengers of hydroxyl radicals (e.g., ethanol, mannitol, or benzoate). In addition to scavengers of hydroxyl radicals, molecules that are thought to specifically scavenge singlet oxygen also tend to prevent lipid peroxidation in microsomes [*95*]. To achieve control over agents that initiate lipid peroxidation, the cell can muster a variety of defense mechanisms. Antioxidants and scavenging agents can quench the toxic agents and radicals before lipid peroxidation can proceed. Vitamin E, cholesterol, and free sulfhydryl groups of proteins represent scavengers that can limit membrane oxidation [*96*]. As noted earlier in Section III, SOD [*83*] and catalase [*66*] serve to catabolize O_2^- and H_2O_2 to water.

Although the chemical events in lipid peroxide formation *in vivo* and *in vitro* are not satisfactorily understood, an initial step may involve the formation of an organic free radical by hydrogen abstraction of an allylic hydrogen, subsequently rearranging to a conjugated *cis,trans*-diene concomitant with the incorporation of oxygen to yield the peroxyl radical [*69, 73, 74, 96*] (Fig. 2). The abstraction of the allylic hydrogen is thought to be due to a radical-like intermediate $(X \cdot)$ whose transient identity is unknown and whose existence may depend on complexed metals. Aust and co-workers [*74, 97*] have proposed a scheme in which the complexed metal undergoes a univalent reduction driven by O_2^-, H_2O_2, or NADPH reductase. The putative metal complex is expected to be a strong oxidant and able to circumvent the spin restriction in a reaction between oxygen and the fatty acid. It also imparts the free-radical characteristic to the reaction. Metals such as cobalt, copper, iron, and manganese that undergo a univalent redox reaction can participate in the promotion of autoxidation and enzymatic peroxidation. Of this group, iron has been found to be the most active catalyst for lipid peroxidation, with iron–heme compounds being the best promoters of lipid peroxidation [*98, 99*]. The newly formed peroxyl radical leads to additional lipid

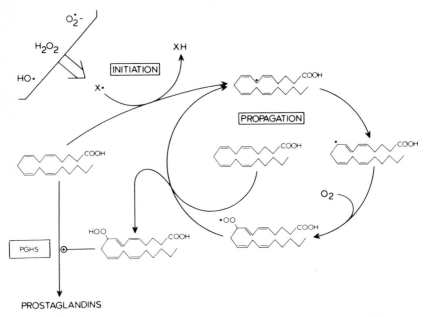

Fig. 2.   Nonenzymatic formation of the lipid hydroperoxide activator of PHGS.

peroxidation, and this new reaction quickly supplants the original lipid peroxidation reaction that was initiated by the active oxygen products [90]. Thus, lipid peroxidation can progress as a chain reaction. It has been estimated that each radical formed goes through 8 to 14 propagation cycles before being terminated [99a, 99b].

Lipid hydroperoxides can also be formed by a controlled, enzyme-catalyzed process in the presence of lipoxygenases [100, 101]. This type of enzyme has been found along with PGHS in leukocytes and platelets, and its hydroperoxy fatty acid product could well serve as an activator for either the lipoxygenase itself [31, 59] or PGHS. An impairment of lipoxygenase activity might manifest in a decreased capacity to form the peroxide needed for PHGS activity. Such an event may be involved in the relationship between platelet lipoxygenase deficiency and bleeding disorders [102, 103]. Patients whose platelets are deficient in lipoxygenase show a marked increase in bleeding time. Their platelets do not aggregate *in vitro* when appropriately stimulated even though assays with exogenous arachidonic acid have indicated that PGHS is present [103]. These results can be explained by a deficiency in lipoxygenase that might result in a failure to form sufficient lipid hydroperoxide activator; as a consequence PHGS remains inactive, thromboxane A_2 is not synthesized, and platelet aggregation does not proceed.

V. IMPAIRMENT OF HYDROPEROXIDE FORMATION BY VITAMIN E

The addition of antioxidants such as vitamin E and butylated hydroxytoluene is routinely used to prevent the oxidation of polyunsaturated fatty acid. However, the use of these antioxidants for investigating the role of peroxides in prostaglandin synthesis has yielded seemingly contradictory results. Several studies have been performed in which the dietary intake of vitamin E has been manipulated. With skeletal muscle microsomes prepared from vitamin E-deficient rabbits, PGE_2 and $PGF_{2\alpha}$ synthesis in the presence of arachidonate was less than with similar preparations from rabbits fed vitamin E [104]. Okuma et al. noted that PGI synthesis in rat aorta was reduced greater than 50% in rats maintained on a vitamin E-deficient diet for 4 to 10 months [105]. Lipid peroxidation as measured by the thiobarbituric acid assay increased 250% in those vitamin E-deficient rats [105]. Supplementing the deficient diet with 20 U of vitamin E per 100 g basal diet for 2 months restored PGI synthesis to normal and also lowered the lipid peroxidation levels. In that study [105] the effects of lipid peroxidation on PGI synthesis most likely reflected an inactivation of PGIS rather than an effect on PGHS. Okuma et al. reported that 15-hydroperoxyarachidonic acid inhibits PGI synthesis half-maximally at 1 μM, which is a similar concentration reported earlier for the inactivation of PGIS [62]. However, Hope et al. [106] reported an inverse relationship between the levels of vitamin E and PGE_2 in rat serum. This group found that the PGE_2 concentration was 7-fold greater in serum from rats fed a vitamin E-deficient diet and that as the dietary level of vitamin E increased the amount of PGE_2 generated during clotting decreased. Machlin et al. [107] observed that platelet aggregation was more responsive during dietary vitamin E depletion (perhaps reflecting increased prostaglandin and thromboxane synthesis). Aggregation followng collagen stimulation was 50% greater for platelets obtained from rats maintained on a vitamin E-deficient diet for 9 weeks. This increase in aggregating capacity may reflect an increase in prostaglandin synthesis resulting from slightly elevated levels of hydroperoxide activators. By the maintenance of adequate vitamin E levels, lipid peroxidation can be held in check. Once a dietary restriction in vitamin E has been imposed, lipid peroxidation can proceed, and hydroperoxide activator levels can increase. Conversely, if vitamin E levels were increased, the likelihood of lipid peroxidation would be decreased, as would the potential for increased platelet aggregation. Vitamin E, at elevated levels (IC_{50} = 0.5 mM), can directly inhibit cyclooxygenase [33]. In vitro addition of vitamin E to human platelets inhibits aggregation [56, 108] perhaps by direct action or by impairing the formation of lipid peroxides within the platelet. White et al. [56] indicated that the combination of two agents, nitroblue tetrazolium and vitamin E, prevents free-radical generation and inhibits the activation, secretion, and stickiness of platelets stimulated with epinephrine

(5.5 μM), collagen (30 $\mu g/ml$), thrombin (0.2 U/ml), or ADP (9 μM). The inhibitory influence of nitroblue tetrazolium and vitamin E on platelet activation seems to be similar to the pattern of suppression obtained with aspirin and indomethacin (both potent inhibitors of platelet prostaglandin synthesis). These results led White *et al.* to propose that the generation of an activative oxygen product is necessary for prostaglandin synthesis and that nitroblue tetrazolium and vitamin E quench the production of this active oxygen product.

Biphasic actions of vitamin E were noted by Goetzl for lipoxygenase [*109*]. The generation of lipoxygenase products (e.g., 5-HETE, 11-HETE, and 5,12-HETE) in human neutrophils *in vitro* was enhanced up to 250% by 30 to 60 μM vitamin E, whereas the lipoxygenase activity was inhibited when the concentration was raised above 0.25 to 1.0 mM [*109*]. Goetzl proposed that these results were not due to interference with the mobilization of arachidonate from phospholipids because the identical biphasic response was seen in the presence of exogenous arachidonic acid.

VI. REMOVAL OF HYDROPEROXIDES

In light of the consequences of lipid peroxide action on the enzymes of arachidonate metabolism and on other nonspecific processes, the suppression of lipid peroxide formation and the removal of peroxides after formation are of considerable importance in the maintenance of normal cell function. The synthesis of prostaglandins in cells can be characterized as being under continual limitation by the events that reduce the available peroxide [*26*]. Two important enzyme activities for the removal of lipid peroxides are glutathione peroxidase and glutathione transferase. By controlling the concentration of available lipid hydroperoxide, these enzymes influence the potential of the cell for prostaglandin synthesis.

Glutathione peroxidase was the first major enzyme identified in the catabolism of lipid peroxides [*75, 110, 110a, 111*]. Glutathione peroxidase catalyzes the reduction of hydroperoxides to the corresponding alcohols and is present in various amounts in mitochrondria and cytosol [*111*]. This selenium-containing, soluble enzyme is composed of four subunits (19,000–23,000 MW per subunit) and has a maximal velocity of 74 $\mu M/min$ *in vitro* [*65*]. Flohe speculated that this enzyme can react with any hydroperoxide regardless of its nature and that the only restraint on its activity appears to be due to steric hindrance or lack of accessibility, as would be the case for membrane-bound peroxides [*65*]. The activity of glutathione peroxidase is also dependent on the availability of glutathione, the cosubstrate in the reaction ($K_m = 0.2$ mM) [*65*]. In living systems the relatively abundant glutathione (1–5 mM) and the excess reducing capacity of glutathione reductase guarantee that sufficient reduced glutathione exists to maintain the glutathione peroxidase in an active form. In this regard Flohe [65] hypothesized that moderate variations in cellular glutathione concentration and/or glutathione reductase capability would not significantly affect the glutathione-dependent hydroperoxide catabolism. This hypothesis was in part substantiated by the findings of Benohr and Waller [112], who demonstrated that a substantial decrease in glutathione reductase does not make human erythrocytes more suspectible to lipid peroxide stress.

In contrast, impairment of glutathione peroxidase activity either by dietary manipulation or by drastic reduction of glutathione levels may lead to an elevated accumulation of lipid peroxides with a subsequent stimulating effect on PGHS. For example, a diet deficient in selenium, an essential cofactor for glutathione peroxidase, has been demonstrated to lead to a decrease in glutathione peroxidase activity and a resultant increase in lipid peroxidation [113–115]. Certain metal ions such as cadmium (K_i = 1.1 μM) and zinc (K_i = 3.7 μM) can inhibit glutathione peroxidase activity by competing with glutathione [116, 116a]. A homozygous deficiency in glutathione biosynthesis or in glutathione reductase activity may result in glutathione levels below 10% of normal [117]. Using isolated hepatocytes, Anundi et al. [118] have shown that a nearly complete depletion of glutathione evokes lipid peroxidation. A reduction in glutathione levels has been demonstrated to correlate with lipid peroxidation induced by mixed-function oxidase activity [119]. Younes and Siegers [120] depleted hepatic glutathione levels by drug treatment and quantified the level of cellular glutathione necessary to protect against microsomal lipid peroxidation. Using liver microsomes from phenobarbital-treated rats, they reported that only after the hepatic glutathione concentration had decreased to a critical value of approximately 20% of the initial level was enhanced lipid peroxidation detected. They also observed that inhibitors of mixed-function oxidases or exogenous glutathione (ED_{50} = 1μM) blocked the peroxidation.

The importance of glutathione peroxidase in regulating peroxide levels and controlling prostaglandin synthesis was indicated by Kawaguchi et al. [121]. Using platelets from hypercholesterolemic rabbits, they noted a 92% decrease in glutathione peroxidase activity and a 70% decrease in glutathione correlated with a 5-fold increase in thromboxane B_2 synthesis and a 2-fold increase in PGH synthesis. They concluded that glutathione and glutathione peroxidase regulate the activity of PGH synthase in platelets.

A second pathway for lipid peroxide disposal, catalyzed by glutathione S-transferase, has been identified in the rat [122–124]. Glutathione S-transferase appears to be a family of multifunctional enzymes that play a role in the catabolism of both endogenous and exogenous substances [125, 126]. These cytosolic enzymes also utilize glutathione (K_m ≃ 0.1 mM) [122] to detoxify the hydroperoxide moiety but differ from glutathione peroxidase in several features. As isolated, glutathione S-transferases are not selenoproteins, and they appear to be more selective in their peroxide substrate requirement (H_2O_2 is not a substrate) [122]. However, a transferase has been reported to catalyze the oxidation of glutathione when acting on organic nitrates [127] and the conjugation of the sulfhydryl group of reduced glutathione with a variety of electrophilic compounds, leading to the formation of water-soluble derivatives [128, 129]. Glutathione S-transferase has also been associated with isomerization of the prostaglandin endoperoxide PGH_2 to PGD_2 [130].

The K_m of glutathione S-transferase for hydroperoxide appears to be higher than that of glutathione peroxidase, and the turnover rate is only 1/300 that of glutathione peroxidase [122]. From a consideration of these kinetic parameters it would appear that glutathione S-transferase is not as effective as glutathione peroxidase on a mole basis in reducing hydroperoxides and may become significant in the removal of hydroperoxides only when it has greater access to the substrate peroxide or when its activity greatly exceeds that of glutathione peroxidase.

Lawrence and Burk [123] measured the tissue distribution of glutathione peroxidase and glutathione S-transferase activities. Glutathione S-transferase activity was measurable in 11 tissues and ranged from 23% of the total peroxide-removing activity in adipose to 91% in testis. Species differences were also noted: Glutathione S-transferase activity accounted for 35% of the total peroxide-removing activity in rat liver, whereas it accounted for 100% of the activity in guinea pig liver.

Increases in glutathione and glutathione S-transferase can occur when there is a decrease in glutathione peroxidase activity. When glutathione peroxidase activity was virtually abolished by a selenium-deficient diet, Lawrence et al. found that glutathione S-transferase activity in rat liver was increased [131]. Hill and Burk [132] examined the combined effects of selenium and vitamin E deficiencies on glutathione levels in rats. The plasma glutathione level was increased 2-fold in the selenium-deficient rat, and the glutathione released from isolated perfused liver was increased 4-fold. Using rat hepatocytes, they found no change in cell viability. After 5 h in culture, selenium-deficient hepatocytes increased both their cellular glutathione levels (1.4-fold) and their capacity to biosynthesize glutathione (1.8-fold). These combined observations support the hypothesis [131] that the decline in glutathione peroxidase can partially be compensated for by accelerated glutathione synthesis and increased glutathione S-transferase activity to maintain protection against lipid peroxidation.

McCay and co-workers [133] reported the presence of a glutathione-dependent, heat-labile factor in rat liver cytosol, which effectively protects against lipid peroxidation. It is clear from both malondialdehyde formation and gas chromatographic analysis of the polyunsaturated fatty acids that peroxidative attack on membranes is prevented by this glutathione-dependent cytosolic factor. Partially purified glutathione peroxidase added to the rat liver microsomes does not mimic the protective effect of the cytosolic factor, nor does it reduce the level of peroxidized microsomal lipids. Burk *et al.* [124] also demonstrated that a cytosolic, glutathione-dependent enzyme in rat liver protects against lipid peroxidation in the NADPH microsomal lipid peroxidation system and that this protection is in part due to glutathione *S*-transferase. These results indicate that the molecular structure of the microsomal lipids may preclude access by glutathione peroxidase and that the lipid hydroperoxide must first be released by hydrolases before it can be reduced by glutathione peroxidase. However, the glutathione-dependent cytosolic factor as described by McCay *et al.* [133] is not limited in this manner and appears to prevent microsomal peroxidation directly. The observation that glutathione peroxidase does not inhibit microsomal peroxidation [133] is an unexpected one, and it indicates that the current concepts of the roles played by glutathione peroxidase and glutathione *S*-transferase in metabolizing membrane lipid peroxides must be reevaluated. Because PGHS is located in microsomal membranes, the mechanisms that control the peroxides in the membrane may have major importance in pathophysiological responses.

VII. INFLUENCE OF HYDROPEROXIDE LEVELS ON DRUG ACTIONS

The mechanism of cyclooxygenase activity indicated in Fig. 1 illustrates the requirement for hydroperoxide in activating the synthetic reaction. This requirement provides an additional means by which control of the biosynthesis of prostanoids can be achieved. Inhibition or antagonism of the hydroperoxide activation of this reaction provides an alternate mode of control to accompany the inhibition or antagonism of substrate access by competitive agents. Interference with peroxide activation may occur either by preventing the binding of peroxide to the catalytic site or by terminating the fatty acid oxygenase radical chain reaction with antioxidant radical-trapping agents. These two modes of interaction tend to have a similar kinetic manifestation even though the mechanism is different. Thus, the inhibition of cyclooxygenase activity by a noncompetitive phenolic agent, acetamidophenol, has been successfully quantified in terms of a competitive binding at the peroxide activation site rather than binding solely at the substrate binding site [27]. Reiterative calculations of a simplified algebraic formulation assuming equilibrium binding provided a close fit to the experimental data and supported the concept that the inhibition of prostaglandin formation by acetamidophenol may include antagonism of peroxide activation in addition to interference with substrate binding. The characterization of this type of inhibition provided a third mechanism of drug action to accompany the more familiar forms of competitive reversible and competitive irreversible inhibition described earlier [134, 135].

At present, the major nonsteroidal antiinflammatory agents can be classified according to these three categories [136]. Although the competitive reversible and competitive irreversible agents are relatively unaffected by a difference in the ambient peroxide levels [137], the antioxidant radical-trapping type of non-competitive reversible agent is very sensitive to the steady-state levels of hydroperoxide in the vicinity of the enzyme active site. Thus, phenolic agents such as acetamidophenol become increasingly potent inhibitors of the prostaglandin-forming cyclooxygenase action when the level of hydroperoxide is lowered. This form of behavior is shared with phenylbutazone and the fenamic acid derivatives [137]. The capacity of these drugs to stimulate *in vitro* assays of cyclooxygenase activity appears to reflect the high ambient levels of lipid hydroperoxide that accumulate in the standard *in vitro* assay systems. Stimulation of this type is not evident when ambient peroxide levels are low [20]. When the assay conditions are changed by the addition of peroxidase activity to keep the hydroperoxide concentration level at a reduced level [137], the effectiveness of these compounds in inhibiting prostaglandin biosynthesis increases.

It seems likely that the paradoxical kinetic behavior of such phenolic agents reflects in part the two separate forms of catalytic activities exhibited by PGHS: cyclooxygenase and peroxidase (illustrated in Fig. 1). Hemler and Lands [20] suggested that the radical-trapping capacity of phenolic agents is the basis for the inhibition of the cyclooxygenase that occurs in low ambient peroxide conditions. An alternate hypothesis was presented by Kuehl *et al.* [138] to account for the paradoxical behavior of these phenolic agents. This hypothesis proposes that the principal inflammatory mediator is a radical generated by PGHS-peroxidase activity on the PGG_2 intermediate in prostaglandin biosynthesis. The hypothesis thus shifts attention away from the generation of prostaglandin per se and proposes new mediators for inflammatory events. Although one cannot disprove the additional putative role for radicals produced during peroxidase catalysis or deny the existence of alternate mediators in inflammation, the primary driving force for this hypothesis is to reconcile the lack of inhibition of prostaglandin biosynthesis of the agent *in vitro* with its observed effectiveness *in vivo*. The demonstration that all of these phenolic agents are effective inhibitors *in vitro* under conditions of moderate ambient hydroperoxide eliminates the need for an alternate hypothesis and places attention once again on the activity of the cyclooxygenase in synthesizing prostaglandins and the inhibitory action of the phenolic agents in preventing prostaglandin biosynthesis.

The decreased capacity of these agents to inhibit prostaglandin formation under conditions of elevated hydroperoxide levels may provide a partial explanation for the weak antiinflammatory activity of several of these agents relative to their widely accepted analgesic activity. It is possible that a number of hyperalgesic states may have only very slightly elevated hydroperoxide levels so that phenolic agents such as acetamidophenol can still effectively inhibit prostaglandin biosynthesis. In the case of severe chronic or acute inflammatory conditions in which phagocytic leukocytes generate appreciable amounts of peroxides, we might expect the phenolic agents to be less effective inhibitors of prostaglandin formation. This rationale may underly the effectiveness of acetamidophenol as an analgesic agent even though it is relatively ineffective as an antiinflammatory agent [*136, 139*]. A clear understanding of the way that hydroperoxide levels alter the effectiveness of certain anticylcooxygenase agents might help in the development of analgesic agents that would not have the ulcerogenic property of strong irreversible inhibitors such as aspirin or indomethacin.

VIII.   SUMMARY

Lipid hydroperoxides play a role in activating the biosynthesis of prostaglandins from arachidonic acid. In general, the biosynthesis can be suppressed by decreasing the level of either the hydroperoxide activator or the nonesterified substrate fatty acid. These two suppressive events can be aided by pharmacological antagonism of substrate binding or antagonism of the activation by hydroperoxide. In the case of eicosapentaenoic acid, the requirement for hydroperoxide activator is sufficiently high that the hydroperoxide levels that occur *in vivo* are insufficient to promote rapid prostaglandin biosynthesis, making this fatty acid a competitive inhibitor of prostaglandin formation from arachidonate. The physiological suppression of prostaglandin formation by peroxidases, which maintain relatively low steady-state concentrations of lipid hydroperoxide within cells, may be overcome when biosynthesis of peroxides is more rapid than removal. The resulting elevation in ambient peroxides facilitates more rapid prostaglandin biosynthesis and provides a rationale for the elevated prostaglandin biosynthesis in inflammatory conditions that have elevated peroxide concentrations.

REFERENCES

1. T. Miyamoto, N. Ogino, S. Yamamoto, and O., Hayaishi, *J. Biol. Chem.* **251**, 2629 (1976).
2. M. Hemler, W. E. M. Lands, and W. L. Smith, *J. Biol. Chem.* **251**, 5575 (1976).
3. F. J. van der Ouderaa, M. Buytenhek, D. H. Nugteren, and D. A. Van Dorp, *Biochim. Biophys. Acta* **437**, 315 (1977).
4. P. P. K. Ho, R. D. Towner, and M. A. Esterman, *Prep. Biochem.* **10**, 597 (1980).
5. F. J. van der Ouderaa, M. Buytenhek, F. J. Slikkerveer, and D.A. Van Dorp, *Biochim. Biophys. Acta* **572**, 29 (1979).
6. S. Ohki, N. Ogino, S. Yamamoto, and O. Hayaishi, *J. Biol. Chem.* **254**, 829 (1979).
7. R. Ueno, T. Shimizu, K. Kondo, and O. Hayaishi, *J. Biol. Chem.* **257**, 5584 (1982).
8. M. Hamberg, B. Samuelsson, I. Bjorkhem, and H. Danielsson, *in* "Molecular Mechanisms of Oxygen Activation" (O. Hayaishi, ed.), p. 29. Academic Press, New York, 1974.
9. C. J. Sih and C. A. Takeguchi, *Prostaglandins* **1**, 83 (1973).

10. N. A. Porter, in "Free Radicals in Biology" (W. A. Pryor, ed.), Vol. 4, p. 261. Academic Press, New York, 1980.
11. M. Hamberg and B. Samuelsson, *J. Biol. Chem.* **242**, 5336 (1967).
12. M. Hamberg and B. Samuelsson, *Proc. Natl. Acad. Sci. U.S.A.* **70**, 899 (1973).
13. D. H. Nugteren, R. K. Beerthuis, and D. A. Van Dorp, *Recl. Trav. Chim. Pays-Bas* **85**, 405 (1966).
14. R. W. Egan, P. H. Gale, E. M. Baptista, K. L. Kennicott, W. J. A. VandenHeuvel, R. W. Walker, P. G. Fagerness, and F. A., Kuehl, Jr., *J. Biol. Chem.* **256**, 7352 (1981).
15. R. W. Egan, P. H. Gale, and F. A. Kuehl, Jr., *J. Biol. Chem.* **254**, 3295 (1979).
16. Y. Ishimura, M. Nozaki, O. Hayaishi, T. Nakamura, M. Tamura, and I. Yamazaki, *J. Biol. Chem.* **245**, 3593 (1970).
17. S. Yamamoto and O. Hayaishi, *J. Biol. Chem.* **242**, 5260 (1967).
18. F. Hirata, T. Ohnishi, and O. Hayaishi, *J. Biol. Chem.* **252**, 4637 (1977).
19. F. J. van der Ouderaa, M. Buytenhek, and D. A. Van Dorp, *Adv. Prostaglandin Thromboxane Res.* **6**, 139 (1980).
20. M. E. Hemler and W. E. M. Lands, *J. Biol. Chem.* **255**, 6253 (1980).
21. L. Que, Jr., J. D., Lipscomb, E. Munck, and J. M. Wood, *Biochim. Biophys. Acta* **485**, 60 (1977).
22. M. E. Hemler, H. W. Cook, and W. E. M. Lands, *Arch. Biochem. Biophys.* **193**, 340 (1979).
23. M. E. Hemler, G. Graff, and W. E. M. Lands, *Biochem. Biophys. Res. Commun.* **85**, 1325 (1978).
24. W. E. M. Lands and M. J. Byrnes, *Prog. Lipid Res.* **20**, 287 (1981).
25. W. E. M. Lands, R. Lee, and W. Smith, *Ann. N.Y. Acad. Sci.* **180**, 107 (1971).
26. H. W. Cook and W. E. M. Lands, *Nature (London)* **260**, 630 (1976).
27. W. E. M. Lands, H. W. Cook, and L. H. Rome, *Adv. Prostaglandin Thromboxane Res.* **1**, 7 (1976).
28. W. E. M. Lands and M. E. Hemler, in "Biochemical and Clinical Aspects of Oxygen" (W. Caughey, ed.), p. 213. Academic Press, New York, 1979.
29. N. Ogino, S. Ohki, S. Yamamoto, and O. Hayaishi, *J. Biol. Chem.* **253**, 5061 (1978).
30. R. J. Kulmacz and W. E. M. Lands, *Prostaglandins* **25**, 531 (1983).
31. W. E. M. Lands, *Prog. Lipid Res.* **20**, 875 (1981).
32. W. E. M. Lands and A. Hanel, in "Prostaglandins and Related Substances" (E. Granström and C. Page-Asciak, eds.), p. 203. Elsevier, Amsterdam/New York, 1983.
33. W. E. M. Lands, P. R. LeTellier, L. H. Rome, and J. V. Vanderhoek, *Adv. Biosci.* **9**, 15 (1973).
34. B. R. Culp, B. G. Titus, and W. E. M. Lands, *Prostaglandin Med.* **3**, 269 (1979).
35. P. Needleman, A. Raz, M. S. Minkes, J. A. Ferrendelli, and H. Sprecher, *Proc. Natl. Acad. Sci. U.S.A.* **76**, 944 (1979).
36. J. Dyerberg and H. O. Bang, *Lancet* **1**, 152 (1978).
37. W. Siess, B. Scherer, B. Bohlig, P. Roth, I. Kurzmann, and P. C. Weber, *Lancet* **2**, 441 (1980).
38. D. E. Morse, H. Duncan, N. Hooker, and A. Morse, *Science* **196**, 298 (1977).
39. R. M. Stewart, E. K. Weir, M. R. Montgomery, and D. E. Niewoehner, *Respir. Physiol.* **45**, 333 (1981).
40. A. Schaefer, M. Komlos, and A. Seregi, *Biochem. Pharmacol.* **27**, 213 (1978).
41. M. Komlos, A. Seregi, and A. Schaefer, *J. Pharm. Pharmacol.* **32**, 592 (1980).
42. A. Seregi, P. Serfozo, Z. Mergl, and A. Schaefer, *J. Neurochem.* **38**, 20 (1980).
43. P. Polgar and L. Taylor, *Prostaglandins* **19**, 693 (1980).
44. P. Polgar and L. Taylor, *Adv. Prostaglandin Thromboxane Res.* **6**, 225 (1980).

45. L. Baud, M.-P. Nivez, D. Chansel, and R. Ardaillou, *Kidney* **20**, 332 (1981).
46. S. L. Hofman, S. M. Prescott, and P. W. Majerus, *Arch. Biochem. Biophys.* **215**, 237 (1982).
47. J. Y. Vanderhoek and M. B. Feinstein, *Mol. Pharmacol.* **16**, 171 (1979).
48. J. J. Adcock, L. G. Garland, S. Moncada, and J. A. Salmon, *Prostaglandins* **16**, 163 (1978).
49. H. S. Mickel and J. Horbar, *Lipids* **9**, 68 (1973).
50. S. Diamant, R. Kissilevitz, and Y. Diamant, *Biol. Reprod.* **23**, 776 (1980).
51. B. B. Vargaftig, Y. Tranier, and M. Chignard, *Prostaglandins* **8**, 133 (1974).
52. B. B. Vargaftig, Y. Tranier, and M. Chignard, *Eur. J. Pharmacol.* **33**, 19 (1975).
53. R. V. Panganamala, H. M. Sharma, H. Sprecher, J. C. Geer, and D. G. Cornwell, *Prostaglandins* **8**, 3 (1974).
54. A. Rahimtula and P. J. O'Brien, *Biochem. Biophys. Res. Commun.* **70**, 893 (1976).
55. R. V. Panganamala, H. M. Sharma, R. E. Heikkila, J. C. Geer, and D. G. Cornwell, *Prostaglandins* **11**, 599 (1976).
56. J. G. White, G. H. R. Rao, and J. M. Gerrard, *Am. J. Pathol.* **88**, 387 (1977).
57. W. Smith and W. E. M. Lands, *Biochemistry* **11**, 3276 (1972).
58. R. W. Egan, J. Paxton, and F. A. Kuehl, Jr., *J. Biol. Chem.* **251**, 7329 (1976).
59. M. I. Siegel, R. T. McConnell, S. L. Abrahams, N. A. Porter, and P. Cuatrecasas, *Biochem. Biophys. Res. Commun.* **89**, 1273 (1979).
60. M. E. Hemler and W. E. M. Lands, *Arch. Biochem. Biophys.* **201**, 586 (1980).
61. Y. Fujimoto and T. Fujita, *Biochim. Biophys. Acta* **710**, 82 (1982).
62. S Moncada, R. J. Gryglewski, S. Bunting, and J. R. Vane, *Prostaglandins* **12**, 715 (1976).
63. E. A. Ham, R. W. Egan, D. D. Soderman, P. H. Gale, and F. A. Kuehl, Jr., *J. Biol. Chem.* **254**, 2191 (1979).
64. J. R. Beetens, M. Claeys, and A. G. Herman, *Biochem. Pharmacol.* **30**, 2811 (1981).
65. L. Flohe, *Ciba Found. Symp. [N.S.]* **65**, 95 (1978).
66. B. Chance, H. Sies, and A. Boveris, *Physiol. Rev.* **59**, 527 (1979).
67. I. Nishigaki, M. Hagihara, H. Tsunekawa, M. Maseki, and K. Yagi, *Biochem. Med.* **25**, 373 (1981).
68. R. P. Mason *in* "Free Radicals in Biology" (W. A. Pryor, ed.), Vol. 5, p. 161. Academic Press, New York, 1982.
69. A. A. Barbar and F. Bernheim, *Adv. Gerontol. Res.* **2**, 355 (1967).
70. C. A. Riely, G. Cohen, and M. Lieberman, *Science* **183**, 208 (1974).
71. M. T. Smith, H. Thor, and S. Orrenius, *Science* **213**, 1257 (1981).
72. C. J. Dillard and A. L. Tappel, *Lipids* **14**, 989 (1979).
73. W. A. Pryor, *Fed. Proc., Fed. Am. Soc. Exp. Biol.* **32**, 1862 (1973).
74. S. D. Aust and B. S. Svingen, *in* "Free Radicals in Biology" (W. A. Pryor, ed.), Vol. 5, p. 1. Academic Press, New York, 1982.
75. L. Flohe', *in* "Free Radicals in Biology" (W. A. Pyror, ed.), Vol. 5 p. 223. Academic Press, New York, 1982.
76. M. Carpenter, *Fed. Proc., Fed. Am. Soc. Exp. Biol.* **40**, 189 (1981).
77. W. Smith and W. E. M. Lands, *J. Biol. Chem.* **247**, 1038 (1972).
78. J. L. Haining and B. Axelrod, *J. Biol. Chem.* **232**, 193 (1958).
79. T. Schewe, R. Wiesner, and S. M. Rapoport, *in* "Methods in Enzymology" (J. M. Lowanstein, ed.), Vol. 71, p. 430. Academic Press, New York, 1981.
80. A. J. Paine, *Biochem. Pharmacol.* **27**, 1805, (1978).
81. L. R. DeChatelet, *J. Reticuloendothel. Soc.* **24**, 73 (1978).
82. J. A. Badwey and M. L. Karnovsky, *Annu. Rev. Biochem.* **49**, 695 (1980).
83. I. Fridovitch, *in* "Free Radicals in Biology" (W. A. Pryor, ed.), Vol. 1, p. 277. Academic Press, New York, 1976.

84. C. Beauchamp and I. Fridovitch, *J. Biol. Chem.* **245,** 4641 (1970).
85. K. L. Fong, P. B. McCay, J. L. Poyer, B. B. Keele, and H. Misra, *J. Biol. Chem.* **248,** 7792 (1973).
86. J. M. C. Gutteridge, R. Richmond, and B. Halliwell, *Biochem. J.* **184,** 469 (1979).
87. S. E. Fridovitch and N. E. Porter, *J. Biol. Chem.* **256,** 260 (1981).
88. J. M. McCord and E. D. Day, Jr., *FEBS Lett.* **86,** 139 (1978).
89. D. Halliwell, *FEBS Lett.* **92,** 321 (1978).
90. B. A. Svingen and S. D. Aust, *In* "Molecular Basis of Environmental Toxicity" (R. S. Bhatnagar, ed.), p. 69. Ann Arbor Sci. Publ., Ann Arbor, Michigan, 1980.
91. S. J. Klebanhoff, *in* "Immunology 80: Progress in Immunology IV" (M. Fougereau and J. Dauseset, eds.), p. 720. Academic Press, New York, 1980.
92. N. I. Krinsky, *Science* **186,** 363 (1974).
93. M. Kasha and A. U. Kahn, *Ann. N.Y. Acad. Sci.* **171,** 5 (1970).
94. J. E. Harrison and J. Schultz, *J. Biol. Chem.* **251,** 1371 (1976).
95. M. M. King, E. K. Lai, and P. B. McCay, Jr., *J. Biol. Chem.* **250,** 6496 (1975).
96. H. B. Demopoulos, *Fed. Proc., Fed. Am. Soc. Exp. Biol.* **32,** 1903 (1973).
97. M. Tien, B. A. Svingen, and S. D. Aust, *Fed. Proc., Fed. Am. Soc. Exp. Biol.* **40,** 179 (1981).
98. A. L. Tappel and H. Zalkin, *Arch. Biochem. Biophys.* **80,** 326 (1959).
99. A. L. Tappel, *in* "Lipids and Their Oxidation" (H. W. Schultz, E. A. Day, and R. O. Sinnbaur, eds.), p. 122. Avi Publ., Wesport, Connecticut, 1962.
99a. T. F. Slater, in "Free Radicals Mechanism in Tissue Injury." Pion Ltd., London, 1972.
99b. W. A. Pryor, J. P. Stanley, E. Blair, and G. B. Cullen, *Arch. Environ. Health* **31,** 201 (1976), esp. p. 205.
100. M. Hamberg and B. Samuelsson, *Proc. Natl. Acad. Sci. U.S.A.* **71,** 3400 (1974).
101. J. F. G. Vliegenthart and G. A. Veldink, *in* "Free Radicals in Biology" (W. A. Pryor, ed.) Vol. 5, p. 29. Academic Press, New York, 1982.
102. M. Okuma and H. Uchino, *N. Engl. J. Med.* **297,** 1351 (1977).
103. A. I. Schafer, *N. Engl. J. Med.* **306,** 381 (1982).
104. A. C. Chan, C. E. Allen, and P. V. J. Hegarty, *J. Nutr.* **110,** 66 (1980).
105. M. Okuma, H. Takayama, and H. Uchino, *Prostaglandins* **19,** 527 (1980).
106. W. C. Hope, C. Dalton, L. J. Machlin, R. J. Filipski, and F. M. Vane, *Prostaglandins* **10,** 1557 (1975).
107. L. J. Machlin, R. Filipski, A. L. Willis, D. C. Kuhn, and M. Brin, *Proc. Soc. Exp. Biol. Med.* **149,** 275 (1975).
108. M. Steiner and J. Anastasi, *J. Clin. Invest.* **57,** 732 (1976).
109. E. J. Goetzl, *Nature (London)* **288,** 183 (1980).
110. P. Boivin, C. Galand, and J. F. Bernard, *in* "Glutathione" (L. Flohe, H. C. Benohr, H. Sies, H. D. Waller, and A. Wendel Georg, eds.), p. 146. Thieme, Stuttgart, 1974.
110a. G. C. Mills, *Arch. Biochem. Biophys.* **86,** 1 (1960).
111. T. C. Stadtman, *Annu. Rev. Biochem.* **49,** 93 (1980).
112. H. C. Benohr and H. D. Waller, *in* "Glutathione" (L. Flohe, H. C. Benohr, H. Sies, H. D. Waller, and A. Wendel Georg, eds.), p. 184. Thieme, Stuttgart, 1974.
113. D. G. Hafeman and W. G. Hoekstra, *J. Nutr.* **107,** 666 (1977).
114. C. K. Chow and A. L. Tappel, *J. Nutr.* **104,** 444 (1974).
115. R. F. Burk, *in* "Selenium-Tellurium Environment" (L. Flohe, ed.), p. 194. Industrial Health Found., Inc., Pittsburgh, Pennsylvania, 1976.
116. W. H. Habig, J. H. Keen, and W. B. Jakoby, *Biochem. Biophys. Res. Commun.* **64,** 501 (1975).
116a. A. G. Splittgerber and A. L. Tappel, *Arch. Biochem. Biophys.* **197,** 534 (1979).

117. P. Boivin, C. Galand, and J. F. Bernard, *in* "Glutathione" (L. Flohe, H. C. Benohr, H. Sies, H. D. Waller, and A. Wendel Georg, eds.), p. 146. Thieme, Stuttgardt, 1974.
118. I. Anundi, J. Hogberg, and A. H. Stead, *Acta Pharmacol.* **45**, 45 (1979).
119. A. Wendel and S. Feuerstein, *Biochem. Pharmacol.* **30**, 2513 (1981).
120. M. Younes and C.-P. Siegers, *Chem.-Biol. Interact.* **34**, 257 (1981).
121. H. Kawaguchi, T. Ishibashi, and Y. Imai, *Lipids* **17**, 577 (1982).
122. J. R. Prohaska and H. E. Ganther, *Biochem. Biophys. Res. Commun.* **76**, 437 (1977).
123. R. A. Lawrence and R. F. Burk, *J. Nutr.* **108**, 211 (1978).
124. R. F. Burk, M. J. Trumble, and R. A. Lawrence, *Biochim. Biophys. Acta* **618**, 35 (1980).
125. E. Boyland and L. F. Chasseaud, *Adv. Enzymol.* **32**, 173 (1969).
126. W. B. Jakoby, *Adv. Enzymol.* **46**, 383 (1978).
127. W. H. Habig, J. H. Keen, and W. B. Jakoby, *Biochem. Biophys. Res. Commun.* **64**, 501 (1975).
128. A. W. Wolkoff, R. A. Weisiger, and W. B. Jakoby, *Prog. Liver Dis.* **6**, 213 (1979).
129. D. L. Vander Jagt, S. P. Wilson, V. L. Dean, and P. C. Simons, *J. Biol. Chem.* **257**, 1997 (1982).
130. E. Christ-Hazelhof, D. H. Nugteren, and D. A. Van Dorp, *Biochim. Biophys. Acta* **450**, 450 (1976).
131. R. A. Lawrence, L. K. Parkhill, and R. F. Burk, *J. Nutr.* **108**, 981 (1978).
132. K. E. Hill and R. F. Burk, *J. Biol. Chem.* **257**, 10668 (1982).
133. P. B. McCay, D. D. Gibson, and K. R. Hornbrook, *Fed. Proc., Fed. Am. Soc. Exp. Biol.* **40**, 199 (1981).
134. W. E. M. Lands, P. R. LeTellier, L. Rome, and J. V. Vanderhoek, *in* "Prostaglandin Synthetase Inhibitors" (H. J. Robinson and J. R. Vane, eds.), p. 1. Raven Press, New York, 1974.
135. L. H. Rome and W. E. M. Lands, *Proc. Natl. Acad. Sci. U.S.A.* **72**, 4863 (1975).
136. W. E. M. Lands, *Trends Pharmacol. Sci.* March, 768 (1981).
137. A. M. Hanel and W. E. M. Lands, *Biochem. Pharmacol.* **31**, 3307 (1982).
138. F. A. Kuehl, J. L. Humes, R. W. Egan, E. A. Ham, G. C. Beveridge, and C. G. Arman, *Nature (London)* **265**, 170 (1977).
139. W. E. M. Lands and A. M. Hanel, *Prostaglandins* **24**, 271 (1982).

CHAPTER **3**

Hydroperoxide-Dependent Oxidations during Prostaglandin Biosynthesis

Lawrence J. Marnett

Department of Chemistry
Wayne State University
Detroit, Michigan

FREE RADICALS IN BIOLOGY, VOL. VI
Copyright © 1984 by Academic Press, Inc.
All rights of reproduction in any form reserved.
ISBN 0-12-566506-7

63

I. INTRODUCTION

The biosynthesis of prostaglandins and thromboxanes can be divided conceptually and experimentally into two stages. In the first stage a polyunsaturated fatty acid is oxygenated to a hydroperoxy endoperoxide termed prostaglandin G (PGG), which is then reduced to a hydroxy endoperoxide termed PGH [1–4] (Scheme 1). Both reactions are catalyzed by an enzyme called prostaglandin H synthase (PGHS) [4–7]. The mechanism of oxygenation of polyunsaturated fatty acids by PGHS is discussed in Lands et al., Chapter 2, this volume. In the second stage of biosynthesis the endoperoxide moiety of PGH is transformed via metabolizing enzymes into the functionalities characteristic of prostaglandins and thromboxanes [7–11]. Fragmentation of PGH to a hydroxy acid and malondialdehyde occurs concomitant with thromboxane synthesis [12]. The metabolism of PGH is determined by the levels of the metabolizing enzymes, and these vary with tissue. For example, platelets make mainly thromboxane, whereas arterial endothelial cells make mainly prostacyclin [13, 14]. Regardless of the fate of PGH, any tissue that makes prostaglandins or thromboxanes must possess PGHS in order to biosynthesize PGH. As a result, virtually all mammalian tissues display some PGHS activity.

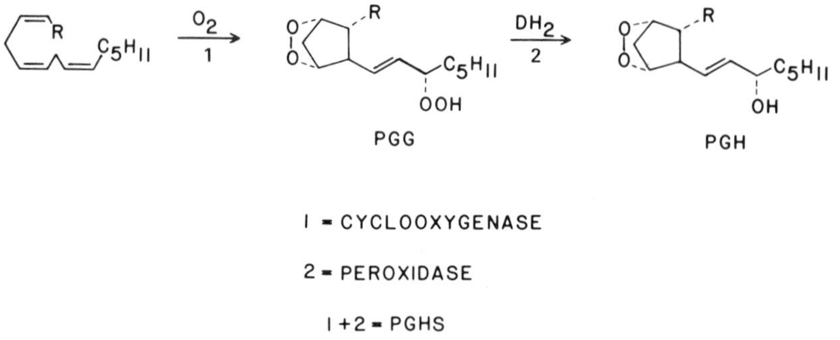

I = CYCLOOXYGENASE

2 = PEROXIDASE

I + 2 = PGHS

Scheme 1

Ram seminal vesicles contain an extremely high level of PGHS and are routinely used for studies of it [15]. The enzyme is membrane bound and appears to be localized in the endoplasmic reticulum and nuclear membranes [16]. It is a glycoprotein with a subunit molecular weight of 70,000 [6]. The protein isolated after a series of chromatographic and electrophoretic steps is inactive but can be reconstituted by the addition of one heme per subunit [17, 18]. Thus, it is believed that the intact enzyme is a heme protein.

In the mid-1970s we discovered that xenobiotics and compounds structurally unrelated to polyunsaturated fatty acids or endoperoxide intermediates undergo oxidation during PGH biosynthesis in ram seminal vesicle microsomes (RSVM)

[*19, 20*]. The oxidizing agent is generated as a result of the interaction of PGG with a microsomal peroxidase. Inhibition by aspirin or indomethacin of the synthesis of PGG_2 from arachidonic acid (AA) abolishes the capacity of RSVM to cooxidize xenobiotics. Addition of PGG_2 to inhibitor-treated microsomes triggers xenobiotic oxidation commensurate with that observed after the addition of AA to uninhibited microsomes [*20*]. Oxidation is not observed after the addition of PGH_2 to intact microsomes or after the addition of PGG_2 to heat-inactivated microsomes [*20*]. These findings suggest that the cooxidations that occur after the addition of AA to RSVM are hydroperoxide-dependent oxidations catalyzed by a microsomal peroxidase with PGG_2 as the substrate [*20*]. A very active peroxidase has been found in RSVM [*21*].

Oxidation, particularly oxygenation, is an important process in xenobiotic metabolism [*22, 23*]. The introduction of an oxygen atom into a drug can cause a loss or an enhancement of the pharmacological activity of the compound. The oxygen atom can serve as a handle for the attachment of polar moieties (sugars, sulfate, or glutathione) that can render a nonpolar molecule water soluble and therefore excretable. Oxygenation also can modify the chemical properties of a poorly reactive substance to an extent that it becomes reactive to cellular constituents. Covalent binding of small molecules to protein and/or nucleic acid can result in toxicity, mutagenicity, or carcinogenicity [*24, 25*]. An enormous amount of literature testifies to the importance of mixed-function oxidases, especially those possessing cytochrome *P*-450 as the terminal oxidase, in xenobiotic oxygenation [*26*]. Considering the dissimilarity between cytochrome *P*-450 and PGHS with respect to physical properties [*6, 26*], response to inhibitors [*27, 28*], and tissue distribution [*15, 26*], it seemed possible that hydroperoxide-dependent oxidations and oxygenations by PGHS might serve as an alternative mechanism for xenobiotic metabolism, particularly in tissues with low mixed-function oxidase activity. Consequently, several laboratories have investigated the cooxidation of drugs and chemical carcinogens during prostaglandin biosynthesis. The scope of the investigations has been reviewed [*29, 30*], and contributions from individuals who are active in the field are contained in this volume. I have restricted the present discussion primarily to work done in my laboratory.

II. ROLE OF PROSTAGLANDIN H SYNTHASE IN UNSATURATED FATTY ACID-DEPENDENT COOXIDATION BY RAM SEMINAL VESICLE MICROSOMES

Unsaturated fatty acid-dependent cooxidations are hydroperoxide-dependent oxidations catalyzed by a peroxidase [*20*]. In RSVM, PGHS and the peroxidase are microsomal [*20*]. An obvious role of PGHS is to synthesize the hydroperoxide substrate for the peroxidase. The importance of the cyclooxygenase component of PGHS in this regard is indicated by the complete inhibition of AA-dependent cooxidation by aspirin or indomethacin [20]. Theoretically, any hydroperoxide-generating system could trigger oxidation if an appropriate peroxidase were present. Prostaglandin H synthase appears to be a major source of hydroperoxides in animals because examples of naturally occurring hydroperoxides are very rare [31]. Considering the potential hazards associated with the generation of hydroperoxides in cells, it is not surprising that most oxygenases have evolved to hydroxylate substrates without the intermediacy of hydroperoxides (e.g., mixed-function oxidases) [32].

The identity of the peroxidase that catalyzes the hydroperoxide-dependent oxidations has been more difficult to establish. Miyamoto *et al.* and Van der Ouderaa *et al.* reported that purified PGHS exhibits a reconstitutable peroxidase activity [4, 6]. Ohki *et al.* subsequently demonstrated that the peroxidase activity is an integral part of PGHS and is contained on the same protein as the cyclooxygenase activity [33]. All of the substrates that were utilized in these studies underwent dehydrogenation rather than oxygenation, so the question of which peroxidase catalyzes hydroperoxide-dependent oxygenations remains unanswered. Several investigators have attempted to identify the peroxidase using inhibitors of other peroxidative enzymes [27, 28, 34–36]. However, there are no specific inhibitors of the peroxidase activity of PGHS, and indomethacin (which exhibits the cyclooxygenase component) does not inhibit the peroxidase component [37]. Therefore, such studies have not been definitive. Egan *et al.* have demonstrated that purified and reconstituted PGHS carries out the hydroperoxide-dependent oxygenation of sulindac sulfide and methyl phenyl sulfide to sulfoxides [38]. It is interesting, however, that sulfide oxygenation was not antagonized by the oxygenatable substrates diphenylisobenzofuran (DPBF) and phenylbutazone (PB) [38].

A. Purification of Peroxidase

We have used two approaches to determine the identity of the major peroxidase that catalyzes hydroperoxide-dependent oxygenation in RSVM. First, we have purified the PB peroxidase activity to apparent electrophoretic homogeneity and compared the fate of the cyclooxygenase and peroxidase activities at various stages of purification [39]. Our results indicate that the activities copurify. Figure 1 displays the elution profile of cyclooxygenase and PB peroxidase from an ion-exchange column (DE 53), the major step in the purification. It is apparent that the two activities coelute. Gel electrophoresis of the pooled active fractions indicates a single major protein of subunit MW 70,000. The cyclooxygenase and PB peroxidase activities are maximally stimulated by exogenous hematin at a ratio of 1 hematin per 70,000 MW subunit. Chromatography of the purified protein on gel filtration, ion-exchange, and chromatofocusing media and also

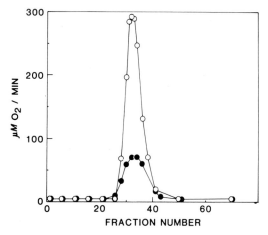

Fig. 1. Elution profile of cyclooxygenase (○) and PB peroxidase (●) activities from a DE 53 ion exchange column.

hydroxylapatite does not resolve the cyclooxygenase from the PB peroxidase activity. These results strongly suggest that the major peroxidase in RSVM that catalyzes the hydroperoxide-dependent oxygenation of PB is contained on the same protein as the cyclooxygenase component of PGH synthase.

B. Immunochemical Identification of the Peroxidase

Our second approach has employed monoclonal antibodies raised against purified cyclooxygenase [40] as reagents to immunoprecipitate peroxidase activities in RSVM [41]. *Staphylococcus aureus* cells that contain protein A on their surface or protein A–Sepharose were used to render the antibody molecules precipitating. Ram seminal vesical microsomes were solubilized with a nonionic detergent (Tween 20) and incubated with *S. aureus* cells or protein A–Sepharose previously treated with the culture medium from secreting (immune) or nonsecreting (nonimmune control) hybridomas. Table I summarizes the amounts of DPBF, PB, and epinephrine peroxidase remaining in the supernatant following precipitations performed with immune and nonimmune preparations. Also presented is the net amount of activity precipitated under the various conditions. Although quantitative differences are observed among the results for each substrate (which are related to differences in assay conditions), the data in Table I indicate that DPBF, PB, and epinephrine peroxidase activities are immunoprecipitated by antibodies raised against the purified cyclooxygenase activity of PGHS. Similar results have been obtained using antibodies secreted by separate clones that recognize different antigenic determinants on the cycloox-

TABLE I   Immunoprecipitation of Peroxidase Activities in RSVM by Anticyclooxygenase

| | Percent activity in supernatant[a] | | |
Compound	Immune	Nonimmune	Percent precipitated
DPBF	31	72	57
PB	17	100	83
Epinephrine	32	62	49

[a]Percent activity remaining in supernatant after centrifugation of incubation mixtures with antibody preparations from secreting (immune) or nonsecreting (nonimmune) hybridomas.

ygenase. Because these antibodies are monospecific for the cyclooxygenase, their capacity to react with most of the peroxidase activity in detergent-solubilized RSVM implies that the cyclooxygenase and cooxygenating peroxidase activities are contained on the same protein. This complements the results of our protein purification studies and provides strong evidence that the major peroxidase activity that catalyzes hydroperoxide-dependent oxidations and oxygenations in RSVM is the peroxidase component of PGHS.

III. HYDROPEROXIDE-DEPENDENT OXIDATIONS

The classical reaction catalyzed by peroxidases is the reduction of a hydroperoxide at the expense of an electron donor [Eq. (1)] [42]. The electron donor is usually dehydrogenated.

$$ROOH + DH_2 \rightarrow ROH + D + H_2O \tag{1}$$

Prostaglandin H synthase catalyzes the dehydrogenation of several compounds including guaiacol, N,N,N^1,N^1-tetramethylphenylenediamine, phenol, acetaminophen, epinephrine, 1-phenyl-3-pyrazolidone (phenidone), 3-amino-1-[m-(triflouromethyl)phenyl]-2-pyrazoline (BW 755C), and aminopyrine [21, 33, 43–46]. Certain substituted phenols and aromatic amines are cyclooxygenase inhibitors, and it has been proposed that their inhibitory capacity is related to their capacity to act as peroxidase cofactors [47]. If this is correct, these compounds are oxidized to a certain extent even when PGG_2 biosynthesis is minimal. This concept has important implications for drug-induced toxicity. For example, a number of antiinflammatory agents that are cyclooxygenase inhibitors exhibit renal toxicity [48]. The toxicity is evident in the proximal tubule after acute administration and in the inner medulla after chronic administration [48]. It has been proposed that the chronic toxicity is due to reduced blood flow, which may be exacerbated by the inhibition of prostaglandin biosynthesis [49]. However, it is also possible that it might be due to the covalent binding of oxidized derivatives of the drugs to protein and/or nucleic acid. This is a commonly invoked mechanism of drug toxicity but is complicated in the kidney by the virtual absence of drug-metabolizing mixed-function oxidases in the inner medulla [50]. It is interesting that this region of the kidney possesses the highest activity of PGHS in the tissue, so that PGHS may play a role in chronic renal toxicity by oxidizing antiinflammatory agents to reactive derivatives [50]. It should be pointed out, however, that aspirin and indomethacin, which cause inner medullary damage following chronic administration, do not appear to be oxidized by the peroxidase of PGHS [37]. Thus, oxidative metabolism may not play a role in the renal toxicity induced by all antiinflammatory agents.

Because oxygenation is an important reaction in drug metabolism and chemical carcinogenesis, we have concentrated our studies on the hydroperoxide-dependent oxygenation of xenobiotics catalyzed by PGHS. We have utilized substrates that give distinct oxygenated derivatives in order to maximize our chances of detecting differences in the mechanisms of oxygenation. Most of the work has been performed with RSVM as the enzyme source, but the oxygenations can be effected with a purified and reconstituted enzyme as well.

A. Diphenylisobenzofuran

Diphenylisobenzofuran is extremely sensitive to oxygenation and is a good molecule with which to test for the enzymatic generation of reactive oxidants. We found that DPBF is oxygenated to dibenzoylbenzene (DBB) [Eq. (2)] if it is present during the incubation of AA or PGG_2 with RSVM [20]. Incubations of PGG_2, DPBF, and RSVM performed under an atmosphere of $^{18}O_2$ lead to the incorporation of significant amounts of ^{18}O into the carbonyl oxygens of DBB (one atom of ^{18}O is incorporated) [51]. Similar experiments indicate that little or no oxygen is incorporated from the hydroperoxide or from H_2O. Thus, the source of the oxygen introduced into DPBF during hydroperoxide-dependent oxygenation by PGHS is atmospheric oxygen.

$$\text{(2)}$$

B. Phenylbutazone

Phenylbutazone is a nonsteroidal antiinflammatory agent that inhibits cyclooxygenase in certain *in vitro* assays. Like its analog, oxyphenylbutazone, PB is hydroxylated at the 4-position by PGHS in the presence of AA or PGG_2 [Eq. (3)] [*52*, *53*]. The hydroxyl oxygen is derived from molecular oxygen rather than the hydroperoxide, and the stoichiometry of 4-hydroxy-PB formed to hydroperoxide reduced is 1 : 1 [*53*].

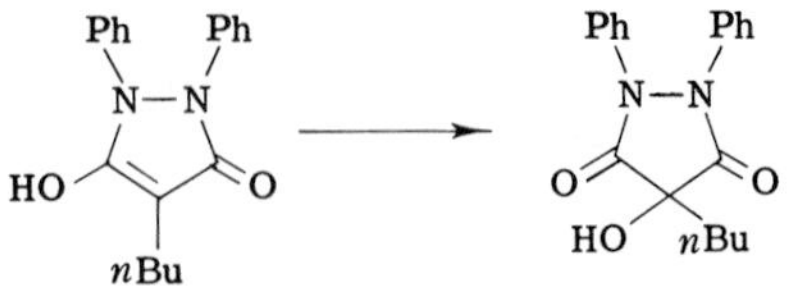

(3)

C. Benzo[*a*]pyrene

Benzo[*a*]pyrene (B[a]P) is the classic example of a noncarcinogenic compound that is oxygenated to a carcinogenic derivative(s) during metabolism. Mixed-function oxidases have primary responsibility for the metabolism of B[a]P *in vivo*, and the oxygenated metabolites formed by the mixed-function oxidases have been extensively studied [*54*]. The ultimate carcinogenic form of B[a]P is believed to be a dihydrodiol-epoxide formed as indicated in Eq. (4) [*55–58*]. The oxygenation steps have been shown to be catalyzed by cytochrome *P*-450 and the hydration step by epoxide hydrase [*59*]. It occurred to us that, if PGHS could trigger the epoxidation of B[a]P or its derivatives, it might be important in the metabolic activation of B[a]P in tissues with low mixed-function oxidase activity. Tissues of particular interest in this regard include lung, skin, and forestomach, which are target organs for B[a]P carcinogenesis, and mammary gland, which is a target organ for dimethylbenzanthracene.

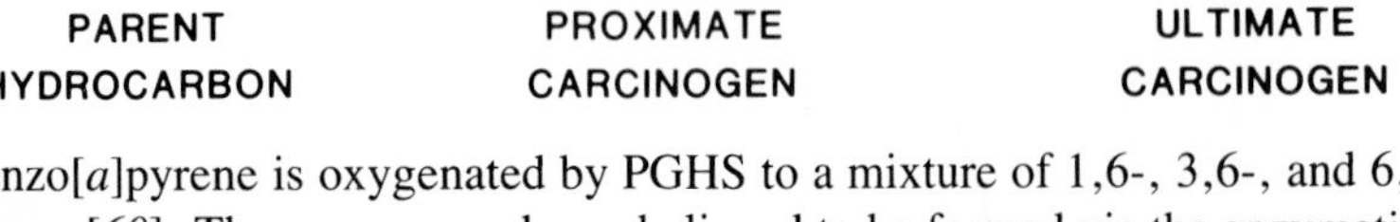

(4)

Benzo[*a*]pyrene   B[a]P-7,8-diol   *anti*-Diol-epoxide

| PARENT HYDROCARBON | PROXIMATE CARCINOGEN | ULTIMATE CARCINOGEN |

Benzo[*a*]pyrene is oxygenated by PGHS to a mixture of 1,6-, 3,6-, and 6,12-quinones [*60*]. These compounds are believed to be formed via the enzymatic or nonenzymatic oxidation of the initial enzymatic oxidation product 6-hydroxy-B[a]P [Eq. (5)] [*61–64*]. We have not identified the source of the oxygen incorporated into B[a]P during hydroperoxide-dependent oxidation by PGHS but, as with DPBF, oxygenation is potently inhibited by antioxidants [*60*]. The stoichiometry of B[a]P oxidized to hydroperoxide added is much less than 1 [*31*].

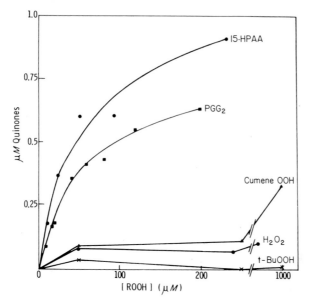

Fig. 2.   Dependence of the RSVM-catalyzed oxidation of B[a]P to quinones on the concentrations of various hydroperoxides.

The hydroperoxide specifiicty of B[a]P oxygenation by RSVM is displayed in Fig. 2 [*31*]. Fatty acid hydroperoxides are clearly superior to hydrogen peroxide and to the nonnaturally occurring *tert*-butyl and cumene hydroperoxides.

$$\text{(5)}$$

D. 7,8-Dihydroxy-7,8-dihydrobenzo[*a*]pyrene

Although 6-hydroxy-B[a]P is mutagenic and possibly carcinogenic, its relevance to B[a]P carcinogenesis is not firmly established [*64, 65*]. The quinones are not mutagenic or carcinogenic, but they can act as cycling redox carriers to generate superoxide anion and H_2O_2 [*66, 67*]. This may be important in tumor promotion [*68, 69*]. We have also investigated the hydroperoxide-dependent metabolism of 7,8-dihydroxy-7,8-dihydrobenzo[*a*]pyrene (B[a]P-7,8-diol) [*70*]. As indicated earlier, B[a]P-7,8-diol is the immediate precursor of the highly mutagenic and carcinogenic epoxides shown in Eq. (6). Thus, this reaction represents the terminal activation step in B[a]P carcinogenesis. Similar reactions appear to be involved in the metabolic activation of several other polycyclic aromatic hydrocarbons [71].

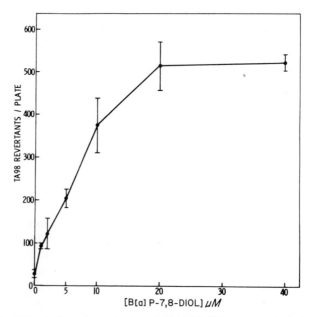

$$\tag{6}$$

The incubation of B[a]P-7,8-diol with Tween 20-solubilized RSVM and AA leads to the formation of a derivative(s) that is strongly mutagenic to *Salmonella typhimurium* strains TA98 and TA100 [72]. The dependence of the mutagenicity on the initial B[a]P-7,8-diol concentration is demonstrated in Fig. 3. An approximately linear dose–response relationship obtains between 1 and 20 μM. In this concentration range the formation of a mutagenic derivative(s) is strongly inhibited by preincubation of the enzyme preparation with indomethacin [72]. No increase in the revertants to histidine independence is observed if 9,10-dihydroxy-9,10-dihydrobenzo[*a*]pyrene or 4,5-dihydroxy-4,5-dihydrobenzo[*a*]pyrene is substituted for B[a]P-7,8-diol. Because neither of the former compounds can be epoxidized to a strongly mutagenic bay-region diol-epoxide, this implies

Fig. 3. Arachidonate-dependent activation of B[a]P-7,8-diol to mutagenic derivatives by detergent-solubilized RSVM.

that the mutagenic derivative formed from B[a]P-7,8-diol during PGH$_2$ biosynthesis is a diol-epoxide.

No one has reported the isolation of a bay-region diol-epoxide of B[a]P following an *in vitro* incubation; all of the experimental evidence for the formation of these compounds is indirect and is based on the trapping of the unstable epoxide at the benzylic carbon by nucleophiles [Eq. (7)]. The identification of the nucleophile adduct then serves as evidence for the intermediacy of the epoxide.

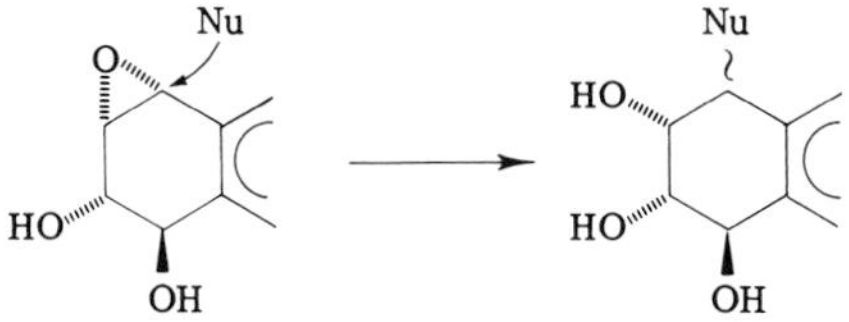

$$\tag{7}$$

Figure 4 summarizes our studies of the products isolated after the incubation of [^{14}C]B[a]P-7,8-diol with RSVM and AA [*74*]. If the reactions are terminated after 15 min, the major products isolated are *trans*- and *cis*-tetrahydrotetraols, presumably formed by hydrolysis of an intermediate epoxide [*27*]. If the reactions are

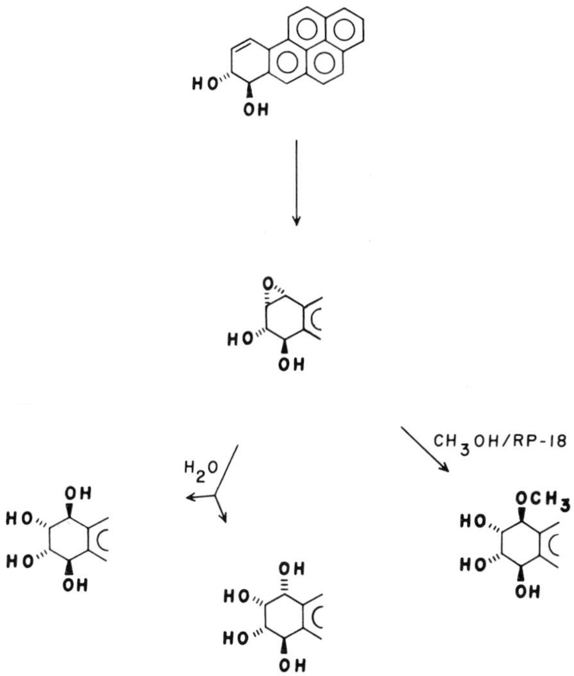

Fig. 4.   Summary of metabolites of B[a]P-7,8-diol formed by PGHS. The tetraols are the sole products observed after a 15-min incubation. The methyl ether is observed after a 3-min incubation.

terminated after 3 min, an additional product, identified as a *trans*-10-methox-ytetrahydrotriol, is isolated [*74*]. We believe this compound is formed by meth-anolysis of the epoxide during reversed-phase HPLC. Thus, the epoxide is extracted from the incubation mixture but decomposes during chromatrographic analysis. The three major products of PGHS-dependent metabolism of B[a]P-7,8-diol are derived from the *anti*-diol-epoxide. No products derived from the *syn*-diol-epoxide are observed.

When incubations of [^{14}C]B[a]P-7,8-diol, AA, and RSVM are carried out in the presence of polyguanylic acid [poly(G)], radioactivity becomes covalently associated with the poly(G). Digestion of the poly(G) generates guanosine ad-ducts derived from the attack of the exocyclic amino group of guanosine on the benzylic carbon of a diol-epoxide (e.g., **1**) [*75*]. Reaction of (±)-*syn*- and (±)-*anti*-diol-epoxides generates eight diastereomeric guanosine adducts analo-gous to **1** by virtue of the *cis*- and *trans*-addition of guanosine to the four enantiomers of the two diol-epoxides. By quantifying the levels of each of the diastereomeric guanosine adducts it is possible to quantify indirectly the levels of each of the enantiomeric diol-epoxides generated as a result of the PGHS-depen-dent metabolism of (±)-B[a]P-7,8-diol. The validity of this approach for deter-mining the stereochemical course of B[a]P-7,8-diol epoxidation has been demon-strated using the mixed-function oxidase in rat liver microsomes [*76*]. By using this technique, we have found that the hydroperoxide-dependent metabolism of B[a]P-7,8-diol by RSVM is nonstereospecific; that is, both enantiomers of the *anti*-diol-epoxide are generated in equal amounts from the two enantiomers of B[a]P-7,8-diol [*75*]. This differs from the stereochemistry of the (±)-B[a]P-7,8-diol metabolism by the mixed-function oxidase in rat liver microsomes [*77, 78*].

1

E. 7,8-Dihydrobenzo[*a*]pyrene

We have completed a study of the cooxygenation of 7,8-dihydroben-zo[*a*]pyrene (H$_2$B[a]P) by RSVM [*79*]. A strongly mutagenic derivative is formed that decomposes to tetrahydrodiols during the incubation [79] [Eq. (8)]. The ratio of *cis*- to *trans*-tetrahydrodiols is sensitive to the level of activity of epoxide hydrase [79]. Unlike the diol-epoxides, 9,10-epoxy-7,8,9,10-tetrahydrobenzo[*a*]pyrene is a substrate for epoxide hydrase [80]. This enzyme hydrates epoxides regiospecifically to *trans*-diols, whereas the epoxide in Eq. (8) hydrolyzes nonenzymatically predominantly to the *cis*-diol [81]. Thus, by modulating the level of activity of epoxide hydrase in the incubations, we have been able to alter the ratio of *cis*- to *trans*-diols, providing additional evidence for the intermediacy of an epoxide. Analysis of the yields of individual guanosine adducts formed in incubations of $H_2B[a]P$, AA, and RSVM in the presence of poly(G) indicates that the epoxidation is not stereospecific [76]. This is analogous to the results of the experiments with B[a]P-7,8-diol described in the preceding section. The study of $H_2B[a]P$ cooxygenation indicates that the AA-dependent epoxidation of B[a]P-7,8-diol is not limited to the latter compound.

$$(8)$$

Guthrie *et al.* have shown that 3,4-dihydroxy-3,4-dihydrobenzo[*a*]anthracene and 1,2-dihydroxy-1,2-dihydrochrysene are metabolized to strongly mutagenic derivatives by RSVM in the presence of AA [73]. Although the products of metabolism were not identified, it is likely that they are the respective bay-region epoxides. In their entirety, these observations indicate that PGHS-dependent metabolism of appropriate dihydroaromatic compounds to strongly mutagenic and carcinogenic diol-epoxides may be general and that PGHS can trigger the metabolic activation of chemical carcinogens *in vitro*.

F. Miscellaneous

Many other compounds are oxidized or oxygenated by PGHS. Most of these are described in reviews [29, 30] or in chapters by Gale and Egan (Chapter 1), Cornwell and Morisaki (Chapter 4), and Kalyanaraman and Sivarajah (Chapter 5), this volume. As a class, aromatic amines appear to be excellent substrates for demethylation or N-hydroxylation. A complete series of primary amines and mono- and dimethylamines has been investigated [45, 82]. Of particular interest are the potent carcinogens 2-aminoflourene, benzidine, and dimethylaminoazobenzene [83–85]. In addition to aromatic amines, diethylstilbestrol, a transplacental uterine carcinogen, and *N*-[4-(5-nitro-2-furyl)-2-thiazolyl]formamide, a bladder carcinogen, are excellent substrates [86, 87].

IV. MECHANISMS OF HYDROPEROXIDE-DEPENDENT OXIDATION AND OXYGENATION

A. Peroxidase Reducing Cofactors

The classical reaction catalyzed by peroxidases is the reduction of a hydroperoxide by an electron donor [Eq. (1)] [42]. The reaction depicted in Eq. (1) is a dehydrogenation, but peroxidases can also catalyze hydroperoxide reduction by transfer of the hydroperoxide oxygen to an acceptor [Eq. (9)] [38, 88]. These reactions are not concerted but occur through the formation of discrete enzymatic intermediates [89, 90].

$$\text{ROOH} + \text{A} \rightarrow \text{ROH} \rightarrow \text{AO} \qquad (9)$$

The catalytic trinity of heme-containing peroxidases is the resting enzyme and two higher oxidation states called compound I and compound II. The interrelationship of these species has been elucidated by numerous studies performed with peroxidases from horseradish and other sources [91]. The resting enzyme, which contains heme iron in the $+3$ oxidation state, reduces the hydroperoxide to an alcohol, generating an iron–oxo complex (compound I) in which the iron is in the formal oxidation state of $+5$ and the oxygen is derived from the hydroperoxide. Compound I can be reduced by a single electron to form an iron–oxo complex (compound II) in which the iron is in the $+4$ oxidation state, or it can transfer the oxygen atom to an acceptor to regenerate the resting enzyme. Compound II is reduced by a single electron to the resting enzyme. Thus, the iron–oxo intermediates of peroxidase catalysis can effect electron transfer or oxygen transfer. Because most of the electron transfers involve a single electron, electron-deficient forms of the donors (DH·) are generated [42]. The chemistry of these electron-deficient forms, either on the enzyme or in solution, determines the overall products of the peroxidase-catalyzed reaction. This is important to remember in attempting to understand the range of hydroperoxide-dependent reactions catalyzed by PGHS.

1. Aromatic Amines

Phenidone and BW 755C are oxidized to radical-cations by RSVM or purified PGHS in the presence of AA or fatty acid hydroperoxides [46]. Hydrogen peroxide is approximately equipotent to fatty acid hydroperoxides in triggering oxidation. The radical-cations are short-lived and decompose rapidly under the incubation conditions. When phenidone is incubated with PGHS and H_2O_2, there is a rapid increase in absorbance at 514 nm due to the formation of its radical-cation. After reaching maximal intensity, the absorbance decreases rapidly. The disappearance of the signal is most likely due to disproportionation of the radical-

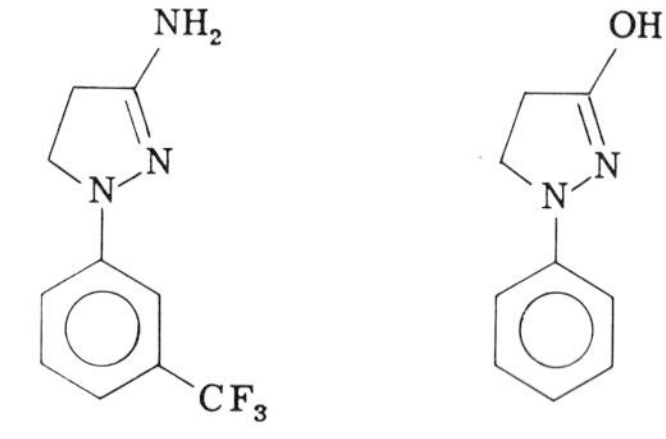

BW 755C

Phenidone
(Lactim tautomer)

2

cation to phenidone and 1-phenyl-3-hydroxypyrazole [Eq. (10)] [92], although we cannot rule out the possibility that the radical-cation is directly oxidized by the peroxidase. The net reaction of phenidone with PGHS is therefore dehydrogenation to 1-phenyl-3-hydroxypyrazole. A radical-cation is produced as an intermediate, and its chemistry determines the course of the overall reaction.

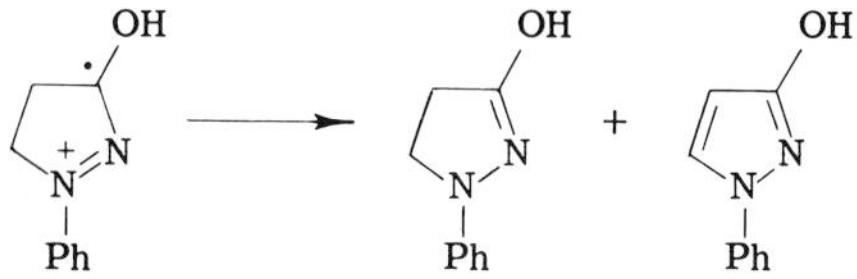

$$\qquad\qquad\qquad\qquad\qquad\qquad\qquad\qquad (10)$$

An interesting variation of this reaction is provided by the PGHS-dependent oxidation of aminopyrine [93]. Lasker *et al.* have demonstrated that the initial oxidation product, the aminopyrine radical-cation, disproportionates to an iminium salt that hydrolyzes to formaldehyde and *N*-methyl-4-aminoantipyrine [Eq. (11)]. The net reaction is the hydroperoxide-dependent oxygenation of the methyl group of aminopyrine, but the actual enzymatic reaction is the same as that observed with phenidone electron transfer. The subsequent steps in the reaction are analogous to those observed with phenidone, but the disproportionation product is unstable and hydrolyzes to the observed product.

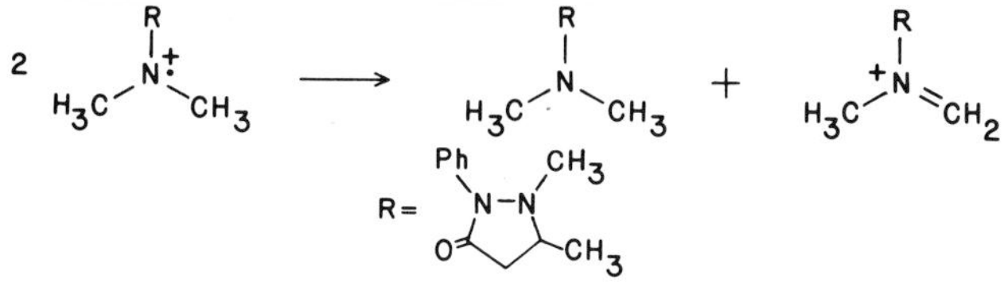

$$\qquad\qquad\qquad\qquad\qquad\qquad\qquad\qquad (11)$$

2. Phenylbutazone

Phenylbutazone is hydroxylated to 4-hydroxy-PB by PGHS and is a cofactor for the enzymatic reduction of hydroperoxides to alcohols [53]. The stoichiometry of PB oxidized to hydroperoxide reduced is approximately 1 : 1 [53]. Incubations with ^{18}O-labeled hydroperoxide, oxygen, and water indicate that the hydroxyl oxygen introduced into PB is derived from molecular oxygen rather than hydroperoxide oxygen. One can utilize precedents similar to those discussed earlier to construct a hypothetical mechanism that explains the key experimental observations. If one assumes that PB acts as a typical peroxidase substrate, the initial reaction should be electron transfer to a higher oxidation state of the peroxidase, generating an electron-deficient derivative of PB [Eq. (12)]. In contrast to the radical-cations of phenidone and aminopyrine, the neutral radical derived from PB can be scavenged by O_2 to form a peroxyl radical because it is a carbon-centered radical [Eq. (13)]. Although we have not elucidated the steps in the conversion of the peroxyl radical to the alcohol, Eqs. (12) and (13) are consistent with the major experimental observations and with the fact that PB is a reducing cofactor for the peroxidase of PGHS. The first step in PB oxidation is analogous to the first step in aromatic amine metabolism and in β-diketone autoxidation [94], but the chemistry of the initial oxidation product alters the overall course of reaction.

$$(12)$$

$$(13)$$

3. Aromatic Sulfides

Egan *et al.* have found that the aromatic sulfides, sulindac sulfide and methyl phenyl sulfide, are reducing cofactors for the peroxidase activity of PGHS and are oxygenated to sulfoxides [37, 38]. The stoichiometry of hydroperoxide reduction is sulfide oxygenation is 1 : 1, and the sulfoxide oxygen is derived from the hydroperoxide oxygen [Eq. (14)]. These observations can be explained by the hypothesis that the initial iron–oxo complex derived from the peroxidase (analogous to compound I) transfers oxygen to the sulfide. Whether the transfer is concerted or stepwise is unknown, but oxygen transfer from iron–oxo complexes has been reported for other peroxidases [88, 95].

$$(14)$$

It is clear that the peroxidase activity of PGHS can catalyze several kinds of hydroperoxide-dependent reactions, including oxidation and oxygenation. Examples of the latter demonstrate that the oxygen can be derived from molecular oxygen, hydroperoxide oxygen, and water. Although insufficient data are available to enable us to propose detailed mechanisms for many of the reactions, literature precedents derived from the study of other peroxidases seem to provide a framework that is useful for understanding the fundamental steps in the oxidations. The examples cited here have been chosen to illustrate this.

4. Active Site for Oxidation of Peroxidase Cofactors

The current knowledge of the physical properties of PGHS raises several interesting questions about the location of the active sites of the cyclooxygenase and peroxidase activities. As stated earlier both activities reside on a single polypeptide chain [33, 39, 41]. Studies have indicated that the stoichiometry of heme binding to the apoprotein is 1 : 1 [17]. This implies that a common active site exists for the cyclooxygenase and peroxidase activities and that electron transfer and oxygen transfer reactions occur at that site. Indeed, many cyclooxygenase inhibitors are reducing cofactors for the peroxidase, but there are important exceptions. Aspirin, indomethacin, and 5,8,11,14-eicosatetraynoic acid are irreversible inhibitors of cyclooxygenase activity but have no inhibitory effect on peroxidase activity and are not oxidized by it [37, 96]. However, because the molecular basis for the inhibition of cyclooxygenase by these three compounds has not been established, their lack of an effect on the peroxidase activity may be irrelevant to the question of the relationship between the two activities.

Equally intriguing is the question of whether electron transfer and oxygen transfer occur at the same site. The oxidations of all of the substrates considered here have been demonstrated with purified enzyme and require the intact, fully reconstituted protein. Because there is only a single heme per polypeptide, one would expect that all the hydroperoxide-dependent reactions would occur at the same site. However, Egan *et al.* have found that a 7-fold excess of PB (an electron transfer substrate) does not inhibit the oxygenation of methyl phenyl sulfide to methyl phenyl sulfoxide [38]. This prompted them to suggest that oxygen transfer reactions occur at a different site than do electron transfer reactions [38]. We have confirmed this observation, but we have found that methyl phenyl sulfide inhibits the hydroperoxide-dependent oxidation of PB and several other electron transfer substrates by purified and reconstituted PGHS (Fig. 5). Because methyl phenyl sulfoxide and methyl phenyl sulfone do not inhibit PB oxidation, the inhibition by methyl phenyl sulfide is most likley due to its preferential oxidation by the peroxidase. Figure 5 indicates that methyl phenyl sulfide does compete with electron transfer substrates for an enzyme-derived oxidant. The nonreciprocal inhibitory relationship between PB and methyl phenyl sulfide is reminiscent of the relationship between guaiacol and BW 755C

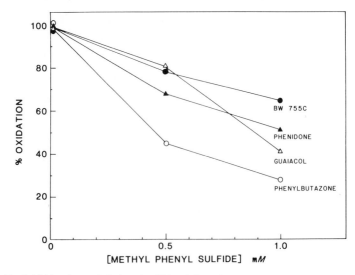

Fig. 5. Inhibition by methyl phenyl sulfide of the oxidation of electron transfer substrates by PHGS and H_2O_2.

[46]. A 20-fold excess of guaiacol does not inhibit the H_2O_2-dependent oxidation of BW 755C by either PGHS or horseradish peroxidase [46]. Yet, at the same concentrations, BW 755C inhibits guaiacol oxidation. A close inspection of the horseradish peroxidase-catalyzed reaction profile indicates that all the horseradish peroxidase-catalyzed reaction profile indicates that all of the BW 755C is consumed by oxidation before any of the guaiacol begins to be oxidized [46]. Because horseradish peroxidase, as well as PGHS, has only a single active site per molecule [91] and because the mechanisms of BW 755C and guaiacol oxidation involve electron transfer [42, 46], it is likely that the preferential oxidation of BW 755C before guaiacol is due either to tighter binding of BW 755C to the enzyme or to its higher reactivity toward the enzyme-generated oxidant. This suggests that methyl phenyl sulfide either binds more tightly to PGHS or is more reactive than PB and that this accounts for the failure of PB to inhibit methyl phenyl sulfide oxidation. However, it is still possible that methyl phenyl sulfide reacts with a form of the peroxidase that is distinct from the form of the enzyme that reacts with electron transfer substrates. Further studies are required to distinguish among these possibilities.

B. Compounds That Are Not Peroxidase Reducing Cofactors

The substrates of PGHS-catalyzed oxidation discussed so far are all reducing cofactors for the peroxidase activity. This is an important observation because it implies that the cosubstrates react with a catalytically functional moiety generated by the enzyme. Because a good deal is known about catalysis by peroxidases, hypotheses can be advanced regarding the nature of the oxidant. We have observed that several compounds that are oxygenated by PGHS, including DPBF, B[a]P, and B[a]P-7,8-diol, are not cofactors for the enzymatic reduction of hydroperoxides to alcohols. This implies that these compounds may be oxidized by unique mechanisms and that the oxidizing agent(s) may be distinct from those that function in peroxidase turnover.

1. Diphenylisobenzofuran

As stated earlier the source of the oxygen incorporated into DPBF is molecular oxygen, not hydroperoxide oxygen [51]. Hydroperoxide-dependent DPBF oxygenation is abolished under strictly anaerobic conditions. The stoichiometry of DPBF oxygenated to hydroperoxide added varies from 2 : 1 to 3000 : 1, indicative of a chain reaction [51]. Furthermore, DPBF oxygenation is extremely sensitive to inhibition by a variety of antioxidants (most of which are also peroxidase reducing cofactors). This suggests that the PGHS-catalyzed oxygenation of DPBF occurs by a free-radical chain process. We have considered a free-radical mechanism for DPBF oxygenation that was originally proposed by Howard and Mendenhall (Scheme 2) [51, 97]. The reaction is initiated by the addition of a radical to the furan followed by scavenging of the incipient carbon-centered radical with molecular oxygen to generate a peroxyl radical. The peroxyl radical then adds to another molecule of DPBF, and the cycle is repeated.

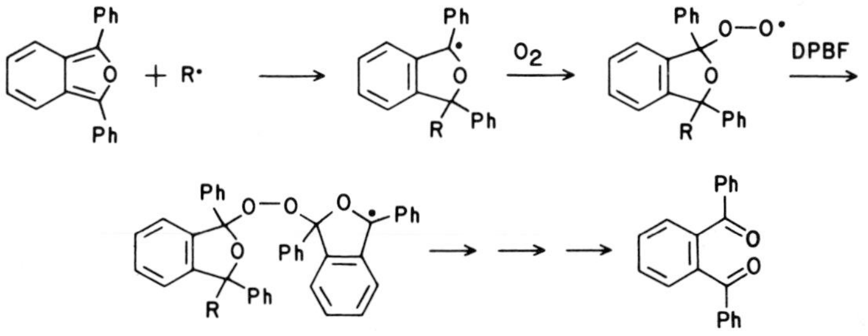

Scheme 2

Eventually, a polymeric peroxide is generated and is reduced or decomposes to DBB. This mechanism is consistent with the source of the oxygen and the chain stoichiometry. Because the chain-propagating agent is a peroxyl radical derived from DPBF, relatively few enzyme-generated radicals are necessary to initiate the oxygenation of a large number of DPBF molecules. This may explain the extreme sensitivity of DPBF oxygenation to inhibition by antioxidants.

2. Benzo[a]pyrene

The oxygenation of B[a]P by PGHS is also potently inhibited by antioxidants, but the stoichiometry of B[a]P oxidized to hydroperoxide added is low (<0.1). Therefore, if the oxygenation of B[a]P to 6-hydroxy-B[a]P (the putative first step) occurs by a radical process, B[a]P is not reactive to its own oxygen-containing radicals. The oxygenation of B[a]P to 6-hydroxy-B[a]P and the derived quinones by the mixed-function oxidases has been studied extensively by Ts'o and Nagata, and they have proposed mechanisms to explain their observations [61, 62, 64]. The 6-oxo radical can be detected by ESR spectroscopy following the incubation of B[a]P with mixed-function oxidases, and the precursor of the radical, the B[a]P cation-radical, can be detected in purely chemical systems [6, 62, 64, 98]. It seems likely that these species are produced during the peroxidatic metabolism of B[a]P by PGHS, but direct experimental evidence is lacking. Likewise, it is not known whether the oxidizing agent is enzyme-derived or hydroperoxide-derived.

3. 7,8-Dihydroxy-7,8-dihydrobenzo[a]pyrene

We have concentrated our studies of the mechanism of polycyclic hydrocarbon oxidation on B[a]P-7,8-diol. This reaction is of interest not only because it represents a metabolic activation, but also because it is an example of a hydroperoxide-dependent epoxidation in which the source of the epoxide oxygen is molecular oxygen [74]. The cleavage of the O—O bond is a four-electron process, but B[a]P-7,8-diol is a two-electron donor, suggesting that another component of the incubation mixture must be concomitantly oxidized. As in the case of B[a]P oxidation the stoichiometry of B[a]P-7,8-diol oxygenated to hydroperoxide added is low (<0.1), and the reaction is potently inhibited by antioxidants [74]. Taken with the source of the oxygen, these observations suggest that the epoxidation is free radical in nature.

C. Epoxidation of Dihydroaromatic Compounds by Hematin and Fatty Acid Hydroperoxides

We have developed a model system that we believe will yield insights into the mechanism of B[a]P-7,8-diol oxidation. Hematin, in the presence of detergents, catalyzes the epoxidation of B[a]P-7,8-diol by unsaturated fatty acid hydroperoxides [99]. The detergent is required at concentrations above its critical micellar concentration and solubilizes B[a]P-7,8-diol in addition to stimulating the catalytic activity of hematin. Epoxidation is saturated at 0.5 μM hematin, and the rates approach those obtained with microsomal PGHS. The products and stereochemistry of epoxidation are very similar to those of the enzymatic system, an exception being that significant amounts of the *syn*-diol-epoxide (~20%) are detected. The epoxide oxygen is derived from molecular oxygen, and the reaction is inhibited by rigorous deoxygenation. The phenolic antioxidants butylated hydroxyanisole and butylated hydroxytoluene inhibit epoxidation at 10^{-6} and 10^{-5} M, respectively. Thus, there is a good correlation between the enzymatic epoxidation of B[a]P-7.8-diol and the hematin-dependent epoxidation.

The incubation of unsaturated fatty acid hydroperoxides with hematin in the *absence* of B[a]P-7,8-diol causes the uptake of molecular oxygen from solution. Experiments with superoxide dismutase and catalase reveal that the molecular oxygen uptake is not due to the formation of O_2^- or H_2O_2. The capacity of some fatty acid methyl esters and hematin to trigger B[a]P-7.8-diol epoxidation and molecular oxygen uptake is closely correlated (Table II). The striking aspect of the data is the requirement for double bonds in the vicinity of the hydroperoxide in order for both reactions to occur. We have proposed a mechanism that is consistent with all of the experimental observations to explain the unsaturated fatty acid-dependent epoxidation of B[a]P-7,8-diol by hematin (Scheme 3) [99]. Hematin reduces the hydroperoxide to an alkoxyl radical and hydroxide ion. The alkoxyl radical cyclizes to an epoxide and a carbon-centered radical derived from the unsaturated portion of the hydroperoxide. This radical is scavenged by molecular oxygen to generate a peroxyl radical that acts as the epoxidizing agent. Epoxidation of aliphatic double bonds by peroxyl radicals is well precedented,

TABLE II   Fatty Acid Hydroperoxide Methyl Esters

Hydroperoxide	O_2 uptake (μM)	7,8-Diol oxidation v_i (μM/min)
COOCH$_3$ ... OOH	0	0
COOCH$_3$ ... OOH	65 ± 1	7.0 ± 1.1
COOCH$_3$ ... OOH	240 ± 15	16 ± 1.8

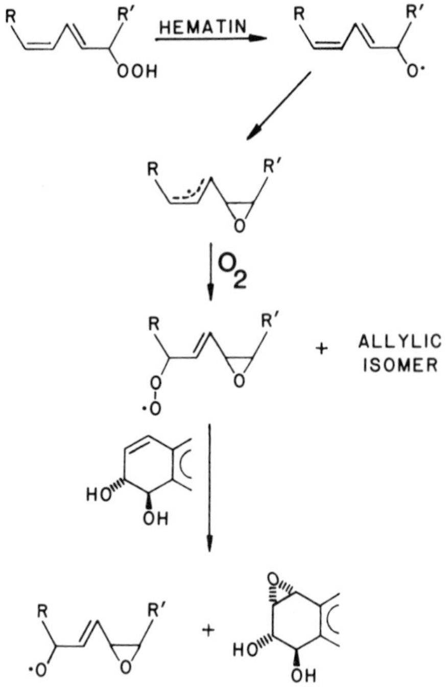

Scheme 3

and the stereochemistry of epoxidation (mainly anti) is consistent with that observed in other peroxyl radical-dependent epoxidations [100–102]. Pryor predicted several years ago that peroxyl radicals can epoxidize polycyclic hydrocarbons [103]. We have found that peroxyl radicals epoxidize dihydroaromatic derivatives of polycyclic hydrocarbons and oxidize the fully aromatic hydrocarbons to quinones. Peroxyl radicals may only be sufficiently reactive to add to nonaromatic double bonds, which may explain the difference in the chemistry observed between B[a]P and B[a]P-7,8-diol.

The peroxyl radical that serves as the epoxidizing agent in the hematin system is generated in a novel fashion by intramolecular cyclization and molecular oxygen scavenging of the incipient carbon radical. Hamberg has reported that 61% of the products isolated after the reaction of 13-hydroperoxy-9,11-octadecadienoic acid with methemoglobin are epoxy alcohols that are presumably derived from reduction of the initially formed epoxy hydroperoxides [104]. Gardner et al. isolated the epoxy hydroperoxides in low yield after the reaction of 13-hydroperoxy-9,11-octadecadienoic acid with ferrous ion and showed that they are reduced to epoxy alcohols under the conditions of the reaction [105]. In addition, Gardner et al. trapped the carbon-centered radical that is the precursor of the peroxyl radical by carrying out the reactions in the presence of tocopherol [106]. Preliminary results from our laboratory indicate that epoxy alcohols formed as indicated in Scheme 3 are major products of the hematin-catalyzed decomposition of 13-hydroperoxy-9,11-octadecadienoic acid.

These observations indicate that cyclization as indicated in Scheme 3 is a major fate of the alkoxyl radical generated by heme- or heme-protein-catalyzed reduction of 13-hydroperoxy-9,11-octadecadienoic acid. The exact mechanism of the hematin-catalyzed generation of the initial alkoxyl radical is unknown. A direct one-electron reduction would oxidize the iron to the +4 oxidation state and yield a hematin derivative analogous to peroxidase compound II. This complex would have to be reduced by an electron in order to catalyze the generation of another molecule of alkoxyl radical. The electron donor might be one of the numerous radical species that are undoubtedly formed in this system. It is also possible that hematin is first reduced to a ferrous derivative by the hydroperoxide (generating a peroxyl radical), which then reduces the hydroperoxide to an alkoxy radical. This would be directly analogous to the decomposition of hydroperoxide by simple iron salts in which the metal cycles between the +3 and +2 oxidation states [107]. It should be pointed out that $FeCl_3$ and $FeSO_4$ do not catalyze B[a]P7,8-diol oxidation even at 1 mM and that, if peroxyl radicals are initial hydroperoxide *oxidation* products, the ^{18}O labeling studies indicate that they do not epoxidize B[a]P-7,8-diol. Nevertheless, Hamberg and Gardner's work along with our own indicates that the pattern of hydroperoxide degradation products is complex, suggesting that multiple reaction pathways probably obtain. In addition, the detergent requirement and the hydrophobic nature of all the components imply that surface effects may be major determinants of the chemistry. Therefore, it seems premature to propose a mechanism for the hematin-catalyzed generation of an alkoxyl radical.

The mechanism outlined in Scheme 3 adequately explains all of the experimental observations made during the hematin-catalyzed epoxidation of B[a]P-7,8-diol by unsaturated fatty acid hydroperoxides. The similarity between the chemistry of the hematin- and PGHS-catalyzed epoxidations makes it tempting to speculate that similar mechanisms are involved. In fact, because other heme proteins catalyze hydroperoxide-dependent epoxidation of B[a]P-7,8-diol, PGHS may trigger the reaction simply because it has an accessible heme group. This would underscore the importance of the hydroperoxide-synthesizing activity of PGHS—the cyclooxygenase—in the cooxygenation of B[a]P-7,8-diol by unsaturated fatty acids. We believe that it is premature to infer mechanisms of the enzyme-catalzyed epoxidation from the results obtained with the model system. Although the latter has provided important insights and raised the exciting possibility that PGHS catalyzes oxidations in which the oxidizing agent is derived from the hydroperoxide, detailed studies of the effect of apo-PGHS on the hematin-dependent reaction should precede further speculation.

V. PEROXIDASE PARTICIPATION IN PATHOLOGICAL RESPONSES

Research in our laboratory has concentrated on the scope and mechanism of hydroperoxide-dependent oxidation by PGHS. Figure 6 is an attempt to integrate our work with current concepts of the control of unsaturated fatty acid metabolism in cellular systems. The rate-limiting step in prostaglandin and thromboxane biosynthesis is the release of substrate fatty acid from phospholipid stores [108, 109]. This occurs in response to any of a number of stimuli that interact with plasma membranes and is dependent on the action of phospholipases. Naturally, all cells do not respond to the same stimuli, but a given cell releases unsaturated fatty acid in response to any agent that binds to or perturbs its plasma membrane,

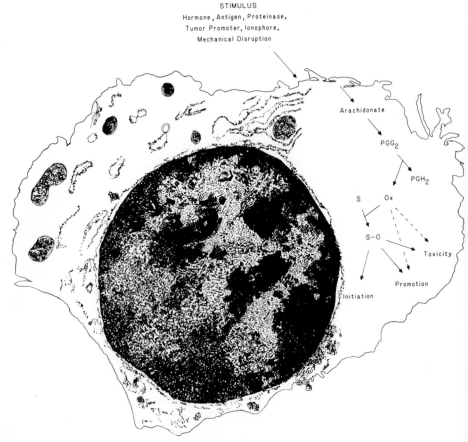

Fig. 6.   Interrelationship of arachidonate metabolism and xenobiotic oxidation.

including agents that cause membrane damage [*110*]. The released fatty acid (usually arachidonate) is then converted by the cyclooxygenase activity of PGHS to PGG, which is the hydroperoxide substrate for the peroxidase. An oxidizing agent that may either be a higher oxidation state of the peroxidase or a hydroperoxide-derived radical is generated and triggers the oxidation or oxygenation of an endogenous substrate or xenobiotic. Very little is known about what the endogenous substrates for the peroxidase might be, although Ogino *et al.* have shown that uric acid is the reducing cofactor for PGHS in bovine seminal vesicles [*111*]. Because the substrate specificity of the peroxidase is rather broad, it is possible that the endogenous reducing cofactors varies with the cell type.

A. Carcinogenesis

The oxidized compounds may play a role in normal cellular metabolism and/or regulation, or they may induce pathological responses such as toxicity and carcinogenicity. Several laboratories, including our own, have studied the oxidative activation of carcinogens by PGHS because oxidation plays a clearly defined role in tumor initiation. It is clear from this research that PGHS can trigger metabolic activation of carcinogens *in vitro,* and relatively recent results suggest it may play a role *in vivo* [*112, 113*].

Studies with polymorphonuclear leukocytes and mouse skin suggest that oxidation, probably by free radicals, may also play an important role in tumor promotion [*68, 114, 115*]. The emphasis of the studies to date has been on the involvement of superoxide or superoxide-derived radicals in tumor promotion, but there is no reason that the reactions triggered by this species could not be effected by alkoxyl or peroxyl radicals derived from unsaturated fatty acid hydroperoxides. Furthermore, there is a direct relationship between the capacity of a series of compounds to promote tumorigenesis and their capacity to stimulate AA release and PGH_2 biosynthesis in cells and in mouse skin *in vivo* [*116*]. This suggests that PGHS may be important in tumor promotion as well as tumor initiation by virtue of its capacity to synthesize hydroperoxides and generate oxidizing agents from them. This possibility is under investigation in our laboratory.

B. Thrombosis and Metastasis

The presence of a peroxidase activity in PGHS suggests that the reduction of PGG_2 and other fatty acid hydroperoxides is an important function. Because hydroperoxides can stimulate or inhibit cyclooxygenase activity, modulation of their concentration can have a dramatic effect on the flux of AA into endoperoxides [*43, 117*]. As expected, peroxidase reducing cofactors can stimulate or inhibit cyclooxygenase activity *in vitro* [*47*]. Another enzyme of the AA cascade is PGI_2 synthase (PGI_2S). Moncada *et al.* [118] first reported that fatty acid hydroperoxides are extremely potent irreversible inhibitors of PGI_2S, and Ham *et al.* [119] demonstrated that peroxidase reducing cofactors ameliorate this effect. Whether cofactors reduce the level of inhibitory hydroperoxides or scavenge the oxidizing agent that reacts with PGI_2 is unclear. Nevertheless, reducing cofactors for PGHS stimulate the conversion of AA to PGI_2 [120, 121].

2,4-Dihydro-5-methyl-2-[2-(2-napthyloxy)ethyl]-3*H*-pyrazole-3-one (nafazatrom) exhibits significant antithrombotic and antimetastatic activity *in vivo* [122, 123]. Its apparent lack of toxic side effects suggests that it may exhibit a high therapeutic ratio in clinical investigations [122]. Nafazatrom is not a thromboxane synthase or phosphodiesterase inhibitor, and it does not exhibit a direct antiaggregatory effect on platelets [122]. Nafazatrom has been reported to elevate the levels of bioassayable PGI_2 after administration to rats [124]. Because PGI_2 is a potent inhibitor of platelet aggregation [125] and of tumor cell metastasis [123], it is possible that the antithrombotic and antimetastatic activities of nafazatrom are derived from its capacity to elevate PGI_2 levels *in vivo* [123, 124].

Nafazatrom
3

We have found, in collaboration with Eling and Honn, that nafazatrom is a potent reducing cofactor for the peroxidase activity of PGHS [126]. It inhibits the hydroperoxide-dependent oxygenation of PB and stimulates the reduction of 15-hydroperoxyeicosatetraenoic acid (15-HPETE) by microsomal or purified PGHS. In addition, nafazatrom causes an elevation in the levels of 6-keto-$PGF_{1\alpha}$, the nonenzymatic hydrolysis product of PGI_2, biosynthesized from AA in RSVM. Nafazatrom does not stimulate the conversion of PGH_2 to PGI_2, indicating that it does not have a direct effect on PGI_2S. However, it protects microsomal PGI_2S from inactivation by 15-HPETE. Thus, nafazatrom stimulates PGI_2 biosynthesis by acting as a reducing cofactor for the peroxidase activity of PGHS. If this is the mechanism of its stimulatory effect on PGI_2 levels *in vivo*, it indicates a critical role for the peroxidase activity of PGHS in the modulation of intracellular fatty acid hydroperoxide levels. This implies that reducing cofactors for PGHS could have important pharmacological activities by virtue of their protection of PGI_2S from inactivation.

VI. RELATED OXIDIZING SYSTEMS

The rate-limiting step in PGHS-dependent cooxidation is the synthesis of hydroperoxide [31]. A unique aspect of this protein is that it contains both a hydroperoxide-synthesizing activity and a peroxidase activity. Theoretically, other hydroperoxide-generating systems could trigger peroxidase-mediated oxidations. In addition, evidence presented in Section IV,C suggests that peroxyl radicals derived from unsaturated fatty acid hydroperoxides will epoxidize B[a]P-7,8-diol. This type of epoxidation may occur as a result of the generation of peroxyl radicals in other biochemical systems.

A. Lipoxygenase

The number of hydroperoxide-generating systems in animals appears to be quite restricted [31]. The main example besides PGHS is lipoxygenase. There are several lipoxygenases that oxygenate fatty acids at different positions to generate hydroperoxy acids [127]. We have found that hydroperoxy acids are equivalent to PGG_2 in triggering cooxidation [31]. Wong et al. have reported that the cyclooxygenase inhibitor indomethacin does not lower the rate of oxidation of N-hydroxyacetylaminoflourene by rat mammary parenchymal cells in the presence of AA [128]. Oxidation is completely abolished by eicosatetraynoic acid, which inhibits both cyclooxygenase and lipoxygenase activities. This implies that a lipoxygenase in the mammary cells triggers the cooxidation. If true, this raises a question about the identity of the peroxidase. It is particularly intriguing in this case because it has been reported that the 12-lipoxygenase of platelets can be separated from the protein that catalyzes hydroperoxide reduction [129]. Therefore, the relation of the lipoxygenase and peroxidase activities may be different from that observed for PGHS.

B. Lipid Peroxidation

The most likely mechanism for the generation of peroxyl radicals of unsaturated fatty acids in biochemical systems independent of PGHS and lipoxygenase is lipid peroxidation. Although the molecular events that initiate lipid peroxidation are incompletely defined, it is clear that propagation reactions generate significant quantities of peroxyl radicals [Eqs. (15) and (16)].

$$ROO \cdot + RH \rightarrow ROOH + R \cdot \tag{15}$$

$$R \cdot + O_2 \rightarrow ROO \cdot \tag{16}$$

We have shown that ascorbate-induced lipid peroxidation causes the epoxidation of B[a]P-7,8-diol to a diol-epoxide [130]. The time course of epoxidation follows the time course of the generation of thiobarbituric acid-reactive material, and the stereochemistry of epoxidation appears to be identical to that of the hematin–fatty acid hydroperoxide system. These observations are consistent with the epoxidation of B[a]P-7,8-diol by peroxyl radicals generated during lipid peroxidation. The implications of this finding for mechanisms of chemically induced toxicity and carcinogenicity are obvious.

VII. CONCLUSION

The enormous range and potency of biological effects exerted by prostaglandins and thromboxanes have prompted the initiation of many studies of their biosynthesis and metabolism and of the factors that regulate these processes. This chapter, along with others in this volume, demonstrates that during the generation of these local hormones oxidizing agents are released, and these may have profound effects on cellular growth and metabolism. The exquisite responsiveness of phospholipases to a variety of membrane stimuli coupled with the explosive oxidative potential of PGHS provides a link between events at the cell surface and oxidations within the cell.

The development of an entire research area devoted to the study of PGHS-catalyzed cooxidation illustrates the uncertain course of science and the direct link between basic and applied research. What began during studies of the mechanism of prostaglandin endoperoxide biosynthesis [*19, 29*] has evolved into a multifaceted investigation of enzyme-mediated free-radical biochemistry. The diversity of the reactions catalyzed is mechanistically challenging, and the pathological implications for toxicological studies are exciting.

ACKNOWLEDGMENTS

Financial support from the American Cancer Society (BC 244) and the National Institutes of Health (GM 23642) is gratefully acknowledged. Lawrence J. Marnett is the recipient of an American Cancer Society Faculty Research Award (FRA 243). This manuscript was skillfully typed by Alice J. Lietz. Robert Sachs assisted in the composition of the illustrations.

REFERENCES

1. M. Hamberg and B. Samuelsson, *Proc. Natl. Acad. Sci. U.S.A.* **70,** 899–903 (1973).
2. D. H. Nugteren and E. Hazelhof, *Biochim. Biophys. Acta* **326,** 448–461 (1973).
3. M. Hamberg, J. Svensson, T. Wakabayashi, and B. Samuelsson, *Proc. Natl. Acad. Sci. U.S.A.* **71,** 345–349 (1974).
4. T. Miyamoto, N. Ogino, S. Yamamoto, and O. Hayaishi, *J. Biol. Chem.* **251,** 2629–2636 (1976).
5. M. Hemler, W. E. M. Lands, and W. L. Smith, *J. Biol. Chem.* **251,** 5575–5579 (1976).
6. F. J. Van der Ouderaa, M. Buytenhek, D. H. Nugteren, and D. A. Van Dorp, *Biochim. Biophys. Acta* **487,** 315–331 (1977).
7. W. E. M. Lands and W. L. Smith, eds., "Methods in Enzymology," Vol. 86. Academic Press, New York, 1982.
8. N. Ogino, T. Miyamoto, S. Yamamoto, and O. Hayaishi, *J. Biol. Chem.* **252,** 890–895 (1977).
9. T. Yoshimoto, S. Yamomoto, M. Okuma, and O. Hayaishi, *J. Biol. Chem.* **252,** 5871–5874 (1977).
10. S. Hammarström and P. Falardeau, *Proc. Natl. Acad. Sci. U.S.A.* **74,** 3691–3695 (1977).
11 S. Yamamoto, S. Ohki, N. Ogino, T. Shimizu, T. Yoshimoto, K. Watanabe, and O. Hayaishi, *Adv. Prostaglandin Thromboxane Res.* **6,** 27–34 (1980).

12. U. Diczfalusy, P. Falardeau, and S. Hammarström, *FEBS Lett.* **84**, 271–274 (1977).
13. M. Hamberg, J. Svensson, and B. Samuelsson, *Proc. Natl. Acad. Sci. U.S.A.* **72**, 2994–2998 (1975).
14. R. A. Johnson, D. R. Morton, J. H. Kinner, R. R. Gorman, J. C. McGuire, F. F. Sun, N. Whittaker, S. Bunting, J. Salmon, S. Moncada, and J. R. Vane, *Prostaglandins* **12**, 915–928 (1976).
15. E. J. Christ and D. A. Van Dorp, *Biochim. Biophys. Acta* **270**, 537–545 (1972).
16. T. E. Rollins and W. L. Smith, *J. Biol. Chem.* **255**, 4872–4875 (1980).
17. G. J. Roth, E. Machuga, and P. Strittmatter, *J. Biol. Chem.* **256**, 10018–10022.
18. R. Ueno, T. Shimizu, K. Kondo, and O. Hayaishi, *J. Biol. Chem.* **257**, 5584–5588 (1982).
19. L. J. Marnett, P. Wlodawer, and B. Samuelsson, *Biochem. Biophys. Res. Commun.* **60**, 1286–1294 (1974).
20. L. J. Marnett, P. Wlodawer, and B. Samuelsson, *J. Biol. Chem.* **250**, 8510–8517 (1975).
21. A. D. Rahimtula and P. J. O'Brien, *Biochem. Biophys. Res. Commun.* **62**, 268–275 (1975).
22. R. T. Williams, "Detoxication Mechanisms. The Metabolism and Detoxication of Drugs, Toxic Substances, and Other Organic Compounds," 2nd ed. Wiley, New York, 1959.
23. D. H. Hutson, *Foreign Compd. Metab. Mamm.* **1**, 314–380 (1970).
24. J. A. Miller, *Cancer Res.* **30**, 559–576 (1970).
25. C. Heidelberger, *Annu. Rev. Biochem.* **44**, 79–121 (1975).
26. R. Sato and T. Omura, eds., "Cytochrome P-450." Academic Press, New York, 1978.
27. L. J. Marnett, J. T. Johnson, and M. J. Bienkowski, *FEBS Lett.* **106**, 13–16 (1979).
28. K. Sivarajah, H. Mukhtar, and T. Eling, *FEBS Lett.* **106**, 17–20 (1979).
29. L. J. Marnett, *Life Sci.* **29**, 531–546 (1981).
30. T. Eling, J. Boyd, G. Reed, R. Mason, and K. Sivarajah, *Drug Metab. Rev.* (in press).
31. L. J. Marnett and G. A. Reed, *Biochemistry* **18**, 2923–2929 (1979).
32. G. Hamilton, *in* "Molecular Mechanisms of Oxygen Activation" (O. Hayaishi, ed.), pp. 405–451. Academic, New York, 1974.
33. S. Ohki, N. Ogino, S. Yamamoto, and O. Hayaishi, *J. Biol. Chem.* **254**, 829–836 (1979).
34. T. V. Zenser, M. B. Mattammal, and B. B. Davis, *J. Pharmacol. Exp. Ther.* **208**, 418–421 (1979).
35. T. V. Zenser, M. B. Mattammal, and B. B. Davis, *J. Pharmacol. Exp. Ther.* **211**, 460–464 (1979).
36. T. V. Zenser, M. B. Mattammal, and B. B. Davis, *J. Pharmacol. Exp. Ther.* **214**, 312–317 (1980).
37. R. W. Egan, P. H. Gale, W. J. A. Vanden Heuvel, E. A. Baptista, and F. A. Kuehl, Jr., *J. Biol. Chem.* **255**, 323–326 (1980).
38. R. W. Egan, P. H. Gale, E. M. Baptista, K. L. Kennicott, W. J. A. Vanden Heuvel, R. W. Walker, P. E. Fagerness, and F. A. Kuehl, Jr., *J. Biol. Chem.* **256**, 7352–7361 (1981).
39. W. R. Pagels, R. J. Sachs, M. Leithauser, and L. J. Marnett, in preparation.
40. D. L. Dewitt, T. E. Rollins, J. S. Day, J. A. Gauger, and W. L. Smith, *J. Biol. Chem.* **256**, 10375–10382 (1981).
41. W. R. Pagels, R. J. Sachs, L. J. Marnett, D. L. Dewitt, J. S. Day, and W. L. Smith, *J. Biol. Chem.* **258**, 6517–6523 (1983).
42. B. C. Saunders, *in* "Inorganic Biochemistry" (G. L. Eichhorn, ed.), pp. 988–1021. Am. Elsevier, New York, 1973.
43. R. W. Egan, J. Paxton, and F. A. Kuehl, Jr., *J. Biol. Chem.* **251**, 7329–7335 (1976).
44. P. Moldeus and A. Rahimtula, *Biochem. Biophys. Res. Commun.* **96**, 469–475 (1980).
45. K. Sivarajah, J. M. Lasker, T. E. Eling, and M. B. Abou-Donia, *Mol. Pharmacol.* **21**, 133–141 (1982).
46. L. J. Marnett, P. H. Siedlik, and L. W. M. Fung, *J. Biol. Chem.* **257**, 6957–6964 (1982).

47. R. W. Egan, P. H. Gale, G. C. Beveridge, L. J. Marnett, and F. A. Kuehl, Jr., *Adv. Prostaglandin Thromboxane Res.* **6**, 153–155 (1980).
48. V. G. Stygles and J. D. Iuliucci, in "Toxicology of the Kidney" (J. B. Hook, ed.), pp. 151–178. Raven Press, New York, 1981.
49. P. Kincaid-Smith, B. M. Saker, I. F. C. McKenzie, and K. D. Muriden, *Med. J. Aust.* **1**, 203–206 (1968).
50. T. V. Zenser, M. B. Mattammal, and B. B. Davis, *J. Pharmacol. Exp. Ther.* **207**, 719–725 (1978).
51. L. J. Marnett, M. J. Bienkowski, and W. R. Pagels, *J. Biol. Chem.* **254**, 5077–5082 (1979).
52. P. S. Portoghese, K. Svanborg, and B. Samuelsson, *Biochem. Biophys. Res. Commun.* **63**, 748–755 (1975).
53. L. J. Marnett, M. J. Bienkowski, W. R. Pagels, and G. A. Reed, *Adv. Prostaglandin Thromboxane Res.* **6**, 149–151 (1980).
54. R. I. Freudenthal and P. W. Jones, eds., "Carcinogenesis," Vol. 1. Raven Press, New York, 1976.
55. P. Sims, P. L. Grover, A. Swaisland, K. Pal, and A. Hewer, *Nature (London)* **252**, 326–328 (1974).
56. P. G. Wislocki, A. W. Wood, R. L. Chang, W. Levin, H. Yagi, O. Hernandez, D. M. Jerina, and A. H. Conney, *Biochem. Biophys. Res. Commun.* **68**, 1006–1012 (1976).
57. E. Huberman, L. Sachs, S. K. Yang, and H. V. Gelboin, *Proc. Natl. Acad. Sci. U.S.A.* **73**, 607–611 (1976).
58. J. Kapitulnik, P. G. Wislocki, W. Levin, H. Yagi, D. M. Jerina, and A. H. Conney, *Cancer Res.* **38**, 354–358 (1978).
59. W. Levin, A. W. Wood, A. Y. H. Lu, D. Ryan, S. West, A. H. Conney, D. R. Thakker, H. Yagi, and D. M. Jerina, *ACS Symp. Ser.* **44**, 99–126 (1977).
60. L. J. Marnett, G. A. Reed, and J. T. Johnson, *Biochem. Biophys. Res. Commun.* **79**, 569–576 (1977).
61. W. Caspary, B. Cohen, S. Lesko, and P. O. P. Ts'o, *Biochemistry* **12**, 2649–2656 (1973).
62. S. Lesko, W. Caspary, R. Lorentzen, and P. O. P. Ts'o, *Biochemistry* **14**, 3978–3984 (1975).
63. L. Jeftic and R. Adams, *J. Am. Chem. Soc.* **92**, 1332–1337 (1970).
64. C. Nagata, Y. Tagashira, and M. Dodama, *Biochem. Dis.* **4**, 87–111 (1974).
65. P. G. Wislocki, A. W. Wood, R. L. Chang, W. Levin, H. Yagi, O. Hernandez, P. M. Dansette, D. M. Jerina, and A. H. Conney, *Cancer Res.* **36**, 3350–3357 (1976).
66. R. J. Lorentzen and P. O. P. Ts'o, *Biochemistry* **16**, 1467–1473 (1977).
67. J. Capdevila and S. Orrenius, *FEBS Lett.* **119**, 33–37 (1980).
68. B. D. Goldstein, G. Witz, M. Amoruso, and W. Troll, *Biochem. Biophys. Res. Commun.* **88**, 854–860 (1979).
69. G. Witz, B. D. Goldstein, M. Amoruso, D. S. Stone, and W. Troll, *Biochem. Biophys. Res. Commun.* **97**, 883–888 (1980).
70. L. J. Marnett, A. Panthananickal, and G. A. Reed, *Drug Metab. Rev.* **13**, 235–247 (1982).
71. D. M. Jerina, R. Lehr, M. Schaefer-Ridder, H. Yagi, J. M. Karle, D. R. Thakker, A. W. Wood, A. Y. H. Lu, D. Ryan, S. West, W. Levin, and A. H. Conney, *Cold Spring Harbor Conf. Cell Proliferation* **4**, 639–658 (1977).
72. L. J. Marnett, G. A. Reed, and D. J. Dennison, *Biochem. Biophys. Res. Commun.* **82**, 210–216 (1978).
73. J. Guthrie, I. G. C. Robertson, E. Zeiger, J. A. Boyd, and T. E. Eling, *Cancer Res.* **42**, 1620–1623 (1982).
74. L. J. Marnett and M. J. Bienkowski, *Biochem. Biophys. Res. Commun.* **96**, 639–647 (1980).
75. A. Panthananickal and L. J. Marnett, *Chem.-Biol. Interact.* **33**, 239–252 (1981).
76. A. Panthananickal, P. Weller, and L. J. Marnett, *J. Biol. Chem.* **258**, 4411–4418 (1983).

77. D. R. Thakker, H. Yagi, H. Akagi, M. Koreeda, A. Y. H. Lu, W. Levin, A. W. Wood, A. H. Conney, and D. M. Jerina, *Chem.-Biol. Interact.* **16,** 281–300 (1977).
78. J. Deutsch, K. P. Vatsis, M. J. Coon, J. C. Leutz, and H. V. Gelboin, *Mol. Pharmacol.* **16,** 1011–1018 (1979).
79. G. A. Reed and L. J. Marnett, *J. Biol. Chem.* (in press).
80. A. W. Wood, W. Levin, A. Y. H. Lu, D. Ryan, S. B. West, H. Yagi, H. D. Mah, D. M. Jerina, and A. H. Conney, *Mol. Pharmacol.* **13,** 1116–1125 (1977).
81. J. F. Waterfall and P. Sims, *Biochem. J.* **128,** 265–277 (1972).
82. A. Rahimtula, P. Moldeus, B. Andersson, and M. Nordenskjöld, *in* "Prostaglandins and Cancer: First International Conference" (T. J. Powles, R. S. Bockman, K. V. Honn, and P. Ramwell, eds.), pp. 159–162. Alan R. Liss, Inc., New York, 1982.
83. J. A. Boyd, D. S. Harvan, and T. E. Eling, *J. Biol. Chem.* **258,** 8246–8254 (1983).
84. T. V. Zenser, M. B. Mattammal, H. J. Armbrecht, and B. B. Davis, *Cancer Res.* **40,** 2839–2845 (1980).
85. S. Vasdev, Y. Tsuruta, and P. J. O'Brien, *in* "Prostaglandins and Cancer: First International Conference" (T. J. Powles, R. S. Bockman, K. V. Honn, and P. Ramwell, eds.), pp. 155–158, Alan R. Liss, Inc., New York, 1982.
86. G. H. Degen, T. E. Eling, and J. A. McLachlan, *Cancer Res.* **42,** 919–923 (1982).
87. T. V. Zenser, M. B. Mattammal, and B. B. Davis, *Cancer Res.* **40,** 114–118 (1980).
88. A. Ishimaru and I. Yamazaki, *J. Biol. Chem.* **252,** 6118–6124 (1977).
89. B. Chance, *Arch. Biochem. Biophys.* **41,** 416–424 (1952).
90. P. George, *Biochem. J.* **54,** 266–276 (1953).
91. H. B. Dunford, *Coord. Chem. Rev.* **19,** 187–251 (1976).
92. W. E. Lee and D. W. Miller, *Photogr. Sci. Eng.* **10,** 192–201 (1966).
93. J. M. Lasker, K. Sivarajah, R. P. Mason, B. Kalyanaraman, M. B. Abou-Donia, and T. E. Eling, *J. Biol. Chem.* **256,** 7764–7767 (1981).
94. G. A. Russell and A. G. Bemis, *J. Am. Chem. Soc.* **88,** 5491–5497 (1966).
95. G. D. Nordblom, R. E. White, and M. J. Coon, *Arch. Biochem. Biophys.* **175,** 524–533 (1976).
96. P. Siedlik, unpublished results.
97. J. A. Howard and G. D. Mendenhall, *Can. J. Chem.* **53,** 2199–2201 (1975).
98. P. D. Sullivan, L. M. Calle, K. Shafer, and M. Nettleman, *in* "Carcinogenesis" (P. W. Jones and R. I. Freudenthal, eds.), Vol. 3; pp. 1–8, Raven Press, New York, 1978.
99. T. A. Dix and L. J. Marnett, *J. Am. Chem. Soc.* **103,** 6744–6746 (1981).
100. K. U. Ingold and B. P. Roberts, eds. "Free Radical Substitution Reactions," pp. 148–193. Wiley (Interscience), New York, 1970.
101. A. Padwa and L. Brodsky, *Tetrahedron Lett.* pp. 1045–1048 (1973).
102. H. Hart and P. B. Lavrik, *J. Org. Chem.* **39,** 1793–1794 (1974).
103. W. A. Pryor, *in* "Environmental Health Chemistry - The Chemistry of Environmental Agents as Potential Human Hazards" (J. D. McKinney, ed.), pp. 445–466. Ann Arbor Sci. Publ. Ann Arbor, Michigan, 1980.
104. M. Hamberg, *Lipids* **10,** 87–92 (1975).
105. H. W. Gardner, D. Weisleder, and R. Kleiman, *Lipids* **13,** 246–252 (1978).
106. H. W. Gardner, K. Eskins, G. W. Grams, and G. E. Inglett, *Lipids* **7,** 324–334 (1972).
107. G. Sosnowtky and D. J. Rawlinson, *in* "Organic Peroxides" (D. S. Swern, ed.), Vol. 2, pp. 153–268. Wiley (Interscience), New York, 1971.
108. W. E. M. Lands and B. Samuelsson, *Biochim. Biophys. Acta* **164,** 426–429 (1968).
109. H. Vonkeman and D. A. Van Dorp, *Biochim. Biophys. Acta* **164,** 430–432 (1968).
110. C. Galli, G. Galli, and G. Porcellati, eds., "Advances in Prostaglandin and Thromboxane Research," Vol. 3. Raven Press, New York, 1978.

111. N. Ogino, S. Yamamoto, O. Hayaishi, and T. Tokuyama, *Biochem. Biophys. Res. Commun.* **87,** 184–191 (1979).

112. S. M. Cohen, T. V. Zenser, G. Murasaki, S. Fukushima, M. Mattammal, N. S. Rapp, and B. B. Davis, *Cancer Res.* **41,** 3355–3359 (1981).

113. J. A. Boyd, J. C. Barrett, and T. E. Eling, *Cancer Res.* **42,** 2628–2632 (1982).

114. G. Witz, B. D. Goldstein, M. Amoruso, D. S. Stone, and W. Troll, *Biochem. Biophys. Res. Commun.* **97,** 883–888 (1980).

115. T. J. Slaga, A. J. P. Klein-Szanto, L. L. Triplett, L. P. Yotti, and J. E. Trosko, *Science* **213,** 1023–1025 (1981).

116. L. Levine, *Adv. Cancer Res.* **35,** 49–79 (1981).

117. M. E. Hemler, H. W. Cook, and W. E. M. Lands, *Arch. Biochem. Biophys.* **193,** 340–345 (1979).

118. S. Moncada, R. J. Gryglewski, S. Bunting, and J. R. Vane, *Prostaglandins* **12,** 715–737 (1976).

119. E. A. Ham, R. W. Egan, D. D. Soderman, P. H. Gale, and F. A. Kuehl, Jr., *J. Biol. Chem.* **254,** 2191–2194 (1979).

120. M. P. Carpenter, *Fed. Proc., Fed. Am. Soc. Exp. Biol.* **40,** 189–194 (1981).

121. J. R. Beetens, M. Claeys, and A. G. Herman, *Biochem. Pharmacol.* **30,** 2811–2815 (1981).

122. F. Seuter, W.-D. Busse, K. Meng, F. Hoffmeister, E. Möller, and H. Horstman, *Arzneim.-Forsch.* **29,** 54–59 (1979).

123. K. V. Honn, J. Meyer, G. Neagos, T. Henderson, C. Westley, and V. Ratanatharathorn, *in* "Interaction of Platelets and Tumor Cells" (G. A. Jamieson, ed.), pp. 295–331. Alan R. Liss, Inc., New York, 1982.

124. J. Vermylen, D. A. F. Chamone, and M. Verstraete, *Lancet,* 518–520 (1979).

125. S. Bunting, R. Gryglewski, S. Moncada, and J. R. Vane, *Prostaglandins* **12,** 897–913 (1976).

126. T. E. Eling, K. V. Honn, W. D. Busse, F. Seuter, and L. J. Marnett, *in* "Prostaglandins and Cancer: First International Conference" (T. J. Powles, R. S. Bockman, K. V. Honn, and P. Ramwell, eds.), pp. 783–787. Alan R. Liss, Inc., New York, 1982.

127. M. Hamberg and B. Samuelsson, *Proc. Natl. Acad. Sci. U.S.A.* **71,** 3400–3404 (1974).

128. P. K. Wong, M. J. Hampton, and R. A. Floyd, *in* "Prostaglandins and Cancer: First International Conference" (T. J. Powles, R. S. Bockman, K. V. Honn, and P. W. Ramwell, eds.), pp. 167–179. Alan R. Liss, Inc., New York, 1982.

129. M. I. Siegel, R. T. McConnell, N. A. Porter, and P. Cuatrecasas, *Proc. Natl. Acad. Sci. U.S.A.* **77,** 308–312 (1980).

130. T. A. Dix and L. J. Marnett, *Science* **221,** 77–79 (1983).

CHAPTER **4**

Fatty Acid Paradoxes in the Control of Cell Proliferation: Prostaglandins, Lipid Peroxides, and Cooxidation Reactions

David G. Cornwell and Nobuhiro Morisaki*

Department of Physiological Chemistry
The Ohio State University
Columbus, Ohio

*Present address: Second Department of Internal Medicine, Chiba University, Chiba, Japan.

FREE RADICALS IN BIOLOGY, VOL. VI
Copyright © 1984 by Academic Press, Inc.
All rights of reproduction in any form reserved.
ISBN 0-12-566506-7

Our interest's on the dangerous edge of things.
The honest thief, the tender murderer,
The superstitious atheist,. . .
We watch while these in equilibrium keep
The giddy line midway:. . .

Robert Browning in *Bishop Blougram's Apology*

I. INTRODUCTION

Studies from many laboratories have shown that polyunsaturated fatty acids are involved in the control of cell proliferation. Studies with some cells and tissues demonstrate stimulatory effects, whereas studies with other cells and tissues demonstrate inhibitory effects. We propose that these contradictory effects, which we call *fatty acid paradoxes,* occur naturally and sometimes in the same cells and tissues. There are many paradoxes. Essential fatty acids support growth, wound healing, and tumorigenesis, and yet an essential fatty acid deficiency leads to hyperplasia. Oleic acid is both a highly toxic growth inhibitor and a stimulator of tumorigenesis. Prostaglandins enhance growth and DNA synthesis in epidermal cells, and yet a prostaglandin deficiency leads to epidermal hyperplasia. Prostaglandins have direct and indirect (immunosurveillance) effects that both enhance and inhibit tumorigenesis, and prostaglandins both enhance and inhibit the proliferation of cells in culture. Lipid peroxides inhibit cell proliferation, yet lipid peroxides may promote both tumorigenesis and metastasis. Antioxidants inhibit initiation and enhance promotion in tumorigenesis. Antioxidants both enhance and inhibit cell proliferation by a variety of direct and indirect effects involving lipid peroxides, the elaboration of growth factors, and cellular differentiation.

Fatty acids are oxidized by a varety of free-radical reactions to cyclic endoperoxides (prostaglandin pathway) and acyclic hydroperoxides (lipid peroxide pathway). The different peroxides, their derivatives, and radicals formed during their synthesis and metabolism can generate either positive or negative signals for cell proliferation. These effects are illustrated in studies involving skin, tumor, and tissue culture systems. We show that intervention strategies for the control of cell proliferation affect both positive and negative signals in ways that sometimes lead to unanticipated results in biological systems.

II. POLYUNSATURATED FATTY ACIDS AND CELL PROLIFERATION

A. Essential Fatty Acids: A Historical Analysis

Burr and Burr [1] first developed the concept of essential fatty acids (EFA) when they found that growth and ovulation in the young rat were greatly retarded by the rigid exclusion of fat from the diet. They soon noted that growth was restored by the addition of a small amount of linoleic acid [18:2 (n-6)]* to the diet [2] and thus identified the EFA.

The amount of an EFA required for growth is small and remarkably consistent. Weight gains in depleted male rats are a linear function of the log dose of the 18:2 (n-6) or 20:4 (n-6) supplied in the diet [3–5]. Weight gains in female rats are a function of fatty acid concentration, but high levels of linoleic acid actually suppress growth in the female [3].

Only two fatty acids, 18:2 (n-6) and its desaturation–chain elongation metabolite 20:4 (n-6) [6, 7], fulfill all the criteria for EFA [2–5]. These fatty acids are classified as members of the linoleic acid or n-6 family of fatty acids. Linolenic acid [18:3 (n-3)] and its desaturation–chain elongation metabolites [6, 7], the n-3 family of fatty acids, support growth somewhat, but their effects on growth are much smaller than the effects of equivalent amounts of n-6 fatty acids [2–5]. Oleic acid [18:1 (n-9)] and its desaturation–chain elongation metabolite 20:3 (n-9) [6, 7], the n-9 family of fatty acids, were elevated in an EFA deficiency [5, 8, 9]. An increase in the ratio of 20:3 (n-9) to 20:4 (n-6), the triene/tetraene ratio, actually became an important criterion for the evaluation of an EFA deficiency [5–7,10]. However, 18:1 (n-9) itself intensified an EFA deficiency [11], and trace amounts of 18:1 (n-9) were unusually toxic, diminishing both growth and survival in EFA-deficient animals [12, 13].

An EFA deficiency has profound effects both on overall growth and on specific organ systems where cell proliferation and differentiation are unusually rapid. Testicular degeneration was an early anatomical observation in deficient experimental animals [14]. Impaired spermatogenesis was found unequivocally in a number of later studies [5, 15]. Female animals maintained on low-fat diets showed irregular ovulation and fetal resorption [2, 5, 15, 16]. Impaired spermatogenesis and fetal resorption in EFA deficiency resembled the effects of vitamin E deficiency on the reproductive systems of male and female animals [17, 18]. The early vitamin E experiments led Mason [17] to propose in 1933 that vitamin E is a cell proliferation factor. No one suggested that EFA or their metabolites are involved in cell proliferation until 1940, when Smedley-Mac-

---

*Fatty acids are designated by the number of carbon atoms:number of double bonds, and the position of the first double bond from the methyl terminus of the acyl chain is noted in parentheses: 18:1 (n-9), 9-octadecenoic acid; 18:2 (n-9), 6,9-octadecadienoic acid; 18:2 (n-6), 9,12-octadecadienoic acid; 18:3 (n-6), 6,9,12-octadecatrienoic acid; 18:3 (n-3), 9,12,15-octadecatrienoic acid; 20:2 (n-9), 8,11-eicosadienoic acid; 20:3 (n-9), 5,8,11-eicosatrienoic acid; 20:3 (n-6), 8,11,14-eicosatrienoic acid; 20:3 (n-3), 11,14,17-eicosatrienoic acid; 20:4 (n-6), 5,8,11,14-eicosatetraenoic acid; 20:5 (n-3), 5,8,11,14,17-eicosapentaenoic acid; 22:4 (n-6), 7,10,13,16-docosatetraenoic acid; 22:6 (n-3), 4,7,10,13,16,19-docosahexaenoic acid. Fatty acid families are designated by the position of the first double bond from the methyl terminus: n-9, oleic acid or 18:1 (n-9) family; n-6, linoleic acid or 18:2 (n-6) family; n-3, linolenic acid or 18:3 (n-3) family. More unsaturated members of each family are formed by enzymatic desaturation–chain elongation reactions.

Lean and Nunn [*19*] noted that arachidonic acid was mobilized in animals with growing ascites tumors.

It first appeared that structural and functional changes in EFA deficiency could be explained by changes in the fatty acid composition of phospholipids [*5*]. Phospholipids, the *élément constant* of lipids in biological systems [*20*], exist in an anisotropic liquid crystal phase that determines the ultimate biological properties of membrane fluidity, ionization, and counterion binding affinity [*21–26*]. However, the physical properties of fatty acids in lipids vary with unsaturation in a step function rather than a continuous function [*22*]. Saturated fatty acids exist in a solid phase, whereas all unsaturated fatty acids exist in a liquid crystal phase at body temperature. Unsaturated fatty acids in this phase have similar physical properties. For example, all unsaturated fatty acids have the same cross-sectional area in the liquid crystal phase (Fig. 1). It is difficult to imagine how the substitution of 18:1 (n-9) and 20:3 (n-9) for 20:4 (n-6) could alter the physical properties of membrane phospholipids sufficiently to explain the physiological effects of an EFA deficiency. Essential fatty acids must achieve their biological effects through the synthesis of a variety of metabolites [*22*]. In fact, the fundamental relationship between EFA and their prostaglandin metabolites was proposed as early as 1968 [*27*].

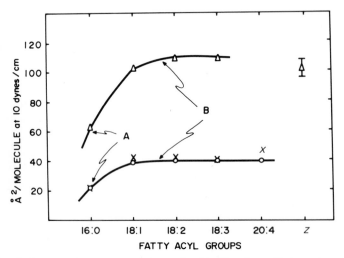

Fig. 1.   Surface areas (square angstroms per molecule) of free fatty acids (x — x), methyl esters (○ — ○), and triglycerides (△ — △) measured at constant surface pressure (10 dynes/cm) and ambient temperature. The fatty acyl groups in simple lipids are designated on the abscissa by number of carbon atoms:number of double bonds. Naturally occurring triglycerides are designated on the abscissa by Z. The solid phase with saturated acyl groups (A) and the first member of the liquid crystal phase with unsaturated acyl groups (B) are noted on the curves. (Data were obtained from refs. *22* and *23*.)

The effects of both polyunsaturated fatty acids and vitamin E on cell proliferation are readily explained in retrospect by the synthesis of prostaglandins and lipid peroxides from polyunsaturated fatty acids. One EFA, 20:4 (n-6), is the specific precursor of many different prostaglandins. Small amounts of 20:4 (n-6) are evidently sufficient for the synthesis of the prostaglandins in quantities that promote cell proliferation. We show in this chapter that all polyunsaturated fatty acids synthesize lipid peroxides and that these compounds inhibit cell proliferation. Large amounts of EFA or polyunsaturated fatty acids derived from 18:1 (n-9) (so-called oleic acid toxicity) inhibit growth through the synthesis of excessive amounts of lipid peroxides. Because vitamin E blocks the synthesis of lipid peroxides, cell proliferation is inhibited by excessive amounts of lipid peroxides formed in a vitamin E deficiency.

B. Dermal Lesions

Essential fatty acids evidently have significant effects on epidermal cell proliferation and growth, and dermal lesions were the first observable changes in the EFA-deficient rat [1, 2, 5, 10]. This concept is supported by studies on wound healing. In 1950 Decker et al. [28] noted that EFA-deficient mice did not heal when their tails were cut to obtain peripheral blood. Other investigators found that the rate of healing of partial thickness burns is prolonged in rats with EFA deficiency [29].

Essential fatty acid deficiencies characterized by elevated triene/tetraene ratios are found in infants receiving parenteral alimentation [30, 31]. Clinical observations suggested that poor wound healing occurred in one infant [31]. However, the effect of EFA deficiency on wound healing could not be confirmed unequivocally in a large group of postsurgical patients who received parental alimentation [32].

Caffrey and Jonsson [33] measured wound healing in rats with an EFA deficiency (elevated triene/tetraene ratios). These investigators found that wound closure was significantly retarded. Furthermore, the topical application of 20:4 (n-6) improved the rate of wound closure. These systematic experiments show clearly that 20:4 (n-6) enhances epidermal cell proliferation.

Cell proliferation is inhibited by a deficiency in vitamin E [17, 18]. Indeed, one early study found that wound healing in the rat was inhibited by a vitamin E deficiency and promoted by a diet rich in vitamin E [34]. However, a more recent study showed that intramuscular injections of a large amount of vitamin E inhibited wound healing [35]. Indomethacin [33], aspirin [36], and cortisone [37] also inhibit wound healing. Vitamin E and these agents may function by diminishing prostaglandin biosynthesis from EFA because they block either fatty acid release or cyclooxygenase activity. Specific prostaglandins and lipid peroxides may be involved in wound healing. One study found that aspirin, which blocks both PGE_2 and $PGF_{2\alpha}$ synthesis, had no effect on wound healing, whereas prednisolone, which blocks only $PGF_{2\alpha}$ synthesis, retarded wound healing [38]. Dipyridamole, an agent that inhibits lipid peroxidation [39] and enhances PGI_2 synthesis [39, 40],promotes the survival of experimental skin flaps [41].

Dermal lesions are complex. Wound healing data suggst that epidermal cell proliferation is inhibited by an EFA deficiency, and yet a number of other studies have shown that epithelial cell proliferation is promoted by a long-term EFA deficiency. Mice maintained on a fat-free diet for 10 to 12 weeks showed enhanced mitotic activity [42, 43]. Thymidine incorporation was increased in both rats [44] and mice [45] maintained for a long period on diets deficient in EFA. These animals had reduced cyclic AMP levels [45] that may have been related to reduced prostaglandin biosynthesis [46–48]. Alternatively, long-term EFA deficiencies may deplete the total polyunsaturated fatty acid pool that is available for the synthesis of inhibitory lipid peroxides because increases in 20:3 (n-9) probably do not compensate for decreases in the total polyunsaturated fatty acid pool [5]. In summary, contradictory wound healing and epidermal cell proliferation experiments support our working hypothesis that EFA both promote and inhibit cell proliferation through the synthesis of prostaglandins and lipid peroxides.

C. Tumorigenesis

In 1930 Watson and Mellanby [49] found an increased number of skin tumors in mice treated with coal tar and fed butter fat. Several laboratories subsequently found that both skin tumors [50, 51] and mammary tumors [51, 52] were enhanced by high-fat diets. These experiments were carried out with a variety of fats, and the early investigators concluded that the amount of fat was more important than the type of fat [53–55].

Shortly after the identification of EFA, other investigators began to publish studies relating tumorigenesis to specific EFA rather than the level of dietary fat. Early studies showed that EFA were mobilized when rats were implanted with Walker sarcoma [19, 56], and Ehrlich tumors apparently grew more slowly in EFA-deficient mice [28, 57]. Systematic studies on the role of EFA in carcinogenesis were begun by Gammal, Carroll, and Plunkett in 1967 [58]. These studies, which have continued to the present time, are summarized in several reviews [55, 59, 60]. They show that fats rich in EFA are more effective than saturated fats in promoting the capacity of 7,12-dimethylbenz[a]anthracene (DMBA) to produce mammary tumors in rats. Essential fatty acids evidently act at the promotional stage of carcinogenesis because they are effective when they are administered several weeks after the DMBA [55, 59, 60]. Furthermore, EFA promote growth and proliferation in transplantable tumors [61, 62].

In addition to EFA, other unsaturated fatty acids such as 18:1 (n-9) [63–65] and members of the n-3 fatty acid family [66] also promoted tumorigenesis when they were included in the diet. These data have been used as evidence against the obligatory role of prostaglandins; however, there are other explanations for these fatty acid effects. Fatty acids in the n-3 family support growth somewhat [2–5], and 20-carbon members of the n-3 family also synthesize prostaglandins in some *in vitro* systems [67]. Early studies of EFA deficiency showed that 18:1 (n-9) is a toxic fatty acid [11–13]. We have suggested (see Section II,A) that 18:1 (n-9) toxicity is related to the synthesis of a large amount of 20:3 (n-9) during EFA deficiency. This intercoversion may be important in tumorigenesis. Isomers of other monoene fatty acids such as erucic acid [22:1 (n-9)] do not synthesize 20:3 (n-9), and these isomers do not promote tumorigenesis [64]. In contrast, 20:3 (n-9) both promotes and inhibits prostaglandin biosynthesis. This fatty acid stimulates the synthesis of immunoreactive PGE and PGI_2 metabolites when a large amount is added to smooth muscle cells in culture, probably by releasing arachidonic acid from phospholipids for prostaglandin biosynthesis [68]. This fatty acid may diminish prostaglandin synthesis when it is incorporated into phospholipids [69] or when it has accumulated in differentiated cells [70]. The fatty acid 20:3 (n-9) also generates lipid peroxides [68]. The 18:1 (n-9) metabolite 20:3 (n-9) may affect tumorigenesis through the synthesis of prostaglandins (indirect effect) and lipid peroxides (direct effect).

The paradoxical effects of EFA on cell growth (Section II,A) and dermal lesions (Section II,B) are also found in tumorigenesis. Rats maintained for as long as 100 weeks on EFA-deficient diets had a significant number of papillary transitional cell tumors of the urinary tract [71]. These animals also exhibited typical and atypical hyperplasias. Further studies showed that the mitotic index was increased in the epithelium of the tongue, esophagus, forestomach, and gastric glands of EFA-deficient animals [72]. These phenomena may again be explained by diminished cyclic AMP or lipid peroxide levels.

Studies of tumorigenesis demonstrate unequivocally that EFA and other polyunsaturated fatty acids both promote and inhibit cell proliferation. We discuss in subsequent sections how the EFA effects on tumorigenesis can be explained by the synthesis of prostaglandins (Section III) and lipid peroxides (Section IV).

D. Tissue Culture

Tissue culture provides greatly simplified systems for studies on fatty acid metabolism and its effects on cell proliferation, but even these systems have many variables. Fatty acid metabolism varies widely in different cell lines [73–75]. It is a function both of the cell density and of the population doubling level in the tissue culture. Fatty acid metabolism depends on the structure and concentration of the exogenous fatty acid and its administration as a single dose or long-term infusion. It depends on the highly variable biological properties of fetal bovine sera (FBS). Fatty acid metabolism and its effects on cell proliferation are modified by the accumulation of metabolic products in cells and media. Nevertheless, whole-animal and tissue culture data are remarkably consistent.

Fatty acid uptake and accumulation in tissue culture follows the classic *élément constant* and *élément variable* pattern that was first described for tissue lipids in whole animals [20]. Exogenous fatty acids accumulated as triglycerides in the *élément variable,* or neutral lipid fraction, of cells in culture [76–79]. Triglyceride droplets were formed rapidly in cells when exogenous fatty acids were added to media, and these droplets disappeared rapidly from cells when exogenous fatty acids were removed from media [79]. Fewer droplets were formed when the fatty acids were infused at a constant rate over a long time interval [80]. The infusion technique diminished the inhibitory effects that higher concentrations of fatty acids had on cell proliferation [81]. However, the inhibitory effects of fatty acids on cell growth were not caused by the formation of cytoplasmic lipid droplets [82]. Confluent cells containing lipid droplets grew normally when they were seeded at low density and challenged with cloning media.

Exogenous fatty acids accumulated in the *élément variable.* The *élément constant,* or phospholipid fraction, did not change in total amount even when cells were challenged with a large amount of fatty acid administrated as a single dose [83]. The fatty acyl groups in the phospholipid fraction of cellular lipids were labile, and it was possible to change significantly the phospholipid composition of cells in culture by the addition of fatty acids to, or the exclusion of fatty acids from, the growth media [7, 80, 83–98]. However, phospholipids tended to conserve their fatty acyl groups [80, 99–101] unless exogenous fatty acids were supplied to the media [75].

Tissue culture experiments have established unequivocally that EFA are not required to maintain the physiological membrane structure that is essential for cell growth. Cells from a number of lines grow either in EFA-deficient media or lipid-free media [76, 84, 102–110]. Essential fatty acids are not detected in some of these growing cultures, and yet neither energy metabolism nor cell morphology is impaired [73]. Cells in culture retain the biosynthetic pathways of chain elongation and retroconversion [83, 95, 98], but many cell lines that grow in tissue culture lose specific enzyme systems for the desaturation reactions that are necessary for EFA interconversions [75, 83, 111]. Thus, the conversion of 18:2 (n-6) to 20:4 (n-6) is not a requirement for cell growth in tissue culture.

The aforementioned studies demonstrate that cells will grow in the absence of EFA. However, there are many other studies that show that low concentrations of EFA promote cell growth (Table I). Both 18:2 (n-6) and 20:4 (n-6) promote cell proliferation in many cell lines when they are added to cultures at concentrations below 30 μM. Also, 18:3 (n-3) promotes growth in L_6 myoblast [102] and

TABLE I   Upper Limit of the Concentration Range in Which Unsaturated Fatty Acids Promote Cell Proliferation

Cell line	Concentration (μM)	References
Linoleic acid [18:2 (n-6)]		
CHD-3	1	112
Macrophage	10	113
HeLa	1	86,114
XS 63.5	20	115
L_6 myoblast	30	93
DMBA tumor	20	116,117
Arachidonic acid [20:4 (n-6)]		
HeLa	1	86,114
XS 63.5	20	115
DMBA tumor	4	116,117
Smooth muscle	10	118,119
Oleic acid [18:1 (n-9)]		
XS 63.5	20	115
L_6 myoblast	40	93
DMBA tumor	4	116,117
WI 38	30	120
Novikoff hepatoma	70	121
Smooth muscle	90	68,122
Fibroblast	20	122

DMBA tumor [116, 117] lines just as it promotes growth [2–5] and tumorigenesis [66] in animals.

The concentrations of free fatty acids (FFA) vary in different lots of FBS [117, 122, 128, 129]; FBS tends to contain lower concentrations of FFA and of EFA than other sera [128–131]. Variations in the composition of FBS and the amount of FBS added to growth media undoubtedly contributed to the differences noted in Table I for the upper limit of the EFA concentration that promoted cell proliferation. Fatty acid additions were substantial. For example, 30 μM EFA more than doubled the EFA supplied to the media as FFA by 20% FBS [117, 122, 128–131].

The 18:1 (n-9) data are particularly interesting. This fatty acid at concentrations as high as 90 μM promoted proliferation in a number of cell lines (Table I). The 18:1 (n-9) could not be replaced in tissue culture by other 18:1 positional isomers [121]. These tissue culture studies resemble experiments on tumorigenesis in which 18:1 (n-9) [63–65] but not its positional isomers [64] promotes cell growth. Thus, tissue culture data also support the hypothesis that a desaturation–chain elongation metabolite of 18:1 (n-9), probably 20:3 (n-9), is involved in the promotion of cell growth.

A number of studies have shown that higher concentrations of EFA and 18:1 (n-9) inhibit cell proliferation (Table II). The lower limit of the fatty acid concentration that inhibited cell proliferation varied widely in different studies. These differences again reflect differences among cell lines, the population doubling level of the cells in the culture, the method of fatty acid addition to the culture media, and both the composition of FBS and the amount of FBS added to the media.

Inhibitory fatty acids are not limited to EFA and 18:1 (n-9). Other studies showed that 18:3 (n-6), 20:3 (n-6), 18:3 (n-3), and 20:3 (n-3) also inhibit cell proliferation [81, 95, 117, 122, 124–126]. The inhibitory effects on cell proliferation of a variety of n-9, n-6, and n-3 fatty acids have been examined in a study [68] in which the cell line, population doubling level, and FBS lot were held constant (Fig. 2). The inhibition of cell proliferation depends both on the number of double bonds in the fatty acid and on the structure of the fatty acid. The inhibitory effect decreases in the order n-9 > n-6 > n-3 for the three fatty acid families. Fatty acids with three double bonds are the most effective inhibitors in each family, and 20:3 (n-9) is the most inhibitory trienoic acid. The effect of 20:3 (n-9), the 18:1 (n-9) metabolite, on cells in tissue culture may explain why 18:1 (n-9) is unusually toxic in animals with an EFA deficiency [11-13].

Fatty acids do not inhibit cell growth in tissue culture simply by lowering the plating efficiency or diminishing clonal growth. In one study (Fig. 3) the same

TABLE II   Lower Limit of the Concentration Range in Which Unsaturated Fatty Acids Inhibit Cell Proliferation

Cell line	Concentration (μM)	References
Linoleic acid [18:2 (n-6)]		
CHD-3	1	112
MB III lymphoblast	60	123
Smooth muscle	60	68,124
Arachidonic acid [20:4 (n-6)]		
Epithelial	35	117
IMR-90	90	81
Fibroblast	20	93,122,125
Smooth muscle	30	68,82,83,118
		122,124,125
Fetal neural	60	126
Glioma	120	126
HTC	35	127
Oleic acid [18:1 (n-9)]		
Epithelial	35	117
DMBA tumor	35	117
IMR-90	160	81
Fibroblast	90	122

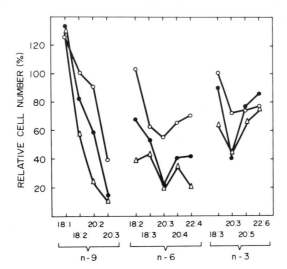

Fig. 2. Proliferation of smooth muscle cells seeded at 40 cells per square centimeter and incubated for 8 days with different concentrations of a specific fatty acid ($\bigcirc$ — $\bigcirc$, 30 μM; $\bullet$ — $\bullet$, 60 μM; $\triangle$ — $\triangle$, 90 μM). Relative cell number was calculated from the mean of total cells in treatment cultures divided by the mean of total cells in control cultures. Each mean was obtained from a minimum of seven Falcon plates. (Adapted from ref. *68.*)

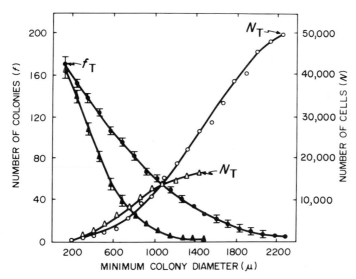

Fig. 3. Colony frequency-size distribution for smooth muscle cells seeded at 200 cells per square centimeter and grown for 8 days with and without 120 μM 20:3 (n-6). Frequency (*f*) data are displayed as the complement of a cumulative distribution function. Vertical lines show SEM. Derived cell number (*n*) data are displayed as a cumulative distribution function. The total number of colonies and cells are designated f_T and N_T, respectively. Control *f*, $\bullet$; control *N*, $\bigcirc$; 120 μM 20:3 (n-6) *f*, $\blacktriangle$; *120 μM 20:3 (n-6) N,* $\triangle$. (Adapted from ref. *132.*)

number of colonies were formed when smooth muscle cells were grown in the presence and absence of 20:3 (n-6), although the total number of cells was much lower in cultures treated with the fatty acid. The number of population doublings decreased from 8.21 to 6.62 when 20:3 (n-6) was added to the culture. Thus, fatty acids lowered the rate of cell division without killing the cells.

E. Summary

Studies with experimental animals demonstrate that EFA and other unsaturated fatty acids promote a weight gain or overall growth. The animal experiments do not prove that EFA promote cell proliferation, because there are other explanations for growth in animals; however, these effects are consistent with enhanced cell proliferation. Strong support for the direct effect of fatty acids on cell proliferation is provided by studies on specific tissues. Fatty acids are needed by tissues where cell proliferation is unusually rapid. For example, spermatogenesis, ovulation, wound healing, and tumorigenesis are promoted by these fatty acids. Finally, EFA and other unsaturated fatty acids are not required for cell proliferation in tissue culture, but these fatty acids promote the proliferation of many cell lines.

A number of studies have shown that higher concentrations of EFA and other unsaturated fatty acids inhibit cell proliferation. Animal studies provide less evidence for inhibitory effects than for promoting effects, but fatty acid toxicities have been reported, and an EFA deficiency was found in some studies to promote both epithelial cell proliferation and tumorigenesis. The most compelling

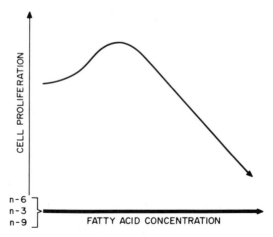

Fig. 4.   Schematic diagram of the fatty acid paradox. Increasing concentrations of unsaturated fatty acids first increase and then decrease cell proliferation.

evidence for the inhibitory effects if fatty acids on cell proliferation is supplied by studies in tissue culture, in which EFA and other unsaturated fatty acids inhibited cell proliferation in a concentration-dependent manner. Fatty acids increase the population doubling times for cells in culture.

These fatty acid studies are summarized in a schematic diagram (Fig. 4). Essential fatty acids and other unsaturated fatty acids show both concentration-dependent stimulation and concentration-dependent inhibition of cell proliferation. These fatty acid effects are related in Sections III and IV to the positive and negative effects of prostanoids and lipid peroxides on cell proliferation.

III. PROSTAGLANDINS AND CELL PROLIFERATION

A. Dermal Lesions

Several experiments with EFA that are summarized in Section II,B suggested that prostaglandins promote the cell proliferation associated with wound healing. Agents that block fatty acid release [37], cyclooxygenase activity [33, 36], and the synthesis of speciifc prostaglandins [38] retard wound healing. A relationship between prostaglandins and epidermal cell growth was first suggested by Kischer [133, 134]. Later investigators found that PGE_1 and PGE_2 stimulate DNA synthesis in human skin [135] and experimental cutaneous wounds [136]. These studies were confirmed by other investigators who also showed that prostaglandins stimulate DNA synthesis and epidermal cell proliferation [137–140].

Epidermal wounding and full skin wounding alter the steady-state equilibrium in epidermis and lead to hyperplasia [141]. The tumor promoter 12-O-tetradecanoylphorbol 13-acetate (TPA) induces epidermal hyperplasia similar to the hyperplasia found in full skin wounds [141]. It stimulates biphasic increases in the prostaglandin content of epidermis [141–144] that can result in as much as a 10-fold increase in PGE_2 and a 3-fold increase in $PGF_{2\alpha}$. Indomethacin and other inhibitors of prostaglandin biosynthesis block TPA-induced stimulation of DNA synthesis and mitotic activity [145–147]. The inhibitory effect of indomethacin on TPA-induced cell proliferation can be overcome by the topical application of PGE_1 and PGE_2 [145], but PGE_2 has negligible mitogenic activity in the absence of TPA [141, 145]. Thus, both prostaglandins and TPA are required for enhanced cell proliferation [141, 148]. Prostaglandins evidently promote cell proliferation in wounds because biochemical events related to wound repair minic some of the effects of TPA on the epidermis.

The prostaglandin paradox is exhibited in studies that show that an EFA deficiency promotes epidermal cell proliferation [42–45]. The PGE_2 level is diminished in the skin of EFA-deficient animals [149]. Scaly dermatosis (hyperplasia) in EFA-deficient animals is improved by the topical application of PGE_2

[*69*]. Prostaglandin levels in EFA deficiency may reflect either the diminished availability of precursor fatty acids or the toxicity of 18:1 (n-9) and its desaturation–chain elongation metabolite 20:3 (n-9), the fatty acid that accumulates in the skin lipids of EFA-deficient animals [*69*]. The topical application of 20:3 (n-9) and its incorporation into phospholipids lead to scaly dermatosis, diminished PGE_2 synthesis, and increased mitotic activity even in the presence of an increased level of 20:4 (n-6) in skin phospholipids [*69*]. These studies support the concept [*70*] that 20:3 (n-9) inhibits the biosynthesis of prostaglandins from 20:4 (n-6).

The epidermal hyperplasia associated with a reduction either in the absolute prostaglandin level or in the effective prostaglandin level probably occurs through a reduction in cyclic AMP [*45–48*]. This mechanism was first suggested as the biochemical basis for epidermal hyperplasia in psoriasis [*150, 151*]. This tissue, like other damaged tissue, synthesizes increased amounts of prostaglandins, but the stimulatory effect of prostaglandins on the cyclic AMP cascade is lower in psoriatic epidermis than in other tissues [*152, 153*]. Indomethacin blocks PGE_2 formation in psoriatic tissue [*152*], and indomethacin exacerbates psoriatic lesions [*154*]. Thus, epidermal cell proliferation is enhanced by lowered levels of cyclic AMP that occur either through diminished prostaglandin synthesis (EFA deficiency) or the diminished capacity of prostaglandins to stimulate the (adenyl cyclase–cyclic AMP)-dependent protein kinase cascade [*150, 151*]. The balance between the stimulatory and inhibitory effects of prostaglandins on epidermal hyperplasia may explain why indomethacin was found in one study to enhance the effect of TPA as a promoter of tumorigenesis [*155*].

B. Tumorigenesis

A number of studies which are summarized in Section II,B, show that tumorigenesis is promoted by both EFA and an EFA deficiency. Thus, tumorigenesis provides additional evidence for the prostaglandin paradox. Prostaglandins and tumorigenesis are discussed in two exhaustive reviews [*148, 156*]. The subject is complex and is reviewed here only in relation to cell proliferation.

Prostaglandin levels are elevated in experimental and human tumors [*148, 156–167*]. Both PGE_2 and $PGF_{2\alpha}$ enhanced the development of tumors induced by 3-methylcholanthrene when the prostaglandins were injected at a site distant from the topical application of the carcinogen [*168*]. These studies suggest that prostaglandins act directly in promoting tumor cell proliferation. However, elevated prostaglandin levels may also act indirectly in promoting tumorigenesis by suppressing immunosurveillance systems [*148, 156, 166, 169, 170*], including macrophage-mediated cytotoxicity [*171*] and nonspecific macrophage-mediated cytotoxicity [*172*]. Cyclosporin A promotes prostanoid synthesis by stimulating fatty acid release, and this effect may explain, in part, how cyclosporin A functions as an immunosuppressive agent to prevent tissue rejection after transplantation [173].

Prostaglandins appear to act directly as inhibitors in some tumors. For example, $PGF_{2\alpha}$ injections induce regression in the MTW9 rat mammary tumor [174]. Levels of PGE_2 and $PGF_{2\alpha}$ are markedly elevated in regressing mammary tumors that were induced by either nitrosomethylurea [175] or DMBA [176]. Finally, PGE_2 injections reduced growth in malignant melanomas [177]. These inhibitory effects may be related to the stimulatory role of prostaglandins in the cyclic AMP cascade (see Section III,A).

Because prostaglandin levels are elevated in some tumors, cyclooxygenase inhibitors might be expected to function as antitumor agents. This hypothesis, first explored by Sykes and Moddox [178], has been supported by a number of studies. Indomethacin, flurbiprofen, and hydrocortisone all inhibit prostaglandin biosynthesis and tumor growth [164, 179–193]. Conversely, diazepam stimualtes PGE_2 synthesis and intrarenal growth in the Walker 256 tumor [194]. If prostaglandins suppress immunosurveillance, cyclooxygenase inhibitors may again act indirectly through the immune system. Cyclooxygenase inhibitors like as prostaglandins have paradoxical effects on cell proliferation. Flurbiprofen significantly enhances the mitotic rate in adenocarcinomas of rat colon that are induced by dimethylhydrazine [195] and blocks the inhibitory effect of 20:4 (n-6) on these tumors [196]. In other studies indomethacin enhanced and PGE_2 inhibited papilloma formation in mice treated with DMBA and repeated applications of the promoters TPA and mezerein [197]. Thus, studies with cyclooxgenase inhibitors add further support to the prostaglandin paradox.

C. Tissue Culture

Prostanoid biosynthesis has been studied in tissue cultures grown from normal, transformed, and tumor cell lines (Table III). Tissue culture has certain advantages in studies on cell proliferation and the prostaglandin paradox because prostaglandin effects may be related directly to the target cell in the culture. However, variables that affect fatty acid metabolism in tissue culture (Section II,D) have profound effects on prostaglandin metabolism. These variables must be considered in the interpretation of cell proliferation data.

Specific cell lines tend to have characteristic prostaglandin profiles. For example, subcultures of porcine endothelial cells synthesize more PGI_2 than PGE_2, whereas subcultures of porcine smooth muscle cells synthesize more PGE_2 than PGI_2 [207]. Endothelial cells from arteries synthesize more PGI_2 than endothelial cells from veins [211]. The history of the cell culture is important. One laboratory reported that smooth muscle cells grown from explant cultures of rat aorta synthesized only PGI_2 [237]. Another laboratory reported that smooth

TABLE III   Tissue Cultures Studied for Prostaglandin and Prostacyclin Biosynthesis

Cell line or tissue	References
Normal	
Adrenal (Y-1), mouse	198
Articular chondrocyte, bovine	199
Endometrial, human	
Stroma	200
Glandular	200
Endothelial	
Arterial, bovine	198,201–205
Arterial, human	206
Arterial, porcine	207–209
Umbilical cord, human	208–216
Venus, human	211,216,217
Epithelial	
Kidney glomerular, rat	218–219a
Fibroblasts	
Adventitia, porcine	208
3T3, mouse	70,220–222
3T3-L1 adipocyte, mouse	70
Embryonic lung, human	222–224
Skin, human	217,225
Glial (C-6), rat	198
HeLa	226
HEp-2	226
Interstitial	
Renal medullary, rabbit	227

(continued)

muscle cells cultured from the aorta of old rats synthesized less PGI_2 and more PGE_2 than smooth muscle cells cultured from the aorta of young rats [235]. Fibroblasts cultured from older human donors synthesized less PGI_2 than fibroblasts cultured from younger human donors [224]. Several investigators have reported that less prostaglandin is synthesized as cell density increases and cultures become confluent [70, 120, 241]. Changes during cell growth may be related to a decrease in the concentration of a precursor fatty acid [241] or the accumulation of an inhibitor in the culture medium [70]. Quiescent cells maintained for a short time in serum-free media actually synthesized more prostaglandin than growing cell in response to either exogenous 20:4 (n-6) or endogenous 20:4 (n-6) released through bradykinin [223]. Finally, the rate of PGI_2 synthesis decreases as cultures age [203], and as a consequence PGI_2 levels are diminished in senescent cultures [222].

Prostaglandin synthesis in cell cultures depends on the origin and concentration of specific fatty acid substrates. Cultured mesothelial cells synthesize both

TABLE III (*Continued*)

Cell line or tissue	References
Macrophage, mouse	*228*
Mandin–Darby canine kidney (MDCK)	*198,229–231*
Mesothelial	
Omentum majus, rabbit	*232*
Smooth muscle, arterial	
Bovine	*198,201*
Guinea pig	*39,68,119*
Human	*217*
Porcine	*207–209*
Rabbit	*233*
Rat	*234–238*
Synovial, human	*239,240*
Transformed or tumor	
Fibroblasts	
SV 3T3, mouse	*220*
Methylcholanthrene, mouse	*241*
Polyoma virus, hamster	*242*
Polyoma virus, 3T3 mouse	*243–245*
Fibrosarcoma, mouse	*246,247*
Hepatoma, rat	*117,225,248*
Leukemia	
L1210 S_6 and R_3, mouse	*249*
Myeloid, mouse	*250*
Murine melanoma (Cloudman S-91)	*251*
Neuroblastoma (41A3), mouse	*198*

PGI_2 and PGE_2 from exogenous 20:4 (n-6), and these cells synthesize only PGI_2 from the endogenous substrate [232]. Human endothelial cells that are enriched with 18:2 (n-6) synthesize less PGI_2 than control cells [212, 213]. These cells are unable to convert 18:2 (n-6) to 20:4 (n-6), and as a consequence 18:2 (n-6) replaces 20:4 (n-6) in the *élément constant,* or phospholipid precursor pool [20, 83]. In contrast, 3T3 cells convert 18:2 (n-6) to 20:4 (n-6), and these cells synthesize an increased amount of PGE_2 when they are grown in the presence of added 18:2 (n-6) [221]. These studies and studies with growing cell cultures [241] emphasize the importance of the size of the precursor pool in regulating prostaglandin synthesis. Several precursor pools that are coupled to specific enzymes may exist in the cell [217]. Exogenous FFA and endogenous phospholipid represent 20:4 (n-6) pools with different rates of prostaglandin synthesis in cultured mesothelial cells [232]. Furthermore, exogenous 20:4 (n-6) is incorporated into Ca^{2+}-sensitive and Ca^{2+}-insensitive lipid pools with different acylhydrolase activities coupled to prostaglandin synthase [252]. The fatty acid

TABLE IV   Agents That Enhance Prostaglandin Biosynthesis by Cells in Culture

Agent	Cell line or tissue	References
A23187	Endothelial, bovine	*203,204*
	Endothelial, human	*212,215,253*
	Epithelial, rat	*219*
	Fibroblast, 3T3 mouse	*221*
	Fibroblast, 3T3-L1 mouse	*70*
	Lymphoma, human	*254*
Angiotensin II and III	Endothelial, human	*206*
	Epithelial, rat	*218,219,219a*
	Fibroblast, human	*224*
	Interstitial, rabbit	*227*
Ascorbic acid	Fibroblast, human	*223,224*
Bradykinin	Endothelial, bovine	*202,203*
	Endothelial, human	*216*
	Fibroblast, mouse	*241*
	Fibroblast, human	*223,224*
	Interstitial, rabbit	*227*
	MDCK	*230*
Cyclosporin A	Smooth muscle, guinea pig	*173*
Dipyridamole	Smooth muscle, guinea pig	*39*
Estradiol	Smooth muscle, rat	*234*
Histamine	Endothelial, human	*216*
Hydralazine	Smooth muscle, guinea pig	*119*
Interleukin 1	Synovial, human	*240*
Mellitin	Endothelial, bovine	*255*
	Fibroblast, human	*255*

(continued)

composition of the precursor pool may also affect prostaglandin synthesis. For example, 22:4 (n-6) and 22:6 (n-3) both inhibit PGE_2 and PGI_2 synthesis from 20:4 (n-6) in cultured cells [68].

Fetal bovine serum itself contains small amounts of several prostaglandins. For example, Hyclone 100348 (Sterile Systems, Logan, Utah) contained 14 nM PGE and 3.9 nM 6-keto-$PGF_{1\alpha}$ [N. Morisaki and D. Cornwell, unpublished observations]. Prostaglandin biosynthesis in cell cultures depends on the properties of the FBS added to the growth media. Two lots of Hyclone FBS generated different amounts of PGI_2 (measured as 6-keto-$PGF_{1\alpha}$) when confluent cultures of smooth muscle cells were incubated with media containing 20% FBS. Hyclone 100364 yielded 200 nM PGI_2, and Hyclone 100348 yielded 100 nM PGI_2 [N. Morisaki and D. Cornwell, unpublished observation]. Prostaglandin synthesis varies directly with the fatty acid content of the media [68]. The differences between Hyclone lots may have reflected their 20:4 (n-6) concentrations, because cultures generated similar amounts of PGI_2, 870 nM and 820 nM,

TABLE IV (*Continued*)

Agent	Cell line or tissue	References
	Fibroblast, MC5-5 mouse	255
	MDCK	255
Propranolol	Adrenal, mouse	198
	Kidney, dog	198
	Leukemia, rat	198
	Neuroblastoma, mouse	198
Quinidine	Kidney, dog	198
	Leukemia, rat	198
Quipazine	Smooth muscle, bovine	201
Serotonin	Smooth muscle, bovine	201
Thrombin	Endothelial, human	212,213,253
	Fibroblast, human	224
	Fibroblast, mouse	241
	Smooth muscle, rat	236
Trypsin	Endothelial, human	253
TPA	Adrenal, mouse	256
	Epidermis, mouse	144,257
	Fibroblast, human	258
	Fibroblast, mouse	259
	Fibrosarcoma, mouse	256
	HeLa	260
	Lymphoma, mouse	261
	MDCK	229,231,256,258,259,262
	Smooth muscle, bovine	256
Vasopressin	Interstitial, rabbit	227

when both media preparations were enriched with 120 μM 20:4 (n-6) before the incubation.

A number of agents stimulate prostaglandin biosynthesis when they are added to cells in culture. Studies with various agents and cell lines are listed in Table IV. Different agents may show specificity for either a prostaglandin or a cell line. Angiotension II stimulates PGE_2 synthesis almost exclusively in rat glomerular epithelial cells [219]. Propranolol stimulates prostaglandin synthesis in dog kidney, mouse adrenal, rat basophil leukemia, and mouse neuroblastoma cells, but this agent does not stimulate prostaglandin synthesis in bovine smooth muscle or rat glial cells [198]. Serotonin and quipazine stimulate PGI_2 synthesis in smooth muscle cells, but these agents have no effect on endothelial cells [201]. Hydralazine stimulates a greater relative increase in PGI_2 than PGE in confluent smooth muscle cells [119].

The agents listed in Table IV promote prostaglandin synthesis in several ways. The calcium ionophore A23187 apparently stimulates release of the endogenous fatty acid substrate from phospholipid [*253, 254*]. For example, A23187 causes the deacylation of platelet phospholipid by activating a phospholipase [*261–263*]. Other effectors that release 20:4 (n-6) through deacylation include angiotensin II [*219, 227*], ascorbic acid [*224, 264*], bradykinin [*224, 264*], and the active peptide of bee venom, melittin [*255*]. These agents do not stimulate prostaglandin synthesis when they are added to cultures together with large amounts of FFA such as 20:4 (n-6) [*224*]. The tumor promoter TPA causes the release of a large amount of 20:4 (n-6) [*231, 256–259, 262*] and enhances FFA conversion to prostaglandins [*259*]. Thus, TPA promotes prostaglandin biosynthesis even when 120 μM 20:4 (n-6) is added to the media [N. Morisaki, L. Tomei, G. Milo, and D. Cornwell, unpublished observation]. Other effectors such as dipyridamole [*39*] function as antioxidants in promoting PGI_2 synthesis. These agents evidently prevent lipid peroxides and lipoproteins that contain lipid peroxides from inactivating PGI_2 synthetase [*265–267*].

The effects of viral transformations on prostaglandin synthesis have been studied with BALB/3T3 fibroblasts [*220, 236–245*] and baby hamster kidney (BHK) fibroblasts [*242, 268*]. The 3T3 cells transformed with simian virus 40 and the BHK cells transformed with polyoma virus synthesize larger amounts of prostaglandins than 3T3 and BHK cells in cultures with comparable cell densities [*220, 268*]. Transformed 3T3 cells have a larger amount of 20:4 (n-6) in media FFA [*243*], and these cells release more labeled 20:4 (n-6) from prelabeled cells [*245*] when they are compared with 3T3 cells. Hydrocortisone, an agent that blocks fatty acid release through phospholipase A_2 activity [*255, 256, 269, 270*], inhibits prostaglandin synthesis in cells transformed with either polyoma virus [*242*] or methylcholanthrene [*271*]. These data strongly suggest that transformed cells have enhanced phospholipase A_2 activity. In addition to viral transformation, tumor promoters such as TPA enhance fatty acid release [*231, 256–259, 262*], and tumors themselves promote fatty acid mobilization from lipid stores [*19, 56, 57, 272, 273*]. Fatty acid mobilization for prostaglandin biosynthesis appears to be an important common property of tumors, their promoters, and transforming agents (see Section V).

The wide variety of effectors listed in Table IV strongly suggests that prostaglandin biosynthesis is not a final common pathway for the physiological activity of all these agents. It is difficult to establish a causal relationship leading from an effector through prostaglandin biosynthesis to cell proliferation. Both hydralazine and vasopressin stimulate prostaglandin synthesis in cell cultures [*119, 227*], and these agents promote proliferation in smooth muscle cells and 3T3 fibroblasts [*119, 274*]. Nafazatrom, like hydralazine [*119*], stimulates PGI_2 synthesis in several tissues [*275, 276*], and yet nafazatrom inhibits the replication of Blba melanoma cells in culture [*277, 178*]. The agent TPA stimulates fatty acid release and prostaglandin synthesis in a number of cell lines (Table IV) and promotes thymidine incorporation and cell division in cultures [*257, 279, 180*], particularly in density-inhibited cells [281, 282]. However, TPA inhibits cell growth in freshly seeded fibroblasts [281] and smooth muscle cells [N. Morisaki, L. Tomei, G. Milo and D. Cornwell, unpublished observations]. Unlike many other effectors, TPA induces profound morphological changes that reduce cell size. It may promote growth in density-inhibited cultures by reducing cell size. It is apparent that prostaglandin effects on cell proliferation must be demonstrated first by the addition of prostaglandins or their immediate fatty acid precursors.

The direct addition of prostaglandins to cells in culture provides vivid evidence for the prostaglandin paradox. In 1975 Jimenez de Asua et al. [283] found that $PGF_{2\alpha}$ promotes cell growth in quiescent 3T3 fibroblasts. Their initial observation was confirmed and extended in a series of papers that have been summarized in two reviews [284, 285]. The growth-promoting effect of $PGF_{2\alpha}$ was later found in experiments with smooth muscle cells [118, 124], human diploid fibroblasts [222, 286], and a lymphoid cell line [287]. Both $PGF_{1\alpha}$ and $PGF_{2\alpha}$ promote cell growth at concentrations as high as 10 μM [124, 286], a level beyond the concentration range for prostaglandins generated in growing cultures by endogenous synthesis. At concentrations that do not exceed 1 μM, PGE_1 and PGE_2 also enhance proliferation in a variety of cell lines [118, 124, 281, 284, 286, 287]. Unlike $PGF_{1\alpha}$ and $PGF_{2\alpha}$, PGE_1 and PGE_2 inhibit cell proliferation at 1 to 10 μM [118, 124, 286] and even lower [222] concentrations.

Endogenous prostaglandin synthesis also promotes cell proliferation. A number of studies have shown that cell proliferation is enhanced when 10–20 μM 20:4 (n-6) is added to growing cultures (Table I). The 20:4 (n-6) in this range enhances PGE synthesis and proliferation when it is added to growing cultures of smooth muscle cells (Fig. 5). Indomethacin both blocks PGE synthesis and diminishes cell growth (Fig. 5). Fetal bovine serum evidently contains sufficient amounts of prostaglandins for cell proliferation at a basal level. It is interesting that hydralazine stimulates PGI_2 synthesis and cell proliferation in growing cultures [119]. Indomethacin again both blocks PGI_2 synthesis and diminishes cell proliferation in these cultures [119].

A number of studies have shown that prostaglandins in both the A series [251, 286–294] and the E series [118, 124, 181, 222, 224, 242, 251, 286, 294–296] and PGI_2 [277, 278] inhibit cell proliferation when they are added to cultures in relatively high (1 μM or greater) concentrations. These prostaglandin concentrations are well beyond the levels achieved in tissues and tissue cultures. Inhibition at these concentrations is difficult to interpret [156]. However, several other studies summarized later suggest that prostaglandins inhibit cell proliferation at the upper end of their physiological concentration range.

Studies with nonsteroidal antiinflammatory drugs provide evidence for the physiological role of prostaglandins in the control of cell growth. These agents both inhibit and enhance cell proliferation when they are added to cells in culture [124, 127, 222, 225, 226, 242, 286, 295, 297]. Their inhibitory effect is ex-

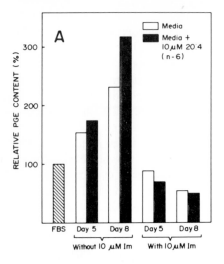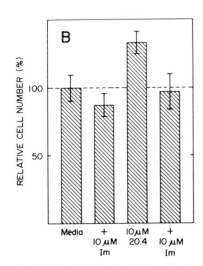

Fig. 5.   Effects of 10 μ*M* 20:4 (n-6) and 10 *M* indomethacin (Im) on PGE synthesis (A) and cell proliferation (B) in growing cultures of smooth muscle cells. Ten micromolar 20:4 (n-6) differs significantly from media ($p < .05$). (Adapted from ref. *119* and unpublished observations.)

plained by the observation that prostaglandins in low concentrations promote cell proliferation. Their enhancing effect is explained by the observation that prostaglandins in high concentrations inhibit cell proliferation. For example, indomethacin inhibits cell proliferation when cells are seeded at low densities, where they generate small amounts of prostaglandins [*119, 124, 222, 226*]. Indomethacin promotes cell proliferation when cells are seeded at high densities, where they generate large amounts of prostaglandins [*222, 226, 295*]. Studies with indomethacin should be limited to the 1–10 μ*M* concentration range, in which indomethacin completely blocks prostaglandin synthesis [*118, 119*]. Higher concentrations of indomethacin are themselves cytotoxic [*124, 127, 286*]. It is difficult with higher concentrations to separate indomethacin cytotoxicity from the effect of indomethacin as a prostaglandin synthase inhibitor.

The physiological role of prostaglandins in the inhibition of cell proliferation is shown clearly in studies with the committed macrophage precursor cell, CFU-C. This precursor cell undergoes clonal proliferation when it is stimulated with colony stimulation factors [*148*]. Added PGE_1 and PGE_2 inhibit CFU-C proliferation [*298*]. The colony stimulation factor stimulates PGE synthesis by macrophages, and the PGE formed in this manner inhibits cell proliferation [*299*]. The inhibitory effects of PGE on the proliferation of a committed granulocyte macrophage culture (CFU-GM) is related to the expression of Ia-like antigens [*300, 301*]. Evidently, PGE functions as a S-phase selective inhibitor of CFU-

GM proliferation [301]. Studies with committed hematopoietic precursor cells are difficult to interpret because prostaglandins may affect proliferation and/or differentiation. The effects of prostaglandins on hematopoietic cells have been reviewed in detail [148, 302, 303]. These studies show that prostaglandins, generated by endogenous synthesis, participate in the control of cell proliferation.

Cell proliferation may be controlled either by the prostaglandins themselves or by oxidation reactions that occur during prostaglandin biosynthesis. Studies with interferon highlight the two mechanisms. Interferon stimulates PGE production in human fibroblasts [304]. Prostaglandin synthase is required for the interferon-induced antiviral state [305]. In fact, an interferon-resistant cell line lacks prostaglandin synthase activity [249]. These studies lead to the hypothesis that interferon inhibits cell proliferation by stimulating the synthesis of prostaglandins [306]. However, added prostaglandins do not restore the interferon effect when prostaglandin synthase is inhibited [307]. Agents such as TPA that promote prostaglandin synthesis also stimulate superoxide production in cells [308, 309]. The interferon effect is inhibited by diethyl dithiocarbamate (DDC) [307]. Cyclooxygenase is blocked by DDC [310], and the radical-generating enzyme prostaglandin hydroperoxidase utilizes DDC as a cooxidation substrate [311]. These studies lead to the hypothesis that interferon inhibits cell proliferation through cooxidation reactions generating superoxide that occur during prostaglandin synthesis. Cooxidation during prostaglandin synthesis may generate inhibitory oxygen-centered radicals just as the xanthine–xanthine oxidase reaction generates oxygen-centered radicals that inhibit cell proliferation in a lymphoblastoid cell line [312]. The DDC effect is paradoxical because this agent blocks superoxide dismutase [307] and might be expected to enhance the inhibitory effect of superoxide.

A number of studies stress the regulatory effects of cyclic nucleotides on cell proliferation [313–316]. Cyclic AMP levels are elevated when PGI_2 synthesis is stimulated [314]. Cyclic nucleotide levels may also be regulated by cooxidation reactions that occur during prostaglandin synthesis and lipid peroxidation. For example, cyclic GMP is synthesized through the action of guanylate cyclase. This enzyme is apparently activated by a sulfhydryl–disulfide interconversion involving either endoperoxides [317–321] or hydroperoxides [313, 321, 322] in the generation of an oxygen-centered radical [323–325]. Cyclic AMP is synthesized through the action of adenylate cyclase, and dopamine-sensitive adenylate cyclase requires a sulfhydryl group [326]. Thus, cooxidation may elevate cyclic GMP levels and suppress cyclic AMP by effects that are not, as has been suggested, mediated by PGI_2 [314]. In contrast, the phosphodiesterase that hydrolyzes cyclic nucleotides is a sulfhydryl enzyme that is stabilized by reducing agents such as dithiothreitol [327, 328]. Cooxidation may explain why high concentrations of the prostaglandin precursor 20:3 (n-6) potentiate the effect of phosphodiesterase inhibitors, caffeine and papaverine, on the inhibition of cell proliferation [*124*].

Cooxidation during prostaglandin synthesis may be important in the control of cell proliferation, but studies with prostaglandin synthtase inhibitors and antioxidants show that other oxidation reactions are also involved in the control of cell proliferation. These oxidation reactions, which are particularly important at higher fatty acid concentrations, are discussed in Section IV.

D. Summary

Many studies in experimental animals have shown that prostaglandins promote cell proliferation. Prostaglandin E stimulates DNA synthesis and proliferation in epidermal cells. TPA and prostaglandins synthesized in response to TPA work together in promoting epidermal hyperplasia. Prostaglandins are elevated in many tumors, and prostaglandins promote tumorigenesis in some model systems. Nonsteroidal antiinflammatory agents suppress cell growth in these model systems.

However, there are contradictory effects, and the prostaglandin paradox is found in animal systems where prostaglandin deficiencies promote cell proliferation. For example, EFA deficiencies lead to dermal hyperplasia. Added PGE diminishes hyperplasia in these animals. Psoriasis, a disease in which epidermal cells have lost the capacity to respond to prostaglandins, is characterized by hyperplasia. Prostaglandins suppress tumorigenesis in some model systems, and blocking agents enhance tumorigenesis in thse systems.

The paradoxical effects of prostaglandins on dermal lesions and tumors are also found in tissue cultures. Prostaglandin levels are elevated in tumor cell lines. Tranforming agents such as viruses promote prostaglandin biosynthesis in some cell lines. Prostaglandins are synthesized in response to TPA, and this agent both promotes and inhibits proliferation in different cell lines. Prostaglandin effects on cells in culture depend on the specific prostaglandin and its concentration. Added $PGF_{1\alpha}$ and $PGF_{2\alpha}$ enhance cell growth over a wide concentration range. Low concentrations of PGE_1 and PGE_2 promote cell proliferation in some cell lines. Cyclooxygenase inhibitors block the enhancing effect of prostaglandins derived from endogenous synthesis. High concentrations of a variety of prostaglandins such as PGA, PGE_1, and PGE_2 inhibit cell proliferation. Cyclooxygenase inhibitors block the inhibitory effect of prostaglandins derived from endogenous synthesis.

The prostaglandin paradox is summarized in a schematic diagram (Fig. 6). Prostaglandin synthesis increases as fatty acid concentration increases. Prostaglandin levels eventually plateau because the capacity of the cyclooxygenase complex for prostaglandin synthesis is exceeded. Cell proliferation does not

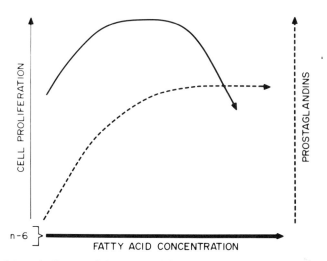

Fig. 6. Schematic diagram of the prostaglandin paradox. Increasing concentrations of prostaglandins first increase and then decrease cell proliferation. Note that prostaglandin levels reach a plateau at higher fatty acid concentrations (see text).

require prostaglandin synthesis. However, prostaglandins show concentration-dependent stimulation and concentration-dependent inhibition of cell proliferation. The fatty acid paradox (Fig. 4) is generated in part of these effects.

Prostaglandins can act either directly on cells or indirectly by mechanisms such as the suppression of immunosurveillance. Prostaglandin synthesis also enhances surveillance mechanisms through the generation of cytotoxic oxygen-centered radicals. In many systems, prostaglandins appear to act through their effect on cyclic nucleotide levels. These effects are sometimes related to the prostaglandins themselves and may be related to changes in cellular redox states through the generation of oxygen-centered radicals during prostaglandin biosynthesis. Direct prostaglandin effects are exhibited by added prostaglandins, whereas indirect prostaglandin effects are exhibited only during endogenous prostaglandin synthesis.

IV. LIPID PEROXIDES AND CELL PROLIFERATION

A. Historical Perspective

Although studies on lipid peroxides and cell proliferation have a long history, much less work has been undertaken with lipid peroxides than with prostaglandins. The reasons are obvious. All prostaglandins are synthesized from common endoperoxide precursors, and these endoperoxides are the products of specific microsomal enzyme reactions. Lipid peroxides include a number of hydroperoxy acids and many other compounds that are formed both through several enzymatic reactions and through a variety of nonenzymatic oxidation reactions.

The first indications that lipid peroxides are involved in the control of cell proliferation were found in experiments that led to the discovery of vitamin E. In 1922 Evans and Bishop [329] showed that a previously unrecognized dietary factor was required for normal gestation. Early investigators were impressed with the observation that this dietary factor, later designated vitamin E [330], exerted its most significant effects on proliferating tissues (Section I,A), Juhász-Schäffer [331] suggested in 1931 that vitamin E is not merely an antisterility factor but a factor necessary for cellular growth and differentiation. Mason [17] noted that Juhász-Schäffer considered the term *antisterility* a misnomer because reproductive tissues were affected only because they are areas of rapid cell proliferation. Mason supported the concept of vitamin E as a cell proliferation factor in a masterful synthesis of the early literature [17], p. 220 that concluded, "It [vitamin E] is essential for some very specific and vital cellular function related, perhaps, to the processes involved in mitotic division."

Shortly after the discovery of vitamin E, Mattill [332] began a series of investigations that established that vitamin E is a lipid antioxidant. The relationship between lipid peroxidation and the physiological action of vitamin E was shown immediately by Evans and Burr [333] in studies that led Burr and Burr to the discovery of EFA [1]. The roles of vitamin E in cell proliferation and in lipid peroxidation were linked in the work of Bernheim, the developer of the thiobarbituric acid (TBA) test for lipid peroxidation [334]. Bernheim [335] reported that tissues with a high mitotic index contain smaller amounts of lipid peroxides than nonregenerating tissues. This concept was largely ignored, although occasional studies have related vitamin E and cell proliferation in mammals [17, 18, 34], and a number of studies with the rotifer *Asplanchna sieboldi* have shown that vitamin E promotes cell growth through an increased number of nuclear divisions and an increased level of DNA duplication [336–339].

B. Tissues and Tumors

Lipid peroxides are formed by a number of enzymatic and nonenzymatic reactions. These reactions have been reviewed in great detail [340–352]. Lipid peroxidation reactions are often initiated by superoxide or the hydroxyl radical; in addition, there is some evidence in special systems for the involvement of singlet oxygen. Specific lipoxygenase enzymes catalyze the synthesis of fatty acids with a hydroperoxyl group at the 5-, 11-, 12- or 15-position. These acyclic hydroperoxides contain a conjugated diene group that has a characteristic UV absorbance maximum, and these acylic hydroperoxides undergo radical rearrangement reactions forming cyclic endoperoxides, which decompose to yield MDA in the TBA reaction [343, 353–355]. Enzymatic and nonenzymatic peroxidation reactions are catalyzed by transition metal ions such as Fe^{2+}. Transition metal ion complexes also catalyze the aerobic decomposition of hydroperoxy fatty acids to oxodiene, hydroxydiene, oxoepoxyene, hydroxyepoxyene, and oxohydroxyene derivatives [345]. Fetal calf aorta converts 12-hydroperoxy-8,10-heptadecadienoic acid to 11, 12-epoxy-10-hydroxy-8-heptadecanoic acid [346]. Additional polyhydroxy derivatives are formed from either an allylic carbonium ion or the S_N2 solvolysis of hydroxyepoxyenes [345]. Many of these compounds do not contain conjugated diene groups, and it is not certain which derivatives, if any, will decompose to yield MDA. Classic TBA and UV assays for lipid peroxidation undoubtedly underestimate significantly the degree of lipid peroxidation in tissues and tumors.

Microsomal lipid peroxidation also involves cytochrome P-450 monooxygenase. These enzymes catalyze ω- and (ω-1) oxidations and epoxidations at ω-6, ω-9, ω-12, and ω-15 [356–358]. The epoxides are decomposed to dihydroxy and trihydroxy derivatives by epoxide hydrolases [356]. Trihydroxy derivatives have been found in other tissues such as platelets and fetal calf aorta [346, 359, 360]. Products of cooxidation reactions again lead to the underestimation of lipid peroxidation by UV analysis or the TBA test.

Microsomes contain the cyclooxygenase complex that generates hydroperoxy fatty acids, which are converted to cyclic endoperoxides in prostaglandin biosynthesis. These complex reactions, which are reviewed elsewhere [343, 351], use a hydroperoxidase reaction [311, 361] for the generation of oxygen-centered radicals that may occur during a burst of metabolic activity [362]. These radicals lead to the cooxidation of many substrates [363] in lipoxygenase-like reactions that are inhibited by indomethacin [363, 364]. Superoxide is produced when TPA promotes prostaglandin biosynthesis [308, 309], and this may explain why TPA also enhances lipoxygenase activity [365].

Although few investigations of tissues and tumors have considered the diversity of fatty acid oxidation reactions and the limitations this diversity places on the interpretation of data, some generalizations are possible.

Several studies have shown that rapidly proliferating cells in regenerating liver, bone marrow, and intestinal mucosa are resistant to lipid peroxidation [332, 335, 340, 366–368]. In contrast, a nonproliferating tissue such as brain generates large amounts of lipid peroxides that are detected by the TBA test [335, 340, 369–371]. Increments in oxygen tension enhance lipid peroxidation in the lung [372], and continuous oxygen exposure inhibits cell proliferation in this tissue [373–375]. Antioxidants such as vitamin E (Section IV,A) and butylated hydroxytoluene (BHT) [376] promote cell proliferation. These data lead to the hypothesis that cell growth and proliferation are blocked by the synthesis of lipid peroxides and the overproduction of free radicals [377–380]. This hypothesis is consistent with the observation that cells are sensitive to oxygen inhibition of proliferation after DNA synthesis but before metaphase [381].

Tumors are unusually deficient in the lipid peroxides that are detected by the TBA test [335, 340, 382]. Homogenates of Ehrlich ascites tumor cells produce only small amounts of lipid peroxides [383–385]. Homogenates of solid tumors produce less lipid peroxide than homogenates of normal tissue [377, 386, 387]. One study reported that lipid peroxides were reduced in tumor-bearing animals and almost absent from the tumors themselves [388]. Other studies have shown that lipid peroxides are absent from tumors but that lipid peroxide levels are normal or even increased in other tissues of the host [375, 389, 390]. Ascorbate-dependent lipid peroxidation is inhibited in many tissues of the host during the terminal stages of tumorigenesis [391]. As would be anticipated from the preceding discussion, a diet deficient in antioxidants causes the shrinkage of Ehrlich ascites tumors in experimental animals [392].

The enzymatic and nonenzymatic oxidation reactions that generate lipid peroxides are suppressed in some (tumor-containing) cell fractions. For example, nonenzymatic lipid peroxidation induced by ascorbate is very low in mitochondria isolated from tumors [387, 393]. NADPH-Dependent lipid peroxidation is diminished in mitochondria isolated from hepatomas [380, 394]. The oxidation rates of drugs by liver microsomes are lowered in rats with Walker tumors [395]. Drug-metabolizing enzymes are suppressed in animals with a variety of tumors including ascites tumors [396, 397], Walker tumors [398], Furth's mammotropic tumor [399], and Guérin carcinoma [400]. Microsomal NADPH-cytochrome c reductase, cytochrome b_5, and cytochrome P-450 are reduced in Morris hepatoma 44 and 3924A and in Novikoff ascites hepatoma [401, 402]. The cooxidation of epinephrine to adrenochrome and NADPH-dependent lipid peroxidation also are diminished in these tumors [402]. Drug-metabolizing enzymes are suppressed in regenerating liver after partial hepatectomy [403], and drug-induced cytochrome P-450 levels are lower in proliferating hepatocyte cultures than in resting cultures [404].

Lipid peroxidation may be suppressed in tumors for several reasons. It has long been recognized that hypoxia results in radioresistance [405] and that tumors are resistant to radiation because they are hypoxic [406]. A number of compounds, called O_2 mimetics, have been developed as radiosensitizers for hypoxic cells [407]. These compounds include nitroxyl free radicals such as triacetoneamine-N-oxyl and electron-affinity compounds such as the substituted (nitro-1-imidazolyl) alkanols [407, 408]. Hypoxia may help to explain why tumor cell fractions retain the capacity to produce superoxide even though cells produce less superoxide under physiological conditions [409]. The effect of O_2 mimetics on lipid peroxidation in tumors deserves study.

Hypoxia does not fully explain why lipid peroxidation is suppressed in tumors because transformed cells in culture are less susceptible to oxygen than normal cells [409]. Lipid peroxidation also may be suppressed in tumors by a deficiency in polyunsaturated fatty acids. A fatty acid deficiency could result either from a decrease in total phospholipids or from a decrease in the concentration of specific polyunsaturated fatty acids. For example, the total phospholipid content of a variety of brain tumors is about 20% of the phospholipid content of normal brain, and as a consequence the polyunsaturated fatty acid pool is diminished even though the relative concentration of 20:4 (n-6) is more than doubled in the phospholipids of neoplastic brain [410]. The relative composition of polyunsaturated fatty acids is lowered in a number of tumors and transformed cells. These data are summarized in Table V. Tumors and transformed cells show a striking increase in 18:1 (n-9) and a striking decrease in 20:4 (n-6). Also, 18:1 (n-9) promotes proliferation in a number of cell lines (Table I), whereas 20:4 (n-6) inhibits this process (Table II). When 18:1 (n-9) is included in the diet, it also promotes tumorigenesis [63–65]. These effects may be related in part to the suppression of lipid peroxidation. Thus, microsomal drug metabolism varies directly with the polyunsaturated fatty acid content of the microsomal membranes [418]. It is important to note that 20:4 (n-6) inhibits the proliferation of both a glioma cell clone and fetal neural cells in tissue cultures [126], and higher fatty acid concentrations are needed to inhibit cell growth in the glioma clone. Fatty acid inhibitory effects are blocked in both cell lines by an antioxidant. The tumor cell line evidently retains the capacity for some lipid peroxidation when it is provided with the appropriate fatty acid substrate.

We have proposed that both cell proliferation in tumors and smooth muscle cell proliferation in atherosclerosis are promoted by an intracellular deficiency in EFA [419–421]. These relationships are discussed in Section V.

C. Tissue Culture

A number of studies from our laboratory have shown that polyunsaturated fatty acids at concentrations greater than 30 μM (Table II) inhibit the proliferation of smooth muscle cells, fibroblasts, fetal neural cells, and a glioma cell clone [39, 68, 82, 118, 122, 124–126, 132, 422]. Indomethacin blocks prostaglandin synthesis in these cultures, but indomethacin does not block the inhibitory effects of the polyunsaturated fatty acids on cell proliferation in the same cultures [39, 118, 124]. Several antioxidants, including the natural antioxidant vitamin E and the synthetic antioxidants BHT, α-naphthol, and 6-hydroxy-2,5,7,8-tetramethylchrom-2-carboxylic acid (Trolox C), block the inhibitory effects of polyunsaturated fatty acids on cell proliferation [82, 124, 126, 132, 419]. One antioxidant, α-naphthol, inhibits prostaglandin synthesis in the cultures [124]. As would be expected from studies with microsomal prostaglandin synthase [423], the other antioxidants have no effect on prostaglandin synthesis. The indomethacin and antioxidant data suggest that the inhibitory effects of the polyunsaturated fatty acids involve lipid peroxidation reactions not related to prostaglandin synthesis.

TABLE V Relative Fatty Acid Composition in Tumors and Transformed Cells

Tissue or cell line	Fatty acid composition (%)					Reference
	18:1 (n-9)	18:2 (n-6)	20:3 (n-6)	20:4 (n-6)	18:1/20:4	
Liver microsomes	18.9	—	—	11.9	1.6	402
Hepatoma 44 microsomes	14.9	—	—	4.9	3.0	
Hepatoma 3924A microsomes	13.3	—	—	4.9	2.7	
Liver mitochondria	10.1	25.8	0.5	21.5	0.5	411
Tumor 7777 mitochondria	27.0	22.2	1.2	9.0	3.0	
Liver microsomes	9.7	19.2	0.9	24.9	0.4	
Tumor 7777 microsomes	24.3	14.9	0.9	9.7	2.5	
Liver microsomes, phosphatidylcholines	8.3	19.1	2.1	22.0	0.4	412
Tumor 7777 microsomes, phosphatidylcholines	19.9	13.4	10.4	7.6	2.6	
Liver (Wistar rat) phospholipid	7.6	15.9	1.2	14.7	0.5	413
Yoshida hepatoma (AH 130) phospholipid	25.1	18.9	0.5	5.9	4.3	
Liver (Buffalo rat) phospholipid	13.3	21.3	1.5	12.6	1.1	413
Morris hepatoma phospholipid	24.6	13.2	0.4	7.6	3.2	
Liver lecithin	10.4	14.6	—	27.1	0.4	414
Hepatoma 27 lecithin	34.6	14.1	—	7.4	4.7	
Kidney lecithin	14.5	15.9	—	25.8	0.6	
Nephroma RA lecithin	26.3	20.9	—	7.9	3.3	
Liver plasma membranes	8.4	19.4	1.0	17.0	0.5	415
Hepatoma 484A plasma membranes, phosphatidylcholines	16.6	20.8	0.5	6.1	2.7	
WI-38 phospholipid	31.0	6.9	—	18.0	1.7	416
WI-38VA13A phospholipid	36.0	9.2	—	12.0	3.0	
Fibroblast (rat) phospholipid	20.3	2.7	2.0	15.00	1.4	417
Fibrosarcoma (rat) phospholipid	32.6	1.9	1.4	4.7	6.9	

Antioxidants such as vitamin E function either as scavengers for free radicals [*340, 341, 418, 424–426*] or, in high concentrations, as surfactants that disrupt lipid configurations in the cell membrane [*353, 421, 427, 428*]. Radical scavenger and surfactant effects are usually distinguished by comparing vitamin E (scavenger and surfactant) and vitamin E acetate (surfactant only) in the oxidation system [*429–431*]. Because cells in tissue culture may contain enzymes that hydrolyze vitamin E acetate to vitamin E, the effect of vitamin E quinone, the surfactant oxidation product of vitamin E [*432*], was studied with cells in tissue culture. Both vitamin E and vitamin E quinone enhance cell proliferation in smooth muscle cells (Fig. 7). Vitamin E quinone is actually more effective than vitamin E in promoting cell proliferation.

All of the polyunsaturated fatty acids that inhibit cell proliferation generated lipid peroxides when they were added to cells in culture [*39, 68, 119, 126, 422*]. Lipid peroxides in tissue culture were estimated by the TBA assay for malondialdehyde (MDA). Vitamin E, vitamin E quinone, and BHT block both lipid peroxide formation and the decomposition of lipid peroxides to MDA [*422*]. Vitamin E quinone is more effective than vitamin E in diminishing MDA, just as vitamin E quinone is more effective than vitamin E in promoting cell proliferation [*422*].

A number of controversial studies beginning shortly after the discovery of vitamin E [*433, 434*] and continuing to the present time [*435–438*] have ascribed antioxidant properties to both vitamin E and vitamin E quinone. Vitamin E quinone not only blocks lipid peroxidation in cells but also blocks lipid peroxida-

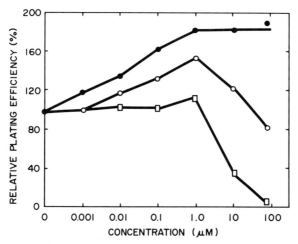

Fig. 7.   Extent of cell proliferation (relative plating efficiency) in smooth muscle cells treated with increasing concentrations of vitamin E (○), vitamin E quinone (●), and menadione (□). (Adapted from ref. *422*.)

tion when fatty acids are incubated with cumeme hydroperoxide in simple aqueous mixtures [422]. It is unlikely that vitamin E quinone acts as a surfactant in unstructured systems containing dissolved acids and micelles. The 1,4-benzoquinones are rapidly reduced to semiquinones by radicals such as aqueous superoxide [426, 439]. Vitamin E quinone may function as a radical scavenger through a quinone–semiquinone cycle involving aqueous superoxide (Fig. 8). These studies show that the efficiency of vitamin E as a radical scavenger in biological systems may be explained in part by a quinone–semiquinone cycle [341, 420]. In tissues, NADPH-cytochrome P-450 reductase may also function as the electron donor for a quinone–semiquinone cycle involving substituted 1,4-benzoquinones such as vitamin E quinone [440]. The antitumor quinones adriamycin and daunomycin react very slowly with NADPH-cytochrome P-450 reductase [440, 441], and these differences in reactivity may help to explain why vitamin E quinone promotes cell growth, whereas antitumor quinones apparently have little effect on this process.

As discussed in the previous section, MDA is generated from bicyclic endoperoxides that are formed both in prostaglandin biosynthesis [343, 421, 422, 442] and in radical rearrangement reactions that convert acyclic hydroperoxides to 2,3-deoxynorboranes [343, 353–355, 421]. Indomethacin blocks prostaglandin biosynthesis, but this agent actually enhances MDA formation in cultured cells [39, 422]. Acyclic hydroperoxides are formed in a number of nonenzymatic reactions and enzymatic reactions catalyzed by a variety of lipoxygenases [340–352]. Malondialdehyde yields from polyunsaturated fatty acids with three or more double bonds are similar when the fatty acids are incubated with either

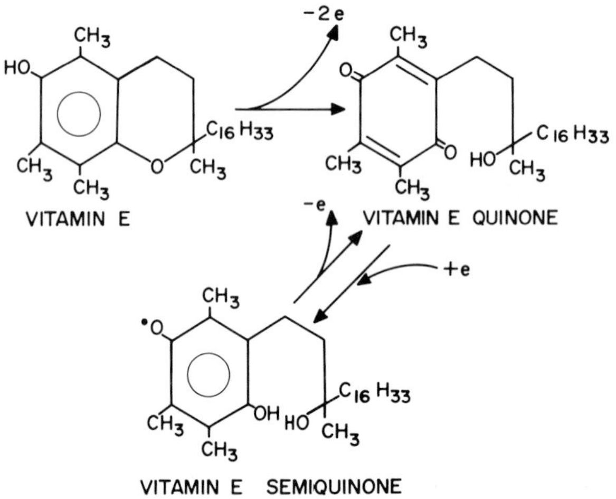

Fig. 8. Hypothetical redox cycle in which vitamin E quinone is reduced to the semiquinone by aqueous superoxide.

cumene hydroperoxide or cells in culture [68]. Lipoxygenases are inhibited by 5,8,11,14-eicosatetraynoic acid (ETYA) [443, 444] and 6,7-dihydroxycoumarin (esculetin) [445]; however, neither ETYA nor esculetin has a significant effect on MDA formation by cells in culture [J. Lindsey, H. Zhang, N. Morisaki, and D. Cornwell, unpublished observations]. The indomethacin, ETYA, and esculetin data suggest that exogenous fatty acids form MDA in smooth muscle cells by oxidation reactions that do not involve either cyclooxygenase or lipoxygenase enzymes directly.

Nonenzymatic lipid peroxidation in smooth muscle cells challenged with high concentrations of polyunsaturated fatty acids might conceivably occur because H_2O_2 is synthesized during β-oxidation in peroxisomes and related microbodies [446]. Böck et al. [446] suggested that peroxisomal β-oxidation works as a security valve for fatty acids when the cell is provided with an abundant supply of fatty acids, and indeed a high-fat diet provokes a marked proliferation of hepatic peroxisomes in the rat [447]. Saturated fatty acids and 18:1 (n-9) are excellent substrates for the fatty acyl-CoA oxidase found in peroxisomes [448]. The generation of H_2O_2 from palmitic acid oxidation in peroxisomes may explain the previously unexplained observation that antioxidants such as α-naphthol and Trolox C block the inhibitory effect of 180 μM palmitic acid on cell proliferation [124].

Peroxisomes are increased in rabbit aorta smooth muscle cells that are lipid laden by cholesterol feeding [449, 450]. Here again, fatty acid metabolism creates the potential for paradoxical effects. Cholesterol feeding may diminish peroxidation by sequestering substrate EFA in lipid droplets [419–421], and cholesterol feeding may promote peroxidation through peroxisomal oxidation reactions.

Fatty acid oxidation through the lipoxygenase pathway occurs in cultured cells [365], and several studies indicate that fatty acid oxidation reactions catalyzed by lipoxygenases also are involved in the control of cell proliferation. Macrophages generate superoxide and hydrogen peroxide, and these agents suppress cell proliferation [451–454]. Vitamin E stimulates the response of murine spleen cells to the T-cell mitogens concanavalin A and phytohemagglutinin [455, 456]. Lipoxygenase products such as 15-hydroperoxyeicosatetraenoic acid (15-HPETE) inhibit the mitogenic response of splenic B cells to lipopolysaccharide [457]. Platelet lipoxygenase activity is deficient in myeloproliferative disorders [458, 459]. Platelet lipoxygenase may control proliferation in distant cells through either the formation of hydroperoxy fatty acids or the generation of metastable scission products of these acids [460, 461]. The 2-alkenals and 4-hydroxyl-2-alkenals formed by β-scission of alkoxyl radicals are cytotoxic. One compound, trans-4-hydroxyl-2-pentenal, has been described as a carcinostatic agent [462].

The complexity of lipid peroxidation effects on cell proliferation is suggested by a few studies with inhibitors of hydroxy fatty acid synthesis and T-lymphocytes. 15-Hydroxyeicosatetraenoic acid (15-HETE) inhibits lipoxygenase reactions in human lymphocytes and mouse spleen lymphocytes [*463, 464*]. Because 15-HPETE inhibits the mitogenic response of B-cells to lipopolysaccharide [*457*], agents that block hydroperoxy fatty acid synthesis such as 15-HETE might be expected to enhance lymphocyte mitogenesis. In fact, 15-HETE inhibited phytohemagglutinin-induced mitogenesis in T-lymphocytes and had almost no effect on lipopolysaccharide-induced mitogenesis in B cells [*464*]. T-Lymphocytes do not synthesize prostaglandins [*465, 466*], but these cells release 20:4 (n-6), which is used for prostaglandin biosynthesis by monocytes and macrophages [*466*]. T-Cell colony formation is unusually sensitive to PGE_1 and PGE_2 levels [*303*], and T-cell mitogenesis is promoted by cyclooxygenase inhibitors [*464*], whereas B-cell mitogenesis is unaffected by these agents [*457*]. 15-Hydroperoxyeicosatetraenoic acid is a potent inhibitor of cyclooxygenase [*467*]. Indomethacin blocks both cyclooxygenase activity and the hydroperoxidase reductive enzyme that converts HPETE to HETE [*467*]. Thus, 15-HETE may inhibit T-cell mitogenesis by blocking the synthesis of 15-HPETE and other HETEs that are required to control prostaglandin levels.

D. Summary

Proliferating tissues and tumors have a diminished capacity for microsomal oxidation reactions, and these tissues generate smaller amounts of lipid peroxides than nonproliferating tissues. Polyunsaturated fatty acids undergo conversion to

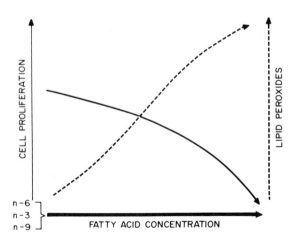

Fig. 9.   Schematic diagram of the inhibitor effects of lipid peroxidation on cell proliferation. Note that lipid peroxides, synthesized by enzymatic and nonenzymatic reactions, vary directly with substrate concentration throughout the concentration range.

lipid peroxides that inhibit proliferation when they are added to cells in culture. Hydroperoxy fatty acids and their scission products are cytostatic. Natural and synthetic antioxidants block both lipid peroxidation and the inhibitory effect of peroxidation reactions on cell proliferation. The natural antioxidants vitamin E and vitamin E quinone promote cell proliferation. Current studies reemphasize the classical description of vitamin E as an agent that promotes the proliferation of many cell types rather than the more limited description of vitamin E as an antisterility factor.

Oxidation and studies with antioxidants demonstrate that lipid peroxides and radicals generated during peroxidation reactions inhibit cell proliferation. The effect of lipid peroxidation on cell growth is summarized in Fig. 9. We suggest that the fatty acid paradox (Fig. 4) is the vector sum of prostaglandin (Fig. 6) and lipid peroxide (Fig. 9) effects on cell proliferation.

V. PROBLEMS WITH FATTY ACID PARADOXES

A. Tumorigenesis: Initiation–Promotion Effects

Free-radical reactions are involved in the conversion of many procarcinogens to their proximate carcinogens [461, 468–471]. Compounds that generate free radicals, benzoyl peroxide [472] and perhaps hydroperoxy fatty acid [197], are involved in tumorigenesis. Tumorigenesis in experimental animals is inhibited by butylated hydoxyanisole, BHT, DDC, ethoxyquin [469, 470], and many other antioxidants. Antioxidants are not always effective inhibitors of tumorigenesis, even when the antioxidants are administered at high levels to experimental animals. For example, vitamin E did not inhibit DMBA-induced neoplasia of the forestomach of Ha/ICR mice [468]. When BHT was administered after urethan, the number of pulmonary adenomas was increased [473]. Similarly, the number of hepatic tumors actually increased when BHT was administered after 2-acetylaminofluorene [474].

We suggest that there are two explanations for enhanced tumorigenesis with antioxidants. Many studies discussed in Section IV show that cell proliferation is inhibited by lipid peroxidation. Antioxidants undoubtedly diminish tumor initiation when they block free-radical reactions that convert procarcinogens to ultimate carcinogens; however, antioxidants may promote tumorigenesis by diminishing the regulatory effect of lipid peroxidation on cell proliferation.

A second mechanism for enhanced tumorigenesis could involve prostaglandin biosynthesis. Studies discussed in Section III show that tumorigenesis in many systems is enhanced by prostaglandin biosynthesis. Prostaglandin biosynthesis is regulated by the oxidative deactivation of the prostaglandin synthase complex [67, 475]. Antioxidants in low concentrations protect prostaglandin synthase and enhance the overall prostaglandin yield [*67, 475–478*], and these prostaglandins may promote cell proliferation and suppress immunosurveillance. Some antioxidants in high concentrations inhibit prostaglandin biosynthesis [*421, 423, 476, 479*], and large amounts of antioxidants in defined diets could suppress prostaglandin levels. Low levels of the inherently variable amounts of antioxidants in free populations may conceivably promote prostaglandin synthesis in some individuals.

The contradictory effects that antioxidants have on initiation and promotion create a dilemma that is complicated even more by the role of antioxidants in cellular differentiation. Mason [*17*] suggested that antioxidants stimulate both growth and differentiation, and these ideas are supported by experiments on differentiation in the rotifer *Asplanchna sieboldi* [*336–339*]. A number of cryoprotective agents, including dimethyl sulfoxide (DMSO), dimethylformamide, tetramethylurea, dimethylurea, and monomethylurea, induce differentiation in Friend leukemia cells [*480, 481*], and one agent, DMSO, is an antioxidant [*482, 483*] that blocks prostaglandin biosynthesis [*476*]. The other cryoprotective agents also function as radical scavengers, and their scavenger efficiency follows the same sequence as their efficiency in inducing differentiation [*484*]. Thus, antioxidants may control tumorigenesis, in part, by inducing differentiation. A detailed analysis of differentiation is beyond the scope of this review [*485*].

Because EFA promote cell proliferation and tumorigenesis (Section II), diets restricted in EFA content have been suggested as therapeutic regimens for the control of tumorigenesis. These diets may also have the contradictory effects on initiation and promotion. Tumors are deficient in EFA (Table V), and as a consequence tumors are deficient in the lipid peroxidation mechanisms that control cell proliferation (Section IV). Unmonitored fat restrictions in free populations may actually exacerbate the relative EFA deficiency in tumors and promote uncontrolled cell proliferation. It may be possible to supplement diets with the cytotoxic n-9 and n-3 fatty acids that generate lipid peroxides but do not serve as prostaglandin precursors [*68*]. This strategy may have significant clinical usefulness and should be explored.

B. Metastasis: Growth Factors and Antioxidants

Early studies with growing ascites tumors indicated that 20:4 (n-6) was mobilized in animals with these tumors [*19, 56*], and studies with tumor-bearing animals have confirmed and extended these observations [*57, 272*]. Fatty acids are mobilized in cells transformed by viruses and chemical agents (Section III,C). Tumor promoters such as TPA mobilize fatty acids (Sections III,B and III,C). Many factors that stimulate growth in eukaryotic cells have been identified [*486*]; a number of these factors [including epidermal growth factor, sar-

coma transforming growth factor, kidney transforming factors, bladder carcinoma transforming growth factor, and interleukin 1 (a monokine) and
interleukin 2 (a lymphokine)] stimulate arachidonic acid release in cells [256].
Agents that block the elaboration of growth factors would be expected to inhibit
tumor promotion.

Platelet aggregation is induced by tumors, and aggregating platelets elaborate
growth factors that promote tumorigenesis [487]. A number of investigators have
suggested that antiaggregatory compounds will break this cycle and inhibit tumor
metastasis [148, 277, 278, 488]. Indeed, PGI_2 was shown to be a potent antimetastatic agent [489]. Dipyridamole stimulates PGI_2 biosynthesis and inhibits
platelet aggregation [39, 490, 491], yet dipyridamole does not inhibit lung colony formation in mice treated with Lewis lung carcinoma cells [492]. We have
suggested that dipyridamole functions as an antioxidant, thus protecting PGI_2
synthase and causing both increased PGI_2 synthesis and increased cell growth
[39]. Thus, antiaggregatory agents may paradoxically have no effect on metastasis. Many studies cited elsewhere in this chapter have shown that lipid peroxides
inhibit cell proliferation, yet 15-HPETE enhances metastasis after the injection
of viable B16a melanoma cells [278]. This effect has been explained by the
inhibition of PGI_2 biosynthesis. The effect may also be explained by diminished
immunosurveillance because 15-HPETE inhibits the mitogenic response of B-
cells [457].

C. Atherosclerosis: Cholesterol and Essential Fatty Acids

The fatty acid paradox creates dilemmas that complicate a rational therapy for
atherosclerosis, just as the fatty acid paradoxes create dilemmas in the treatment
of carcinogenesis and metastasis. One model for atherosclerosis suggests that the
process is initiated by injury to the endothelium. Endothelial injury itself and the
platelet aggregation that occurs in response to endothelial injury lead to increased
endothelial permeability, free cholesterol influx, the accumulation of cholesterol
esters, and ultimately smooth muscle cell proliferation [421, 493]. This model is
summarized in Fig. 10.

Endothelial injury may be prevented by antioxidants such as vitamin E [420,
421]. Sections of the vascular endothelium are almost totally destroyed in animals maintained on a vitamin E-deficient diet [494]. Autoxidation and the microsomal oxidation of sterols and fatty acids are enhanced in a vitamin E deficiency
[495]. Hydroperoxy fatty acids and oxygenated sterols damage endothelial cells
and initiate the process of inflammation and repair [496, 497]. In contrast,
vitamin E and other antioxidants promote the proliferation of smooth muscle
cells (Section IV,C). In an early study on vitamin E and atherosclerosis, Iwakami
[498] noted that cholesterol diets supplemented with vitamin E were less athe-

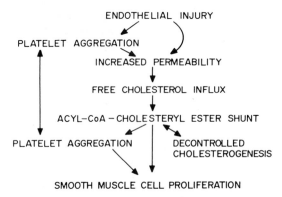

Fig. 10.   Model for the development of the atherosclerotic lesion. (Adapted from ref. *420.*)

rogenic in the rabbit than cholesterol diets alone. However, Iwakami used a diet that did not contain sufficient vitamin E to block lipid peroxidation in the aorta. He may have selected an antioxidant level that blocked endothelial injury without promoting smooth muscle cell proliferation. Such a diet is difficult to maintain in a free population.

Platelet aggregation occurs as a consequence of endothelial injury, and platelets themselves contribute to vessel injury and to changes in vessel permeability [*421, 493*]. Aggregating platelets also release a growth factor that stimulates the proliferation of aorta smooth muscle cells [*493, 499–501*]. Platelet effects are summarized in Fig. 10. The platelet-derived growth factor stimulates 20:4 (n-6) release and prostaglandin biosynthesis [*201, 256*]. Because both of the antioxidants vitamin E and dipyridamole block platelet aggregation [*420, 421, 490, 491*], these agents might be expected to diminish experimental atherosclerosis. Dipyridamole actually enhances experimental plaque formation in rabbits and monkeys maintained on cholesterol-rich atherogenic diets [*502–505*]. The antiaggregatory agent, again paradoxically, promotes plaque formation probably by functioning as an antioxidant and therefore stimulating smooth muscle cell proliferation [*39*].

Triglycerides and cholesterol esters accumulate as intracellular lipid droplets in tissues. Triglyceride droplets in the renal medulla and cholesterol ester droplets in the fatty streak lesion of the aorta contain large amounts of 20:3 (n-6), 20:4 (n-6), and 22:4 (n-6) [*419, 420, 506–509*]. We have suggested that the unusual fatty acid composition of neutral lipid droplets is characteristic of tissues in which microsomes are stimulated to shunt fatty acyl-CoA intermediates into neutral lipids [*83, 419, 420*]. This shunt appears in Fig. 10. Cellular phospholipids, the *élément constant,* do not change in total amount, whereas cellular neutral lipids, the *élément variable,* change dramatically in total content (Section II,D). The shunt converting polyunsaturated fatty acids to cholesterol esters explains why phospholipids in aorta plaques resemble tumors in that they contain low amounts of polyunsaturated fatty acids [419, 421, 510–512]. Several observations suggest that cholesterol, through conversion to cholesterol fatty esters, decreases the amounts of polyunsaturated fatty acids available in phospholipids for the synthesis of prostaglandins and lipid peroxides. For example, hypercholesterolemia in liver disease leads to a decrease in the phospholipid 20:4 (n-6) available for prostaglandin biosynthesis [513]. The recovery of PGI_2 synthesis from exogenous 20:4 (n-6) by injured rabbit aorta is inhibited in hypercholesterolemic animals [514, 515]. Diminished PGI_2 synthesis in these animals is explained if exogenous 20:4 (n-6) is diluted by saturated fatty acids from the endogenous phospholipid pool. In this model a relative EFA deficiency in aorta lesions decreases substrate availability for prostaglandin biosynthesis and lipid peroxidation and enhances the vulnerability of these lesions to the growth-promoting effects of antioxidants that also function as antiaggregatory agents. The many fatty acid paradoxes suggest at least one alternative explanation for diminished PGI_2 synthesis in hypercholesterolemic animals. Peroxisome activity, stimulated in lipid-laden cells [449, 450], may generate H_2O_2 (Section IV,C), which inactivates PGI_2 synthase.

ACKNOWLEDGMENTS

This study was supported in part by Research Grant HL-11897 from the National Institutes of Health. We appreciate many continuing discussions with Dr. Rao V. Panganamala and the assistance of Ms. Dorothy Ferguson.

REFERENCES

1. G. O. Burr and M. M. Burr, *J. Biol. Chem.* **82,** 345 (1929).
2. G. O. Burr, *Fed. Proc., Fed. Am. Soc. Exp. Biol.* **1,** 224 (1942).
3. S. M. Greenberg, C. E. Calbert, E. E. Savage, and H. J. Deuel, Jr., *J. Nutr.* **41,** 473 (1950).
4. S. M. Greenberg, C. E. Calbert, H. J. Deuel, Jr., and J. B. Brown, *J. Nutr.* **45,** 521 (1951).
5. R. T. Holman, *Prog. Chem. Fats Other Lipids* **9,** 279 (1969).
6. H. Sprecher, *in* "Polyunsaturated Fatty Acids" (W.-H. Kunau and R. T. Holman, eds.), pp. 1–18. Am. Oil Chem. Soc., Champaign, Illinois, 1977.
7. H. Sprecher, *Prog. Lipid Res.* **20,** 13 (1981).
8. I. G. Rieckehoff, R. T. Holman, and G. O. Burr, *Arch. Biochem.* **20,** 331 (1949).
9. A. J. Fulco and J. F. Mead, *J. Biol. Chem.* **234,** 1411 (1959).
10. R. T. Holman, *in* "Polyunsaturated Fatty Acids" (W.-H. Kunau and R. T. Holman, eds.), pp. 163–182. Am. Oil Chem. Soc., Champaign, Illinois, 1977.
11. G. A. Dhopeshwarkar and J. F. Mead, *J. Am. Oil Chem. Soc.* **38,** 297 (1961).
12. D. K. Bosshardt, M. Kryvokulsky, and E. E. Howe, *J. Nutr.* **69,** 185 (1959).
13. H. J. Thomasson, G. J. von Beers, J. J. Gottenbos, H. deIongh, J. G. Keppler, and S. Sparreboom, *in* "Biochemistry of Lipids" (G. Popjak, ed.), pp. 194–202. Pergamon, Oxford, 1960.
14. H. M. Evans, S. Lepkovsky, and E. A. Murphy, *J. Biol. Chem.* **106,** 431 (1934).

15. J. P. Funch, E. Aaes-Jorgensen, and H. Dam, *Br. J. Nutr.* **11**, 426 (1957).
16. E. C. Maeder, *Anat. Rec.* **70**, 73 (1937).
17. K. E. Mason, *Am. J. Anat.* **52**, 153 (1933).
18. J. S. Nelson, *in* "Vitamin E. A Comprehensive Treatise" (L. J. Machlin, ed.), pp. 397–402. Dekker, New York, 1980.
19. I. Smedley-MacLean and L. C. A. Nunn, *Biochem. J.* **35**, 983 (1941).
20. E. F. Terroine and P. Bélin, *Bull. Soc. Chim. Biol.* **9**, 12 (1927).
21. G. S. Patil, R. H. Matthews, and D. G. Cornwell, *Adv. Chem. Ser.* **144**, 44 (1975).
22. D. G. Cornwell and G. S. Patil, *in* "Polyunsaturated Fatty Acids" (W.-H. Kunau and R. T. Holman, eds.), pp. 105–137. Am. Oil Chem. Soc., Champaign, Illinois, 1977.
23. L. I. Burke, G. S. Patil, R. V. Panganamala, J. C. Geer, and D. G. Cornwell, *J. Lipid Res.* **14**, 9 (1973).
24. G. S. Patil, R. H. Matthews, and D. G. Cornwell. *J. Lipid Res.* **17**, 197 (1976).
25. G. S. Patil and D. G. Cornwell, *J. Lipid Res.* **18**, 1 (1977).
26. G. S. Patil, N. J. Dorman, and D. G. Cornwell, *J. Lipid Res.* **20**, 663 (1979).
27. R. K. Beerthuis, D. H. Nugteren, H. J. J. Pabon, and D. A. van Dorp, *Recl. Trav. Chim. Pays-Bas* **87**, 461 (1968).
28. A. B. Decker, D. L. Fillerup, and J. F. Mead, *J. Nutr.* **41**, 507 (1950).
29. T. K. Hulsey, S. J. Burnham, W. W. Neblett, J. A. O'Neill, Jr., and H. C. Meng, *Surg Forum* **28**, 31 (1977).
30. J. R. Paulsrud, L. Pensler, C. F. Whitten, S. Stewart, and R. T. Holman, *Am. J. Clin. Nutr.* **25**, 897 (1972).
31. M. D. Caldwell, H. T. Jonsson, and H. B. Othersen, Jr., *J. Pediatr.* **81**, 894 (1972).
32. J. A. O'Neill, Jr., M. D. Caldwell, and H. C. Meng, *Ann. Surg.* **185**, 535 (1977).
33. B. B. Caffrey and H. T. Jonsson, Jr., *Prog. Lipid Res.* **20**, 641 (1981).
34. E. Pegreffi, *Pathologica* **27**, 543 (1935).
35. H. P. Ehrlich, H. Tarver, and T. K. Hunt, *Ann. Surg.* **175**, 235 (1972).
36. K. H. Lee, *J. Pharm. Sci.* **57**, 1238 (1968).
37. E. L. Howes, C. M. Plotz, J. W. Blunt, Jr., and C. Ragan, *Surgery* **28**, 177 (1950).
38. M. H. McGrath, *Plast. Reconstr. Surg.* **69**, 74 (1982).
39. N. Morisaki, J. M. Stitts, L. Bartels-Tomei, G. E. Milo, R. V. Panganamala, and D. G. Cornwell, *Artery* **11**, 88 (1982).
40. J. Mehta, P. Mehta, and D. Hay, *Prostaglandins* **24**, 751 (1982).
41. C. Arnander, G. Jurell, G. Tornling, and G. Unge, *Scand. J. Plast. Reconstr. Surg.* **13**, 261 (1979).
42. A. N. Nasr and S. Shostak, *Nature (London)* **207**, 1395 (1965).
43. R. L. Snipes, *Anat. Rec.* **159**, 421 (1967).
44. J. L. McCullough, S. H. Schreiber, and V. A. Ziboh, *J. Invest. Dermatol.* **70**, 318 (1978).
45. N. J. Lowe, E. Duell, and J. J. Voorhees, *Br. J. Dermatol.* **101**, 33 (1979).
46. V. A. Ziboh and S. L. Hsia, *J. Lipid Res.* **13**, 458 (1972).
47. J. J. Voorhees, E. A. Duell, M. Stawiski, and E. R. Harrell, *Adv. Cyclic Nucleotide Res.* **4**, 117 (1974).
48. C. Prottey, *Br. J. Dermatol.* **94**, 579 (1976).
49. A. F. Watson and E. Mellanby, *Br. J. Exp. Pathol.* **11**, 311 (1930).
50. C. A. Baumann and H. P. Rusch, *Am. J. Cancer* **35**, 213 (1939).
51. A. Tannerbaum, *Cancer Res.* **2**, 468 (1942).
52. H. Silverstone and A. Tannenbaum, *Cancer Res.* **10**, 448 (1950).
53. H. P. Jacobi and C. A. Baumann, *Am. J. Cancer* **39**, 338 (1940).
54. F. L. Haven and W. R. Bloor, *Adv. Cancer Res.* **4**, 237 (1956).
55. K. K. Carroll and H. T. Khor, *Prog. Biochem. Pharmacol.* **10**, 308 (1975).

56. I. Smedley-MacLean and E. M. Hume, *Biochem. J.* **35,** 996 (1941).
57. J. F. Mead, *Prog. Lipid Res.* **20,** 1 (1981).
58. E. B. Gammal, K. K. Carroll, and E. R. Plunkett, *Cancer Res.* **27,** 1737 (1967).
59. K. K. Carroll, *J. Environ. Pathol. Toxicol.* **3,** 253 (1980).
60. K. K. Carroll, G. J. Hopkins, T. G. Kennedy, and M. B. Davidson, *Prog. Lipid Res.* **20,** 685 (1981).
61. G. J. Hopkins and C. E. West, *JNI, J. Natl. Cancer Inst.* **58,** 753 (1977).
62. G. A. Rao and S. Abraham, *JNCI, J. Natl. Cancer Inst.* **56,** 431 (1976).
63. J. Benson, M. Lev, and C. G. Grand, *Cancer Res.* **16,** 135 (1956).
64. K. K. Carroll and H. T. Khor, *Lipids* **6,** 415 (1971).
65. S. Dayton, S. Hashimoto, and J. Wollman, *J. Nutr.* **107,** 1353 (1977).
66. G. J. Hopkins, T. G. Kennedy, and K. K. Carroll, *JNCI,* **66,** 517 (1981).
67. W. E. M. Lands and M. J. Byrnes, *Prog. Lipid Res.* **20,** 287 (1981).
68. N. Morisaki, H. Sprecher, G. E. Milo, and D. G. Cornwell, *Lipids* **17,** 893 (1982).
69. V. A. Ziboh, T. T. Nguyen, J. L. McCullough, and G. D. Weinstein, *Prog. Lipid Res.* **20,** 857 (1981).
70. B. T. Hyman, L. L. Stoll, and A. A. Spector, *Biochim. Biophys. Acta* **713,** 375 (1982).
71. B. Monis and A. R. Eynard, *JNCI, J. Natl. Cancer Inst.* **64,** 73 (1980).
72. B. Monis and A. R. Eynard, *Prog. Lipid Res.* **20,** 691 (1981).
73. J. M. Bailey and L. M. Dunbar, *Exp. Mol. Pathol.* **18,** 142 (1973).
74. B. V. Howard and W. J. Howard, *Adv. Lipid Res.* **12,** 51 (1974).
75. A. A. Spector, S. N. Mathur, T. L. Kaduce, and B. T. Hyman, *Prog. Lipid Res.* **19,** 155 (1980).
76. R. P. Geyer, *in* "Lipid Metabolism in Tissue Culture Cells" (G. H. Rothblat and D. Kritchevsky, eds.), pp. 33–44. Wistar Inst. Press, Philadelphia, Pennsylvania, 1967.
77. M. S. Moskowitz, *in* "Lipid Metabolism in Tissue Culture Cells" (G. H. Rothblat and D. Kritchevsky, eds.), pp. 49–59. Wistar Inst. Press, Philadelphia, Pennsylvania, 1967.
78. C. G. Mackenzie, J. B. Mackenzie, and O. K. Reiss, *in* "Lipid Metabolism in Tissue Culture Cells" (G. H. Rothblat and D. Kritchevsky, eds.), pp. 63–81. Wistar Inst. Press, Philadelphia, Pennsylvania, 1967.
79. E. E. Schneeberger, R. D. Lynch, and R. P. Geyer, *Exp. Cell Res.* **69,** 193 (1971).
80. J. J. Mulligan, R. D. Lynch, E. E. Schneeberger, and R. P. Geyer, *Biochim. Biophys. Acta* **470,** 92 (1977).
81. R. D. Lynch, *Lipids* **15,** 412 (1980).
82. J. S. Miller, V. C. Gavino, G. A. Ackerman, H. M. Sharma, G. E. Milo, J. C. Geer, and D. G. Cornwell, *Lab. Invest.* **42,** 495 (1980).
83. V. C. Gavino, J. S. Miller, J. M. Dillman, G. E. Milo, and D. G. Cornwell, *J. Lipid Res.* **22,** 57 (1981).
84. R. P. Geyer, A. Bennett, and A. Rohr, *J. Lipid Res.* **3,** 80 (1962).
85. A. A. Spector and D. Steinberg, *J. Biol. Chem.* **242,** 3057 (1967).
86. L. E. Gerschenson, J. F. Mead, I. Harary, and D. F. Haggerty, Jr., *Biochim. Biophys. Acta* **131,** 42 (1967).
87. A. F. Horwitz, M. E. Hatten, and M. M. Burger, *Proc. Natl. Acad. Sci. U.S.A.* **71,** 3115 (1974).
88. R. E. Williams, B. J. Wisnieski, H. G. Rittenhouse, and C. F. Fox, *Biochemistry* **13,** 1969 (1974).
89. K. A. Ferguson, M. Glaser, W. H. Bayer, and P. R. Vagelos, *Biochemistry* **14,** 146 (1975).
90. A. B. Awad and A. A. Spector, *Biochim. Biophys. Acta* **426,** 723 (1976).
91. M. D. Rosenthal and R. P. Geyer, *Biochim. Biophys. Acta* **441,** 465 (1976).
92. C. P. Burns, D. G. Luttenegger, S.-P. L. Wei, and A. A. Spector, *Lipids* **12,** 747 (1977).

93. A. F. Horwitz, A. Wight, P. Ludwig, and R. Cornell, *J. Cell Biol.* **77**, 334 (1978).
94. C. G. Mackenzie, E. Moritz, J. A. Wisneski, O. K. Reiss, and J. B. Mackenzie, *Mol. Cell. Biochem.* **19**, 7 (1978).
95. A. A. Spector, R. E. Kiser, G. M. Denning, S.-W. M. Koh, and L. E. DeBault, *J. Lipid Res.* **20**, 536 (1979).
96. A. A. Spector, G. M. Denning, and L. L. Stoll, *In Vitro* **16**, 932 (1980).
97. M. D. Rosenthal, *Lipids* **15**, 838 (1980).
98. B. T. Hyman, L. L. Stoll, and A. A. Spector, *J. Biol. Chem.* **256**, 8863 (1981).
99. R. D. Lynch, E. E. Schneeberger, and R. P. Geyer, *Biochemistry* **15**, 193 (1976).
100. P.-Y. Tsai and R. P. Geyer, *Biochim. Biophys. Acta* **489**, 381 (1977).
101. P.-Y. Tsai and R. P. Geyer, *Biochim. Biophys. Acta* **528**, 344 (1978).
102. V. J. Evans, J. C. Bryant, H. A. Kerr, and E. L. Schilling, *Exp. Cell Res.* **36**, 439 (1964).
103. J. M. Bailey and J. Menter, *Proc. Soc. Exp. Biol. Med.* **125**, 101 (1967).
104. G. H. Rothblat, M. K. Buchko, and D. Kritchevsky, *Biochim. Biophys. Acta* **164**, 327 (1968).
105. Y. Kagawa, T. Takaoka, and H. Katsuta, *J. Biochem. (Tokyo)* **68**, 133 (1970).
106. J. M. Bailey and L. M. Dunbar, *Cancer Res.* **31**, 91 (1971).
107. T. Takaoka and H. Katsuta, *Exp. Cell Res.* **67**, 295 (1971).
108. J. M. Bailey, B. V. Howard, L. M. Dunbar, and S. F. Tillman, *Lipids* **7**, 125 (1972).
109. E. Lengle and R. P. Geyer, *Biochim. Biophys. Acta* **260**, 608 (1972).
110. G. H. Rothblat, L. Arbogast, D. Kritchevsky, and M. Naftulin, *Lipids* **11**, 97 (1976).
111. L. M. Dunbar and J. M. Bailey, *J. Biol. Chem.* **250**, 1152 (1975).
112. R. G. Ham, *Science* **140**, 802 (1963).
113. I. N. Dubin, B. Czernobilsky, and B. Herbst, *JNCI, J. Natl. Cancer Inst.* **34**, 43 (1965).
114. I. Harary, L. E. Gerschenson, D. F. Haggerty, Jr., W. Desmond, and J. F. Mead, *in* "Lipid Metabolism in Tissue Culture Cells" (G. H. Rothblat and D. Kritchevsky, eds.), pp. 17–30. Wistar Inst. Press, Philadelphia, Pennsylvania, 1967.
115. R. W. Holley, J. H. Baldwin, and J. A. Kiernan, *Proc. Natl. Acad. Sci. U.S.A.* **71**, 3976 (1974).
116. W. R. Kidwell, M. E. Monaco, M. S. Wicha, and G. S. Smith, *Cancer Res.* **38**, 4091 (1978).
117. M. S. Wicha, L. A. Liotta, and W. R. Kidwell, *Cancer Res.* **39**, 426 (1979).
118. J. J. Huttner, E. T. Gwebu, R. V. Panganamala, G. E. Milo, and D. G. Cornwell, *Science* **197**, 289 (1977).
119. N. Morisaki, J. Lindsey, G. E. Milo, and D. G. Cornwell, *Lipids* **18**, 349 (1983).
120. R. Cornell, G. L. Grove, G. H. Rothblat, and A. F. Horwitz, *Exp. Cell Res.* **109**, 299 (1977).
121. D. E. Wennerström and H. M. Jenkin, *Biochim. Biophys. Acta* **431**, 469 (1976).
122. J. J. Huttner, G. E. Milo, R. V. Panganamala, and D. G. Cornwell, *In Vitro* **14**, 854 (1978).
123. J. M. Bailey, *in* "Lipid Metabolism in Tissue Culture Cells" (G. H. Rothblat and D. Kritchevsky, eds.), pp. 85–109. Wistar Inst. Press, Philadelphia, Pennsylvania, 1967.
124. D. G. Cornwell, J. J. Huttner, G. E. Milo, R. V. Panganamala, H. M. Sharma, and J. C. Geer, *Lipids* **14**, (1979).
125. V. C. Gavino, J. S. Miller, J. M. Dillman, G. E. Milo, and D. G. Cornwell, *Prog. Lipids Res.* **20**, 323 (1981).
126. V. C. Liepkalns, C. Icard-Liepkalns, and D. G. Cornwell, *Cancer Lett.* **15**, 173 (1982).
127. M. C. F. DeMello, B. M. Bayer, and M. A. Beaven, *Biochem. Pharmacol.* **29**, 311 (1980).
128. C. W. Roone, N. Mantel, T. D. Caruso, Jr., E. Kazam, and R. E. Stevenson, *In Vitro* **7**, 174 (1972).
129. S. Ruggieri, R. Roblin, and P. H. Black, *J. Lipid Res.* **20**, 772 (1979).
130. V. P. Skipski, *in* "Blood Lipids and Lipoproteins:Quantitation, Composition and Metabolism" (G. J. Nelson, ed.), pp. 539–551. Wiley (Interscience), New York, 1972.

131. D. G. Cornwell, F. A. Kruger, G. J. Hamwi, and J. B. Brown, *Metab., Clin. Exp.* **11**, 840 (1962).
132. V. C. Gavino, G. E. Milo, and D. G. Cornwell, *Cell Tissue Kinet.* **15**, 225 (1982).
133. C. W. Kischer, *Dev. Biol.* **16**, 203 (1967).
134. C. W. Kischer, *Am. J. Anat.* **124**, 491 (1969).
135. W. H. Eaglstein and G. D. Weinstein, *J. Invest. Dermatol.* **64**, 386 (1975).
136. A. Lupulescu, *Prostaglandins* **10**, 573 (1975).
137. C. B. Bentley-Phillips, H. Paulli-Jorgensen, and R. Marks, *Arch. Dermatol. Res.* **257**, 233 (1977).
138. N. J. Lowe and R. B. Stoughton, *J Invest. Dermatol.* **68**, 134 (1977).
139. V. A. Ziboh, J. T. Lord, and N. S. Penney, *J. Lipid Res.* **18**, 37 (1977).
140. F. Marks, G. Fürstenberger, and E. Kownatzki, *Cancer Res.* **41**, 696 (1981).
141. G. Fürstenberger, M. Gross, and F. Marks, *in* "Prostaglandins and Cancer" (T. J. Powles, R. S. Bockman, K. V. Honn, and P. Ramwell, eds.), pp. 239–254. Alan R. Liss, Inc., New York, 1982.
142. E. Bresnick, P. Meunier, and M. Lamden, *Cancer Lett.* **7**, 121 (1979).
143. C. L. Ashendel and R. K. Boutwell, *Biochem. Biophys. Res. Commun.* **90**, 623 (1979).
144. G. Fürstenberger and F. Marks. *Biochem. Biophys. Res. Commun.* **92**, 749 (1980).
145. G. Fürstenberger and F. Marks, *Biochem. Biophys. Res. Commun.* **84**, 1103 (1978).
146. G. Fürstenberger, M. de Bravo, S. Bertsch, and F. Marks, *Res. Commun. Chem. Pathol. Pharmacol.* **24**, 533 (1979).
147. A. K. Verma, C. L. Ashendel, and R. K. Boutwell, *Cancer Res.* **40**, 308 (1980).
148. K. V. Honn, R. S. Bockman, and L. J. Marnett, *Prostaglandins* **21**, 833 (1981).
149. D. A. van Dorp, *Ann. N.Y. Acad. Sci.* **180**, 181 (1971).
150. . J. J. Voorhees and E. A. Duell, *Arch. Dermatol.* **104**, 352 (1971).
151. J. J. Voorhees, *Arch. Dermatol.* **118**, 869 (1982).
152. S. Hammarström, M. Hamberg, B. Samuelsson, E. A. Duell, M. Stawiski, and J. J. Voorhees, *Proc. Natl. Acad. Sci. U.S.A.* **72**, 5130 (1975).
153. S. Hammarström, J. A. Lindgren, C. Marcelo, E. A. Duell, T. F. Anderson, and J. J. Voorhees, *J. Invest. Dermatol.* **73**, 180 (1979).
154. H. Katayama and A. Kawada, *J. Dermatol.* **8**, 323 (1981).
155. S. M. Fischer, G. L. Gleason, G. D. Mills, and T. J. Slaga, *Cancer Lett.* **10**, 343 (1980).
156. R. A. Karmali, *Prostaglandins Med.* **5**, 11 (1980).
157. A. H. Tashjian, Jr., E. F. Voelkel, P. Goldhaber, and L. Levine, *Prostaglandins* **3**, 515 (1973).
158. W. C. Tan, O. S. Privett, and M. E. Goldyne, *Cancer Res.* **34**, 3229 (1974).
159. B. M. Jaffe and S. Condon, *Ann. Surg.* **184**, 516 (1976).
160. A. H. Tashjian, Jr., E. F. Voelkel, and L. Levine, *Prostaglandins* **14**, 309 (1977).
161. A. Bennett, E. M. Charlier, A. M. McDonald, J. S. Simpson, I. F. Stanford, and T. Zebro, *Lancet* **2**, 624 (1977).
162. A. Maiorana and P. M. Gullino, *Cancer Res.* **38**, 4409 (1978).
163. I. Alam, E. F. Voelkel, A. H. Tashjian, Jr., and L. Levine, *Prostaglandins Med.* **4**, 227 (1980).
164. A. Trevisani, E. Ferretti, A. Capuzzo, and V. Tomasi, *Br. J. Cancer* **41**, 341 (1980).
165. A. H. Tashjian, Jr., E. F. Voelkel, and L. Levine, *in* "Prostaglandins and Cancer" (T. J. Powles, R. S. Bockman, K. V. Honn, and P. Ramwell, eds.), pp. 513–523. Alan R. Liss, Inc., New York, 1982.
166. O. J. Plescia, *in* "Prostaglandins and Cancer" (T. J. Powles, R. S. Bockman, K. V. Honn, and P. Ramwell, eds.), pp. 619–631. Alan R. Liss, Inc., New York, 1982.

138 David G. Cornwell and Nobuhiro Morisaki

167. T. Nagasaka, W. D. Winters, O. M. Soriero, and M. J. K. Harper, *in* "Prostaglandins and Cancer" (T. J. Powles, R. S. Bockman, K. V. Honn, and P. Ramwell, eds.), pp. 697–700. Alan R. Liss, Inc., New York, 1982.
168. A. Lupulescu, *JNCI, J. Natl. Cancer Inst.* **61,** 97 (1978).
169. J. S. Goodwin and D. R. Webb, *Clin. Immunol. Immunopathol.* **15,** 106 (1980).
170. Z. L. Olkowski, M. B. Waitzman, and M. J. Skeen, *Prog. Lipid Res.* **20,** 759 (1981).
171. J. L. Murray, *in* "Prostaglandins and Cancer" (T. J. Powles, R. S. Bockman, K. V. Honn, and P. Ramwell, eds.), pp. 713–718. Alan R. Liss, Inc., New York, 1982.
172. S. M. Taffet, T. E. Eurell, and S. W. Russell, *Prostaglandins* **24,** 763 (1982).
173. J. A. Lindsey, N. Morisaki, J. M. Stitts, R. A. Zager, and D. G. Cornwell, *Lipids* **18,** 566 (1983).
174. W. Jubiz, J. Frailey, and J. B. Smith, *Cancer Res.* **39,** 998 (1979).
175. S. C. Liu and R. A. Knazek, *Prostaglandins, Leukotrienes Med.* **8,** 191 (1982).
176. M. K. Foecking, R. V. Panganamala, H. Abou-Issa, and J. P. Minton, *in* "Prostaglandins and Cancer" (T. J. Powles, R. S. Bockman, K. V. Honn, and P. Ramwell, eds.), pp. 657–662. Alan. R. Liss, Inc., New York, 1982.
177. C. M. Popescu, *Prostaglandins Med.* **7,** 321 (1981).
178. J. A. Sykes and I. S. Moddox, *Nature (London), New Biol.* **237,** 59 (1972).
179. H. R. Strausser and J. L. Humes, *Int. J. Cancer* **15,** 724 (1975).
180. V. Hial, Z. Horakova, F. E. Shaff, and M. A. Beaven, *Eur. J. Pharmacol.* **37,** 367 (1976).
181. M. G. Santoro, G. W. Philpott, and B. M. Jaffe, *Nature (London)* **263,** 777 (1976).
182. A. Bennett, J. Houghton, D. J. Leaper, and I. F. Stamford, *Br. J. Pharmacol.* **63,** 356 (1978).
183. A. Bennett, J. Houghton, D. J. Leaper, and I. F. Stamford, *Prostaglandins* **17,** 179 (1979).
184. N. R. Lynch and J.-C. Salomon, *JNCI, J. Natl. Cancer Inst.* **62,** 117 (1979).
185. J. S. Goodwin, G. Husby, and R. C. Williams, Jr., *Cancer Immunol. Immunother.* **8,** 3 (1980).
186. T. Kudo, T. Narisawa, and S. Abo, *Gann* **71,** 260 (1980).
187. T. Narisawa, M. Sato, M. Tani, T. Kudo, T. Takahashi, and A. Goto, *Cancer Res.* **41,** 1954 (1981).
188. W. R. Waddell and R. E. Gerner, *J. Surg. Oncol.* **15,** 85 (1980).
189. A. Bennett, *Prog. Lipid Res.* **20,** 677 (1981).
190. M. Pollard and P. H. Luckert, *Science* **214,** 558 (1981).
191. A. Bennett, *in* "Prostaglandins and Cancer" (T. J. Powles, R. S. Bockman, K. V. Honn, and P. Ramwell, eds.), pp. 759–766. Alan R. Liss, Inc., New York, 1982.
192. K. Hellmann and B. A. Pym, *in* "Prostaglandins and Cancer" (T. J. Powles, R. S. Bockman, K. V. Honn, and P. Ramwell, eds.), pp. 767–774. Alan R. Liss, Inc., New York, 1982.
193. N. R. Lynch and J. C. Salomon, *in* "Prostaglandins and Cancer" (T. J. Powles, R. S. Bockman, K. V. Honn, and P. Ramwell, eds.), pp. 775–781. Alan. R. Liss, Inc., New York, 1982.
194. R. A. Karmali, A. Volkman, W. Spivey, P. Muse, and T. M. Louis, *Prostaglandins Med.* **4,** 239 (1980).
195. P. J. M. Tutton and D. H. Barkla, *Br. J. Cancer* **41,** 47 (1980).
196. P. J. M. Tutton, F. M. Petry, and D. H. Barkla, *in* "Prostaglandins and Cancer" (T. J. Powles, R. S. Bockman, K. V. Honn, and P. Ramwell, eds.), pp. 753–766. Alan R. Liss, Inc., New York, 1982.
197. S. M. Fischer and T. J. Slaga, *in* "Prostaglandins and Cancer" (T. J. Powles, R. S. Bockman, K. V. Honn, and P. Ramwell, eds.), pp. 255–264. Alan R. Liss, Inc., New York, 1982.
198. A. Rigas and L. Levine, *Prostaglandins Med.* **7,** 217 (1981).
199. D. Mitrovic, E. McCall, and F. Dray, *Prostaglandins* **23,** 17 (1982).
200. D. Gal, M. L. Casey, J. M. Johnston, and P. C. MacDonald, *J. Clin. Invest.* **70,** 798 (1982).

201. S. R. Coughlin, M. A. Moskowitz, H. N. Antoniades, and L. Levine, *Proc. Natl. Acad. Sci. U.S.A.* **78,** 7134, (1981).
202. G. L. Hahn, M. Menconi, and P. Polgar, *in* "Prostaglandins and Cancer" (T. J. Powles, R. S. Bockman, K. V. Honn, and P. Ramwell, eds.), pp. 381–384. Alan R. Liss, Inc., New York, 1982.
203. S. L. Hong, T. Carty, and D. Deykin, *J. Biol. Chem.* **255,** 9538 (1980).
204. T. L. Kaduce, A. A. Spector, and R. S. Bar, *Arteriosclerosis (Dallas)* **2,** 380 (1982).
205. A. Eldor, I. Vlodavsky, E. Hyam, R. Atzmon, and Z. Fuks, *Prostaglandins* **25,** 263 (1983).
206. M. A. Gimbrone, Jr. and R. W. Alexander, *Science* **189,** 219 (1975).
207. A. Ager, J. L. Gordon, S. Moncada, J. D. Pearson, J. A. Salmon, and M. A. Trevethick, *J. Cell. Physiol.* **110,** 9 (1982).
208. W. Siess, F. Dray, C. Seillan, C. Ody, and F. Russo-Marie, *Biochem. Biophys. Res. Commun.* **99,** 608 (1981).
209. T. Neichi, W.-C. Chang, Y. Mitsui, and S. Murota, *Artery* **11,** 47 (1982).
210. A. J. Marcus, B. B. Weksler, E. A. Jaffe, and M. J. Broekman, *J. Clin. Invest.* **66,** 979 (1980).
211. A. R. Johnson, *J. Clin. Invest.* **65,** 841 (1980).
212. A. A. Spector, J. C. Hoak, G. L. Fry, G. M. Denning, L. L. Stoll, and J. B. Smith, *J. Clin. Invest.* **65,** 1003, (1980).
213. A. A. Spector, J. C. Hoak, G. L. Fry, L. L. Stoll, C. T. Tanke, and T. L. Kaduce, *Prog. Lipid Res.* **20,** 471 (1981).
214. A. A. Spector, T. L. Kaduce, J. C. Hoak, and G. L. Fry, *J. Clin. Invest.* **68,** 1003 (1981).
215. J. M. F. Thomas, H. Chap, and L. Douste-Blazy, *Biochem. Biophys. Res. Commun.* **103,** 819 (1981).
216. F. Alhenc-Gelas, S. J. Tsai, K. S. Callahan, W. B. Campbell, and A. R. Johnson, *Prostaglandins* **24,** 723 (1982).
217. N. L. Baenziger, P. R. Becherer, and P. W. Majerus, *Cell* **16,** 967 (1979).
218. J. Sraer, J. Foidart, D. Chansel, P. Mahieu, and R. Ardaillou, *Int. J. Biochem.* **12,** 203 (1980).
219. A. S. Petrulis, M. Aikawa, and M. J. Dunn, *Kidney Int.* **20,** 469 (1981).
219a. R. Ardaillou, J. Foidart, P. Mahieu, and J. Sraer, *in* "Prostaglandin Synthetase Inhibitors: New Clinical Applications" (P. Ramwell, ed.), p. 315. Alan R. Liss, Inc., New York, 1980.
220. E. M. Ritzi and W. A. Stylos, *JNCI, J. Natl. Cancer Inst.* **56,** 529 (1976).
221. G. M. Denning, P. H. Figard, and A. A. Spector, *J. Lipid Res.* **23,** 584 (1982).
222. L. Taylor and P. Polgar, *FEBS Lett.* **79,** 69 (1977).
223. L. Taylor and P. Polgar, *Prostaglandins* **22,** 723 (1981).
224. L. Taylor, E. Schneider, J. Smith, and P. Polgar, *Mech. Ageing Dev.* **16,** 311 (1981).
225. V. Hial, M. C. DeMello, Z. Horakova, and M. S. Beaven, *J. Pharmacol. Exp. Ther.* **202,** 446 (1977).
226. D. R. Thomas, G. W. Philpott, and B. M. Jaffe, *Exp. Cell Res.* **84,** 40 (1974).
227. R. M. Zusman and H. R. Keiser, *J. Clin. Invest.* **60,** 215 (1977).
228. R. S. Bockman, *Prostglandins* **21,** 9 (1981).
229. L. Levine and A. Hassid, *Biochem. Biophys. Res. Commun.* **76,** 1181 (1977).
230. M. G. Lewis, T. L. Kaduce, and A. A. Spector, *Prostaglandins* **22,** 747 (1981).
231. L. W. Daniel, L. King, and M. Waite, *J. Biol. Chem.* **256,** 12830 (1981).
232. M.-C. Coene, C. Van Hove, M. Claeys, and A. G. Herman, *Biochim. Biophys. Acta* **710,** 437 (1982).
233. J. Larrue, C. Leroux, D. Daret, and H. Bricaud, *Biochim. Biophys. Acta* **710,** 257 (1982).
234. W.-C. Chang, J. Nakao, H. Orimo, and S.-I. Murota, *Biochim. Biophys. Acta* **619,** 107 (1980).

235. W.-C. Chang, S.-I. Murota, J. Nakao, and H. Orimo, *Biochim. Biophys. Acta* **620**, 159 (1980).
236. J. Whiting, K. Salata, and J. M. Bailey, *Science* **210**, 664 (1980).
237. R. L. Tansik, D. H. Namm, and H. L. White, *Prostaglandins* **15**, 399 (1978).
238. J. Grünwald, W. Schäper, J. Mey, and W. H. Hauss, *Artery* **11**, 1 (1982).
239. J.-M. Dayer, S. M. Krane, R. G. G. Russell, and D. Robinson, *Proc. Natl. Acad. U.S.A.* **73**, 945 (1976).
240. S. B. Mizel, T.-M- Dayer, S. M. Krane, and S. E. Mergenhagen, *Proc. Natl. Acad. Sci. U.S.A.* **78**, 2474 (1981).
241. S. L. Hong, G. Patton, and D. Deykin, *Prostaglandins* **17**, 53 (1979).
242. S. Hammarström, *in* "Prostaglandins and Cancer" (T. J. Powles, R. S. Bockman, K. V. Honn, and P. Ramwell, eds.), pp. 297–307. Alan R. Liss, Inc., New York, 1982.
243. S. Hammarström, *Eur. J. Biochem.* **74**, 7 (1977).
244. M. E. Goldyne, J. A. Lindgren, H. E. Claesson, and S. Hammarström, *Prostaglandins* **19**, 155 (1980).
245. P. Roos, J. A. Lindgren, and S. Hammarström, *Eur. J. Biochem.* **108**, 279 (1980).
246. A. H. Tashjian, Jr., E. F. Voelkel, L. Levine, and P. Goldhaber, *J. Exp. Med.* **136**, 1329 (1972).
247. L. Levine, P. M. Hinkle, E. F. Voelkel, and A. H. Tashjian, Jr., *Biochem. Biophys. Res. Commun.* **47**, 888 (1972).
248. B. M. Bayer and M. A. Beaven, *Biochem. Pharmacol.* **28**, 441 (1979).
249. K. A. Chandrabose, P. Cuatrecasas, R. Pottathil, and D. J. Lang, *Science* **212**, 329 (1981).
250. Y. Honma, T. Kasukabe, M. Hozumi, and Y. Koshihara, *J. Cell. Physiol.* **104**, 349 (1980).
251. M. D. Bregman, D. Sander, and F. L. Meyskens, Jr., *Biochem. Biophys. Res. Commun.* **104**, 1080 (1982).
252. A. Erman, R. Azuri, and A. Raz, *Lipids* **17**, 119 (1982).
253. B. B. Weksler, C. W. Ley, and E. A. Jaffe, *J. Clin. Invest.* **62**, 923 (1978).
254. H. R. Knapp, O. Oelz, L. J. Roberts, B. J. Sweetman, J. A. Oates, and P. W. Reed, *Proc. Natl. Acad. Sci. U.S.A.* **74**, 4251 (1977).
255. L. Levine and I. Alam, *Prog. Lipid Res.* **20**, 81 (1981).
256. L. Levine, *in* "Prostaglandins and Cancer" (T. J. Powles, R. S. Bockman, K. V. Honn, and P. Ramwell, eds.), pp. 189–204. Alan R. Liss, Inc., New York, 1982.
257. G. Fürstenberger, H. Richter, N. E. Fusenig, and F. Marks, *Cancer Lett.* **11**, 191 (1981).
258. L. Levine, I. Alam, and J. J. Langone, *Prostaglandins Med.* **2**, 177 (1979).
259. L. Levine and A. Hassid, *Bichem. Biophys. Res. Commun.* **79**, 477 (1977).
260. D. J. Crutchley, L. B. Conanan, and J. R. Maynard, *Cancer Res.* **40**, 849 (1980).
261. W. C. Pickett, R. L. Jesse, and P. Cohen, *Biochim. Biophys. Acta* **486**, 209 (1977).
262. K. Ohuchi and L. Levine, *Biochim. Biophys. Acta* **619**, 11 (1980).
263. T. V. Zenser and B. B. Davis, *Am. J. Physiol.* **235**, F213 (1978).
264. P. Polgar, W. H. J. Douglas, L. Terracio, and L. Taylor, *Adv. Prostaglandin Thromboxane Res.* **6**, 225 (1980).
265. J. A. Salmon, D. R. Smith, R. J. Flower, S. Moncada, and J. R. Vane, *Biochim. Biophys. Acta* **523**, 250 (1978).
266. J. Turk, A. Wyche, and P. Needleman, *Biochem. Biophys. Res. Commun.* **95**, 1628 (1980).
267. A. Szczeklik and R. J. Gryglewski, *Artery* **7**, 488 (1980).
268. S. Hammarström, B. Samuelsson, and G. Bjursell, *Nature (London), New Biol.* **243**, 50 (1973).
269. G. J. Blackwell, R. Carnuccio, M. DiRosa, R. J. Flower, L. Parente, and P. Persico, *Nature (London)* **287**, 147 (1980).

270. F. Hirata, E. Schiffmann, K. Venkatasubramanian, D. Salomon, and J. Axelrod, *Proc. Natl. Acad. Sci. U.S.A.* **77,** 2533 (1980).
271. S. L. Hong and L. Levine, *Proc. Natl. Acad. Sci. U.S.A.* **73,** 1730 (1976).
272. S. Kitada, E. F. Hays, and J. F. Mead, *Prog. Lipid Res.* **20,** 823 (1981).
273. R. Wood, A. Zoeller, and M. Matocha, *Lipids* **17,** 771 (1982).
274. E. Rozengurt, A. Legg, and P. Pettican, *Proc. Natl. Acad. Sci. U.S.A.* **76,** 1284 (1979).
275. J. Vermylen, D. A. F. Chamone, and M. Verstraete, *Lancet* **1,** 518 (1979).
276. L. O. Carreras, D. A. F. Chamone, P. Klerckx, and J. Vermylen, *Thromb. Res.* **19,** 663 (1980).
277. K. V. Honn, J. Meyer, G. Neagos, T. Henderson, C. Westley, and V. Ratanatharathorn, *in* "Interaction of Platelets and Tumor Cells" (G. A. Jamieson, ed.), pp. 295–328. Alan R. Liss, Inc., New York, 1982.
278. K. V. Honn, *in* "Prostaglandins and Cancer" (T. J. Powles, R. S. Bockman, K. V. Honn, and P. Ramwell, eds.), pp. 733–752. Alan R. Liss, Inc., New York, 1982.
279. A. Sivak and B. L. Van Duurin, *Cancer Res.* **30,** 1203 (1970).
280. R. Süss, G. Kreibich, and V. Kinzel, *Eur. J. Cancer* **8,** 299 (1972).
281. L. Diamond, S. O'Brien, C. Donaldson, and Y. Shimizu, *Int. J. Cancer* **13,** 721 (1974).
282. L. D. Tomei, J. C. Cheney, and C. E. Wenner, *J. Cell. Physiol.* **107,** 385 (1981).
283. L. Jimenez de Asua, D. Clingan, and P. S. Rudland, *Proc. Natl. Acad. Sci. U.S.A.* **72,** 2724 (1975).
284. L. Jimenez de Asua, A. M. Otto, M.-O. Ulrich, J. Martin-Perez, and G. Thomas, *in* "Prostaglandins and Cancer" (T. J. Powles, R. S. Bockman, K. V. Honn, and P. Ramwell, eds.), pp. 309–331. Alan R. Liss, Inc., New York, 1982.
285. A. M. Otto, *in* "Prostaglandins and Cancer" (T. J. Powles, R. S. Bockman, K. V. Honn, and P. Ramwell, eds.), pp. 391–396. Alan R. Liss, Inc., New York, 1982.
286. W. J. Bettger and R. G. Ham, *Prog. Lipid Res.* **20,** 265 (1981).
287. R. A. Karmali, D. F. Horrobin, J. Menezes, and P. Patel, *Pharmacol. Res. Commun.* **11,** 69 (1979).
288. M. Adolphe, J. P. Giroud, J. Timsit, and P. Lechat, *C.R. Hebd. Seances Acad. Sci., Ser. D* **277,** 537 (1973).
289. G. S. Eisenbarth, D. K. Wellman, and H. E. Lebovitz, *Biochem. Biophys. Res. Commun.* **60,** 1302 (1974).
290. L. J. Kagen, D. T. Neigel, K. Collins, and H. J. Robinson, *In Vitro* **13,** 18 (1977).
291. K. V. Honn, J. R. Dunn, II, L. R. Morgan, M. Bienkowski, and L. J. Marnett, *Biochem. Biophys. Res. Commun.* **87,** 795 (1975).
292. W. A. Turner, D. R. Bennett, K. C. Thompson, J. D. Taylor, and K. V. Honn, *in* "Prostaglandins and Cancer" (T. J. Powles, R. S. Bockman, K. V. Honn, and P. Ramwell, eds.), pp. 365–368. Alan R. Liss, Inc., New York, 1982.
293. W. A. Turner, J. D. Taylor, and K. V. Honn, *in* "Prostaglandins and Cancer" (T. J. Powles, R. S. Bockman, K. V. Honn, and P. Ramwell, eds.) pp. 369–373. Alan R. Liss, Inc., New York, 1982.
294. M. G. Santoro and B. M. Jaffe, *in* "Prostaglandins and Cancer" (T. J. Powles, R. S. Bockman, K. V. Honn, and P. Ramwell, eds.), pp. 425–436. Alan R. Liss, Inc., New York, 1982.
295. J. A. Lindgren, H. E. Claesson, and S. Hammarström, *Exp. Cell Res.* **124,** 1 (1979).
296. D. T. Mayschak, E. Glass, S. Kacy, W. Boerwinkle, S. Barranco, W. Flye, and C. M. Townsend, Jr., *in* "Prostaglandins and Cancer" (T. J. Powles, R. S. Bockman, K. V. Honn, and P. Ramwell, eds.), pp. 385–389. Alan R. Liss, Inc., New York, 1982.
297. R. Ardaillou, J. Foidart, P. Mahieu, and J. Sraer, *in* "Prostaglandin Synthetase Inhibitors:

New Clinical Applications'' (P. Ramwell, ed.), pp. 315–322. Alan R. Liss, Inc., New York, 1980.

298. J. Kurland and M. A. S. Moore, *Exp. Hematol. (Copenhagen)* **5**, 357 (1977).

299. J. I. Kurland, R. S. Bockman, H. E. Broxmeyer, and M. A. S. Moore, *Science* **199**, 552 (1978).

300. L. M. Pelus, S. Saletan, R. Silver, and M. A. S. Moore, *Blood* **59**, 284 (1982).

301. L. M. Pelus, *J. Clin. Invest.* **70**, 568 (1982).

302. L. M. Pelus, *in* "Prostaglandins and Cancer" (T. J. Powles, R. S. Bockman, K. V. Honn, and P. Ramwell, eds), pp. 399–413, Alan R. Liss, Inc., New York, 1982.

303. R. S. Bockman, *in* "Prostaglandins and Cancer" (T. J. Powles, R. S. Bockman, K. V. Honn, and P. Ramwell, eds.), pp. 415–423. Alan R. Liss, Inc., New York, 1982.

304. M. Yaron, I. Yaron, D. Gurari-Rotman, M. Revel, H. R. Lindner, and U. Zor, *Nature (London)* **267**, 457 (1977).

305. R. Pottathil, K. A. Chandrabose, P. Cuatrecasas, and D. J. Lang, *Proc. Natl. Acad. Sci. U.S.A.* **77**, 5437 (1980).

306. F. A. Fitzpatrick and D. A. Stringfellow, *in* "Prostaglandins and Cancer" (T. J. Powles, R. S. Bockman, K. V. Honn, nd P. Ramwell, eds.), pp. 333–343. Alan. R. Liss, Inc., New York, 1982.

307. K. A. Chandrabose, R. Pottathil, E. C. Borden, R. Fox, and P. Cuatrecasas, *in* "Prostaglandins and Cancer" (T. J. Powles, R. S. Bockman, K. V. Honn, and P. Ramwell, eds.), pp. 345–364. Alan R. Liss, Inc., New York, 1982.

308. B. D. Goldstein, G. Witz, M. Amoruso, D. S. Stone, and W. Troll, *Cancer Lett.* **11**, 257 (1981).

309. J. E. Repine, J. G. White, C. C. Clawson, and B. M. Holmes, *J. Lab. Clin. Med.* **83**, 911 (1974).

310. W. E. M. Lands, P. R. LeTellier, L. H. Rome, and J. Y. Vanderhoek, *Adv. Biosci.* **9**, 15 (1973).

311. R. W. Egan, P. H. Gale, E. M. Baptista, and F. A. Kuehl, Jr., *Prog. Lipid Res.* **20**, 173 (1981)..

312. D. K. Klassen, P. R. Conkling, and A. L. Sagone, Jr., *Infect. Immun.* **35**, 818 (1982).

313. I. H. Pastan, G. S. Johnson, and W. B. Anderson, *Annu. Rev. Biochem.* **44**, 491 (1975).

314. M. F. Vesin, D. Leiber, and S. Harbon, *Prostaglandins* **24**, 851 (1982).

315. D. L. Friedman, *Physiol. Rev.* **56**, 652 (1976).

316. N. D. Goldberg and M. K. Haddox, *Annu. Rev. Biochem.* **46**, 823 (1977).

317. D. B. Glass, W. Frey, II, D. W. Carr, and N. D. Goldberg, *J. Biol. Chem.* **252**, 1279 (1977).

318. D. B. Glass, J. M. Gerrard, D. Townsend, D. W. Carr, J. G. White, and N. D. Goldberg, *J. Cyclic Nucleotide Res.* **3**, 37 (1977).

319. M. K. Haddox, J. H. Stephenson, M. E. Moser, and N. D. Goldberg, *J. Biol. Chem.* **253**, 3143 (1978).

320. J. A. Rillema, *Prostaglandins* **15**, 857 (1978).

321. G. Graff, J. H. Stephenson, R. R. Winget, and N. D. Goldberg, *Lipids* **14**, 212 (1979).

322. H. Hidaka and T. Asano, *Proc. Natl. Acad. Sci. U.S.A.* **74**, 3657 (1977).

323. C. K. Mittal and F. Murad, *Proc. Natl. Acad. Sci. U.S.A.* **74**, 4360 (1977).

324. C. K. Mittal, W. P. Arnold, and F. Murad, *J. Biol. Chem.* **253**, 1266 (1978).

325. K. D. Munkres and H. J. Colvin, *Mech. Ageing Dev.* **5**, 99 (1976).

326. E. T. Suen, P. C. K. Kwan, and Y. C. Clement-Cormier, *Mol. Pharmacol.* **22**, 595 (1982).

327. M. Samir Amer and W. E. Kreighbaum, *J. Pharm. Sci.* **64**, 1 (1975).

328. B. Weiss and R. Fertel, *Adv. Pharmacol. Chemother.* **14**, 189 (1977).

329. H. M. Evans and K. S. Bishop, *Science* **56**, 650 (1922).

330. K. E. Mason, *in* "Vitamin E" (L. J. Machlin, ed.), pp. 1–6. Dekker, New York, 1980.
331. A. Juhász-Schäffer, *Virchows Arch. Pathol. Anat. Physiol.* **281**, 35 (1931).
332. H. A. Mattill, JAMA, *J. Am. Med. Assoc.* **89**, 1505, (1927).
333. H. M. Evans and G. O. Burr, *JAMA, J. Am. Med. Assoc.* **89**, 1587 (1927).
334. F. Bernheim, M. L. C. Bernheim, M. L. C. Bernheim, and K. M. Wilbur, *J. Biol. Chem.* **174**, 257 (1948).
335. F. Bernheim, *Radiat. Res., Suppl.* **3**, 17 (1963).
336. C. W. Birky, Jr., *J. Exp. Zool.* **169**, 205 (1968).
337. E. S. Wurdak and J. J. Gilbert, *Cell Tissue Res.* **169**, 435 (1976).
338. P. A. Jones and J. J. Gilbert, *J. Exp. Zool.* **201**, 163 (1977).
339. J. J. Gilbert, *Am. Sci.* **68**, 636 (1980).
340. J. F. Mead, *in* "Free Radicals in Biology" (W. A. Pryor, ed.), Vol. 1, pp. 51–68. Academic Press, New York, 1976.
341. W. A. Pryor, *in* "Free Radicals in Biology" (W. A.Pryor, ed.), Vol. 1, pp. 1–49. Academic Press, New York, 1976.
342. I. Yamazaki, *in* "Free Radicals in Biology" (W. A. Pryor, ed.), Vol. 3, pp. 183–218. Academic Press, New York, 1977.
343. N. A. Porter, *in* "Free Radicals in Biology" (W. A. Pryor, ed.), Vol. 4, pp. 261–294. Academic Press, New York, 1980.
344. J. F. Mead, R. A. Stein, G.-S. Wu, A. Sevanian, and M. Gan-Elepano, *in* "Autoxidation in Food and Biological Systems" (M. G. Simic and M. Karel, eds.), pp. 413–428. Plenum, New York, 1979.
345. H. W. Gardner, *in* "Autoxidation in Food and Biological Systems" (M. G. Simic and M. Karel, eds.), pp. 447–504. Plenum, New York, 1979.
346. C. D. Funk and W. S. Powell, *Prostaglandins* **25**, 299 (1983).
347. H. O. Hultin, *in* "Autoxidation in Food and Biological Systems" (M. G. Simic and M. Karel, eds.), pp. 505–527. Plenum, New York, 1979.
348. J. D. George, G. M. Rosen, and E. J. Rauckman, *in* "Autoxidation in Food and Biological Systems" (M. G. Simic and M. Karel, eds.), pp. 541–562. Plenum, New York, 1979.
349. P. J. O'Brien, *in* "Autoxidation in Food and Biological Systems" (M. G. Simic and M. Karel, eds.), pp. 563–587. Plenum, New York, 1979.
350. J. A. Badwey and M. L. Karnovsky, *Annu. Rev. Biochem.* **49**, 695 (1980).
351. R. A. Sheldon and J. K. Kochi, "Metal-Catalyzed Oxidations of Organic Compounds," pp. 215–268. Academic Press, New York, 1981.
352. J. F. G. Vliegenthart and G. A. Veldink, *in* "Free Radicals in Biology" (W. A. Pryor, ed.), Vol. 5, pp. 29–64. Academic Press, New York, 1982.
353. T. Nakamura and I. Hishinuma, *in* "Tocopherol, Oxygen and Biomembranes," (C. de Duve and O. Hayaishi, eds.), pp. 95–108. Elsevier/North Holland Biomedical Press, Amsterdam, 1978.
354. L. K. Dahle, E. G. Hill, and R. T. Holman, *Arch. Biochem. Biophys.* **98**, 253 (1962).
355. W. A. Pryor, J. P. Stanley, and E. Blair, *Lipids* **11**, 370 (1976).
356. E. H. Oliw, J. A. Lawson, A. R. Brash, and J. A. Oates. *J. Biol. Chem.* **256**, 9924 (1981).
357. E. H. Oliw and J. A. Oates, *Prostaglandins* **22**, 863 (1981).
358. E. H. Oliw, P. Guengerich, and J. A. Oates, *J. Biol. Chem.* **257**, 3771 (1982).
359. R. L. Jones, P. J. Kerry, N. L. Poyser, and I. C. Walker, *Prostaglandins* **16**, 583 (1978).
360. R. W. Bryant and J. M. Bailey, *Prostaglandins* **17**, 9 (1979).
361. L. J. Marnett and M. J. Bienkowski, *Biochem. Biophys. Res. Commun.* **96**, 639 (1980).
362. H. A. Kontos, E. P. Wei, E. F. Ellis, W. D. Dietrich, and J. T. Povlishock, *Fed. Proc., Fed. Am. Soc. Exp. Biol.* **40**, 2326 (1981).

363. L. J. Marnett, *Life Sci.* **29**, 531 (1981).
364. W. C. Hubbard, A. J. Hough, Jr., A. R. Brash, J. T. Watson, and J. A. Oates, *Prostaglandins* **20**, 431 (1980).
365. F. H. Valone, R. Obrist, N. Tarlin, and R. C. Bast, Jr., *Cancer Res.* **43**, 197 (1983).
366. F. Bernheim, A. Ottolenghi, and K. M. Wilbur, *Radiat. Res.* **4**, 132 (1956).
367. N. Wolfson, K. M. Wilbur, and F. Bernheim, *Exp. Cell Res.* **10**, 556 (1956).
368. A. A. Barber and K. M. Wilbur, *Radiat. Res.* **10**, 167 (1959).
369. H. Zalkin and A. L. Tappel, *Arch. Biochem. Biophys.* **91**, 117 (1960).
370. H. J. Abramson, *J. Biol. Chem.* **178**, 179 (1949).
371. A. E. Kitabchi and R. H. Williams, *J. Biol. Chem.* **243**, 3248 (1968).
372. N. Oshino and B. Chance, *in* ''Biochemical and Medical Aspects of Active Oxygen'' (O. Hayaishi and K. Asada, eds.), pp. 191–207. University Park Press, Baltimore, Maryland, 1977.
373. M. J. Evans, J. D. Hackney, and R. F. Bils, *Aerosp. Med.* **40**, 1365 (1969).
374. D. H. Bowden and I. Y. R. Adamson, *Arch. Pathol.* **92**, 279 (1971).
375. W.H. Northway, Jr., R. Petriceks, and L. Shahinian, *Pediatrics* **50**, 67 (1972).
376. H. Witshi and S. Lock, *Toxicology* **9**, 137 (1978).
377. E. B. Burlakova, *Biofizika* **12**, 82 (1967).
378. B. T. Cole, *Proc. Soc. Exp. Biol. Med.* **93**, 290, (1956).
379. N. N. Glushchenko, S. V. Shestakova, and V. S. Danilov, *Biol. Nauki (Moscow)* **1**, 51 (1975).
380. T. J. Player, D. J. Mills, and A. A. Horton, *Biochem. Biophys. Res. Commun.* **78**, 1397 (1977).
381. A. K. Balin, D. B. P. Goodman, H. Rasmussen, and V. J. Cristofalo, *J. Cell Biol.* **78**, 390 (1978).
382. C. A. Apffel, *Prog. Exp. Tumor Res.* **22**, 317 (1978).
383. C. W. Shuster, *Proc. Soc. Exp. Biol. Med.* **90**, 423 (1955).
384. E. D. Lash, *Arch. Biochem.* **116**, 332 (1966).
385. K. Utsumi, N. Goto, Y. Kanemasa, T. Yoshioka, and T. Oda, *Physiol. Chem. Phys.* **3**, 467 (1971).
386. S. K. Donnan, *J. Biol. Chem.* **182**, 415 (1950).
387. E. H. Thiele and J. W. Huff, *Arch. Biochem.* **88**, 203 (1960).
388. T. Ohnishi, *Gann* **49**, 113 (1958).
389. E. A. Neifakh and V. E. Kagan, *Biokhimiya (Moscow)* **34**, 692 (1969).
390. E. A. Neifakh, *Biofizika* **16**, 560 (1971).
391. V. Z. Lankin and L. P. Micheeva, *Tr. Mosk. O-va. Ispyt. Priv.* **52**, 151 (1975).
392. W. A. Baumgartner, V. A. Hill, and E. T. Wright, *Am. J. Clin. Nutr.* **31**, 457 (1978).
393. K. Utsumi, G. Yamamoto, and K. Inaba, *Biochim. Biophys. Acta* **105**, 368 (1965).
394. T. J. Player, D. J. Mills, and A. A. Horton, *Biochem. Soc. Trans.* **5**, 1506 (1977).
395. R. Kato, G. Frontino, and P. Vassanelli, *Experientia* **19**, 30 (1963).
396. R. Rosso, M. G. Donelli, G. Franchi, and S. Garattini, *Eur. J. Cancer* **7**, 565 (1971).
397. H. Yoshida, *Kumamoto Med. J.* **24**, 55 (1971).
398. J. D. Khandekar, D. Dardachti, B. Tuchweber, and K. Kovacs, *Cancer* **29**, 738 (1972).
399. J. T. Wilson, *Endocrinology* **88**, 185 (1971).
400. T. K. Basu, D. V. Parke, and D. C. Williams, *Cytobios* **11**, 71 (1974).
401. G. M. Bartoli, T. Galeotti, G. Palombini, G. Parisi, and A. Azzi, *Arch. Biochem. Biophys.* **184**, 276 (1977).
402. T. Galeotti, G. M. Bartoli, S. Bartoli, and E. Bertoli, *in* ''Biological and Clinical Aspects of Superoxide and Superoxide Dismutase'' (W. H. Bannister and J. V. Bannister, eds.), Vol. 11B, pp. 106–117. Elsevier/North Holland Biomedical Press, Amsterdam, 1980.

403. J. Hilton and A. C. Sartorelli, *J. Biol. Chem.* **245,** 4187 (1970).
404. H. Hirsiger, F. R. Althaus, V. Giger, and U. A. Meyer, *in* "Microsomes, Drug Oxidations, and Chemical Carcinogenesis" (M. J. Coon, A. H. Conney, R. W. Estabrook, H. V. Gelboin, J. R. Gillette, and P. J. O'Brien, eds.), pp. 591–594. Academic Press, New York, 1980.
405. E. J. Hall, *in* "Chemical Modification: Radiation and Cytotoxic Drugs" (R. M. Sutherland, ed.), pp. 323–325. Pergamon, Oxford, 1982.
406. T. L. Phillips, T. H. Wasserman, J. Stetz, and L. W. Brady, *in* "Chemical Modification: Radiation and Cytotoxic Drugs" (R. M. Sutherland, ed.), pp. 327–334. Pergamon, Oxford, 1982.
407. I. J. Stratford, *in* "Chemical Modification: Radiation and Cytotoxic Drugs" (R. M. Sutherland, ed.), pp. 391–398. Pergamon, Oxford, 1982.
408. G. E. Adams, I. Ahmed, E. M. Fielden, P. O'Neill, and I. J. Stratford, *in* "Radiation Sensitizers" (L. W. Brady, ed.), pp. 33–38. Masson, New York, 1980.
409. L. W. Oberley, *in* "Superoxide Dismutase" (L. W. Oberley, ed.), Vol. II, pp. 127–165. CRC Press, Boca Raton, Florida, 1982.
410. H. B. White, Jr., *in* "Tumor Lipids: Biochemistry and Metabolism" (R. Wood, ed.), pp. 75–88. Am. Oil Chem. Soc., Champaign, Illinois, 1973.
411. R. C. Reitz, J. A. Thompson, and H. P. Morris, *Cancer Res.* **37,** 561 (1977).
412. M. Waite, B. Parce, R. Morton, C. Cunningham, and H. P. Morris, *Cancer Res.* **37,** 2092 (1977).
413. S. Ruggieri and A. Fallani, *in* "Tumor Lipids: Biochemistry and Metabolism" (R. Wood, ed.), pp. 89–110. Am. Oil Chem. Soc., Champaign, Illinois, 1973.
414. L. D. Bergelson and E. V. Dyatlovitskaya, *in* "Tumor Lipids: Biochemistry and Metabolism" (R. Wood, ed.), pp. 111–125. Am. Oil Chem. Soc., Champaign, Illinois, 1973.
415. R. P. Van Hoeven, P. Emmelot, J. H. Krol, and E. P. M. Oomen-Meulemans, *Biochim. Biophys. Acta* **380,** 1 (1975).
416. B. V. Howard, J. DeB. Butler, and J. M. Bailey, *in* "Tumor Lipids: Biochemistry and Metabolism" (R. Wood, ed.), pp. 200–214. Am. Oil Chem. Soc., Champaign, Illinois, 1973.
417. L. W. Daniel, L. S. Kucera, and M. Waite, *J. Biol. Chem.* **255,** 5697 (1981).
418. E. D. Wills, *in* "Microsomes, Drug Oxidations, and Chemical Carcinogenesis" (M. J. Coon, A. H. Conney, R. W. Estabrook, H. V. Gelboin, J. R. Gillette, and P. J. O'Brien, eds.), pp. 545–548. Academic Press, New York, 1980.
419. D. G. Cornwell, J. C. Geer, and R. V. Panganamala, *in* "Pharmacology of Lipid Transport and Atherosclerotic Processes" (E. J. Masoro, ed.), pp. 445–483. Pergamon, Oxford, 1975.
420. D. G. Cornwell and R. V. Panganamala, *Prog. Lipid Res.* **20,** 365 (1981).
421. R. V. Panganamala and D. G. Cornwell, *Ann. N.Y. Acad. Sci.* **393,** 376 (1982).
422. V. C. Gavino, J. S. Miller, S. O. Ikharebha, G. E. Milo, and D. G. Cornwell, *J. Lipid Res.* **22,** 763 (1981).
423. R. V. Panganamala, J. S. Miller, E. T. Gwebu, H. M. Sharma, and D. G. Cornwell, *Prostaglandins* **14,** 261 (1977).
424. A. L. Tappel, *in* "Free Radicals in Biology" (W. A. Pryor, ed.), Vol. 4, pp. 1–47. Academic Press, New York, 1980.
425. L. A. Witting, *in* "Free Radicals in Biology" (W. A. Pryor, ed.), Vol. 4, pp. 295–319. Academic Press, New York, 1980.
426. T. F. Slater, *Ciba Found. Symp.* [N.S.] **65,** 143–159 (1979).
427. D. B. Menzel, *in* "Free Radicals in Biology" (W. A. Pryor, ed.), Vol. 2, pp. 181–202. Academic Press, New York, 1976.
428. J. A. Lucy, *in* "Tocopherol, Oxygen and Biomembranes" (C. de Duve and O. Hayaishi, eds.), pp. 109–120. Elsevier/North Holland Biomedical Press, Amsterdam, 1978.

429. J. M. C. Gutteridge, *Res. Commun. Chem. Pathol. Pharmacol.* **22,** 563 (1978).
430. M. Mino and K. Sugita, *in* ''Tocopherol, Oxygen and Biomembranes'' (C. de Duve and O. Hayaishi, eds.), pp. 71–81. Elsevier/North Holland Biomedical Press, Amsterdam, 1978.
431. E. T. Gwebu, R. W. Trewyn, D. G. Cornwell, and R. V. Panganamala, *Res. Commun. Chem. Pathol. Pharmacol.* **28,** 361 (1980).
432. G. S. Patil and D. G. Cornwell, *J. Lipid Res.* **19,** 416 (1978).
433. O. H. Emerson, G. A. Emerson, and H. M. Evans, *J. Biol. Chem.* **131,** 409 (1939).
434. P. Karrer and A. Geiger, *Helv. Chim. Acta* **23,** 455 (1940).
435. J. B. Mackenzie, H. Rosenkrantz, S. Ulick, and A. T. Milhorat, *J. Biol. Chem.* **183,** 655 (1950).
436. T. Epstein and D. Gershon, *Mech. Ageing Dev.* **1,** 257 (1972).
437. A. C. Cox, G. H. R. Rao, J. M. Gerrard, and J. G. White, *Blood* **55,** 907 (1980).
438. G. H. R. Rao, C. A. Cox, J. M. Gerrard, and J. G. White, *Prog. Lipid Res.* **20,** 549 (1981).
439. J. A. Fee and J. S. Valentine, *in* ''Superoxide and Superoxide Dismutases'' (A. M. Michelson, J. M. McCord, and I. Fridovich, eds.), pp. 19–60. Academic Press, New York, 1977.
440. G. Powis and P. L. Appel, *Biochem. Pharmacol.* **29,** 2567 (1980).
441. B. A. Svingen and G. Powis, *Arch. Biochem. Biophys.* **209,** 119 (1981).
442. R. J. Flower, H. S. Cheung, and D. W. Cushman, *Prostaglandins* **4,** 325 (1973).
443. D. T. Downing, J. A. Barve, F. D. Gunsteon, F. R. Jacobsberg, and M. Lie ken Jie, *Biochim. Biophys. Acta* **280,** 343 (1972).
444. J. M. Goetz, H. Sprecher, D. G. Cornwell, and R. V. Panganamala, *Prostaglandins* **12,** 187 (1976).
445. K. Sekiya, H. Okuda, and S. Arichi, *Biochim. Biophys. Acta* **713,** 68 (1982).
446. P. Böck, R. Kramar, and M. Pavelka, ''Peroxisomes and Related Particles in Animal Tissues,'' pp. 31–44. Springer-Verlag, Berlin and New York, 1980.
447. H. Ishii, N. Fukumori, S. Horie, and T. Suga, *Biochim. Biophys. Acta* **617,** 1 (1980).
448. M. Bronfman, N. C. Inestrosa, and F. Leighton, *Biochem. Biophys. Res. Commun.* **88,** 1030 (1979).
449. T. J. Peters and C. de Duve, *Exp. Mol. Pathol.* **20,** 228 (1974).
450. H. Shio, M. G. Farguhar, and C. de Duve, *Am. J. Pathol.* **76,** 1 (1974).
451. B. M. Babior, J. T. Curnutte, and R. S. Kipnes, *J. Lab. Clin. Med.* **85,** 235 (1975).
452. M. S. Meltzer, R. W. Tucker, and A. C. Breuer, *Cell. Immunol.* **17,** 30 (1975).
453. R. Keller, *JNCI, J. Natl. Cancer Inst.* **56,** 369 (1976).
454. Z. Metzger, J. T. Hoffeld, and J. J. Oppenheim, *J. Immunol.* **124,** 983 (1980).
455. L. M. Corwin and J. Shloss, *J. Nutr.* **110,** 916 (1980).
456. L. M. Corwin, R. K. Gordon, and J. Shloss, *Scand. J. Immunol.* **14,** 565 (1981).
457. M. G. Goodman and W. O. Weigle, *J. Supramol. Struct.* **13,** 373 (1980).
458. M. Okuma and H. Uchino, *Blood* **54,** 1258 (1979).
459. A. I. Schafer, *N. Engl. J. Med.* **306,** 381 (1982).
460. R. O. Recknagel, E. A. Glende, Jr., and A. M. Hruszkewycz, *in* ''Free Radicals in Biology'' (W. A. Pryor, ed.), Vol. 3 pp. 97–132. Academic Press, New York, 1977.
461. W. A. Pryor, *Ann. N.Y. Acad. Sci.* **393,** 1 (1982).
462. P. J. Conroy, J. T. Nodes, T. F. Slater, and G. W. White, *Eur. J. Cancer* **11,** 231 (1975).
463. E. J. Goetzl, *Biochem. Biophys. Res. Commun.* **101,** 344 (1981).
464. J. M. Bailey, R. W. Bryant, C. E. Low, M. B. Pupillo, and J. Y. Vanderhoek, *Cell. Immunol.* **67,** 112 (1982).
465. A. D. Bankhurst, E. Hastain, J. S. Goodwin, and G. T. Peake, *J. Lab. Clin. Med.* **97,** 179 (1982).
466. M. E. Goldyne and J. D. Stobo, *Prostaglandins* **24,** 623 (1982).
467. M. I. Siegel, *Headache* **21,** 264 (1981).

468. L. W. Wattenberg, *JNCI, J. Natl. Cancer Inst.* **48,** 1425 (1972).
469. P. O. P. Ts'o, W. J. Caspary, and R. J. Lorentzen, *in* 'Free Radicals in Biology'' (W. A. Pryor, ed.), Vol. 3, pp. 251–303. Academic Press, New York, 1977.
470. L. W. Wattenberg, *J. Environ. Pathol. Toxicol.* **3,** 35 (1980).
471. P. B. McCay, M. King, L. E. Rikans, and J. V. Pitha, *J. Environ. Pathol. Toxicol.* **3,** 451 (1980).
472. T. J. Slaga, A. J. P. Klein-Szanto, L. L. Triplett, and L. P. Yotti, *Science* **213,** 1023 (1981).
473. H. Witschi, D. Williamson, and S. Lock, *JNCI, J. Natl. Cancer Inst.* **58,** 301 (1977).
474. C. Peraino, R. J. Fry, E. Staffeldt, and J. P. Christopher, *Food Cosmet. Toxicol.* **15,** 93 (1977).
475. F. A. Kuehl, Jr., R. W. Egan, and J. L. Humes, *Prog. Lipid Res.* **20,** 97 (1981).
476. R. V. Panganamala, H. M. Sharma, R. E. Heikkila, J. C. Geer, and D. G. Cornwell, *Prostaglandins* **11,** 599 (1976).
477. R. V. Panganamala, V. C. Gavino, and D. G. Cornwell, *Prostaglandins* **17,** 155 (1979).
478. C. Deby and G. Deby-Dupont, *in* "Biological and Clinical Aspects of Superoxide and Super-oxide Dismutase" (W. H. Bannister and J. V. Bannister, eds.), Vol. 11B, pp. 4–97. Elsevier/North Holland Biomedical Press, Amsterdam, 1980.
479. M. P. Carpenter, *Prog. Lipid Res.* **20,** 143 (1981).
480. C. Friend, W. Scher, J. G. Holland, and T. Sato, *Proc. Natl. Acad. Sci. U.S.A.* **68,** 378 (1971).
481. H. D. Preisler, G. Christoff, and E. Taylor, *Blood* **47,** 363 (1976).
482. M. J. Ashwood-Smith, *Ann. N.Y. Acad. Sci.* **243,** 246 (1975).
483. N. R. Brownlee, J. J. Huttner, R. V. Panganamala, and D. G. Cornwell, *J. Lipid Res.* **18,** 635 (1977).
484. J. S. Miller and D. G. Cornwell, *Cryobiology* **15,** 585 (1978).
485. R. W. Ruddon, "Cancer Biology," pp. 99–119. Oxford Univ. Press, London and New York, 1981.
486. R. W. Ruddon, "Cancer Biology," pp. 175–210. Oxford Univ. Press, London and New York, 1981.
487. M. Steiner, *in* "Interaction of Platelets and Tumor Cells" (G. A. Jamieson, ed.), pp. 383–403. Alan R. Liss Inc., New York, 1982.
488. L. R. Zacharski, *in* "Interaction of Platelets and Tumor Cells" (G. A. Jamieson, ed.), pp. 113–129. Alan R. Liss, Inc., New York, 1982.
489. K. V. Honn, B. Cicone, and A. Skoff, *Science* **212,** 1270 (1981).
490. K.-E. Blass, H.-U. Block, W. Forster, and K. Ponicke, *Br. J. Pharmacol.* **68,** 71 (1980).
491. N. Serneri, G. G. Masoti, L. Poggesi, G. Galanti, and A. Morettini, *Eur. J. Clin. Pharmacol.* **21,** 9 (1981).
492. P. Hilgard, *in* "Interaction of Platelets and Tumor Cells" (G. A. Jamieson, ed.), pp. 143–158. Alan R. Liss, Inc., New York, 1982.
493. R. Ross, *Arteriosclerosis* **1,** 293 (1981).
494. I. Nafstad, *Thromb. Res.* **5,** 25 (1974).
495. W. A. Yu, M. C. Yu, and P. A. Young, *Exp. Mol. Pathol.* **21,** 289 (1974).
496. A. W. Sedar, M. J. Silver, J. J. Kocsis, and J. B. Smith, *Atherosclerosis* **30,** 273 (1978).
497. H. Imai, N. T. Werthessen, V. Subramanyam, P. W. LeQuesne, A. H. Soloway, and M. Kanisawa, *Science* **207,** 651 (1980).
498. M. Iwakami, *Nagoya J. Med. Sci.* **28,** 50 (1965).
499. R. Ross, J. Glomset, B. Kariya, and L. Harker, *Proc. Natl. Acad. Sci. U.S.A.* **71,** 1207 (1974).
500. H. N. Antoniades and C. D. Scher, *Proc. Natl. Acad. Sci. U.S.A.* **74,** 1973 (1977).
501. H. N. Antoniades and A. J. Owen, *Annu. Rev. Med.* **33,** 445 (1982).

502. A. Dembínska-Kièc, W. Rücker, and P. S. Schönhöfer, *Atherosclerosis* **33**, 315 (1979).
503. A. Dembínska-Kièc, W. Rücker and P. S. Schönhöfer, *Naunyn-Schmiedeberg's Arch. Pharmacol.* **309**, 59 (1979).
504. W. Hollander, B. Kirkpartrick, J. Paddock, M. Colombo, S. Nagraj, and S. Prusty, *Exp. Mol. Pathol.* **30**, 55 (1979).
505. J. K. Koster, Jr., A. F. Tryka, P. H'Doubler, and J. J. Collins, Jr., *Artery* **9**, 405 (1981).
506. J. C. Geer, R. V. Panganamal, and D. G. Cornwell, *Atherosclerosis* **12**, 63 (1970).
507. I. Bojesen, *Lipids* **9**, 835 (1974).
508. K. Comai, S. J. Farber, and J. R. Paulsrud, *Lipids* **10**, 555 (1975).
509. A. Danon, M. Heimberg, and J. A. Oates. *Biochem. Biophys. Acta* **388**, 318 (1975).
510. C. J. F. Bottcher and C. M. Van Gent, *J. Atheroscler. Res.* **1**, 36 (1961).
511. J. L. Rabinowtiz and R. W. Riemenschneider, *Lab. Invest.* **12**, 549 (1963).
512. Y. Homma, H. Nakamura, and Y, Goto, *Jpn. Heart J.* **13**, 43 (1972).
513. J. S. Owen, R. A. Hutton, R. C. Day, K. R. Bruckdorfer, and N. McIntyre, *J. Lipid Res.* **22**, 423 (1981).
514. D. J. Falcone, D. P. Hajjar, and C. R. Minick, *Am. J. Pathol.* **99**, 81 (1980).
515. A. Eldor, D. J. Falcone, D. P. Hajjar, C. R. Minick, and B. B. Weksler, *Am. J. Pathol.* **107**, 186 (1982).

CHAPTER **5**

The Electron Spin Resonance Study of Free Radicals Formed during the Arachidonic Acid Cascade and Cooxidation of Xenobiotics by Prostaglandin Synthase

B. Kalyanaraman

National Biomedical ESR Center
Department of Radiology
Medical College of Wisconsin
Milwaukee, Wisconsin

and K. Sivarajah

Laboratory of Pulmonary Function and Toxicology
National Institute of Environmental Health Sciences
Research Triangle Park, North Carolina

FREE RADICALS IN BIOLOGY, VOL. VI
Copyright © 1984 by Academic Press, Inc.
All rights of reproduction in any form reserved.
ISBN 0-12-566506-7

I. INTRODUCTION

The arachidonic acid (AA) cascade, which is initiated by the prostaglandin (PG) cyclooxygenase and the lipoxygenase activity of the PG synthase multi-enzyme complex, consists of the transformation of AA to PGs, prostacyclins, thromboxanes, and leukotrienes [1]. Because these biologically active compounds are involved in various physiological and regulatory processes [1, 2], an understanding of the molecular events influencing the biosynthesis of these compounds is of vital importance [2]. The free-radical aspects of the chemical and biological synthesis of PGs and thromboxanes have been reviewed [3]. This chapter focuses on the experimental detection of free-radical intermediates produced during the metabolism of AA, PG endoperoxide (PGG_2), and other xenobiotics by using ESR spectroscopy. The chapter is divided into three parts, as follows:

1. The application of the ESR–spin trapping technique to the detection of carbon-centered and oxygen-centered radicals derived from AA, catalyzed by the PG cyclooxygenase component of PG synthase
2. The ESR study of the "hemoprotein-derived" free radical produced during the interaction of PGG_2 and other hydroperoxides with the PG hydroperoxidase component of PG synthase in crude microsomes
3. The ESR identification of free radicals produced during cooxidation of various xenobiotics by the PG hydroperoxidase component of PG synthase

Because several reviews of the application of ESR and ESR–spin trapping to the detection of free radicals in biological systems have already appeared in preceding volumes of this series [4–6] and elsewhere [7–9], the reader is referred to these for background information.

II. APPLICATION OF ELECTRON SPIN RESONANCE–SPIN TRAPPING TO THE DETECTION OF FREE RADICALS FORMED DURING THE TRANSFORMATION OF ARACHIDONIC ACID TO PROSTAGLANDIN ENDOPEROXIDE

The first isolatable intermediate during the AA cascade is PGG_2 [1, 2]. This transformation, which involves the introduction of 2 mol of oxygen to 1 mol of AA, is catalyzed by the PG cyclooxygenase component of PG synthase [1, 2] (Scheme 1). This specialized type of lipid peroxidation has long been presumed to form free-radical intermediates [10–13].

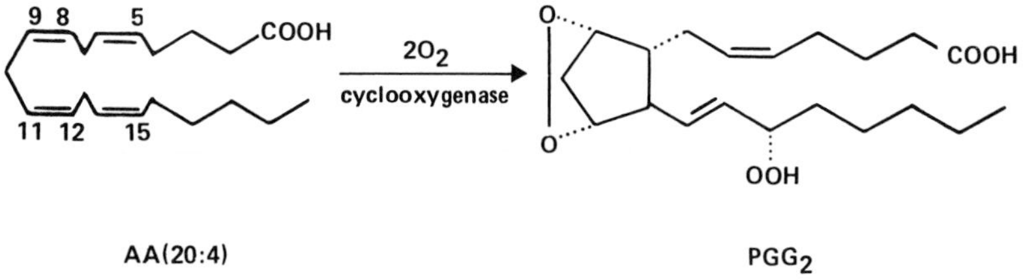

Scheme 1

A. Evidence for Free-Radical Intermediates

The free-radical nature of AA oxygenation is based on the following experimental observations:

1. Various free-radical scavengers (vitamin E, propyl gallate, etc.) inhibit PG production [11].

2. The bisoxygenation of AA is noticeably stimulated in the presence of sulfite, implying that the free radicals generated during AA oxygenation initiate sulfite oxidation, which is accompanied by oxygen uptake [10, 14] (for an alternative explanation of this effect due to cooxidation of sulfite see Section IV,C).

3. The oxidation of linoleic acid by lipoxygenase, an enzyme thought to be analogous to PG cyclooxygenase, is attributed to free-radical mediation [15, 16].

4. An ESR signal attributable to a paramagnetic intermediate is formed from incubations containing PG cyclooxygenase, AA, and air [11] (see also Section III,A).

B. Stereospecifically Controlled Free-Radical Reaction and Peroxyl Radical Cyclization Mechanism

The initial step during the transformation of AA to PGG_2 is the stereospecific removal of the 13-L hydrogen atom [13]. The radical thus formed apparently undergoes an allylic rearrangement, shifting the radical center to the C-11 position, followed by stereospecific addition of oxygen to form a C-11 AA peroxyl radical ($AAO_2 \cdot$). The cyclization of this peroxyl radical, followed by a second cyclization of the C-8 carbon-centered free radical, gives rise to a C-15 carbon-centered radical [3, 11]. This radical undergoes further oxygenation, resulting in a $AAO_4 \cdot$ radical that abstracts a hydrogen atom to form PGG_2 [17]. This proposed radical-mediated transformation of AA to PGG_2 is shown in Fig. 1. Although alternative mechanisms have been proposed [18], only the radical-mediated AA oxygenation is considered here.

C. Detection of a Carbon-Centered Arachidonate Free Radical Formed by Hydrogen Abstraction Catalyzed by Prostaglandin Cyclooxygenase: A Nitroso Spin Trapping Study

Although carbon-centered lipid free radicals have long been proposed to be reactive intermediates during lipoxygenase-catalyzed oxygenation of linoleic acid [15, 16] on the basis of inhibitor studies [15, 16] and other indirect methods [14], absolute spectral evidence has been lacking for quite some time. However, De Groot et al. [19] have clearly demonstrated the involvement of a carbon-centered free radical in the oxidation of linoleic acid by soybean lipoxygenase. These investigators used a water-soluble spin trap, 2-methyl-2-nitrosopropanol, to detect a rather reactive carbon-centered conjugated dienyl linoleic acid free radical.

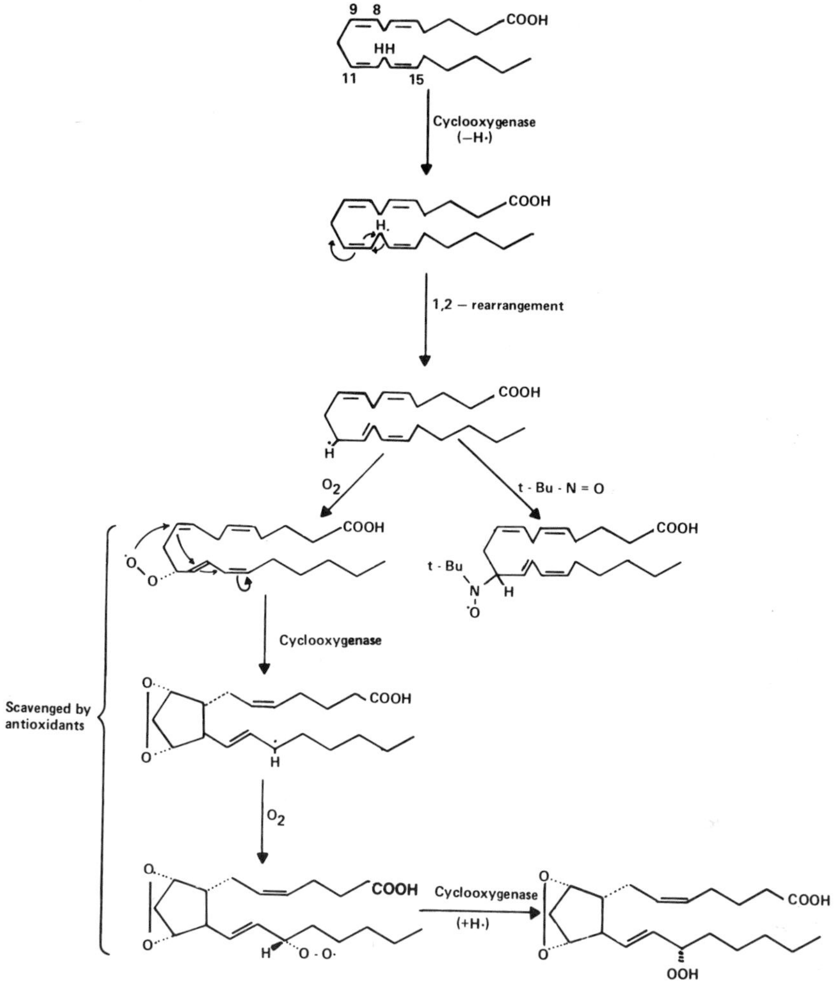

Fig. 1. Free-radical mechanism of transformation of AA to PGG$_2$ catalyzed by the PG cyclooxygenase.

Because the mechanism of soybean lipoxygenase is formally similar to that of PG synthase [12, 17], we used a similar spin trap, 2-methyl-2-nitrosopropane (MNP) to detect the initial free radical involved in the PG synthase-catalyzed conversion of AA to PGG$_2$ [20]. The ESR spectrum of the spin adduct obtained in an incubation mixture containing ram seminal vesicle (RSV) microsomes (a rather rich source of PG synthase), AA, and air clearly shows an unpaired electron interacting with the nitroxide nitrogen and the attached β-proton, as shown in Fig. 2A. The ESR parameters (a_N = 15.7 G; a_H = 2.5 G) are similar to

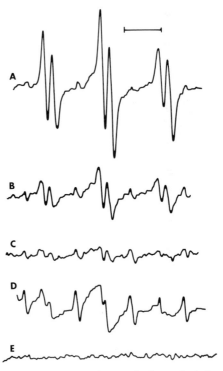

Fig. 2. Electron spin resonance spectra of prostaglandin synthesis incubations. (A) The ESR spectrum of the carbon-centered AA–MNP radical adduct obtained upon the addition of 400 μ*M* AA to an incubation mixture (0.25% ethanol) containing 2.0 mg/ml of RSV microsomal protein in a 1 mg/ml MNP solution of Tris buffer (pH 7.5); (B) ESR spectrum obtained from an incubation mixture (0.3% ethanol) like that described in (A) but that had been preincubated for 2 min with 100 μ*M* indomethacin; (C) ESR spectrum obtained from an incubation mixture (0.5% ethanol) like that described in (A) but that had been preincubated for 2 min with 400 μ*M* indomethacin; (D) ESR spectrum obtained from the incubation of 2.0 mg/ml of RSV microsomal protein in the MNP solution of (A); (E) baseline obtained upon mixing 400 μ*M* AA with a 1 mg/ml MNP solution of Tris buffer (pH 7.5, 0.25% ethanol).

the reported values of the spin adduct formed by trapping of the carbon-centered linoleic acid radical with MNP [*19*]. Indomethacin, a classic PG synthase inhibitor [*21*], considerably decreased the ESR signal intensity at a concentration of 100 μ*M* (Fig. 2B). No ESR signal was detected at a concentration of 400 μ*M* indomethacin (Fig. 2C). Concomitant with the appearance of the ESR signal in Fig. 2A, there is an increased rate of oxygen uptake upon the addition of AA to the incubation mixture containing RSV microsomes (control, Fig. 3). The observed decrease in ESR signal intensity in the presence of indomethacin (Figs. 2B and 2C) also corresponds to a lower incorporation of oxygen into AA (curves

- - - and —— in Fig. 3). These results imply that the trapped carbon-centered radical is responsible for AA oxygenation. In the absence of an enzyme, no ESR signal was observed (Fig. 2E). The presence of a small signal (Fig. 2D) from a mixture of microsomes and MNP alone corresponds to the incorporation of oxygen into endogenous fatty acids present in the microsomes (Fig. 3).

When octadeuterated AA (5,6,8,9,11,12,14,15-[^{2}H]eicosatetraenoic acid) was used with RSV and MNP, the six-line spectrum (Fig. 2A) collapsed to a three-line spectrum with the same nitrogen hyperfine coupling constant [22]. This proves that the β-hydrogen was responsible for the doublet hyperfine interaction seen in Fig. 2A.

Although AA has three methylenic carbon atoms, the hydrogen abstraction has been shown [20] to take place at the C-13 position, forming a carbon-centered radical followed by isomerization of the cis-11-double bond to a trans-12-double bond (Fig. 1). As shown in Fig. 1 the cyclization of the AAO$_2$· peroxyl radical can also generate two other carbon-centered radicals at positions C-8 and C-15 [3]. The spin trapping experiments using AA specifically deuterated at either C-11 and/or C-15 position(s) may be able to reveal the site of hydrogen abstraction.

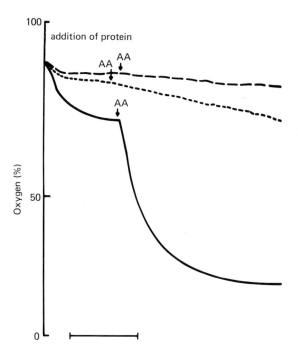

Fig. 3. Oxygen consumption tracings corresponding to Figs. 2A (—, after the addition of AA), 2B (- - -), 2C (- -), and 2D (—, before the addition of AA).

Similar radical-mediated schemes for PG cyclooxygenase-induced oxygenation of AA also have been proposed by other investigators [23, 24].

D. Chemical Verification of the Nitroso Spin Adduct

The chemical identity of the AA–MNP spin adduct obtained during oxygenation of AA was verified by independently synthesizing a similar radical adduct of MNP and AA [20]. After a prolonged incubation or at higher concentrations of AA and MNP, a chemical reaction between AA and MNP occurs, yielding an ESR spectrum very similar to that shown in Fig. 2A. The ESR spectrum of this species (a_N = 15.5 G; a_H = 2.0 G) appears under anaerobic or aerobic conditions and therefore cannot be due to the trapping of free radicals formed by AA autoxidation. Because alkenyl nitroxides have previously been synthesized by a pseudo-Diels–Alder reaction between an alkene and a C-nitroso compound [25, 26], a similar mechanistic scheme can be proposed to account for the chemical formation of the AA–MNP adduct (Scheme 2). For the sake of simplicity, the

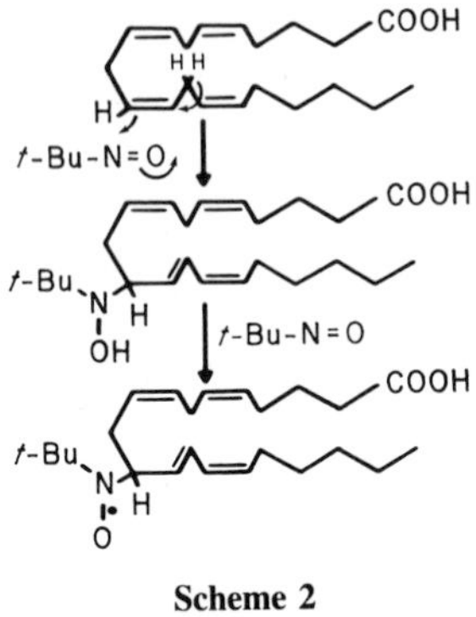

Scheme 2

chemical reaction shown in the scheme is not distinguished from the enzymatic reaction (Fig. 1). However, it is possible that the observed ESR spectrum of the chemical AA–MNP adduct is a composite of spectra of several such adducts.

E. Conflicting Views on the Mechanism of Spin Adduct Formation: Primary Hydrogen Abstraction versus Secondary Radical Addition

De Groot's mechanism [19] and our adapted version of it [20] and that of other investigators [17] (Fig. 1) implicate the abstraction of a methylenic hydrogen atom from the C-11 position of linoleic acid and the C-13 position of AA (Fig. 1). However, Evans [27] and Janzen [6] have suggested that the addition of a peroxyl or an alkoxyl radical across the double bond of linoleic acid could have led to the spin-trapped carbon-centered radical (radical addition to carbon at position 9, 10, 12, or 13 of linoleic acid). If this view is extended to the arachidonate system, it would imply that the trapped carbon-centered arachidonate radical is formed in a secondary rather than a primary oxygenation pathway. Therefore, further studies based on kinetic isotope effects are needed to resolve this question [9, 28].

F. Mechanistic Implications of the Spin Trapping of Lipid Arachidonate Radical: Inhibition of Prostaglandin Biosynthesis

As illustrated in Fig. 1 the presence of nitroso spin traps (radical scavengers) should inhibit the oxygenation of AA, as is indeed the case [20]. The carbon-centered linoleic acid radical intermediate has already been found to react competitively with either oxygen or nitrosobenzene [29], and it has also been reported that nitrosobenzene is a much stronger inhibitor of the oxygenation of AA than is MNP [20]. The fact that nitrosobenzene is a better inhibitor of oxygen uptake than is MNP is consistent with nitrosodurene reacting with secondary alkyl radicals 6·7 times faster than MNP [30, 31]. The natural consequence of trapping the primary arachidonate radical is thus a decreased formation of PGG_2 as well as of other PGs.

In this context it is worth pointing out that other xenobiotics that are not necessarily nitroso spin traps but that possess a nitroso functional group also could trap the lipid arachidonate radicals (Scheme 3). According to Scheme 3 the

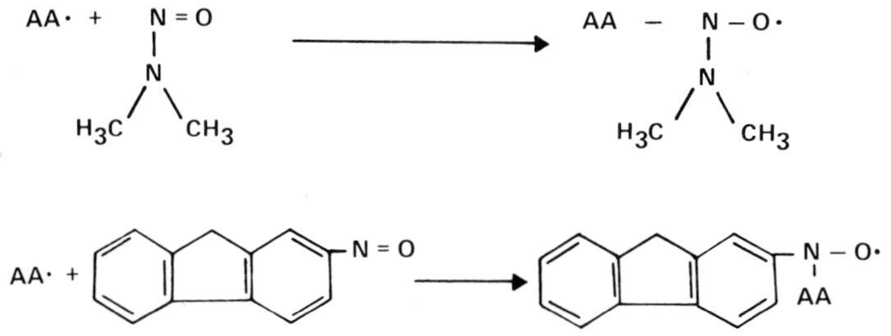

Scheme 3

carcinogens dimethylnitrosoamine and nitrosofluorene could inhibit AA oxygenation and therefore PG synthase merely by acting as "spin traps." However, before one arrives at such a conclusion, the resulting nitroxide must be spectroscopically identified because the nonspecific radical–radical recombination reaction [32] (i.e., between the nitroxide and the lipid radical) has also been reported to inhibit oxygenation during lipid peroxidation [33]. It is also of interest that the ESR spectrum of the lipid–nitrosofluorene spin adduct is well characterized [26]. It is possible that the environmental agents nitrogen dioxide and nitric oxide could also combine with arachidonate radical via a "radical recombination'' reaction, as shown in Scheme 4. Although the observed decrease in prostacyclin synthesis following NO_2 exposure has been related to the formation of lipid hydroperoxides [34], possibly via a radical-initiated major reaction pathway [35], a minor radical recombination reaction [35] cannot be entirely dismissed. The trapping reaction between the arachidonate radical and other xenobiotics may in turn result in an indirect regulation of PG biosynthesis.

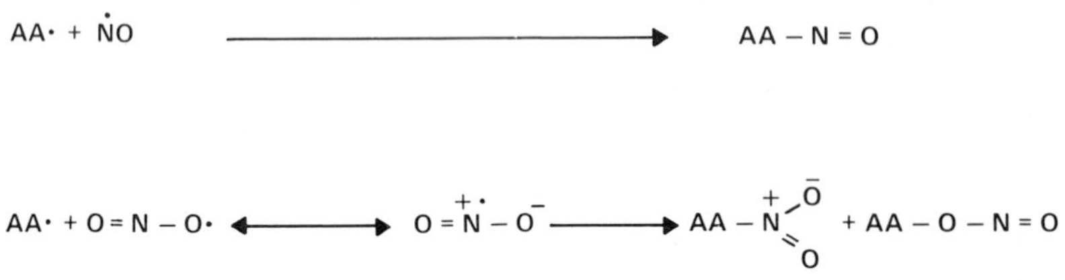

AA· + ṄO ⟶ AA – N = O

AA· + O = N – O· ⟷ O = Ṅ – O⁻ ⟶ AA – N⁺⟨O⁻ / O + AA – O – N = O

Scheme 4

G. Tentative Evidence against the Involvement of Prostaglandin Hydroperoxidase during the Conversion of Arachidonic Acid to Prostaglandin Endoperoxide

Two reports [36, 37] do not concur with the proposed pathway of AA oxygenation (Fig. 1). It is suggested (*a*) that the carbon-centered arachidonate lipid free radicals (AA·) are formed as a consequence of PG hydroperoxidase activity acting nonspecifically on AA and (*b*) that the 11-AA peroxyl radical shown in Fig. 1 is formed apparently as a result of interaction between PG hydroperoxidase and 11-hydroperoxyeicosatetraenoic acid (11-HPETE). It is desirable to design further experiments that will test the involvement of the PG cyclooxygenase component of PG synthase in the formation of lipid arachidonate radical (AA·). One such experiment would be to carry out an analogous spin trapping investigation using Mn(III)–protoporphyrin cyclooxygenase, which has no PG hydroperoxidase activity [38]. It was elegantly demonstrated that 11-HPETE cannot act as a substrate for PG cyclooxygenase, so that once 11-HPETE is formed it cannot be reincorporated into the PG biosynthetic pathway [39]. Thus, the proposed involvement of PG hydroperoxidase in the initial oxygenation of AA [36, 37] must be revised.

H. Possibilities of Spin Trapping the Carbon-Centered Lipid Arachidonate Radical Using Nitrones

In addition to nitroso compounds, nitrones, another well-known class of spin traps [40], can potentially compete with oxygen for trapping the AA· radical, as shown in Scheme 5 [40]. Previously, nitrones were shown to trap carbon-centered lipid dienyl free radicals produced during microsomal [41] and lipoxygenase-catalyzed lipid peroxidations [42]. However, it must be mentioned that the spin adduct in these studies was characterized on the basis of biochemical intuition rather than spectroscopic evidence. In the proposed spin trapping reaction shown in Scheme 5 both AA–α-phenyl-*tert*-butylnitrone (AA–PBN) and AA–5,5-dimethyl-1-pyrroline *N*-oxide (AA–DMPO) spin adducts will exhibit six-line ESR spectra with different hyperfine coupling constants for nitrogen and β-hydrogen [6, 9]. Because most PBN spin adducts and DMPO carbon-centered radical adducts exhibit six-line ESR spectra, one cannot unambiguously attribute a six-line ESR spectrum to a specific radical reaction as shown in the scheme.

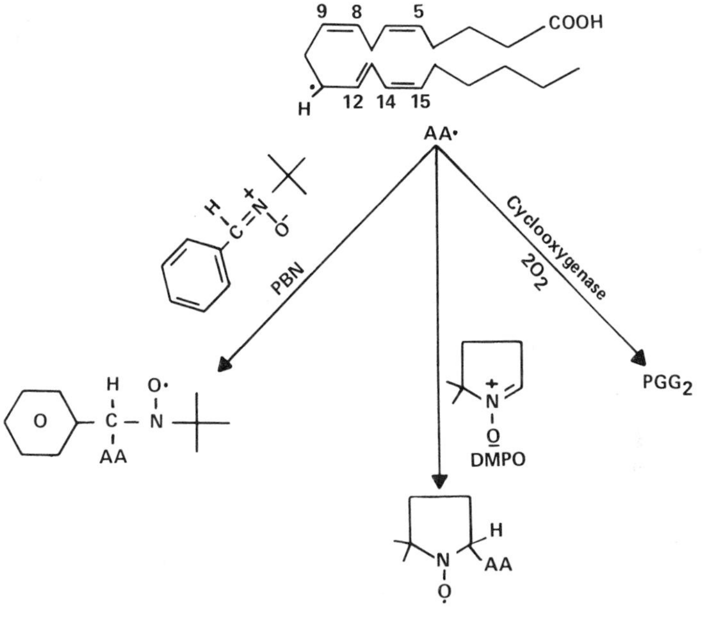

Scheme 5

However, if C-11 in AA is isotopically enriched with the ^{13}C isotope ($I = \frac{1}{2}$), then trapping of this carbon-centered radical, by both PBN and DMPO, should result in an ESR spectrum exhibiting hyperfine coupling from the attached ^{13}C nucleus. Such procedures have previously been employed in biochemical systems for the conclusive demonstration of the presence of reactive intermediates [43, 44]. As mentioned before, one of the natural consequences of trapping the arachidonate radical is the decreased biosynthesis of various PGs, which may also be observed during the PG synthase-catalyzed metabolism of the xenobiotics possessing a nitrone group [28].

I. Experimental Limitation in the Detection of Peroxyl Radicals Formed during Arachidonic Acid Oxygenation

As illustrated in Fig. 1 other key intermediates, which remain speculative thus far, in the conversion of AA to PGG_2 are the postulated 11-AA peroxyl and 15-AA peroxyl radicals. Due to the extremely high reactivity of primary and secondary peroxyl radicals, direct ESR detection of these species involves special experimental techniques such as continuous-flow or rapid freeze-quenching methods, as previously demonstrated in the detection of hydroperoxyl ($\cdot$OOH) and superoxide ($O_2{}^-$) radicals [45, 46]. Because peroxyl radicals have no hyperfine coupling except that due to the ^{17}O nucleus [47], an accurate determination of either the isotropic g value ($g_{iso} = 2.0134$) or axial g values ($g_\perp = 2.0092$ and $g\parallel = 2.0355$) should characterize them unambiguously [47].

Because ESR–spin trapping is yet another viable technique for the study of transient radicals [6, 9], potentialities of trapping the AA peroxyl radicals, using different spin traps, are discussed as follows (Scheme 6). Although nitroso spin

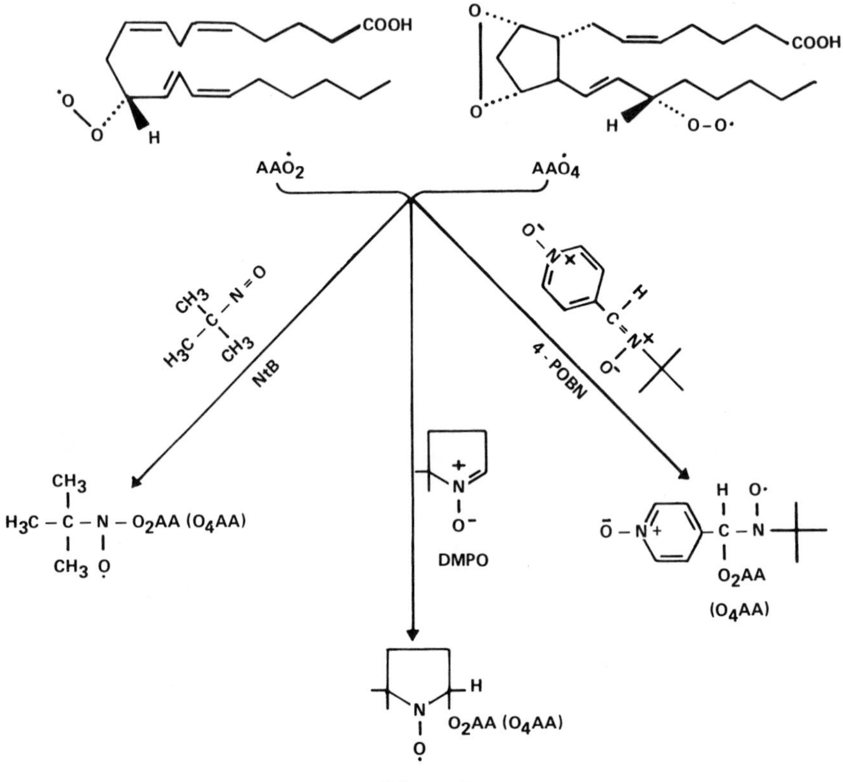

Scheme 6

traps react fairly rapidly with peroxyl radicals [6], MNP–peroxyl radical adducts are too unstable to be easily observed at room temperature [48], decomposing to give the *tert*-butyl radical, which in turn reacts with MNP to from di-*tert*-butyl nitroxide [48, 49]. In our spin trapping experiments, after about 30 min the spectrum shown in Fig. 2A became distorted by a second signal, which was shown to be that of di-*tert*-butyl nitroxide by direct comparison with the ESR spectrum of commercial di-*tert*-butyl nitroxide [20].

There have been reports of spin trapping of lipid peroxyl radicals with a nitrone, α-(4-pyridyl 1-oxide)-*N-tert*-butylnitrone (4-POBN) [50]. As mentioned earlier the ESR spectrum of the resulting 4-POBN–peroxyl adduct, if formed, is not uniquely different from other 4-POBN radical adducts. Also, there exist conflicting reports regarding the stability of such adducts at room temperature [9, 48]. Therefore, the use of phenyl or other structurally related nitrones to detect AA peroxyl radicals should be aided by experiments using ^{17}O-enriched oxygen, which would exhibit hyperfine coupling from the ^{17}O nucleus ($I = 5/2$; $a^{\beta}_{17O} = 2.9$ G) [48]. In contrast, the ESR spectrum of the DMPO–peroxyl adduct is unique and can be differentiated from most other DMPO spin adducts (Although the DMPO–superoxide adduct exhibits a similar spectrum, its sensitivity to superoxide dismutase can distinguish it from the DMPO–peroxyl adduct) [51, 52]. Because the rate constant between DMPO and peroxyl radicals is low, the resulting spin adduct can be detected only at high concentrations of spin trap [51]. Furthermore, this spin trapping reaction also could be hampered by radical cyclization reactions of some peroxyl radicals derived from AA [53].

J. Effect of Radical Scavengers (Antioxidants)

It has been suggested that antioxidants (or free-radical scavengers) regulate PG biosynthesis by modulating the production of free radicals [54]. The most commonly used antioxidants are α-tocopherol, propyl gallate, hydroquinone, simple and substituted phenols, and substituted aromatic amines. The AA peroxyl radicals can act as key propagating intermediates during the enzymatic cyclic transformation of AA to PGG_2 (Fig. 1). The AA peroxyl radical can either propagate the reaction by abstracting a hydrogen atom from enzyme-bound AA [24] or participate in the "rate-limiting termination" step by abstracting of the labile hydrogen atom from antioxidants (i.e., phenols or amines) as follows [55–57]:

$$ArOH \text{ or } ArNHR + AAO_2 \cdot \xrightarrow{k_{inh}} ArO \cdot \text{ or } ArNR \cdot + AAOOH \tag{1}$$

The inhibition rate constants (k_{inh}) for a list of antioxidants [58, 59] in chemical systems are given in Table I. At higher concentrations, nearly all the antioxidants listed in Table I would be inhibitory by participating in the rate-limiting termination reaction [Eq. (1)] and cause inhibition of PG synthesis [54]. The accumula-

TABLE I   Rate Constants for Reactions of Peroxyl Radicals with Antioxidants

Antioxidant	$k_{inh} \times 10^{-4}$ (mol^{-1} sec^{-1})
α-Tocopherol	$\sim$235 $\pm$ 50
2,6-Di-*tert*-butyl-4-methylphenol (BHT)	$\sim$1.2
2,6-Di-*tert*-butyl-4-methoxyphenol (BHA)	$\sim$7.8
N,N-Dimethylphenylenediamine	$\sim$250
Hydroquinone	$\sim$110
Phenol	$\sim$0.5

tion of di-*tert*-butyl nitroxide in Fig. 1, which was formally attributed to decomposition of the nitroso–peroxyl adduct, did not occur in the presence of phenol, probably because of the scavenging reaction shown in Eq. (1). However, at lower concentrations these antioxidants can act in a stimulatory fashion, by acting as hydrogen atom donors during the peroxidatic conversion of PGG$_2$ to PGH$_2$ (see Section IV). But Mn(III)–protoporphyrin cyclooxygenase, which has no peroxidase activity, was inhibited by phenol, even at lower concentrations, without showing any stimulation [24]. In addition to these well-known antioxidants (Table I), other compounds also can inhibit PG synthesis via the radical-scavenging reaction [Eq. (1)]. For instance, 2,2,6,6-tetramethyl-4-piperidonehydroxylamine is an efficient inhibitor of autoxidation ($k_{inh} = 5 \times 10^5 M^{-1}{}_{sec}{}^{-1}$) because of its capacity to donate a hydrogen atom [58]. In addition to these "indirect" effects of antioxidants (radical-scavenging mechanisms) in regulating the biosynthesis of PGs, these compounds also exert "direct" effects by interacting with the enzyme [60, 61].

K. Possible Extension of Spin Trapping to Lipoxygenase-Catalyzed Oxygenation of Arachidonic Acid

It has been proposed that endogenous lipid hydroperoxides may indirectly regulate PG biosynthesis, in addition to the direct regulation of PGs by "radical scavengers" [62]. The lipoxygenase component of PG synthase oxygenates AA to various hydroperoxides [63]. It is well established that soybean lipoxygenase could catalyze the oxygenation of AA to HPETE and hydroxyeicosatetraenoic acid (HETE) in a stereospecific manner [64]. The first step in this reaction is the stereospecific removal of a hydrogen atom attached to the C-13 position, implying the formation of the "pentadienyl" carbon-centered radical, which then undergoes stereospecific oxygenation at C-15 to yield 15-HPETE [64]. The transformation of AA to various biologically important HPETEs and HETEs has also been shown to be catalyzed by lipoxygenase enzymes [65–67]. For exam-

ple, a novel lipoxygenase present in human blood platelets was shown to oxygenate AA to 12-HPETE (and 12-HETE) [65, 66], whereas another lipoxygenase
in neutrophils transforms AA to 5-HPETE [67], an important intermediate leading to leukotriene synthesis (see Scheme 7). The major difference in these

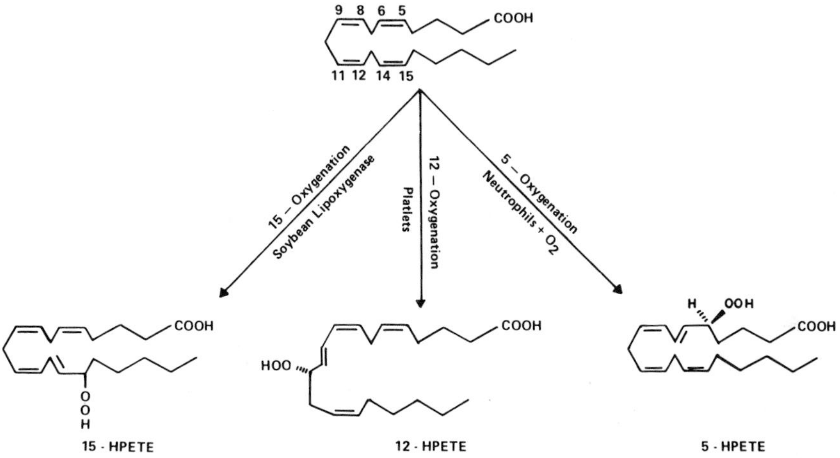

Scheme 7

oxygenations is the positional stereospecificity of the lipoxygenase enzyme [66].
The application of spin trapping has elucidated the role of lipid radicals in 9-
oxygenation as well as 9/13-oxygenations of linoleic acid by soybean [19] and
potato lipoxygenases [68]. The lipoxygenase enzyme derived from potato tubers
was found to oxygenate AA to 5-HETE and other products [69]. It is, perhaps,
possible to use nitroso spin trapping to understand the mechanistic pathway of
AA oxygenation in platelets and neutrophils [65–67]. The fact that indomethacin
inhibits PG cyclooxygenase but not the lipoxygenase can be used to differentiate
between the two similar reactions [65, 67].

III. APPLICATION OF ELECTRON SPIN RESONANCE
TO THE STUDY OF FREE RADICALS FORMED
DURING THE CONVERSION OF PROSTAGLANDIN G$_2$
TO PROSTAGLANDIN H$_2$

So far, we have discussed the critical regulatory role of free radicals derived
from AA during PGG$_2$ biosynthesis. In the following section efforts are made to
understand the free-radical mechanism of interaction of PGG$_2$ and other hydroperoxides with PG hydroperoxidase, an enzyme responsible for the reduction of

PGG$_2$ to PGH$_2$ [*38, 70*], as well as the oxidative intermediates responsible for the free-radical-mediated cooxidation of xenobiotics during the reduction of PGG$_2$ to PGH$_2$.

A. Direct Electron Spin Resonance Study of an Enzyme-Derived Free Radical (Ram Seminal Vesicle Free Radical) in the Microsomal Prostaglandin Synthase–Prostaglandin G$_2$ or ROOH System

By means of ESR spectroscopy, a free-radical signal was detected in incubations containing RSV microsomes and AA in air but not in a nitrogen atmosphere [*23*]; the same radical was detected in incubations containing microsomes and PGG$_2$, in both air and nitrogen atmospheres [*23*]. Both phenol and methional inhibited the ESR signal intensity [*23*]. Both phenol and methional inhibited the ESR signal intensity [*23*]. It was concluded that the free radical obtained during AA and PGG$_2$ metabolism by RSV is an oxygen-centered species and that this oxygen-centered free radical is released during the reduction of the 15-hydroperoxy moiety present in PGG$_2$ to the hydroxy moiety in PGH$_2$ [*23*].

The same ESR signal, with varying intensity, was also observed during the reduction of 15-hydroperoxy-PGE$_1$ to PGE$_1$, 15-HPETE to 15-HETE, and H$_2$O$_2$ to H$_2$O by the RSV microsomes [*71*].

We have reinvestigated the free radical formed during the reaction of RSV with PGG$_2$ and have concluded, from the known ESR parameters for the oxygen-centered free radicals, that the free radical formed by the interaction of hydroperoxide and RSV (henceforth referred to as RSV free radical) is neither a hydroxyl nor any known oxygen-centered radical [*72*]. The g value of the RSV free radical (Fig. 4) was measured to be 2.0054 and compared with the g values of other known oxygen-centered radicals (2.014–2.030) [*72*]. Because there is a considerable difference between the g value of RSV free radical and other oxygen-centered free radicals, it is clear that the RSV free radical is not a freely rotating oxygen-centered radical but is derived from RSV microsomes [*72*], presumably from their protein counterpart. A g value in the range of 2.004 to 2.005 suggests that an organic free radical is responsible for the ESR spectrum of this unknown RSV radical (Fig. 4). However, the absence of hyperfine structure precludes identification of its structure [*72*].

Because PG synthase is thought to be a hemoprotein, we have tried to compare the ESR parameters of the unknown RSV free radical with previous ESR data of free radicals formed by the reaction of H$_2$O$_2$ with methemoglobin [*73, 74*] and metmyoglobin [*75, 76*]. Comparison of the ESR parameters of the RSV free radical and the methemoglobin and metmyoglobin free radicals revealed significant differences in P_{max} and the peak-to-peak linewidth and identical g values of

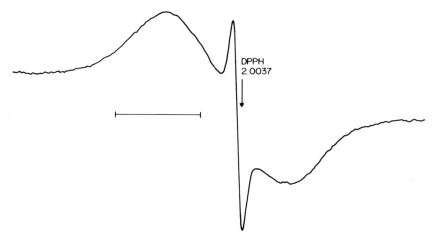

Fig. 4. The ESR spectrum obtained upon the addition of 400 μM (2.5 μl of a solution of 50 mg of AA in 1 ml of ethanol) to 0.5 ml of 30 mg/ml of microsomal protein in Tris buffer (pH 7.5). The incubation mixture was saturated with O_2 for 2 min before the addition of AA. A capillary containing DPPH was then inserted in the 3-mm cylindrical tube. The incubation mixture was freeze-quenched within 20 sec and kept in liquid N_2. The ESR spectrum was recorded at −196°C. The spectrometer conditions were as follows: gain, 1 × 10^4; modulation amplitude, 3.2 G; time constant, 1 sec; scan time, 8 min; microwave power, 60 mw.

the three hemoprotein free radicals, indicating that the RSV free radical is most likely to be derived from a hemoprotein [72]. Moreover, the fact that this RSV free radical saturates only at higher microwave powers indicates that RSV free radical may be formed by the oxidation of an amino acid(s) located near the iron of the heme [72].

B. Possible Mechanisms of Formation of Ram Seminal Vesicle Free Radical

The study of free radicals associated with a higher oxidation state of hydroperoxidases formed upon interaction with H_2O_2 and other hydroperoxides during the peroxidatic enzymatic cycle is an area of active research. Electron spin resonance studies have established that compound I of horseradish peroxidase (HRP) contains 2 oxidizing equivalents per mole as oxyferryl iron [Fe(IV)] and a porphyrin π-cation-radical [77], whereas the isoelectronic enzyme–substrate complex of cytochrome c peroxidase retains its 2 oxidizing equivalents in the form of a ferryl ion [Fe(IV)] and a free radical of an aromatic amino acid present in the protein group [78, 79]. Because the g value of the enzyme–substrate complex of cytochrome c peroxidase is close to that of the RSV free radical [72], it is possible that the RSV free radical is formed by the reaction of hydroperoxides, including H_2O_2, with the hemoprotein, as follows (also see Fig. 5):

$$E[X, Fe(III)] + ROOH \rightarrow E^*[X^+, Fe(IV)] + ROH \qquad (2)$$

The species X^+ may in fact be the RSV free radical, and it is very likely that X^+ is the ultimate site of oxidation of the apoprotein and not the primary free radical. However, before one can implicate the RSV free radical in the peroxidatic activity of PG synthase, it is important to determine the stoichiometry of the heme and RSV radical concentrations [72].

It is also conceivable that some hemoproteins other than PG synthase act as a substrate for the peroxidase activity of RSV, and the RSV free radical could merely be due to adventitious oxidation of a hemoprotein [80]. The fact that the purified PG cyclooxygenase–hydroperoxidase system does not exhibit an ESR signal during AA oxygenation or hydroperoxide interaction [80] is consistent with such a possibility (see Fig. 6). The optical spectroscopic identification of the higher oxidation intermediate(s), which in this case is thought to be analogous to compound I of HRP, has also been attempted [81].

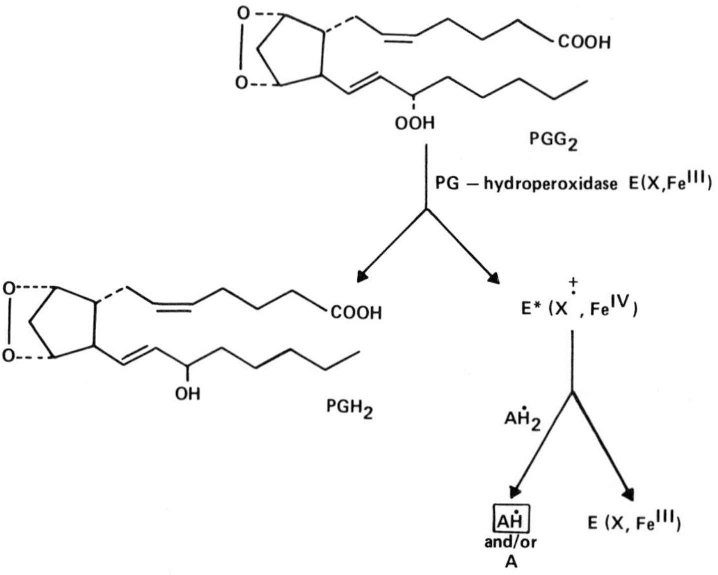

Fig. 5.   Hypothetical scheme showing the mechanism of cooxidation of xenobiotics; $A\dot{H}_2$, AH, and A denote the cooxidizable substrate, the one-electron oxidation intermediate, and the two-electron oxidation product or intermediate, respectively; E* refers to a higher oxidation intermediate possibly analogous to the enzyme–substrate complex of cytochrome c peroxidase; and X^+ is the RSV free radical, also possibly associated with the higher oxidation intermediate E*.

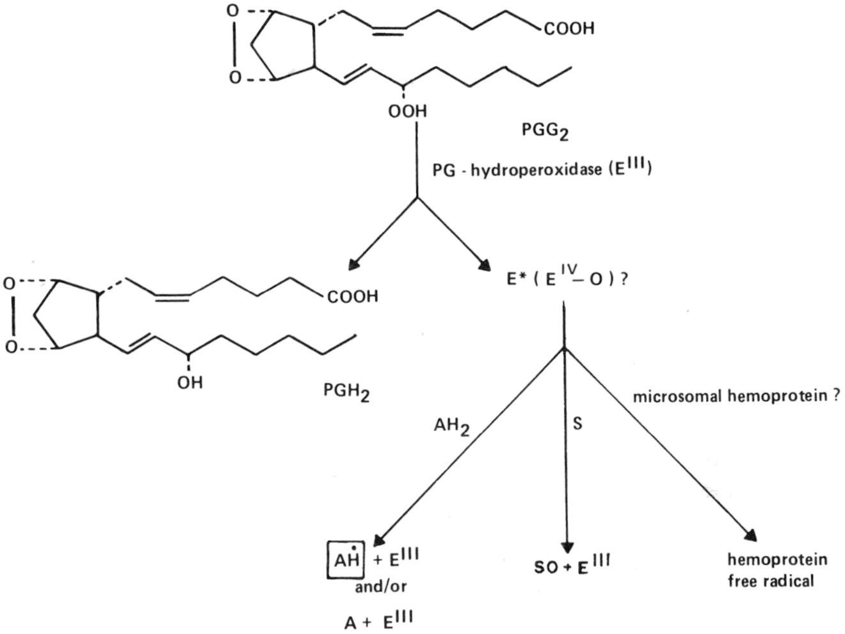

Fig. 6. Hypothetical schme showing the mechanism of cooxidation and cooxygenation of xenobiotics; AH_2, AH, and A are as described in Fig. 5. The E* refers to a higher oxidation intermediate possibly anlaogous to either the compound I of HRP or an oxy–heme complex; S is the substrate, which undergoes cooxygenation to form the product, So. The RSV free radical is possibly associated with the hemoprotein free radical formed by an adventitious oxidation of an endogenous hemoprotein present in the crude microsomes.

C. Implications in Cooxidation

During the PG hydroperoxidase-catalyzed reduction of PGG_2 to PGH_2, several xenobiotics undergo cooxygenation [82]. Efforts to identify the oxidant initiating the cooxidation of substrates have not been definitive. The involvement of superoxide anion as the oxidizing species appears to be unlikely, because superoxide dismutase had no effect either on the cooxidation reaction or on PG biosynthesis [82]. Nor does singlet oxygen appear to play a role in cooxidation [82, 83]. Inhibition studies of luminol chemiluminescence in vesicular gland incubations indicate the possible involvement of the hydroxyl radical [82, 84]. It also has been proposed that a free radical generated as a result of reaction of PG synthase with PGG_2 could be responsible for the cooxidation [82 85]. However, it should be borne in mind that the proposed mechanism(s) must account for cooxidations involving one- and/or two-electron transfer(s) as well as cooxygenations involving an oxygen atom transfer from hydroperoxide, molecular oxygen, or water.

It has been suggested that the mechanism of PG hydroperoxidase-catalyzed one-electron oxidation of substrates is similar to that of the HRP–H_2O_2 system [86], because the majority of compounds cooxidized by PG hydroperoxidase are known substrates of HRP [5]. Therefore, one can visualize that, in the presence of the one-electron donating peroxidase substrates, one electron is transferred from the substrate (AH_2) to the RSV free radical (Fig. 5), forming the AH· radical. This scheme shown in Fig. 5 also accounts for the two-electron oxidation (or dehydrogenation) of AH_2 to A. Indirect experimental support of the scheme shown in Fig. 5 comes from the observed decrease of RSV free-radical signal in the presence of phenol (a one-electron donor) [23] and iodide ion (a two-electron donor) [87]. Because the formation of free-radical intermediates is often predictive of the one-electron-mediated oxidative mechanisms [5], except in a few cases, the ESR technique can be used to study the one-electron-mediated cooxidation of substrates.

Because the scheme shown in Fig. 5 does not explain the incorporation of an oxygen atom from hydroperoxide into a substrate, another scheme involving a modified higher oxidation state (an oxene complex), which is similar to that of "peroxygenases" [80, 88, 89], is proposed (Fig. 6). This scheme explains not only the electron and oxygen atom transfer, but also the fact that the RSV free radical signal is observed in crude microsomal preparations of PG synthase [80].

In the next section, based on these two working models (Fig. 5 and 6) the free-radical mechanism of cooxidation of xenobiotics undergoing one-electron oxidations during the PG hydroperoxidase-catalyzed reduction of PGG_2 to PHG_2 will be discussed. The substrate AH_2 in Fig. 5 and 6 may simply be substituted by the individual substrate" described in Section IV.

It must be said that these two schemes are very hypothetical, and further spectroscopic characterization of the "higher oxidation" intermediate(s) is sorely needed. Although oxidants such as the hydroxyl or an alkoxy radical have been proposed to participate in both the anaerobic metabolism of hydroperoxides by lipoxygenases [90] and the oxidation of some substrates by PG cyclooxygenase and AA [91, 92], it is unlikely that they are responsible for the metabolism of xenobiotics observed during the reduction of PGG_2 to PGH_2.

IV. APPLICATION OF ELECTRON SPIN RESONANCE TO THE STUDY OF FREE-RADICAL-MEDIATED COOXIDATION OF XENOBIOTICS BY PROSTAGLANDIN HYDROPEROXIDASE

A number of structurally different substrates have been shown to undergo cooxidation during the PG hydroperoxidase-catalyzed reduction of PGG_2 to

PGH$_2$ [93]. The cooxidation has been studied by using one or more of the following experimental techniques: (a) polarographic measurement of the increase in oxygen consumption by stimulation of PG synthase, (b) measurement of covalent binding of reactive intermediates to cellular macromolecules, (c) measurement of the stimulation in the conversion of PGG$_2$ to PHG$_2$, (d) chemiluminescence, (e) fluorescence and mass and visible spectroscopy, (f) chromatographic method, (g) measurement of formaldehyde via N-demethylation, and (h) ESR. Of these techniques, ESR is the only physical technique by which one can unambiguously detect free-radical intermediates formed during cooxidation. The identification of free-radical intermediates is often predictive of a one-electron oxidation mechanism, although in some cases when there is an equilibrium between the parent compound, the two-electron oxidation product and the free-radical ESR cannot be used to differentiate between the one- and two-electron-mediated cooxidation [5]. In the following sections the potential application of ESR to the cooxidative metabolism of different classes of compounds is discussed.

A. Electron Spin Resonance Detection of Cation-Radicals from Aminopyrene, Aromatic Amines, Hydrazines, and Benzidines: Possible Intermediates during N-Demethylation

1. Aminopyrene

The facile stoichiometric one-electron oxidation of aminopyrene (AP) to its radical-cation during the N-demethylation pathway has been demonstrated in the HRP–H$_2$O$_2$ system [94]. Because the peroxidatic activity of the HRP–H$_2$O$_2$ system often models that of mammalian peroxidases [28], it was felt that AP could act as an electron donor during PG hydroperoxidase-catalyzed conversion of PGG$_2$ to PGH$_2$ [95]. The multiline ESR spectrum obtained from incubations containing AP, RSV microsomes, and AA, was found under aerobic conditions to be similar to the ESR spectrum of the AP radical-cation obtained either during controlled electrochemical oxidation [96] or during other enzymatic one-electron oxidations [97]. Inhibitors of PG synthase (indomethacin and flufenamic acid) caused a reduction in the ESR signal intensity of the AP cation-radical. An identical ESR spectrum was obtained when either *tert*-butyl hydroperoxide or 15-HPETE was substituted for AA and air. This observation suggested the involvement of a PG hydroperoxidase component of PG synthase.

Because it has previously been established that the AP radical-cation exhibits an absorption maximum at 580 nm [94], the rate of free-radical formation was conveniently monitored by visible spectroscopy [95]. It has also been shown that the AP radical-cation is an obligatory intermediate during HRP-catalyzed N-demethylation of AP to formaldehyde [94]. This is found to be the case even for the PG hydroperoxidase-catalyzed oxidation of AP [95]. A good correlation exists between the rates of AP cation-radical and formaldehyde formation as a function of both AA and AP concentrations [95]. This implies that the metabolism of AP to formaldehyde is free-radical-mediated. Moreover, the maximum steady-state concentrations of the AP cation-radical, measured at 580 nm, is linearly related to the square root of enzyme concentration. This square root relationship between the enzyme concentration and the AP radical-cation concentration indicates that AP radical-cation decays via either disproportion or dimerization [95], as shown in Scheme 8. It can be seen that another important

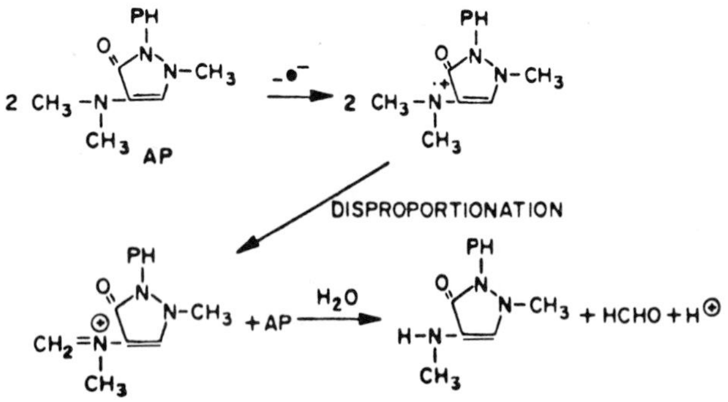

Scheme 8

intermediate, namely, an iminium cation (the two-electron oxidation product), is formed from disproprotination of the radical-cation rather than through subsequent one-electron oxidation of AP radical-cation by PG hydroperoxidase. Furthermore, this free-radical mechanism of N-dealkylation also predicts that the oxygen atom present in formaldehyde is derived from water rather than from oxygen or hydroperoxide [95].

Although, as mentioned earlier, the optical study conveniently measures the rate of formation of AP radical-cation, due to fortuitious spectral characteristics of the radical-cation it is highly desirable that parallel ESR studies be carried out. However, the transient nature of the AP cation-radical at physiological pH would hamper such a study. Nonetheless, becaue most cation-radicals are fairly stable at acidic pH [98], the lifetime of the AP radical-cation can be considerably enhanced by rapidly decreasing the pH of the mixture, after initiating the metabolism at physiological pH values. In such preliminary "pH jump" experiments, in which an identical incubation mixture is used, as previously described [95], the AP radical-cation can persist for up to several minutes [99]. This technique could be usefully employed in ESR studies of other transient radical-cations.

2. Aromatic Amines

Sivarajah *et al.* have shown that microsomes derived from RSV, lung, and kidney medulla can catalyze the N-demethylation of several substituted and unsubstituted mono- and dimethylanilines to formaldehyde when incubated with AA and air [*100*]. The fact that indomethacin inhibited the production of formaldehyde indicated that the hydroperoxide, needed for the metabolism, was catalyzed by PG cyclooxygenase [*21*]. The occurrence of N-demethylation in the presence of lipid hydroperoxides under anaerobic conditions suggested the involvement of PG hydroperoxidase. Although no definite structure–activity relationship could be established from the N-demethylation data, the following aspects were highlighted: (*a*) The ring substitution of both mono- and dimethylanilines with a nitro group inhibited N-demethylation the most; (*b*) the rates of N-demethylation from both unsubstituted mono- and dimethylanilines were greater than those for the corresponding substituted compounds.

Because the PG hydroperoxidase component of PG synthase is proposed to catalyze N-demethylation reactions [*100*], it is quite reasonable to adapt the free-radical-mediated mechanistic schemes proposed for N-demethylations of substituted *N,N'*-dimethylanilines to analogous hemoprotein–hydroperoxide systems [*101*]. In Scheme 9 an alkylamine undergoes two sequential one-electron

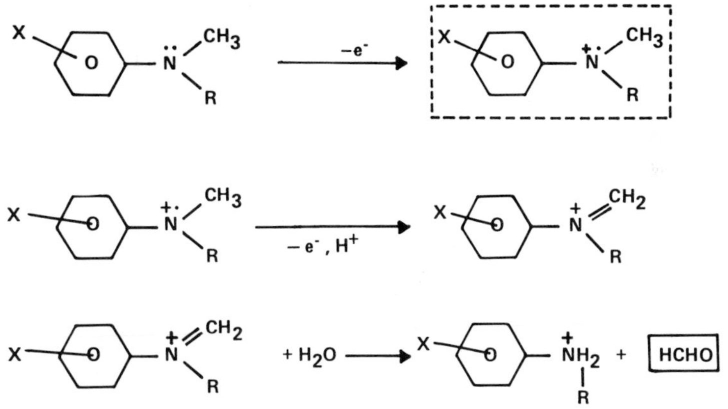

Scheme 9

oxidations, forming the radical and an iminium cation, which is subsequently hydrolyzed to give formaldehyde [*94, 101*]. However, before applying this scheme to the PG synthase-catalyzed N-demethylation of *N*-methylanilines, it is necessary to show that the radical-cation is an intermediate. As mentioned in the previous section these radical-cations (which are not resonance-stabilized) are somewhat persistent only at acidic pH, so that their detection by static ESR, at physiological pH values, is very difficult. The ESR spectra of primary radical-cations of some substituted N,N'-dimethylanilines have been obtained in the HRP–H_2O_2 system, but only under acidic conditions [102].

Two groups of workers [102–104] have attempted to explain the differing rates of N-demethylation found in the HRP–H_2O_2 system by considering factors affecting the oxidation potential of these N-alkylamines, as well as the lifetimes and fates of the corresponding radical-cations [102, 104]. Included among these factors are the electronic, steric, and lipophilic aspects of substituents on the aromatic ring, as well as on the amino nitrogen [103, 104]. The inhibitory effect of a nitro group on the PG synthese-catalyzed N-demethylation of nitro-substituted N,N'-dimethyl- or monomethylanilines could therefore be attributed to its electron-withdrawing effect (negative Hammett factor), lowering the ease of oxidation of the amine to its radical-cation in additon to destabilizing the radical-cation [100]. Another factor influencing the rate of PG synthase-catalyzed N-demethylation may be the nonspecific effect of substituted N-alkylanilines that is exerted during initial AA oxygenation producing the hydroperoxide necessary to drive the cooxidation. For example, N-methyl-N-nitrosoaniline, a rather poor substrate for N-demethylation, may in fact inhibit AA oxygenation by merely acting as a ''spin trap'' (see Section II,F).

Another factor governing the lifetime of these anilinium radical-cations is the ''tail-to-tail'' dimerization reaction [105, 106], as shown in Scheme 10. In this

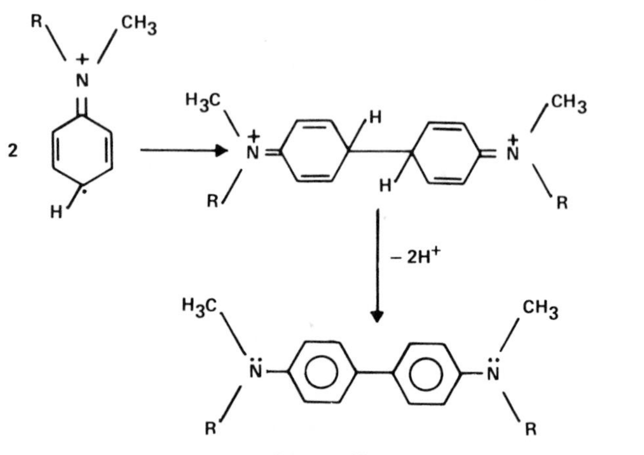

Scheme 10

scheme the one-electron oxidation product, the N,N-dimethylanilinium radical-cation (R = CH$_3$, Scheme 10), undergoes tail-to-tail dimerization to give the corresponding benzidine. The ESR spectrum of N,N,N',N'-tetramethylbenzidine cation-radical (see Section IV,A,4) has been observed during the HRP-catalyzed N-demethylation of N,N-dimethylaniline [101]. This dimerization reaction has been suggested to compete with the N-demethylation reaction [102], so the faster rate of N-demethylation observed during HRP-catalyzed oxidation of p-methyltoluidine as compared with ortho- and meta-isomers, was due partly to the absence of such a dimerizaton reaction [102] for p-methyltoluidine [102]. It is not clear why a similar trend is not observed with PG synthase-catalyzed N-demethylation of ortho-, para-, and meta-substituted monomethylanilines [100].

Studies on HRP-catalyzed N-demethylation of N,N-dimethyl-p-toluidine have demonstrated that more formaldehyde is formed in air than under nitrogen [107]. It was postulated that molecular oxygen increased the rate of N-demethylation by accepting an electron from the anilinium radical-cation to form superoxide [107, 108]. The possibility of such an activation of molecular oxygen during cooxidation of methyl-substituted anilines should also be tested by monitoring oxygen consumption during cooxidation of N,N-dimethyl-p-toluidine, PG hydroperoxidase, and a lipid hydroperoxide (15-HPETE). In any case, the fact that formaldehyde is also produced during the cooxidation of N,N-dimethylanilines by PG hydroperoxidase and a lipid hydroperoxide under anaerobic conditions clearly rules out the incorporation of an oxygen atom from molecular oxygen into formaldehyde.

As mentioned before, the detection of a primary radical-cation of methyl-substituted anilines during PG hydroperoxidase-catalyzed N-demethylation is rather crucial. The stability of an anilinium radical-cation can be increased through resonance stabilization by suitable substituents either on the ring or on the amino nitrogen [109]. Perhaps PG synthase-catalzyed N-demethylation studies using substrates such as N-methyl-di-p-anisylamine (R = C$_6$H$_5$OCH$_3$; X = p-OCH$_3$), N-methyl-di-p-tolylamine (R = C$_6$H$_5$OCH$_3$; X = p-CH$_3$), N-methyl-N-phenyl-p-anisylamine (R = C$_6$H$_5$; X = OCH$_3$), and N,N-dimethyl-p-anisidine (R = CH$_3$; X = OCH$_3$) would make it possible to identify the corresponding radical-cation by ESR at physiological pH as well as the correlation of radical-cation with formaldehyde formation. (For the structures of these substrates, R and X should be substituted correspondingly in Scheme 9.)

In addition to N-substituted mono-and dimethylanilines, the N-methyl-substituted p-phenylenediamines also undergo cooxidation [100, 109a]. Sivarajah et al. have measured formaldehyde during PG synthase-catalyzed oxidation of N,N-dimethyl- and N-monomethyl-p-phenylenediamines [100]. Egan et al. [80] have suggested a possible one-electron-mediated cooxidation of N,N-dimethyl-p-phenylenediamines during the reduction of PGG$_2$ to PGH$_2$. The cooxidation of N,N,N',N'-tetramethyl-p-phenylenediamine to a colored intermediate, Wurter's blue, is used in the measurement of PG hydroperoxidase activity [110]. However, before postulating a mechanism of cooxidation for N-methyl-substituted p-phenylenediamines, it is necessary to consider relevant oxidation–reduction schemes pertaining to the chemistry of p-phenylenediamines.

This class of compounds undergoes sequential one-electron oxidation to semi-quinoneimine and quinoneimine in chemical systems, as shown in Scheme 11.

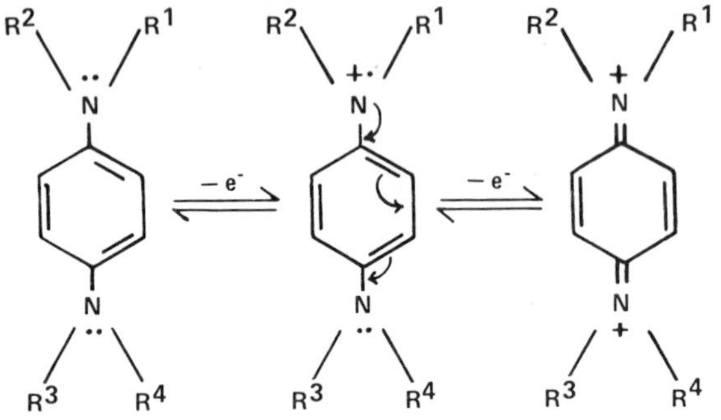

Scheme 11

The one-electron oxidation intermediate, the semiquinoneimine, is analogous to a semiquinone, and its chemical stability results from an increased resonance stabilization [98, 109]. The aminium radicals of p-phenylenediamine are further stabilized by increasing the number of N-methyl groups, due to both inductive and steric effects [98, 109] (Scheme 12). The radical-cations of N,N,N',N'-

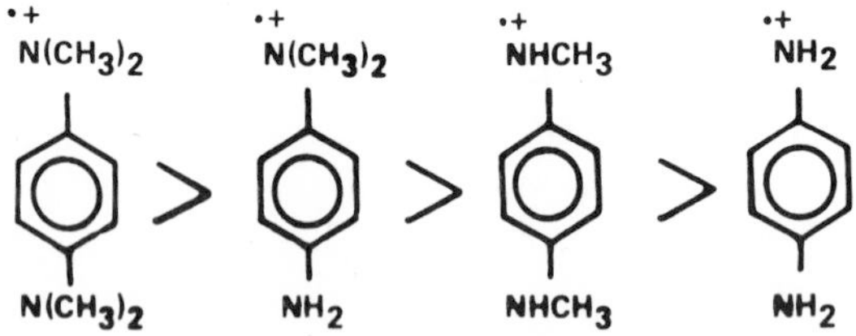

Scheme 12

tetramethyl-p-phenylenediamine ($R^1 = R^2 = R^3 = R^4 = CH^3$, Wurster's blue, Scheme 11) and N,N-dimethyl-p-phenylenediamine ($R^1 = R^2 = CH_3$; $R^3 = R_4$ = H, Wurster's red, Scheme 11) exist as stable paramagnetic salts [109]. The stability of Wurster's salts can be explained in terms of an electron delocalization [109] (Scheme 13). The ESR spectra of these stable radical-cations are well known [98, 109]. For example, the ESR spectrum of Wurster's blue radical-cation is characterized by coupling to two equivalent nitrogens, four equivalent methyl protons, and four equivalent ring hydrogens [109].

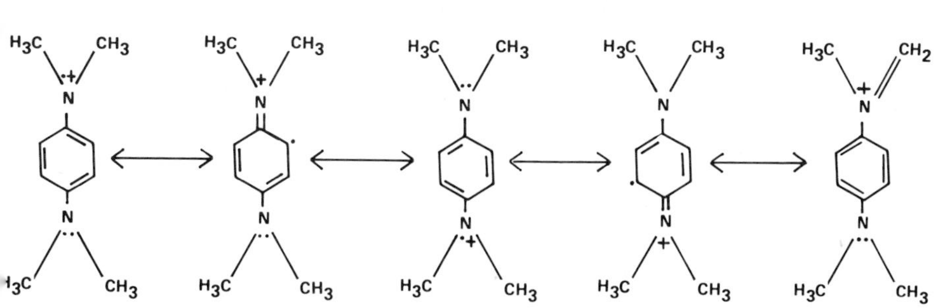

Scheme 13

Increasing the number of methyl substituents at the ortho-position of N-methyl-substituted p-phenylenediamine radical-cations causes destabilization because of steric interactions between the ring and aminomethyl groups, which prevent the coplanarity of these two nitrogen centers, thus disrupting electron delocalization. The degree of destabilization is greater for the N,N,N'N'-tetramethyl-p-phenylenediamine series than for N,N'-dimethyl-substituted series [98, 109] (Scheme 14). The radical-cation of N,N,N',N'-tetramethyldurene is so unstable

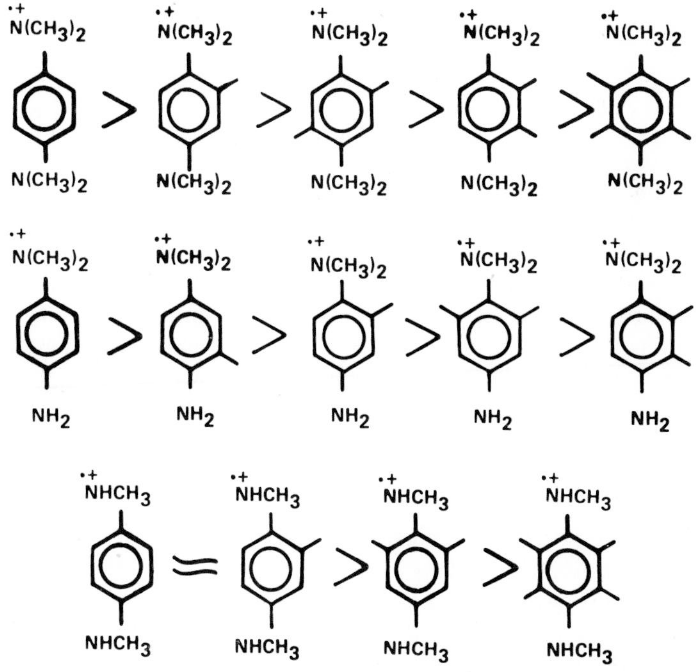

Scheme 14

that it has not been detected [109]. In contrast, aminium radicals of p-phe-nylenediamines are stabilized by increasing the number of o-methyl substituents [98, 109] (Scheme 15). The stability of Wurster's radical-cations is further

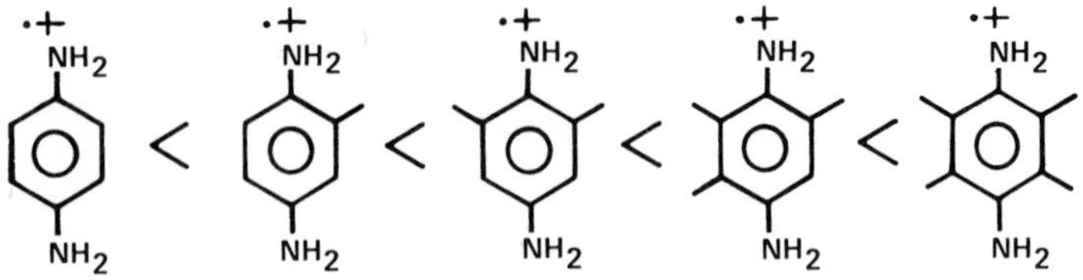

Scheme 15

determined by the existence of an equilibrium between the parent p-phe-nylenediamine, semiquinoneimine, and quinonediimine [111] (Scheme 16).

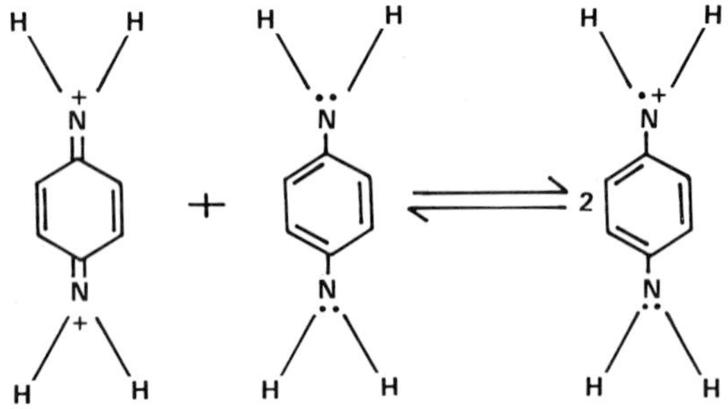

Scheme 16

Given this complicated background of free-radical chemistry of p-phe-nylenediamines, it is obvious that any postulated mechanism is, at best, specula-tive in the absence of concrete ESR data. The formaldehyde formation from PG synthase-catalyzed N-demethylation of N-methyl-substituted p-phenylenedia-mines can be considered to occur by hydrolysis of an iminium cation in a manner similar to that of AP. However, the actual mechanism of formation of an im-inium cation is open to question. It is important to study the effect of molecular oxygen on formaldehyde formation. Finally, due to the existence of a redox equilibrium (Scheme 16), it is possible to observe the ESR spectrum of the radical-cation even if the initial oxidation involves a two-electron mechanism [5, 111].

In related studies on the cooxidation of acridine derivatives, nitroxide radicals were detected during PG synthase-catalyzed metabolism of 9-aminoacridine and quinacrine, which were shown to bind covalently to microsomal membranes [112]. Perhaps the most notable feature of this study [112] is the possible incor-poration of an oxygen atom from molecular oxygen into the NH-containing moiety of acridine derivatives.

The existence of a free-radical intermediate during the cooxidation of methylaminoazobenzene by PG synthase, AA, and air has been postulated [113]. It would be interesting to compare the mechanism of PG synthase-catalyzed oxidation of methylaminoazobenzene with that of NADPH-dependent microsomal metabolism, during which the formation of a nitroxide radical was reported [114].

3. Hydrazines

The toxic, mutagenic, and tumorigenic aspects of several substituted hydrazine derivatives have aroused considerable interest in the metabolic activation reaction of this class of nitrogenous compounds [115, 116]. Although the exact mechanism of activation is not known, it has been suggested that some hydrazines elicit their toxic response due to oxidative metabolic liberation of electrophilic species such as carbonium ions or radicals [117, 118]. Studies on the metabolism of hydrazines have so far been restricted to microsomal monooxygenases [119].

We have previously shown that tetramethylhydrazine (TMH) undergoes oxidative N-demethylation to formaldehyde via a radical-cation by HRP and H_2O_2 [120]. Our interest in a similar oxidative metabolism of TMH by PG synthase and AA originated from a report that PG hydroperoxidase can initiate the metabolism of a number of xenobiotics in tissues containing low mixed-function oxidase activity [93]. Tetramethylhydrazine was chosen as a model compound because of the known chemistry of its oxidation products, in particular its radical-cation [121].

The addition of AA to an incubation mixture containing RSV microsomes and TMH immediately resulted [122] in an ESR spectrum of 15 equally spaced groups of lines (Fig. 7A). On the basis of previous reports of the chemical [123, 124], electrochemical [125], and enzymatic [120] oxidation of TMH in aprotic and protic solvents, the ESR spectrum (Fig. 7A) was identified as that of the radical-cation of TMH. The TMH radical-cation (TMH$^+$) was formed as a result of the one-electron oxidation of TMH. The TMH radical-cation possesses a novel three-electron π-bond between the two nitrogens and is inherently stable due to the favorable electronic interaction. The spectrum of TMH$^+$ shows electronic nuclear hyperfine coupling constants of 2 equivalent nitrogens ($a_N = 13.5$ G) and 12 equivalent protons ($a_H = 12.7$ G), in agreement with the literature values in water [124]. The incubation mixture was bubbled with oxygen for 1 min before the addition of AA in order to increase the concentration of PGG_2. The inhibition of the ESR spectrum by indomethacin (Fig. 7D) demonstrated that the hydroperoxide needed for oxidation of TMH was produced by the oxidative metabolism of AA by PG cyclooxygenase rather than the lipoxygenase component of PG synthase [93].

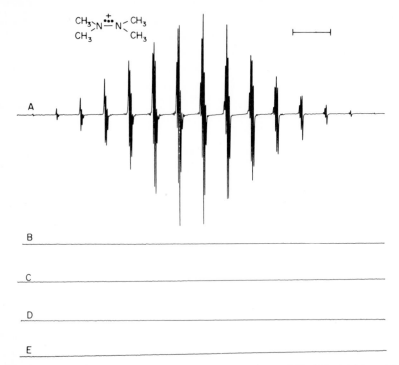

Fig. 7.   An ESR spectrum of an incubation containing 87 mm TMH, 200 μM AA, and 2 mg protein in 1 ml of Tris buffer (pH 7.4). The reaction was initiated at room temperature by the addition of 5 μl AA in ethanol (50 mg/ml). The mixture was bubbled with O_2 for 1 min before the addition of AA. The ESR spectrometer conditions were as follows for all spectra. Scan range, 200 G; time constant, 0.25 sec; scan time, 8 min; modulation amplitude, 0.05 G; spectrometer gain, 1.6×10^3; microwave power, 20 mW. (B) As in (A) except that AA was omitted. (C) As in (A) except that the enzyme was preincubated with 400 μM indomethacin for 2 min before the addition of AA. (D) As in (A) except that the incubation mixture described in (A) was bubbled with N_2 for 5 min before the addition of AA. (E) As in (A) except that the enzyme was denatured by heating for 10 min in boiling water before the addition of AA.

The ESR spectrum obtained upon the addition of TMH to an incubation containing RSV microsomes and H_2O_2 in a buffer (allowing for a negligible amount of preincubation time between RSV microsomes and H_2O_2) yielded the same spectrum (Fig. 7A). Furthermore, the addition of RSV microsomes preincubated with indomethacin to a mixture containing H_2O_2 and TMH did not reduce the ESR spectral intensity of TMH$^+$ to any appreciable extent. These results implicate the involvement of PG hydroperoxidase, which is not inhibited by indomethacin [93].

The participation of cytochrome P-450 acting either as an oxygenase [126] or a peroxidase [127] was ruled out because RSV microsomes do not contain any detectable amount of cytochrome *P*-450 and typical cytochrome *P*-450 inhibitors such as metyrapone and carbon monoxide do not inhibit the reaction [*112, 117*].

The PG synthase-dependent N-dealkylation was determined by measuring the formaldehyde fromation obtained from an incubation containing RSV microsomes, TMH, AA, and air. Formaldehyde formation is inhibited by heat inactivation of RSV microsomes, omission of AA, and preincubation of RSV microsomes wth indomethacin before its addition [*122*]. Therefore, the formation of PGG_2 is vital to the metabolism of TMH.

Because the TMH radical-cation is inherently very stable [*121*], the subsequent peroxidase-catalyzed oxidation of the radical-cation to a dication (Scheme 17) is a reasonable decay mechanism [*120*], as opposed to the decay via disproportionation (see Section IV,A,1).

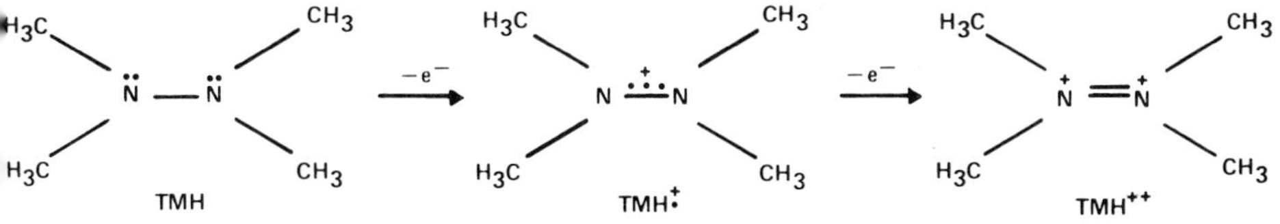

Scheme 17

Previous studies on the electrochemical oxidation of TMH have shown that the removal of a formally antibonding π-electron in the TMH radical-cation leading to the formation of the TMH dication (TMH^{2+}) is fairly easy [*125*]. However, the dication, with two adjacent positive charges, is unstable and was suggested to decay via deprotonation to give an iminium cation with subsequent formation of *N*-methylenemethylamine [*125*]. On this basis, two possible mechanistic schemes can be postulated to account for the formaldehyde formation during oxidative metabolism of TMH (Scheme 18). As indicated in Scheme 18, several

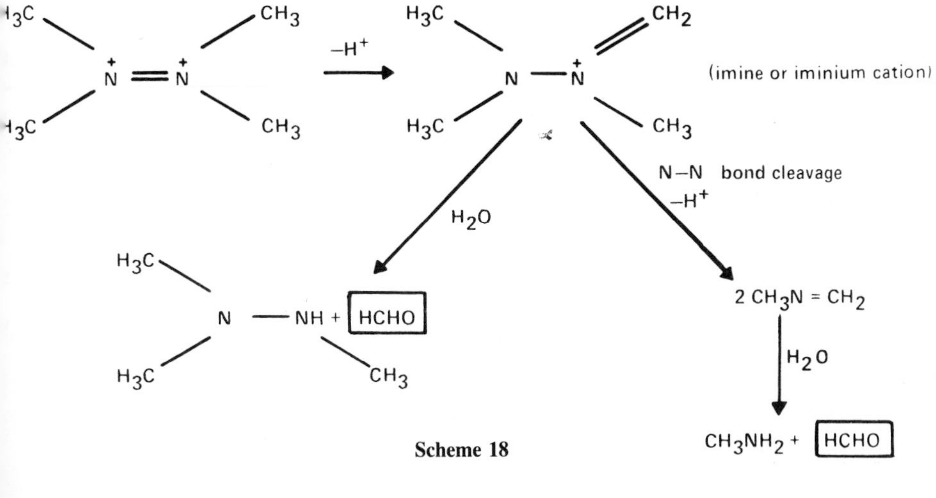

Scheme 18

products other than formaldehyde are possible. However, in order to prove that the TMH radical-cation is an obligatory intermediate in the production of formaldehyde, stoichiometric rate measurements of the TMH radical-cation and formaldehyde formation must be carried out [120].

In view of this work it is suggested that tetrasubstituted arylhydrazines could undergo a similar type of cooxidation. For example, the cation-radical of tetraphenylhydrazine (TPH) is considered to be the prototype of *violenes,* a family of persistent free radicals [109]. A preliminary investigation on HRP–H_2O_2-catalyzed oxidation of TPH yielded a five-line ESR spectrum attributable to coupling from two equivalent nitrogens of TPH$^+$ [128]. In chemical systems the TPH radical-cation has been shown to undergo benzidine rearrangement, as shown in Scheme 19 [129]. Benzidine could undergo a series of free-radical

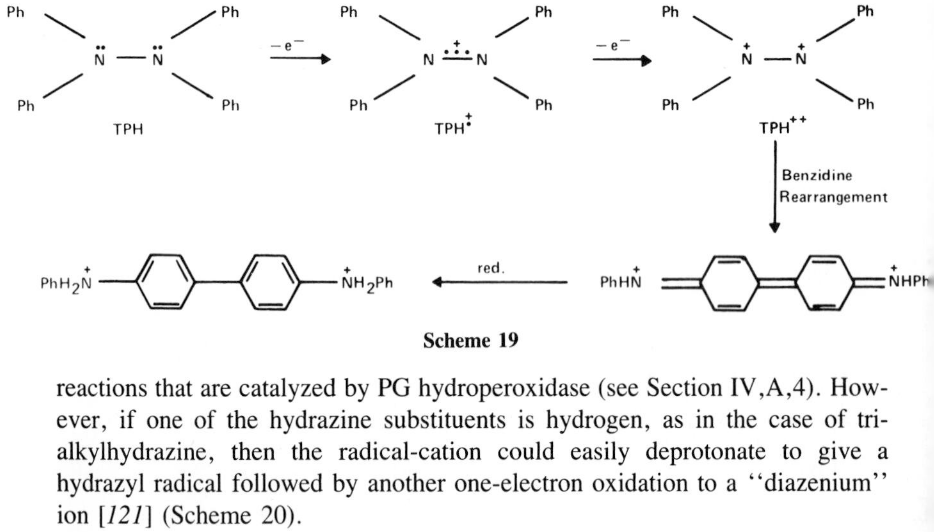

Scheme 19

reactions that are catalyzed by PG hydroperoxidase (see Section IV,A,4). However, if one of the hydrazine substituents is hydrogen, as in the case of trialkylhydrazine, then the radical-cation could easily deprotonate to give a hydrazyl radical followed by another one-electron oxidation to a "diazenium" ion [121] (Scheme 20).

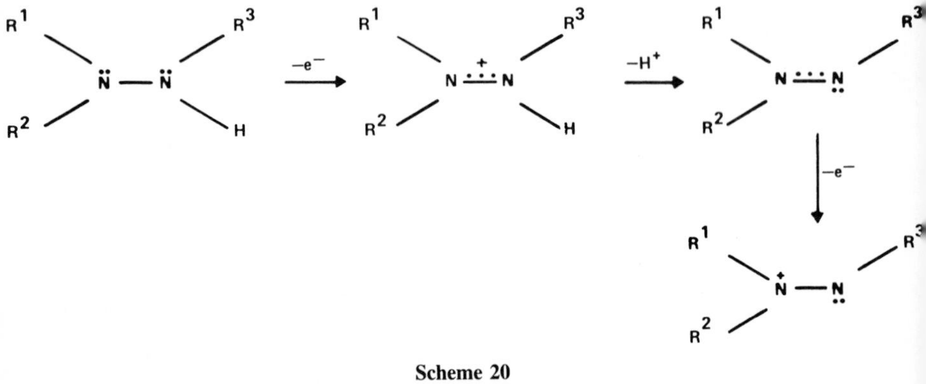

Scheme 20

Because the hydrazyl radical as well as hydrazine radical-cations possessing an N—H bond are too short-lived to be detected by static ESR methods [130], one has to employ either a continuous -flow system or spin trapping methods to detect these species. Both nitrogen-centered and the carbon-centered radicals produced during the oxidative metabolism of hydralazine [131] and other hydrazines [132, 133] have been spin-trapped with DMPO, and the resulting spin adducts have also been verified by mass spectrometry [131, 133]. The stability of the hydrazyl radical is increased by replacing R^1, R^2, and R^3 substituents in Scheme 20 with an aryl or aryl-substituted group [109]. For example, the remarkably stable 2,2-diphenyl-1-picrylhydrazyl (DPPH) free radical could be formed by PG hydroperoxidase-catalyzed cooxidation of the corresponding trisubstituted parent hydrazine.

4. Benzidines

The metabolic activation of benzidine to reactive electrophilic intermediates has been linked to its carcinogenic effects [134]. Early studies on the histochemical localization of PG synthase implicated the peroxidase component of PG synthase in the oxidation of aminobenzidine [135, 136]. Zenser et al. have shown that PG synthase can activate benzidines to reactive products in tissues containing low mixed-function oxidase activity [137, 138]. The PG hydroperoxidase-mediated cooxidation of benzidine was inferred mainly from covalent binding of products derived from [^{14}C]benzidine to nucleic acids [138]. The observed decrease in binding of an activated benzidine in the presence of glutathione with subsequent formation of a benzidine–glutathione conjugate constituted a possible detoxication pathway [138]. A study on the HRP–H_2O_2-catalyzed activation of benzidine confirmed that the quinoneimine, a two-electron oxidation product of benzidine, formed a ring-substituted thio ether-like conjugate with sulfhydryl-containing groups [139]. Zenser et al. also suggested that benzidine may be metabolized via a radical pathway by PG hydroperoxidase, possibly by acting as a peroxidase substrate because is quenched the ESR signal of an enzyme-derived free radical [138].

The oxidation of benzidine and its derivatives to various colored products by peroxidases in the presence of hydroperoxides has long been used in the detection of occult blood [140]. It was first suggested by Saunders [141] that benzidine could undergo a sequential one-electron oxidation to its radical-cation and diimine, corresponding to the appearance of the different colored products. Josephy et al. studied this sequence during the HRP–H_2O_2-catalyzed oxidation of 3,5,3′,5′-tetramethylbenzidine by optical spectroscopy and concluded that the initially formed colored product (which is blue in this case) is in fact a charge-transfer complex of the parent compound and the diimine (the two-electron oxidation product), which exists in a rapid equilibrium with the radical-cation.

The radical-cation is oxidized further to diimine (which in this case is yellow) [142]. Therefore, it was suggested that a free radical-cation could be formed during an enzymatic oxidation of benzidines whether the initial oxidation proceeds by a one- or a two-electron step (142). This excellent study should provide a basis for understanding the peroxidatic metabolism of other benzidines.

As discussed earlier (Section IV,A,2) the stability of the radical-cation is dependent on the pH of the medium as well as on the substituents attached to the aromatic ring and the amino group. Because N-alkyl substitution increases the stability of aromatic aminium radical-cations [109], we chose N,N,N',N'-tetramethylbenzidine (N,N,N',N'-TMB) as a substrate for cooxidation studies. Furthermore, it was also felt that the oxidative metabolism of N,N,N',N'-TMB could be monitored by measuring the formaldehyde formation in a manner similar to that from substituted N,N-di-methylanilines [100]. The chemistry of this compound (N,N,N',N'-TMB) is probably quite similar to that of the ring-substituted tetramethylbenzidine (3,5,3',5'-TMB) except for the fact that the latter compound is not expected to yield formaldehyde. The results on the PG synthase-catalyzed cooxidation of N,N,N',N'-TMB are discussed here in the light of previous findings on the HRP-catalyzed oxidation of 3,5,3',5'-TMB [142, 143].

We employed the pH jump technique (Section IV,A,1) for characterizing the ESR spectrum of transient radical-cations of benzidines. The metabolism of N,N,N',N'-TMB was initiated at physiological pH values, followed by rapid lowering of the pH to acidic values. The resulting ESR spectrum is identified as the radical-cation of N,N,N',N'-TMB, the spectral parameters of which are known ($a_N = 4.88$ G; $a_H^{CH3} = 4.70$ G; $a_{H1}^{ring} = 0.75$ G; $a_{H2}^{ring} = 1.65$ G) [144].

Appropriate inhibitor experiments established that the hydroperoxide needed for the metabolism of N,N,N',N'-TMB was generated by PG cyclooxygenase and not lipoxygenase or cytochrome P-450. The appearance of an identical ESR spectrum from RSV microsomal incubations containing N,N,N',N'-TMB and H_2O_2 indicated that the PG hydroperoxidase component, in conjunction with the hydroperoxide, catalyzed the oxidation of N,N,N',N'-TMB.

In the present work [143] the appearance of an ESR spectrum is associated with the formation of a green-colored compound. The ESR spectrum was found to decay when the color changed to yellow. As judged from previous findings [142], the green color is probably due to a charge-transfer complex between the parent compound and the diimine, which is in a rapid equilibrium with the radical-cation, and the yellow color represents the diamagnetic diimine compound. The simplified overall reaction sequence is shown in Scheme 21. Note that the radical-cation is in equilibrium with the charge-transfer complex (not shown in this scheme). A previous study clearly established that the concentration of the radical-cation is directly proportional to the square root of the concentration of the charge-transfer complex [142].

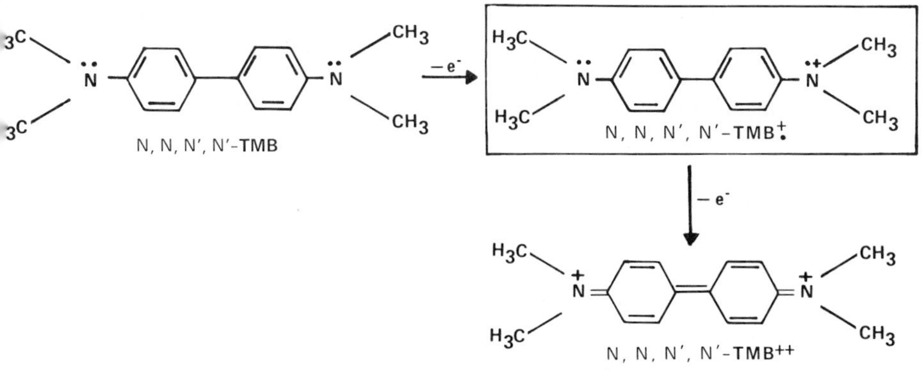

Scheme 21

The reason that an ESR spectrum of radical-cation of N,N,N',N'-TMB can still be observed at low pH, even though the enzymatic activity is greatly inhibited at the low pH, is due to the existence of the equilibrium shown in Scheme 22

$$TMB^{++} + TMB \rightleftharpoons 2\,TMB^{\cdot +}$$
$$(N, N, N', N') \quad (N, N, N', N') \quad (N, N, N', N')$$

Scheme 22

as well as the inherent stability of the diimine and the radical-cation in an acidic medium [98]. Note again that the radical-cation may be in equilibrium with the charge-transfer complex [142]. In view of the equilibrium reaction, the ESR detection of free radicals in these systems does not necessarily mean that the initial enzymatic oxidation is mediated via a one-electron transfer [142].

Prostaglandin synthase also initiated the facile N-dealkylation of N,N,N',N'-TMB to formaldehyde (Scheme 23). The formation of formaldehyde

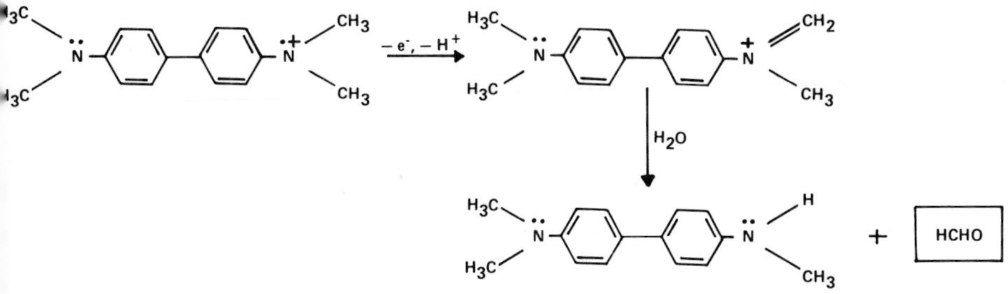

Scheme 23

from incubations containing RSV microsomes, N,N,N',N'-TMB, and AA in air was inhibited by indomethacin; the formation of formaldehyde increased with both RSV microsomes and *N,N,N',N'*-TMB concentrations [*143*]. The cytochrome *P*-450 inhibitors did not have any effect on formaldehyde formation. It is conceivable that the mechanism of formaldehyde formation is similar to that of *N,N*-dimethylaniline and AP. In this context it is also of interest to determine whether the iminium cation is formed by disproprotionation of TMB$^+$ or by oxidation of a *C*-alkyl radical intermediate. The study of a possible activation of molecular oxygen by TMB$^+$ is also of mechanistic importance.

Present ESR methodology also can be employed to detect transient radical-cations formed from *o*-anisidine and other related benzidines [142, 145]. Although the reactive electrophilic intermediate is the diamagnetic diimine compound [*139*], the importance of radical-cation stems from other radical-mediated reactions, including N-demethylation [*143*].

B. Free-Radical Intermediates from Phenols, Catechols, Catecholamines, and Indole Derivatives

1. Phenols

The stimulation of PG synthase by phenol itself [*146*] is considered to be partly due to its protective action during self-deactivation of PG cyclooxygenase [*23*] as well as its capacity to donate hydrogen atoms during the PG hydroperoxidase-catalyzed reduction of PGG$_2$ to PGH$_2$ [*71*]. Although efforts to identify products of PG hydroperoxidase-catalyzed metabolism of [^{14}C]phenol have been unsuccessful [*71, 80*], the isolation of a polymeric product may indicate the involvement of free radicals in a manner similar to that of HRP-catalyzed oxidation of phenols [*147*]. However, because phenoxyl radicals are very reactive [*147*], the spectroscopic detection of these species must involve the use of ESR flow techniques, as previously demonstrated during the oxidation of phenols in other peroxidizing systems [*148*].

Because most substituted phenols (with electron-donating groups) undergo rapid peroxidatic oxidation [*149*], the metabolism of these compounds by PG hydroperoxidase is a viable process. Of the various substituted phenols, the metabolism of acetominophen (also known as paracetamol) by PG hydroperoxidase is of considerable interest due to its widespread therapeutic applications. Studies on metabolic activation of paracetamol by PG hydroperoxidase in the presence of hydroperoxides indicate that paracetamol is metabolized to an electrophilic intermediate, possibly a quinoneimine that was trapped as the glutathione conjugate [*150–152*]. In the absence of glutathione or other sulfhydryl reagents, the electrophilic metabolite was shown to bind covalently to microsomal proteins [*153*]. Previous studies have also suggested that the electrophilic intermediate, the quinoneimine, is probably formed via disproportionation of the semiquinoneimine, a one-electron oxidation intermedaite of paracetamol, as shown in Scheme 24.

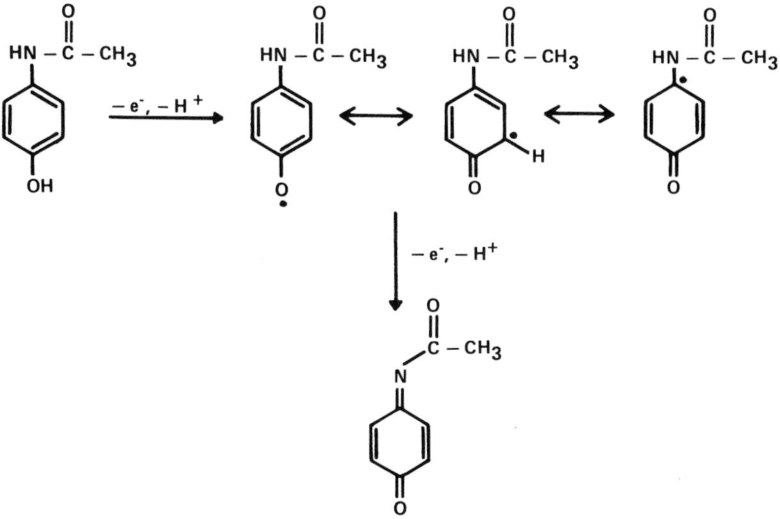

Scheme 24

Although a broad ESR signal characteristic of a polymer has been detected during HRP–H$_2$O$_2$-catalyzed oxidation of paracetamol by means of static ESR [154], the spectroscopic identification of the precursor radical, namely, the semiquinoneimine, formed during cooxidation by PG hydroperoxidase at physiological pH, will probably require the use of continuous-flow techniques.

Phenoxyl radicals of the type mentioned previously are unlikely to be detected by static ESR techniques, except under acidic conditions [109]. However, substituted phenoxyl radicals with ortho- and para-positions blocked by methyl or methoxyl groups have been detected by static ESR methods [155]. Antioxidants that possess the hindered phenolic structure [BHA (butylated hydroxyanisole), BHT (butylated hydroxytolene), and tocopherols] have been known to inhibit the PG synthase-catalyzed cooxidation of several xenobiotics [54]. These compounds act as competitive inhibitors during the HRP–H$_2$O$_2$-catalyzed oxidation of other phenols [156, 157]. It is conceivable that the BHA, BHT, or tocopherol could act as a better substrate for PG hydroperoxidase and inhibit the cooxidation of other xenobiotics. Direct ESR can be used to study the free-radical-mediated metabolism of BHA, BHT, or tocopherol due to the persistence of phenoxyl radicals derived from these hindered phenols [155, 157]. It is again worth mentioning that these antioxidants can inhibit the cooxidation of xenobiotics either by scavenging the propagating AA peroxyl radicals during AA oxygenation or by acting as competitive substrates for PG hydroperoxidase during their reduction of PGG_2 to PGH_2, depending on the concentrations.

In addition to phenols, hydroquinones stimualte the PG biosynthesis [158]. A free-radical mechanism of activation by hydroquinone has been proposed to explain the stimulatory effects of 1,4-hydroquinone [158]. On the basis of existing knowledge of free-radical chemistry of hydroquinones [8, 159], a reaction consisting of sequential one-electron oxidations can be visualized for PG synthase-catalyzed oxidation of hydroquinone (Scheme 25). However, the one-

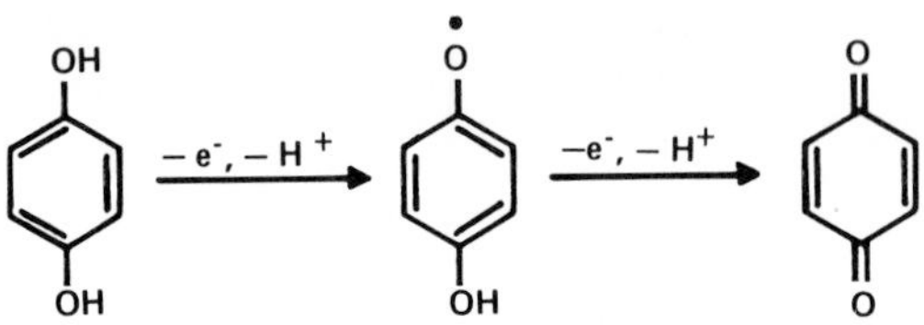

Scheme 25

electron oxidation intermediate, *p*-benzosemiquinone, is not kinetically stable at neutral pH. Thus, only a continuous-flow technique will make it possible to detect *p*-benzosemiquinones [159]. Furthermore, due to the existence of the comproportionation–disproportionation equilibrium between the hydroquinone, *p*-quinone, and *p*-semiquinone [8], detailed kinetic analysis is needed before an actual pathway leading to the *p*-semiquinone free radical can be ascertained [5].

The intermediacy of a diethylstilbestrol-*p*-semiquinone may also be proposed during PG hydroperoxidase-catalyzed oxidative metabolism of diethylstilbestrol (DES) to dienestrol (β-DIES) [160] (Scheme 26). The putative DES-*p*-semi-

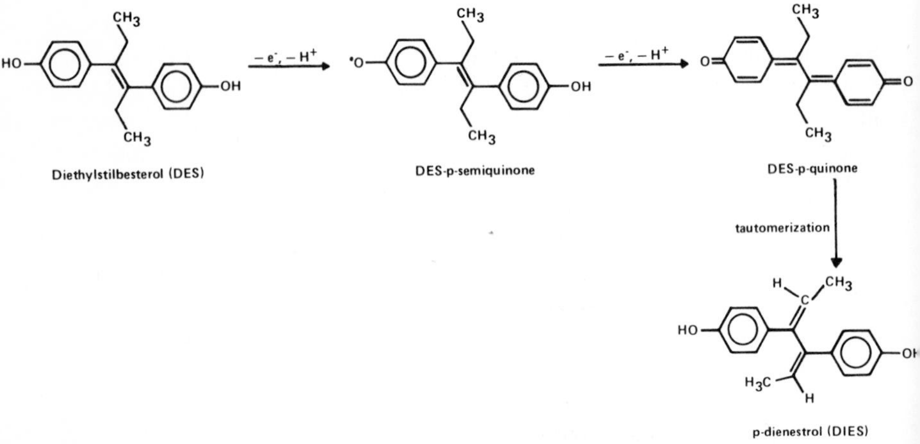

Scheme 26

quinone and/or p-quinone was implicated in the covalent binding to macromolecules [161] in the $HRP-H_2O_2$ system. However, as mentioned earlier, the DES-p-semiquinone is expected to be kinetically unstable, so that ESR confirmation of this free-radical metabolite must again await continuous-flow measurements; the product, β-DIES, itself could undergo further peroxidatic oxidation [8].

2. Catechols and Catecholamines

The addition of catechol derivatives (catechol, pyrogallal, and propyl gallate) and catecholamines (norepinephrine, dopamine, and epinephrine) as cofactors of prostaglandin-producing (exogenous and endogenous) mammalian tissues has been shown to stimulate PG biosynthesis [162–168]. It was originally proposed that these 1,2-dihydroxy compounds stimulated prostaglandin production by providing reducing equivalents for the nonenzymatic reduction of PGG_2 to PGH_2 [169]. However, the AA-dependent PG cyclooxygenase induced oxidation of epinephrine (also known as adrenaline) to adrenochrome (the four-electron oxidation product) was demonstrated by optical spectroscopy [170] (Scheme 27).

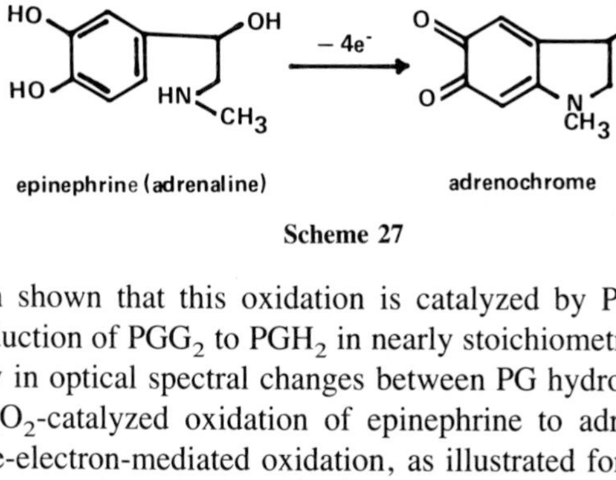

epinephrine (adrenaline)   adrenochrome

Scheme 27

It has been shown that this oxidation is catalyzed by PG hydroperoxidase during the reduction of PGG_2 to PGH_2 in nearly stoichiometric amounts [109a]. The similarity in optical spectral changes between PG hydroperoxidase–PGG_2- and $HRP-H_2O_2$-catalyzed oxidation of epinephrine to adrenochrome is suggestive of one-electron-mediated oxidation, as illustrated for most substrates in the $HRP-H_2O_2$ system [109a]. We have shown that the $HRP-H_2O_2$ system could catalyze the oxidation of a wide variety of catechols and catecholamines to quinones via the respective o-semiquinone radical-anions [171]. On the basis of this work and previous electrochemical data on catecholamine oxidations [172], a tentative reaction scheme for PG hydroperoxidase-catalyzed oxidation of epinephrine to adrenochrome can be proposed, as shown in Scheme 28. Although the occurrence of adrenochrome can easily be demonstrated by optical spectroscopy [170], the ESR detection of the epinephrine semiquinone may involve the use of a spin stabilization technique [173–175]. The ESR–spin stabilization technique in an aqueous solution involves the addition of diamagnetic metal ions to increase the kinetic stability of radical-ions such as o-semiquinones through

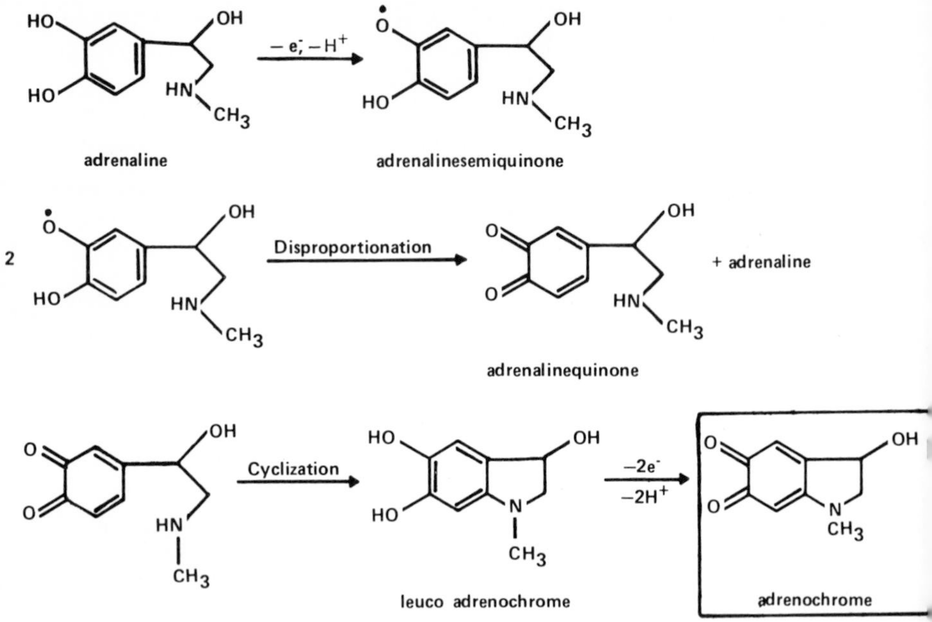

Scheme 28

the formation of complexes, as shown in Scheme 29. This technique can be easily used to distinguish semiquinones from other catecholamines (e.g., dopamine, norepinephrine, and dopa). However, the inhibitory effects of some diamagnetic metal ions on PG synthesis may limit the applicability of this technique. Finally, the findings that the catecholamines could cause stimulation of PG biosynthesis in renal [164] and brain tissues [168] may mean that the PG hydroperoxidase-catalyzed cooxidation of catecholamines is a viable metabolic pathway in these tissues.

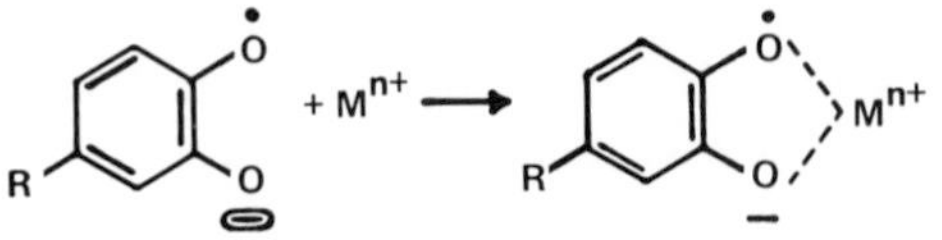

$$(M^{n+} = Zn^{2+}, Cd^{2+}, Mg^{2+}, Ca^{2+}, La^{3+})$$

Scheme 29

3. Indole Derivatives

Sih *et al.* have shown that the addition of substituted indole derivatives to PG-generating (*in vitro*) systems could cause an increase in PG biosynthesis [109a,

162]. It has further been shown that these compounds stimulate PG hydroperoxidase-catalyzed reduction of hydroperoxides [109a]. Because neither the intermediate(s) nor the product(s) is known, the actual mechanism of stimulation by indole derivatives cannot be determined at this time. The stimulatory effects of tryptophan were attributed mainly to its increased heme affinity [109a] rather than to cooxidation due to the failure to detect any appreciable transformation in tryptophan during the conversion of PGG_1 to PGH_1 [109a] However, with other 5-hydroxyindole derivatives the possibility of a one-electron-mediated cooxidation to semiquinoneimine cannot be dismissed (Scheme 30). There is evidence

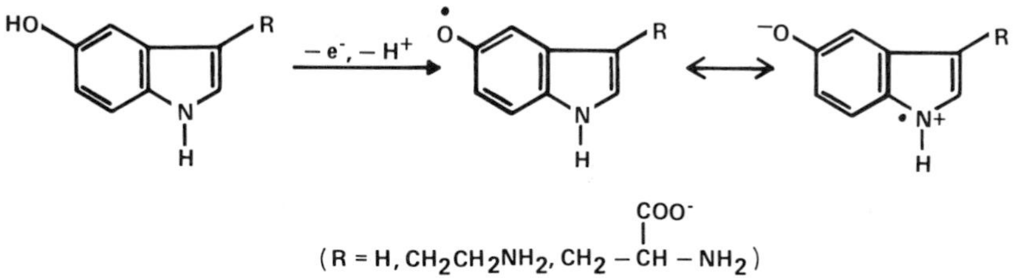

$$(R = H, CH_2CH_2NH_2, CH_2 - \overset{\overset{\displaystyle COO^-}{|}}{C}H - NH_2)$$

Scheme 30

for the importance of a 5-hydroxy group in the resonance stabilization of the resulting semiquinoneimines [176]. Unfortunately, these semiquinoneimines cannot be detected at physiological pH (these are stable only in an alkaline solution), because they undergo a rapid polymerization reaction to yield a melanin-like derivative, so that the direct ESR study of cooxidation of hydroxyindole derivatives may yield a melanin-like spectrum [177]. The proposed scheme of the peroxyl radical-mediated oxidation of indole-3-acetic acid to indole-3-aldehyde [178] may serve as a model system for the PG hydroperoxidase-catalyzed oxidation of 3-substituted indole derivatives such as indole-3-acetic acid, and indole-3-acetamide.

C. Phenylbutazone, Sulfur Compounds, and Miscellaneous Compounds

1. Phenylbutazone

The facile cooxidation of oxyphenbutazone to 4-hydroxyoxyphenbutazone was first demonstrated by Portoghese *et al.* using either PG synthase or lipoxygenase [179]. It was subsequently shown that phenylbutazone, a structurally similar compound, was also cooxygenated to the corresponding 4-hydroxyphenbutazone by the same system [180] (Scheme 31). Furthermore, incubations containing phenbutazone, hydroperoxide, and RSV microsomes under an $^{18}O_2$ atmosphere lead to the formation of 4-hydroxyphenbutazone [80, 180] with

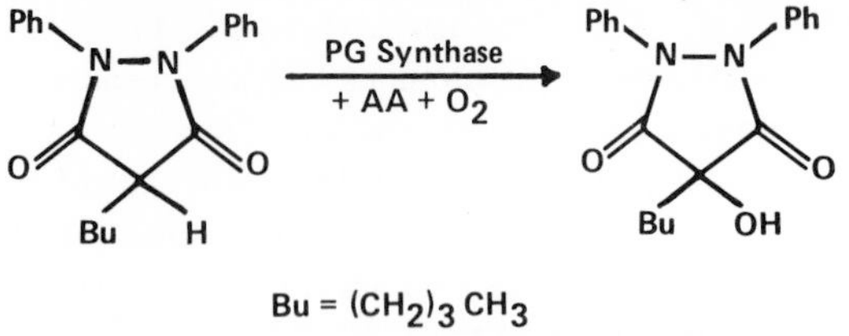

$$Bu = (CH_2)_3 CH_3$$

Scheme 31

substantial incorporation of ^{18}O into the product, showing the involvement of PG hydroperoxidase and molecular oxygen during cooxidation of phenbutazone [180]. In a more recent investigation, phenbutazone was found to stimulate the PG hydroperoxidase-catalyzed reduction of hydroperoxides [80]. A free-radical-mediated mechanism of cooxidation was suggested because antioxidants inhibited the cooxidation of phenbutazone [80, 180]. Portoghese et al. have also suggested the formation of a carbon-centered radical, which would in turn initiate the oxygen uptake observed under peroxidatic conditions [179]. A free-radical mechanism of incorporation of molecular oxygen into the product is therefore proposed (Scheme 32).

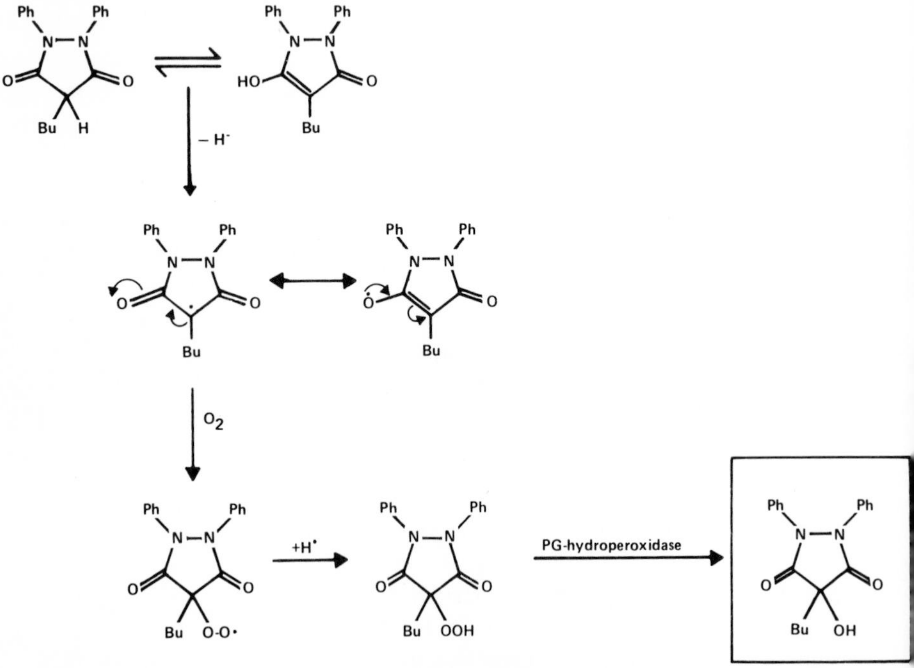

Scheme 32

Because carbon-centered free radicals react with molecular oxygen at diffusion-controlled rates, it is unlikely that they can be detected by ESR in air. A spin trapping experiment with MNP, phenylbutazone, hydroperoxide, and RSV microsomes under nitrogen could lead to the formation of the MNP–carbon-centered radical adduct. Even though this adduct would exhibit only a three-line ESR spectrum, the nitroxide hyperfine coupling may be considerably lower, and isotopic enrichment of the radical carbon center with ^{13}C ($I = \frac{1}{2}$) would enable one to make an unambiguous spectral characterization [8, 9]. The peroxyl adduct could be detected either by direct ESR at high modulation amplitudes or by spin trapping with DMPO [8, 9, 52].

2. Sulfur Compounds

The PG hydroperoxidase component of PG synthase derived from either RSV or guinea pig lung microsomes catalyzes the one-electron oxidation of (bi)sulfite to sulfur trioxide radical-anion (SO_3^{-}), which is immediately spin-trapped by DMPO [181], yielding an ESR spectrum identical to that obtained during the HRP–H_2O_2-catalyzed oxidation of (bi)sulfite in the presence of DMPO [182]. The PG hydroperoxidase-catalyzed reduction of hydroperoxides is also stimulated by other sulfur-containing compounds: sulindac sulfide [80], methyl phenyl sulfide [80], chlorpromazine [183, 184], promethiazine [183], and methional [23, 183].

During the reduction of hydroperoxides, both sulindac sulfide and methyl phenyl sulfide were converted to their corresponding sulfoxides. The oxgyen atom that was incorporated into the sulfoxides originates from the hydroperoxide [80]. Although products of metabolism of chlorpromazine and promethiazine are not known, the sequential one-electron oxidation of chlorpromazine to the radical-cation and dication, respectively, has been established in the HRP–H_2O_2 system [185]. Either the radical-cation or dication of chlorpromazine was proposed to undergo hydrolysis to form the corresponding sulfoxide [186, 187]. Therefore, the oxygen atom in chlorpromazine sulfoxide, if formed during the peroxidatic cooxidation by PG hydroperoxidase, is most likely to originate from water rather than the hydroperoxide. In other oxidizing systems, methional is converted to a sulfonium radical-cation followed by decomposition to ethylene gas [188]. The nonenzymatic reduction of PGG_2 to PGH_2 has been suggested to be the reason for stimulatory actions of lipoic acid [189] and other Cu^{2+}–dithiol complexes [190, 191].

3. Miscellaneous Compounds

In addition to the foregoing list of cooxidizable substrates that possess characteristic functional groups or atoms, as evidenced by their electron–hydrogen donor or oxygen acceptor property during the conversion of PGG_2 to PGH_2, several other compounds such as uric acid [192, 193], luminol [82], diphenylisobenzofuran (DPBF) [82, 86], adrenochrome [109a], 1,4-benzoquinone, and N-[4-(5-nitro-2-furyl)-2-thiazolyl]formamide (FANFT) have been found to increase PG biosynthesis and/or undergo cooxidation [194, 195]. A free-radical chain mechanism was proposed for cooxidation of DPBF on the basis of the tremendous increase in the stoichiometric conversion of DPBF with respect to PGG_2 [86]. It is also of interest that both $HRP-H_2O_2$ [196] and Ce(IV) [197] oxidized DPBF to dibenzoylbenzene, a product of metabolism of DPBF by PG hydroperoxidase [86, 194]. (A free-radical signal has been observed during chemical oxidation of DPBF [197].) In a related peroxidizing system, both uric acid and luminol have been proposed to undergo cooxidation via a nitrogen-centered radical intermediate [198, 199]. The requirement of molecular oxygen for the PG hydroperoxidase-dependent metabolism of FANFT indicates that the incorporation of molecular oxygen into the product possibly occurs through a free radical [194, 195]. Whereas the slight stimulation of PG biosynthesis by 1,4-benzoquinone [109a] remains anomalous, increased conversion of PGG_2 to PGH_2 by adrenochrome can be explained by rapid nonenzymatic reduction of PGG_2 by the adrenochrome tautomer (a hydroxy compound), which exists in an unfavorable equilibrium with the adrenochrome [200, 201].

ACKNOWLEDGMENTS

This work was supported in part by National Institutes of Health Grant RR-01008. The authors wish to thank Dr. William Pryor, Dr. Roger Sealy, Lilian Kalyanaraman, Yvonne Morauski, and Rita Pavelko for their assistance in the preparation of this manuscript.

REFERENCES

1. B. Samuelsson, M. Goldyne, E. Granström, M. Hamberg, S. Hammerström, and C. Malmsten, *Annu. Rev. Biochem.* **47**, 997 (1978).
2. W. E. M. Lands, *Annu. Rev. Physiol.* **41**, 633 (1979).
3. N. Porter, in "Free Radicals in Biology" (W. A. Pryor, ed.), Vol. 4, p. 261. Academic Press, New York, 1980.
4. D. C. Borg, in "Free Radicals in Biology" (W. A. Pryor, ed.) Vol. 1, p. 69. Academic Press, New York, 1976.
5. I. Yamazaki, in "Free Radical in Biology" (W. A. Pryor, ed.), Vol. 3, p. 183. Academic Press, New York, 1977.
6. E. G. Janzen, in "Free Radicals in Biology" (W. A. Pryor, ed.), Vol. 4, p. 115. Academic Press, New York, 1980.
7. J. R. Bolton, D. C. Borg, and H. M. Swartz, in "Biological Applications of Electron Spin Resonance" (H. M. Swartz, J. R. Bolton, and D. C. Borg, eds.), p. 63. Wiley (Interscience), New York, 1980.
8. R. P. Mason, *Rev. Biochem. Toxicol.* **1**, 151 (1979).
9. B. Kalyanaraman, *Rev. Biochem. Toxicol.* **4**, Elsvier/North Holland, Amsterdam, 1982.

10. B. Samuelsson, E. Granström, and M. Hamberg, *in* "Prostaglandins, Proceedings of the Second Nobel Symposium" (S. Bergstrom and B. Samuelsson, eds.), p. 51. Interscience, Stockholm 1967.

11. D. H. Nugteren, R. K. Beerthuis, and D. A. Van Dorp, *in* "Prostaglandins, Proceedings of the Second Nobel Symposium" (S. Bergstrom and B. Samuelsson, eds.)., p. 45 Interscience, Stockholm, 1967.

12. M. Hamberg and B. Samuelsson, *J. Biol. Chem.* **242**, 5344 (1967).

13. B. Samuelsson, *Fed. Proc., Fed. Am. Soc. Exp. Biol.* **31**, 1442 (1972).

14. I. Fridovich and P. Handler, *J. Biol. Chem.* **236**, 1836 (1961).

15. S. Bergström and R. T. Holman, *Adv. Enzymol.* **8**, 425 (1948).

16. A. L. Tappel, *in* "The Enzymes" (P. D. Boyer, H. Lardy, and K. Myrbäck, eds.), 2nd ed., Vol. 8, p. 275. Academic Press, New York, 1963.

17. M. Hamberg, B. Samuelsson, I. Bjorkhem, and H. Danielsson, *in* "Molecular Mechanisms of Oxygen Activation" (O. Hayaishi, ed.), Chapter 2. Academic Press, New York, 1974.

18. R. J. Flower, *Ciba Found. Symp.* [N.S.] **65**, 123 (1979).

19. J. J. M. C. De Groot, G. J. Garssen, J. F. G. Vliegenthart, and J. Boldingh, *Biochim. 8iophys. Acta* **326**, 279 (1973).

20. R. P. Mason, B. Kalyanaraman, B. E. Tainer, and T. E. Eling, *J. Biol. Chem.* **255**, 5019 (1980).

21. H. J. Robinson and J. R. Vane, *in* "Prostaglandin Synthetase Inhibitors" (H. J. Robinson and J. R. Vane, eds.), p. 20. Raven Press, New York, 1974.

22. B. Kalyanaraman and R. P. Mason, unpublished results.

23. R. W. Egan, J. Paxton, and F. A. Kuehl, Jr., *J. Biol. Chem.* **251**, 7329 (1976).

24. M. E. Hemler and W. E. M. Lands, *J. Biol. Chem.* **255**, 6253 (1980).

25. A. B. Sullivan, *J. Org. Chem.* **31**, 2811 (1966).

26. R. A. Floyd, L. M. Soong, M. A. Stuart, and D. L. Reigh, *Arch. Biochem. Biophys.* **185**, 450 (1978).

27. C. A. Evans, *Aldrichimica Acta* **12**, 23 (1979).

28. R. P. Mason and C. F. Chignell, *Pharmacol. Rev.* **33**, 189 (1982).

29. H. Aoshima, T. Kajiwara, A. Hatanaka, and H. Hatano, *J. Biochem. (Tokyo)* **82**, 1559 (1977).

30. Y. Maeda and K. U. Ingold, *J. Am. Chem. Soc.* **101**, 4975 (1979).

31. P. Schmid and K. U. Ingold, *J. Am. Chem. Soc.* **100**, 2493 (1978).

32. S. Nigam, K. D. Asmus, and R. L. Wilson, *J. Chem. Soc., Faraday Trans.* **1**, p. 2324 (1976).

33. M. Hicks and J. M. Gebicki, *Arch. Biochem. Biophys.* **210**, 56 (1981).

34. T. Kobayashi, I. Morita, and S. Murota, *Toxicol. Lett.* **9**, 373 (1981).

35. W. A. Pryor and J. W. Lightsey, *Science* **214**, 435 (1981).

36. A. Rahimtula and P. J. O'Brien, *Biochem. Biophys. Res. Commun.* **70**, 893 (1976).

37. P. J. O'Brien and A. Rahimtula, *Biochem. Biophys. Res. Commun.* **70**, 832 (1976).

38. N. Ogino, S. Ohki, S. Yamamoto, and O. Hayaishi, *J. Biol. Chem.* **252**, 890 (1977).

39. N. A. Porter, R. A. Wolf, W. R. Pagels, and L. J. Marnett, *Biochem. Biophys. Res. Commun.* **92**, 349 (1980).

40. B. Kalyanaraman and R. P. Mason, unpublished work.

41. A. N. Saprin and L. H. Piette, *Arch. Biochem. Biophys.* **180**, 480 (1977).

42. B. Kalyanaraman, R. P. Mason, E. Perez-Reyes, C. F. Chignell, C. R. Wolf, and R. M. Philpot, *Biochem. Biophys. Res. Commun.* **89**, 1065 (1979).

43. J. L. Poyer, P. B. McCay, E. K. Lai, E. G. Janzen, and E. R. Davis, *Biochem. Biophys. Res. Commun.* **94**, 1154 (1980).

44. A. Tomasi, E. Albano, K. A. K. Lott, and T. F. Slater, *FEBS Lett.* **122**, 303 (1980).

45. G. Czapski, *Annu. Rev. Phys. Chem.* **171** (1971).

46. P. F. Knowles, J. F. Gibson, F. M. Pick, and R. C. Bray, *Biochem. J.* **111**, 53 (1969).

47. K. U. Ingold, *Acc. Chem. Res.* **2,** 1 (1969).
48. J. A. Howard and J. C. Tait, *Can. J. Chem.* **56,** 176 (1978).
49. J. Pfab, *Tetrahedron Lett.* **9,** 843 (1978).
50. G. M. Rosen and E. J. Rauckman, *Proc. Natl. Acad. Sci. U.S.A.* **78,** 7346 (1981).
51. E. Finkelstein, G. M. Rosen, and E. J. Rauckman, *J. Am. Chem. Soc.* **102,** 4994 (1980).
52. B. Kalyanaraman, C. Mottley, and R. P. Mason, *J. Biol. Chem.* **258,** 3855 (1983).
53. N. A. Porter, L. S. Lehman, B. A. Weber, and K. J. Smith, *J. Am. Chem. Soc.* **103,** 6447 (1981).
54. M. Carpenter, *Fed. Proc., Fed. Am. Soc. Exp. Biol.* **40,** 189 (1981).
55. L. H. Mahoney and M. A. DaRooge, *J. Am. Chem. Soc.* **92,** 4063 (1970).
56. G. S. Hammond, C. E. Boozer, C. E. Hamilton, and J. N. Sen, *J. Am. Chem. Soc.* **77,** 3238 (1955).
57. C. E. Boozer and G. S. Hammond, *J. Am. Chem. Soc.* **76,** 3861 (1972).
58. J. A. Howard, *Adv. Free-Radical Chem.* **4,** 49 (1972).
59. G. W. Burton and K. U. Ingold, *J. Am. Chem. Soc.* **103,** 6472 (1981).
60. R. W. Egan, P. H. Gale, G. C. Beveridge, L. J. Marnett, and F. A. Kuehl, Jr., in "Advances in Prostaglandins and Thromboxane Research" (P. Samuelsson, P. W. Ramwell, and R. Paoletti, eds.), Vol. 6, p. 153. Raven Press, New York, 1980.
61. J. Y. Vanderhoek and W. E. M. Lands, *Biochim. Biophys. Acta* **296,** 382 (1973).
62. M. E. Hemler, H. W. Cook, and W. E. M. Lands, *Arch. Biochem. Biophys.* **193,** 340 (1979).
63. M. Hamberg and B. Samuelsson, *J. Biol. Chem.* **242,** 5344 (1967).
64. M. Hamberg and B. Samuelsson, *J. Biol. Chem.* **242,** 5329 (1967).
65. M. Hamberg and B. Samuelsson, *Proc. Natl. Acad. Sci. U.S.A.* **71,** 3400 (1974).
66. D. H. Nugteren, *Biochim. Biophys. Acta* **380,** 299 (1975).
67. P. Borgeat, M. Hamberg, and B. Samuelsson, *J. Biol. Chem.* **251,** 7816 (1976).
68. J. Sekiya, H. Aoshima, T. Kajiwara, T. Togo, and A. Hatanaka, *Agric. Biol. Chem.* **41,** 827 (1977).
69. E. J. Corey, J. O. Albright, A. E. Barton, and S. Hashimoto, *J. Am. Chem. Soc.* **102,** 1435 (1980).
70. T. Miyamoto, N. Ogino, S. Yamamoto, and O. Hayaishi, *J. Biol. Chem.* **251,** 2629 (1976).
71. R. W. Egan, P. H. Gale, and F. A. Kuehl, Jr., *J. Biol. Chem.* **254,** 3295 (1979).
72. B. Kalyanaraman, R. P. Mason, B. Tainer, and T. E. Eling, *J. Biol. Chem.* **257,** 4764 (1982).
73. T. Shiga and K. Imaisumi, *Arch. Biochem. Biophys.* **167,** 469 (1975).
74. N. K. King and M. E. Winfield, *J. Biol. Chem.* **238,** 1520 (1963).
75. J. F. Gibson, D. J. E. Ingram, and P. Nicholls, *Nature (London)* **181,** 1398 (1958).
76. T. Yonetani and H. Schleyer, *J. Biol. Chem.* **242,** 1974 (1967).
77. J. E. Roberts, B. M. Hoffman, R. Rutter, and L. P. Hager, *J. Biol. Chem.* **256,** 2118 (1981).
78. B. M. Hoffman, J. E. Roberts, T. G. Brown, C. H. Kang, and E. Margoliash, *Proc. Natl. Acad. Sci. U.S.A.* **76,** 6132 (1979).
79. T. Yonetani, H. Schleyer, and A. Ehrenberg, *J. Biol. Chem.* **241,** 3240 (1966).
80. R. W. Egan, P. H. Gale, E. M. Baptista, K. L. Kennicott, W. J. A. VandenHeuvel, R. W. Walker, P. E. Fagerness, and F. A. Kuehl, Jr., *J. Biol. Chem.* **256,** 7352 (1981).
81. P. J. O'Brien and A. D. Rahimtula, in "Advances in Prostaglandin and Thromboxane Research" (P. Samuelsson, P. W. Ramwell, and R. Paoletti, eds.), Vol. 6, p. 145. Raven Press, New York, 1980.
82. L. J. Marnett, P. Wlodawer, and B. Samuelsson, *J. Biol. Chem.* **250,** 8510 (1975).
83. R. V. Panganamala, N. R. Brownlee, H. Sprechter, and D. G. Cornwell, *Prostaglandins* **7,** 21 (1974).
84. P. J. O'Brien and L. G. Hulett, *Prostaglandins* **19,** 683 (1980).
85. L. J. Marnett and G. A. Reed, *Biochemistry* **18,** 2923 (1979).

86. L. J. Marnett, M. J. Bienkowski, and W. R. Pagels, *J. Biol. Chem.* **254**, 5077 (1979).
87. R. W. Egan, P. H. Gale, G. C. Beveridge, G. B. Phillips, and L. J. Marnett, *Prostaglandins* **16**, 861 (1978).
88. A. Ishimaru and I. Yamazaki, *J. Biol. Chem.* **252**, 6118 (1977).
89. A. Ishimaru, *Bioorg. Chem.* **9**, 472 (1980).
90. G. J. Garssen, J. F. G. Vliegenthart, and J. Boldingh, *Biochem. J.* **130**, 435 (1972).
91. P. J. O'Brien and L. G. Hulett, *Prostaglandins* **19**, 683 (1980).
92. S. J. Weiss, J. Turk, and P. Needleman, *Blood* **53**, 1191 (1979).
93. L. J. Marnett, *Life Sci.* **29**, 531 (1981).
94. B. W. Griffin and P. L. Ting, *Biochemistry* **17**, 2206 (1978).
95. J. M. Laskar, K. Sivarajah, R. P. Mason, B. Kalyanaraman, M. S. Abou-Donia, and T. E. Eling, *J. Biol. Chem.* **256**, 7764 (1981).
96. H. Sayo and M. Masui, *J. Chem Soc., Perkin Trans.* 2 p. 1640 (1973).
97. B. W. Griffin, *FEBS Lett.* **74**, 139 (1977).
98. Y. L. Chow, in *Reactive Intermediates* (R. A. Abrahmowitch, ed.) **1**, 151 Plenum Press, New York, (1980).
99. K. Sivarajah and B. Kalyanaraman, unpublished results.
100. K. Sivarajah, J. M. Laskar, T. E. Eling, and M. B. Abou-Donia, *Mol. Pharmacol.* **21**, 133 (1982).
101. B. W. Griffin, *Arch. Biochem. Biophys.* **190**, 850 (1978).
102. B. W. Griffin, D. K. Davis, and G. V. Bruno, *Bioorg. Chem.* **10**, 342 (1981).
103. G. Galiiani and B. Rindone, *J. Chem. Soc., Perkin Trans.* **1**, 456 (1978).
104. G. Galiiani and B. Rindone, *Bioorg. Chem.* **10**, 283 (1981).
105. E. T. Seo, R. F. Nelson, J. M. Fritsch, L. S. Marcoux, D. W. Leedy, and R. N. Adams, *J. Am. Chem. Soc.* **88**, 3498 (1966).
106. R. N. Adams, *Acc. Chem. Res.* **2**, 175 (1969).
107. P. L. Ashley, D. K. Davis, and B. W. Griffin, *Biochem. Biophys. Res. Commun.* **97**, 660 (1980).
108. G. Galliani and B. Rindone, *J. Chem. Soc., Perkin Trans.* **2**, 1 (1980).
109. A. R. Forrester, J. M. Hay, and R. H. Thompson, "Organic Chemistry of Stable Free Radicals," p. 254. Academic Press, New York, 1968.
109a. S. Ohki, N. Ogino, S. Yamamoto, and O. Hayaishi, *J. Biol. Chem.* **254**, 829 (1979).
110. F. J. Van der Ouderra, M. Buytenhek, D. H. Nugteren, and D. A. Van Dorp, *Biochim. Biophys. Acta* **487**, 315 (1977).
111. L. H. Piette, P. Ludwig, and R. N. Adams, *Anal. Chem.* **34**, 916 (1962).
112. B. K. Sinha, *Biochem. Biophys. Res. Commun.* **103**, 166 (1981).
113. S. Vasder, Y. Tsuruta, and P. J. O'Brien, *Biochem. Pharmacol.* **31**, 607 (1982).
114. A. Stier, R. Clauss, A. Luche, and I. Reitz, *Xenobiotica* **10**, 661 (1980).
115. M. R. Juchau and A. Horita, *Drug Metab.* **1**, 71 (1972).
116. L. B. Colvin, *J. Pharm. Sci.* **58**, 1433 (1969).
117. R. Braun, U. Greeff, and K. J. Netter, *Xenobiotica* **10**, 557 (1980).
118. S. D. Nelson, J. R. Mitchell, J. A. Timbrell, W. R. Snodgrass, and G. B. Corcoran, III, *Science* **194**, 901 (1976).
119. R. A. Prough, M. L. Coomes, and D. L. Dunn, in "Microsomes and Drug Oxidations" (V. Ullrich, ed.), p. 500. Pergamon, Oxford, 1977.
120. B. Kalyanaraman, R. P. Mason, and K. Sivarajah, *Biochem. Biophys. Res. Commun.* **105**, 217 (1982).
121. S. F. Nelson, in "Organic Free Radicals" (W. A. Pryor, ed.), p. 309. ACS Symposium Series, 69, American Chemical Society, Washington, D.C. 1978.
122. B. Kalyanaraman, R. P. Mason, K. Sivarajah, and T. E. Eling, *Carcinogenesis* (in press).

123. S.F. Nelson, *J. Am. Chem. Soc.* **88**, 5666 (1966).
124. W. H. Bruning, C. J. Michejda, and D. Romans, *J. Chem. Soc., Chem. Commun.* p. 11 (1967).
125. P. J. Kinlen, D. H. Evans, and S. F. Nelson, *J. Electroanal. Chem.* **97**, 265 (1979).
126. G. D. Nordblom, R. E. White, and M. J. Coon, *Arch. Biochem. Biophys.* **175**, 524 (1976).
127. E. G. Hrycay and P. J. O'Brien, *Arch. Biochem. Biophys.* **147**, 14 (1971).
128. B. Kalyanaraman, unpublished observations.
129. G. Koga, N. Koga, and J. P. Anselme, *in* "The Chemistry of the Hydrazo, Azo, and Azoxy Groups" (S. Patai, ed.), p. 915. Wiley, New York, 1975.
130. S. F. Nelson, W. P. Parmelee, M. Gobl, K. O. Hiller, D. Veltwisch, and K. D. Astmus, *J. Am. Chem. Soc.* **102**, 5606 (1980).
131. B. K. Sinha and A. G. Motten, *Biochem. Biophys. Res. Commun.* **105**, 1044 (1982).
132. O. Augusto, P. R. Ortiz de Montellano, and A. Quintanilha, *Biochem. Biophys. Res. Commun.* **101**, 1324 (1981).
133. H. A. O. Hill and P. J. Thornalley, *FEBS Lett.* **125**, 235 (1981).
134. D. B. Clayson and R. C. Ganner, *in* "Chemical Carcinogens" (C. E. Searle, ed.), p. 366. Am. Chem. Soc., Washington, D.C., 1976.
135. J. A. Lutwin, *Histochemistry* **53**, 301 (1977).
136. F. H. A. Janszen and D. H. Nugteren, *Histochemie* **27**, 159 (1971).
137. T. V. Zenser, M. B. Mattammal, H. J. Armbrecht, and B. B. Davis, *Cancer Res.* **40**, 2839 (1980).
138. T. V. Zenser, M. B. Mattammal, and B. B. Davis, *J. Pharmacol. Exp. Ther.* **211**, 460 (1979).
139. J. R. Rice and P. T. Kissinger, *Biochem. Biophys. Res. Commun.* **104**, 1312 (1982).
140. T. H. Haley, *Clin. Toxicol.* **8**, 13 (1975).
141. B. C. Saunders, *in* "Inorganic Biochemistry" (G. L. Eichhorn, ed.), Vol. II, p. 988. Elsevier, Amsterdam, 1973.
142. R. D. Josephy, T. E. Eling, and R. P. Mason, *J. Biol. Chem.* **257**, 3669 (1982).
143. B. Kalyanaraman, R. P. Mason, and K. Sivarajah, unpublished results.
144. P. Smejtek, J. Honzl, and V. Metalova, *Collect. Czech. Chem. Commun.* **30**, 3875 (1965).
145. A. Claiborne and I. Fridovich, *Biochemistry* **18**, 2324 (1979).
146. W. L. Smith and W. E. M. Lands, *J. Biol. Chem.* **246**, 6700 (1971).
147. K. U. Ingold, *Spec. Publ.—Chem. Soc.* **24**, 285 (1970).
148. T. Shiga and K. Imaizumi, *Arch. Biochem. Biophys.* **167**, 469 (1975).
149. D. Job and H. B. Dunford, *Eur. J. Biochem.* **66**, 607 (1976).
150. P. Moldeus and A. Rahimtula, *Biochem. Biophys. Res. Commun.* **96**, 469 (1980).
151. P. Moldeus, B. Andersson, A. Rahimtula, and M. Berggren, *Biochem. Pharmacol.* **31**, 1363 (1982).
152. P. Mohandas, G. G. Duggin, J. S. Horvath, and D. J. Tiller, *Toxicol. Appl. Pharmacol.* **61** 252 (1981).
153. J. A. Boyd and T. E. Eling, *J. Pharmacol. Exp. Ther.* **219**, 659 (1981).
154. S. D. Nelson, D. C. Dahlin, E. J. Rauckman, and G. M. Rosen, *Mol. Pharmacol.* **20**, 195 (1981).
155. E. S. Caldwell and C. Steelink, *Biochim. Biophys. Acta* **184**, 420 (1969).
156. W. Boguth and H. Niemann, *Biochim. Biophys. Acta* **248**, 121 (1971).
157. G. Sgaragli, L. D. Corte, R. Puliti, F. De Sarlo, R. Francalanci, and A. Guarna, *Biochem. Pharmacol.* **29**, 763 (1980).
158. A. Yoshimoto, H. Iko, and K. Tomita, *J. Biochem. (Tokyo)* **68**, 487 (1970).
159. T. Ohnishi, H. Yamazaki, T. Iyanagi, T. Nakamura, and I. Yamazaki, *Biochim. Biophys. Acta* **172**, 357 (1969).
160. G. H. Degen, T. E. Eling, and J. A. McLachlan, *Cancer Res.* **42**, 919 (1982).

161. M. Metzler and J. A. Mclachlan, *Biochem. Biophys. Res. Commun.* **85,** 874 (1978).
162. C. J. Sih, C. Takeguchi, and P. Foss, *J. Am. Chem. Soc.* **92,** 6670 (1970).
163. D. A. Van Dorp, *Prog. Biochem. Pharmacol.* **3,** 71 (1967).
164. A. Schaefer, M. Komlos, and A. Seregi, *Biochem. Pharmacol.* **27,** 213 (1978).
165. L. S. Wolfe, K. Rostworowski, and H. M. Pappius, *Can. J. Biochem.* **54,** 629 (1976).
166. H. H. Tai, C. L. Tai, and C. S. Hollander, *Biochem. J.* **154,** 257 (1976).
167. J. Baumann, F. V. Bruchhausen, and G. Wurm, *Naunyn-Schmiedegerg's Arch. Pharmacol.* **307,** 73 (1979).
168. L. S. Wolfe, H. M. Pappius, and J. Marion, *in* "Advances in Prostaglandin and Thromboxane Research" (B. Samuelsson and R. Paoletti, eds.), Vol. 1, p. 345. Raven Press, New York, 1976.
169. J. A. Chan, M. Nagassawa, C. Takeguchi, and C. J. Sih, *Biochemistry* **14,** 2987 (1975).
170. C. Takeguchi and C. J. Sih, *Prostaglandins* **2,** 169 (1972).
171. B. Kalyanaraman and R. C. Sealy, *Biochem. Biophys. Res. Commun.* **106,** 1119 (1982).
172. M. D. Hawley, S. V. Tatawawadi, S. Piekarski, and R. N. Adams, *J. Am. Chem. Soc.* **89,** 447 (1967).
173. H. B. Stegmann, H. U. Bergler, and K. Scheffler, *Angew. Chem.* **20,** 389 (1981).
174. C. C. Felix and R. C. Sealy, *Photochem. Photobiol.* **34,** 423 (1981).
175. C. C. Felix and R. C. Sealy, *J. Am. Chem. Soc.* **103,** 2831 (1981).
176. E. Perez-Reyes and R. P. Mason, *J. Biol. Chem.* **254,** 2427 (1981).
177. T. Uemura, T. Shimazu, R. Miura, and T. Yamano, *Biochem. Biophys. Res. Commun.* **93,** 1074 (1980).
178. I. Yamazaki, *in* "Molecular Mechanisms of Oxygen Activation" (O. Hayaishi, ed.), p. 535. Academic Press, New York, 1974.
179. P. S. Portoghese, K. Svanborg, and B. Samuelsson, *Biochem. Biophys. Res. Commun.* **63,** 748 (1975).
180. L. J. Marnett, M. J. Bienkowski, W. R. Pagels, and G. A. Reed, *in* "Advances in Prostaglandin and Thromboxane Research", Vol. 6, (B. Samuelsson, P. W. Ramwell, and R. Paoletti, eds.), pp. 149–151. Raven, New York, 1980.
181. C. Mottley, R. P. Mason, C. F. Chignell, K. Sivarajah, and T. E. Eling, *J. Biol. Chem.* **257,** 5050 (1982).
182. C. Mottley, T. B. Trice, and R. P. Mason, *Mol. Pharmacol.* **22,** 732 (1982).
183. R. W. Egan, P. H. Gale, W. J. A. Valdenheuvel, and F. A. Kuehl, Jr., *in* "Prostaglandin and Inflammation" (K. D. Rainsford and A. W. Ford-Hutchinson, eds.), p. 39. Birkhaeuser, Basel, 1979.
184. J. Robak, H. Kasperczyk, and A. Chytkowski, *Biochem. Pharmacol.* **28,** 2844 (1979).
185. L. H. Piette, G. Bulow, and I. Yamazaki, *Biochim. Biophys. Acta* **88,** 120 (1964).
186. H. Y. Cheng, P. H. Sackett, and R. L. McCreery, *J. Am. Chem. Soc.* **100,** 962 (1978).
187. D. C. Borg and G. C. Cotzias, *Proc. Natl. Acad. Sci. U.S.A.* **48,** 643 (1962).
188. C. Beauchamp and I. Fridovich, *J. Biol. Chem.* **245,** 4641 (1970).
189. L. J. Marnett and C. L. Wilcox, *Biochim. Biophys. Acta* **487,** 222 (1977).
190. R. E. Lee and W. E. M. Lands, *Biochim. Biophys. Acta* **260,** 203 (1972).
191. N. A. Porter, J. R. Nixon, and D. W. Gilmore, *in* "Organic Free Radicals", (W. A. Pryor, ed.), ACS Symposium Series, No. 9, p. 89, Am. Chem. Soc., Washington, D. C., 1978.
192. N. Ogino, S. Yamamoto, O. Hayaishi, and T. Tokuyama, *Biochem. Biophys. Res. Commun.* **87,** 184 (1979).
193. C. Deby, G. Deby-Dupont, F. X. Noel, and L. Lavergre, *Biochem. Pharmacol.* **30,** 2243 (1981).
194. T. V. Zenser, M. B. Mattammal, W. W. Brown, and B. R. Davis, *Kidney Int.* **16,** 688 (1979).
195. T. V. Zenser, M. B. Mattammal, and B. R. Davis, *J. Pharmacol. Exp. Ther.* **214,** 312 (1980).

196. H. W. S. Chan, *J. Am. Chem. Soc.* **93**, 4632 (1971).
197. D. C. Borg, K. M. Schaich, J. J. Elmore, Jr., and J. A. Bell, *Photochem. Photobiol.* **28**, 887 (1978).
198. M. J. Cormier and P. M. Prichard, *J. Biol. Chem.* **243**, 4706 (1968).
199. R. R. Howell and J. B. Wyngaarden, *J. Biol. Chem.* **235**, 3544 (1960).
200. W. Bors, C. Michel, M. Saran, and E. Lengfelder, *Biochim Biophys. Acta* **540**, 162 (1978).
201. W. Bors, M. Saran, E. Lengfelder, C. Michel, C. Fuchs, and C. Frenzel, *Photochem. Photobiol.* **28**, 629 (1978).

CHAPTER **6**

Phagocytes, Oxygen Radicals, and Lung Injury

Robert M. Tate* and John E. Repine

Webb-Waring Lung Institute and the Departments of Medicine
and Pediatrics
University of Colorado Health Sciences Center
Denver, Colorado

I. INTRODUCTION

Phagocytes in the lung are critical for host defense against microbial invasion. Indeed, both alveolar macrophages (macrophages that normally reside in alveolar air spaces of the lung) and polymorphonuclear leukocytes (neutrophils) have well-defined roles in the defense of the respiratory tract. These functions have been fully discussed [1–3]. The ability of phagocytes to kill microbes involves their capacity to reduce oxygen (O_2) to toxic O_2 metabolites including hydrogen peroxide (H_2O_2) and the free radicals, superoxide anion (O_2^- and hydroxyl radical ($HO\cdot$) [4–6]. However, alveolar macrophages and neutrophils are also capable of releasing these toxic reduced species of O_2 into the extracellular environment. The purpose of this chapter is to examine the evidence supporting the hypothesis that excessive accumulation of phagocytes in lung tissue followed by their release of toxic O_2 products could result in tissue injury.

---

*Present address: Pulmonary Division, Department of Medicine, Denver General Hospital, Denver, Colorado 80204.

FREE RADICALS IN BIOLOGY, VOL. VI

Henson provided the modern groundwork for our understanding of phagocyte-mediated tissue injury by proposing that phagocytes might be "frustrated" in their attempts to engulf impossibly large substances such as damaged epithelial surfaces or epithelial surfaces with adherent immune complexes [7]. The inability to phagocytize such materials might result in spillage of toxic products into the extracellular space and subsequent host tissue injury. It is now well recognized that phagocytes do not have to be "frustrated" in order to release O_2 products extracellularly. In fact, it is possible to detect extracellular reduced O_2 products during a "normal" stimulated respiratory burst *in vitro* [8]. With this phenomenon in mind, selected interactions of the lung with macrophages, neutrophils, and O_2 products will be examined.

II. ALVEOLAR MACROPHAGES AND LUNG INJURY

Alveolar macrophages are derived from circulating blood monocytes and become resident cells of the alveolar space. These cells have been well characterized as phagocytes, secretory cells, and participants in various pulmonary inflammatory processes. Particularly relevant to this discussion is their capacity to secrete complement components, reactive O_2 metabolites, arachidonate metabolites, and neutrophil chemotactic factors [9].

The possible roles of alveolar macrophages in lung injury have been well demonstrated in experimental pulmonary O_2 toxicity. Hyperoxia reproducibly causes an acute, severe lung injury in experimental animals. This injury is characterized by respiratory distress, pulmonary inflammation and edema, and death in a few days in the majority of exposed animals. For example, exposure of rats to hyperoxia ($>95\% O_2$) resulted in a 70% mortality between 66 and 72 h of exposure (Fig. 1) [10].

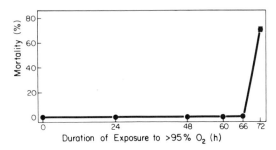

Fig. 1.   Deaths were observed in rats exposed to hyperoxia for 72 h (mean ± SEM). Each point represents at least 24 determinations. (From Fox [10], with permission.)

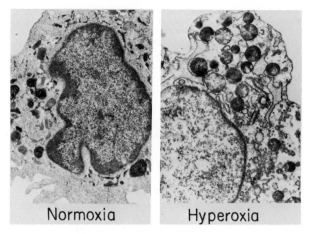

Normoxia   Hyperoxia

Fig. 2. Alveolar macrophages (AM) lavaged from lungs of rats exposed to hyperoxia (>95% O_2) for 66 h (right panel) had nuclear abnormalities, disorganized cytoplasm, and vacuolization not seen in AM from rats exposed to normoxia (left panel). (From Repine [10a], with permission.)

An early event in experimental O_2 toxicity is alveolar macrophage damage. Alveolar macrophages obtained from hyperoxia-exposed rats demonstrated nuclear abnormalities, disorganized cytoplasm, and vacuolization (Fig. 2) [10a]. These changes are very similar to the abnormal histology of rabbit alveolar macrophages exposed in culture to hyperoxia for 48 to 72 h.

In addition to the ultrastructural changes, alveolar macrophages exposed to hyperoxia *in vitro* release an increased amount of cytoplasmic lactate dehydrogenase [11]. Hyperoxia not only damages alveolar macrophages, but also stimulates them to release a variety of proinflammatory factors. For example, cultured rabbit alveolar macrophages exposed to hyperoxia release neutrophil chemotactic factors [11]. Specifically, we have identified two chemotactic factors having molecular weights of less than 1000 and ~12,000 (Fig. 3).

It is interesting that two chemotactic factors with very similar molecular weights were recovered by lung lavage from hyperoxia-exposed rabbits, suggesting that alveolar macrophages might be responsible for the production of neutrophil chemotaxins during *in vivo* O_2 toxicity (Fig. 3). In addition to chemotaxins, cultured alveolar macrophages stimulated by hyperoxia release factors that promote neutrophil adhesivity [12] and neutrophil superoxide anion release [11]. These findings in models of pulmonary oxygen toxicity suggest that a basic mechanism of lung injury may be alveolar macrophage damage and stimulation to release factors that cause neutrophils to adhere to endothelial cells, to migrate out of vascular spaces, and to release toxic neutrophil products.

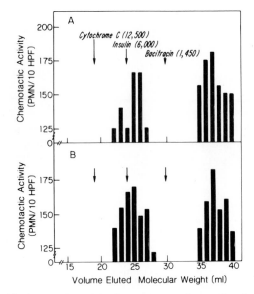

Fig. 3. Neutrophil chemotactic activity in supernatants from cultured rabbit alveolar macrophages exposed to hyperoxia (>95% O_2 for 48 h) and in lung lavage fluid from rabbits exposed to hyperoxia (>95% O_2 for 48 h). The molecular weights of two peaks of chemotactic activity *in vivo* are very similar to those of the two peaks of chemotactic activity present in supernatants of cultured hyperoxia-exposed alveolar macrophages. PMN, Polymorphonuclear neutrophils; HPF, high power field.

III. NEUTROPHILS AND LUNG INJURY

In contrast to alveolar macrophages, there are normally very few neutrophils in lung tissue except for the apparently quiescent neutrophils marginated within the pulmonary vasculature. However, it is clear that neutrophils accumulate in the lung during acute and chronic lung injury. For example, pathological descriptions of noncardiogenic pulmonary edema [now referred to as the adult respiratory distress syndrome (ARDS)], frequently cited pulmonary vascular leukocyte sequestration and neutrophil migration into the interstitial and alveolar spaces as prominent findings [13–16]. An increased number of neutrophils is also present in the lungs of smokers and in various idiopathic interstitial lung diseases [17–19]. Thus, neutrophils are clearly positioned to participate in the pathogenesis of a variety of acute and chronic lung disorders.

The lung disorder that is most frequently hypothesized to involve neutrophils and/or O_2 metabolites is ARDS [20–22]. The clinical evidence to support this premise is limited. As mentioned, neutrophil accumulation in lung biopsies or postmortem specimens from ARDS patients is common [13–16]. Information available from lung lavage fluid from ARDS patients reemphasizes the potential involvement of neutrophils in this syndrome. Several reports document the dramatic appearance of neutrophils or neutrophil products in lavage fluid of ARDS patients early in the course of the syndrome [23–25]. The mechanism(s) responsible for this pulmonary leukocyte accumulation is unclear. Many investigators suspect that complement is involved, and indeed neutrophil-aggregating activity, which appeared to be complement fragment C5a, has been demonstrated in the plasma of ARDS patients [26]. The final piece of clinical evidence relating neutrophils to ARDS is that circulating neutrophils from ARDS patients appear to be in an activated state when analyzed for chemotactic responsiveness, chemiluminescence, and superoxide anion release [27].

In contrast to the limited available clinical evidence supporting this hypothesis of neutrophil-mediated ARDS, there is abundant evidence to support a pathogenetic role for neutrophils in experimental noncardiogenic pulmonary edema (also known as permeability pulmonary edema). The most established findings are derived from studies in the sheep lung lymph fistula model as originally described by Staub and associates [28]. Increased vascular permeability has been demonstrated after the intravenous injection of gram-negative bacteria [29–31] or microemboli [32]. Of particular interest in these experiments is the fact that the increased permeability occurring after these stimuli is attenuated in neutropenic sheep, suggesting that neutrophils are necessary for the development of the permeability changes. However, studies that rely on neutrophil depletion by chemotherapeutic agents remain subject to many criticisms, including the specificity of the agent causing neutropenia (i.e., perhaps other important components of lung injury and/or defense mechanisms are affected) and the completeness and persistence of neutropenia during the experiments. Nevertheless, the findings in the sheep preparation have been substantiated in other animal models of permeability edema. For example, glass bead microembolization in dogs caused increased extravascular lung water that is attenuated in neutropenic animals [33]. Similarly, hyperoxic lung injury in rabbits can be reduced by making the rabbits neutropenic [34]. Taken together, these studies suggest that neutrophils can contribute to the pathogenesis of some acute pulmonary injuries.

Unfortunately, in intact animals one frequently cannot determine whether a given factor plays a minor or major role in the event being studied. Thus, it is difficult to assess the magnitude of the contribution of the neutrophil to the development of pulmonary edema in the aforementioned studies. For this reason investigators have used isolated perfused lungs to provide a more controlled setting for the study of specific mechanisms at pulmonary edema. For example, by washing and then perfusing lungs with a physiological salt solution one can essentially eliminate blood components that could interfere or interact with the experimental test conditions. As noted previously, neutrophils appear to contribute to hyperoxic lung injury in rabbits [34]. Similarly, neutrophils are necessary

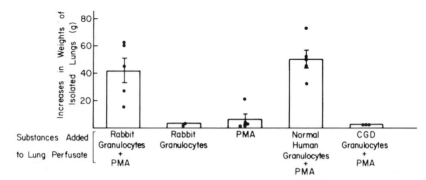

Fig. 4.   Increases in weight of isolated lungs perfused with granulocytes and/or PMA. Increases in weight of lungs perfused with rabbit granulocytes and PMA were greater than when either rabbit granulocytes alone or PMA alone were added to the perfusates. Increases in weight of lungs perfused with healthy human granulocytes and PMA were greater than when chronic granulomatous disease granulocytes and PMA were added to the perfusates. Each point represents data from one rabbit. (From Shasby [*35*], with permission.)

in intact rabbits for the development of pulmonary edema after injection of phorbol myristate acetate (PMA), and agent that increases neutrophil adhesivity, aggregability, degranulation, and a production of reduced O_2 species [*35*]. To establish a major role for neutrophils in hyperoxia and PMA-induced acute lung injury, Shasby and VanBenthuysen perfused isolated, ventilated, salt-perfused rabbit lungs with neutrophils and PMA and found that neutrophils plus PMA caused dramatic lung weight gains and increases in lung lavage albumin concentrations [*35*]. This severe pulmonary edema did not occur with neutrophils or PMA alone (Fig. 4).

This finding supported the hypothesis that neutrophils contribute to the pathogenesis of acute lung injury by releasing toxic neutrophil products that might include proteases, arachidonate products, and/or toxic reduced species of O_2. In support of the involvement of neutrophil-derived O_2 products, neutrophils from a patient with chronic granulomatous disease did not cause pulmonary edema in the aforementioned isolated lung experiments [*35*]. Because neutrophils in chronic granulomatous disease are considered to be normal except for their deficient O_2 radical production, this findings strongly suggests that O_2-derived products were involved in the pathogenesis of neutrophil-mediated pulmonary edema in isolated rabbit lungs.

IV. REDUCED OXYGEN PRODUCTS AND LUNG INJURY

To confirm the possibility that stimulated neutrophils damage the lung by releasing O_2-derived products, Tate and associates examined the effects of O_2

products generated by either xanthine oxidase or glucose oxidase on isolated salt-perfused lungs [*36*]. By means of this method, the effects of O_2 products on the lung could be isolated from the effects of other toxic properties of stimulated neutrophils. Reduced species of O_2 generated by xanthine oxidase caused severe pulmonary edema, increases in alveolar–capillary barrier permeability, and pulmonary vasoconstriction. Additional experiments using O_2 metabolite scavengers suggested that these changes were mediated by H_2O_2 or H_2O_2 products (Fig. 5). For example, superoxide dismutase did not prevent the xanthine oxidase-induced lung changes, suggesting that O_2^- was not the directly toxic species. In contrast, purified catalase or dimethylthiourea (a potent HO· scavenger) did prevent edema. This suggested that either H_2O_2 itself or an H_2O_2-derived product such as HO· (generated by H_2O_2 and trace metal ions present in the perfusate) caused the changes. The role of H_2O_2 was confirmed when the injection of glucose oxidase, an enzymatic reaction that generates H_2O_2 but not O_2^- [*37*], had similar effects. Thus, intravascular generation of H_2O_2 can cause acute pulmonary edema with increased alveolar–capillary barrier permeability in isolated lungs.

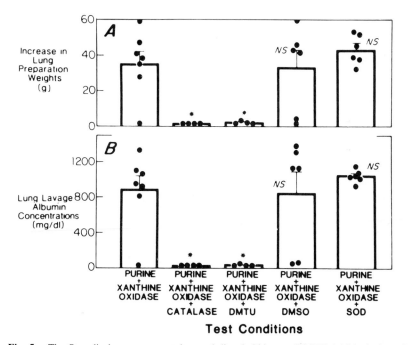

Fig. 5.   The O_2 radical scavengers catalase and dimethylthiourea (DMTU) inhibited edema induced by purine plus xanthine oxidase. Asterisk indicates significant difference ($p < .05$) from purine plus xanthine oxidase group; NS indicates no significant difference; DMSO, dimethyl sulfoxide; SDO, superoxide dismutase. (From Tate [*36*], with permission.)

The effects of O_2 products on vascular permeability and lung tissues have been demonstrated in several other models. In studies by Del Maestro and colleagues using the hamster cheek pouch model, vascular permeability was found to be increased following HO· generation by the xanthine oxidase reaction [38, 39]. In the sheep lung lymph model, Flick and Staub have reported that increased vascular permeability after air emboli can be reduced by the administration of superoxide dismutase [40] or by the inhibition of leukocyte superoxide anion production by phenytoin [41]. Using a rat model to assess the effects of oxygen metabolites on the lung, Johnson and Ward demonstrated that intratracheal instillation of chemical generators of O_2 products caused acute lung injury as determined by histological abnormalities and labeled albumin accumulation in lung tissue [37]. Superoxide anion, H_2O_2, and myeloperoxidase-produced derivatives appeared to cause acute lung injury, and in some cases this injury progressed to fibrosis. In another report these authors demonstrated that airway administration of catalase was protective in a neutrophil-dependent model of immune complex lung injury, again suggesting that acute lung injury may be related to the production of H_2O_2 products [42]. Furthermore, in a neutrophil-independent model, airway instillation of PMA caused an acute lung injury that was inhibited by catalase but not by superoxide dismutase [43]. The source of the toxic H_2O_2 in this model appeared to be PMA-stimulated alveolar macrophages. Finally, this group reported that intravascular activation of complement by cobra venom factor caused an acute lung injury that was attenuated by neutrophil depletion or intravenous administration of superoxide dismutase or catalase [44].

In addition to the organ and whole animal studies cited earlier, it has been shown in cell culture systems that stimulated neutrophils can damage endothelial cells through O_2 metabolite generation. Sacks and associates reported that neutrophils exposed to activated complement induce the release of ^{51}Cr from endothelial cells [45]. The amount of ^{51}Cr released from the "damaged" endothelial cells in this study was small, and therefore, it was uncertain if this neutrophil-mediated injury was physiologically significant. Nevertheless, the ^{51}Cr release was inhibited by the addition of superoxide dismutase plus catalase or catalase alone. These findings suggested that neutrophils might damage endothelial cells by a mechanism involving O_2 metabolites. Weiss and colleagues significantly extended these observations by showing that neutrophils can destroy endothelial cells in culture by generating cytotoxic amounts of hydrogen peroxide [46]. In these studies, neutrophils stimulated by PMA caused more severe cytotoxicity. The cytolysis was prevented by adding catalase, but not by adding superoxide dismutase. In addition, neutrophils derived from patients with chronic granulomatous disease did not damage the endothelial cell targets, whereas myeloperoxidase-deficient neutrophils did. Based on these findings, Weiss proposed that neutrophil-derived O_2 metabolites, specifically hydrogen peroxide, can significantly damage endothelial cells. Other investigators have confirmed the ability of O_2 metabolites to damage pulmonary artery endothelial cells [47], lung epithelial [47] and parenchymal [48] cells, and fibroblasts [49].

An interesting phenomenon related to O_2 radical tissue injury is the production of neutrophil chemotactic factors after exposure of plasma to O_2 radical sources. Petrone and McCord exposed human plasma to xanthine and xanthine oxidase and generated a chemotactic factor that appeared to be a lipid bound to albumin [50]. The nature of the lipid remains to be determined. Similarly, the exposure of arachidonic acid to O_2 radicals elaborates a chemotactic factor [51]. Thus, there appears to be not only direct tissue toxicity from O_2 radicals but also mechanisms by which O_2 radical generation might amplify the inflammatory response.

V. SUMMARY

Phagocytes and their toxic O_2 metabolites are critical for normal host defense. However, the same phagocytes can potentially cause harm to the host when they are inappropriately stimulated to accumulate and release their toxic products. The evidence cited in this chapter clearly suggests that neutrophil-derived O_2 metabolites can injure lung tissue. However, it remains to be demonstrated whether this basic mechanism of lung injury is involved in the pathophysiology of the human lung.

ACKNOWLEDGMENTS

This work was supported in part by the National Institutes of Health, the American Lung Association, the American Heart Association, the Council for Tobacco Research, and the Kroc, Hill, Swan, and Kleberg Foundations. Dr. Tate is a recipient of a Clinician–Scientist Award from the American Heart Association. The editor of the American Review of *Respiratory Diseases* gave permission to reprint Figs. 1, 4, and 5. The editor of the *Lung Biology in Health and Disease* series, published by Marcel Dekker, Inc., gave permission to reprint Fig. 2.

REFERENCES

1. J. D. Brain, D. F. Proctor, and L. M. Reid, *in* "Lung Biology in Health and Disease." Dekker, New York, 1977.
2. T. L. Petty and R. M. Cherniack, *in* "Seminars in Respiratory Medicine" (G. L. Huber, ed.), p. 187. Thieme-Stratton, New York, 1980.
3. S. R. Rehm, G. N. Cross, and A. K. Pierce, *J. Clin. Invest.* **66**, 194 (1980).
4. B. M. Babior, J. T. Curnutte, and R. S. Kipnes, *J. Lab. Clin. Med.* **85**, 235 (1975).
5. B. M. Babior, *N. Engl. J. Med.* **298**, 659 (1978).
6. R. L. Baehner, L. A. Boxer, and L. M. Ingraham, *in* "Free Radicals in Biology" (W. A. Pryor, ed.), Vol. 5, p. 91. Academic Press, New York, 1982.
7. P. M. Henson, *Am. J. Pathol.* **68**, 593 (1972).
8. J. C. Fantone and P. A. Ward, *Am. J. Pathol.* **107**, 397 (1982).
9. C. F. Nathan, H. W. Murray, and Z. A. Cohn, *N. Engl. J. Med.* **303**, 622 (1980).

10. R. B. Fox, J. R. Hoidal, D. M. Brown, and J. E. Repine, *Am. Rev. Respir. Dis.* **123,** 521 (1981).

10a. J. E. Repine, *in* "Acute Respiratory Failure" (W. M. Zapol and C. J. M. Lenfant, eds.). Dekker, New York (in press).

11. R. N. Harada, C. M. Bowman, R. B. Fox, and J. E. Repine, *Chest* **81,** 52S (1982).

12. C. M. Bowman, R. N. Harada, R. B. Fox, G. A. Shibao, and J. E. Repine, *Inflammation* (in press).

13. N. B. Ratlift, J. W. Wilson, E. Mikat, D. B. Hackel, and T. C. Graham, *Am. J. Pathol.* **65,** 325 (1971).

14. M. Bachofen and E. R. Weibel, *Am. Rev. Respir. Dis.* **116,** 589 (1977).

15. P. C. Pratt, *in* "The Lung: Structure, Function, and Disease" (W. M. Thurlbeck and M. R. Abell, eds.), p. 43. Williams & Wilkins, Baltimore, Maryland, 1978.

16. G. Schlag and H. Redl, *Anaesthetist* **29,** 606 (1980).

17. R. G. Crystal, J. E. Gadek, V. J. Ferrans, J. D. Fulmer, B. R. Line, and G. W. Hunninghake, *Am. J. Med.* **70,** 542 (1981).

18. S. E. Weinberger, J. A. Kelman, N. A. Elson, R. C. Young, H. Y. Reynolds, J. D. Fulmer, and R. G. Crystal, *Ann. Intern. Med.* **89,** 459 (1978).

19. P. L. Haslam, C. W. G. Turton, B. Heard, A. Lukoszek, J. V. Collins, A. J. Salsbury, and M. Turner-Warwick, *Thorax* **35,** 9 (1980).

20. K. L. Brigham, *in* "Clinics in Chest Medicine" (R. C. Bone, ed.), p. 9. Saunders, Philadelphia, Pennsylvania, 1982.

21. T. M. Hyers, *Semin. Respir. Med.* **2,** 104 (1981).

22. J. E. Repine, C. M. Bowman, and R. M. Tate, *Chest* **81S,** 47S (1982).

23. C. T. Lee, A. M. Fein, M. Lippman, H. Holtzman, P. Kimbel, and G. Weinbaum, *N. Engl. J. Med.* **304,** 192 (1981).

24. W. W. McGuire, R. G. Spragg, A. B. Cohen, and C. G. Cochrane, *J. Clin. Invest.* **69,** 543 (1982).

25. A. A. Fowler, S. Walchak, P. C. Giclas, P. H. Henson, and T. M. Hyers, *Chest* **81S,** 50S (1982).

26. D. E. Hammerschmidt, L. J. Weaver, L. D. Hudson, P. R. Craddock, and H. S. Jacob, *Lancet* **1,** 947 (1980).

27. G. A. Zimmerman, A. D. Renzetti, and H. R. Hill, *Am. Rev. Respir. Dis.* **127,** 290 (1983).

28. N. Staub, R. Bland, K. Brigham, R. Demling, J. Erdman, and W. Woolverton, *J. Surg. Res.* **19,** 315 (1975).

29. K. Brigham, R. Bowers, and J. Haynes, *Circ. Res.* **45,** 292 (1979).

30. K. Brigham, W. Woolverton, L. Blake, and N. Staub, *J. Clin. Invest.* **54,** 792 (1974).

31. A. J. Heflin and K. L. Brigham, *J. Clin. Invest.* **68,** 1253 (1981).

32. M. R. Flick, G. Perel, and N. C. Staub, *Circ. Res.* **48,** 344 (1981).

33. A. Johnson and A. B. Malik, *Am. Rev. Respir. Dis.* **122,** 561 (1980).

34. D. M. Shasby, R. B. Fox, R. N. Harada, and J. E. Repine, *J. Appl. Physiol.* **52,**1237 (1982).

35. D. M. Shasby, K. M. VanBenthuysen, R. M. Tate, S. S. Shasby, I. F. McMurtry, and J. E. Repine, *Am. Rev. Respir. Dis.* **125,** 443 (1982).

36. R. M. Tate, K. M. VanBenthuysen, D. M. Shasby, I. F. McMurtry, and J. E. Repine, *Am. Rev. Respir. Dis.* **126,** 802 (1982).

37. K. J. Johnson, J. C. Fantone, J. Kaplan, and P. A. Ward, *J. Clin. Invest.* **67,** 983 (1981).

38. R. F. Del Maestro, J. Bjork, and K. E. Arfors, *Microvasc. Res.* **22,** 239 (1981).

39. R. F. Del Maestro, J. Bjork, and K. E. Arfors, *Microvasc. Res.* **22,** 255 (1981).

40. M. R. Flick, J. Hoeffel, and N. C. Staub, *Fed. Proc., Fed. Am. Soc. Exp. Biol.* **40,** 405 (1981).

41. M. R. Flick, J. Hoeffel, and R. O. Webster, *Fed. Proc., Fed. Am. Soc. Exp. Biol.* **41,** 1500 (1982).

42. K. J. Johnson and P. A. Ward, *J. Immunol.* **126,** 2365 (1981).
43. K. J. Johnson and P. A. Ward, *Am. J. Pathol.* **107,** 29 (1982).
44. G. O. Till, K. J. Johnson, R. Kunkel, and P. A. Ward, *J. Clin. Invest.* **69,** 1126 (1982).
45. R. Sacks, C. F. Moldow, P. R. Craddock, T. K. Bowers, and 'I. S. Jacob, *J. Clin. Invest.* **61,** 1161 (1978).
46. S. J. Weiss, J. Young, A. F. LoBuglio, A. Slivka, and N. Nimen, J. Clin. Invest. **68,** 714 (1981).
47. N. Suttorp and L. M. Simon, *J. Clin. Invest.* **70,** 342 (1982).
48. W. J. Martin, J. E. Gadek, G. W. Hunninghake, and R. G. Crystal, *J. Clin. Invest.* **68,** 1277 (1981).
49. R. H. Simon, C. H. Scoggin, and D. Patterson, *J. Biol. Chem.***256,** 7181 (1981).
50. W. F. Petrone, D. K. English, K. Wong, and J. M. McCord, *Proc. Natl. Acad. Sci. U.S.A.* **77,** 1159 (1980).
51. H. D. Perez, B. B. Weksler, and I. M. Goldstein, *Inflammation* **4,** 313 (1980).

CHAPTER 7

Low-Level Chemiluminescence of Biological Systems

Enrique Cadenas

Institut für Physiologische Chemie I, Universität Düsseldorf,
Düsseldorf, Federal Republic of Germany

Alberto Boveris

Departamento de Quimica Biologica, Facultad de Farmacia y Bioquimica,
Buenos Aires, Argentina

Britton Chance

Johnson Research Foundation, School of Medicine,
University of Pennsylvania, Philadelphia, Pennsylvania

FREE RADICALS IN BIOLOGY, VOL. VI
211

I. INTRODUCTION

Luminescence of living organisms has fascinated scientists since antiquity [1, 2]. It was thought to be restricted to organisms with light organs containing luciferin–luciferase systems until 1961, when Tarusov et al. [3] used photon counting to identify a weak blue-green light emission from mouse liver in situ. This observation was later extended to brain, intestine, tissue homogenates, and lipid extracts [3–6]. (For a critical review see Baremboim et al. [7].) Isolated cells, such as amoebas [8], yeast [9, 10], phagocytizing leukocytes [11, 12] and macrophages [13], and hepatocytes [14] exhibit "low-level" [15] or "dark" chemiluminescence [7], thus named to differentiate this phenomenon from the photoemission of the luciferin–luciferase systems, which is 10^3–10^6 times brighter. Low-level chemiluminescence cannot be seen by the dark-adapted human eye, because a retinal illumination of 3×10^3 photons sec^{-1} cm^{-2} is required to perceive a luminous signal [16], and this value is about 10^4 times higher than the spontaneous light emission from living aerobic tissues.

The Russians' work [3–7] on chemiluminescence was largely unnoticed in America and Europe, probably because of earlier reports of the so-called mutagenic radiation [17, 18] from living tissue, which could not be observed with modern photon-counting equipment [19]. Exceptions were Stauff and Ostrowski's [20] and Howes and Steele's reports [21, 22] on rat liver. Reports on light emission during lipid peroxidation [23, 24] in isolated microsomes and during other oxidative reactions [25] revived the interest in chemiluminescence and suggested its use for the investigation of the radical reactions of lipid peroxidation under physiological conditions.

The most important aspect of cell and organ chemiluminescence is that it gives a noninvasive signal of the results of oxidative metabolism that can be readily detected and continuously monitored. Exposure and probing of the organ with optical fibers yields information about the rate of oxidative reactions occurring in vivo. For a correct interpretation of experimental data on organ light emission in vivo, current information on the mechanism of chemiluminescent reactions is necessary. Simplified and chemically well-defined model systems provide the

TABLE I Noninvasive Assays of Lipid Peroxidation

Assay	Applicability	Species detected	Method	Efficiency
Hydrocarbon expiration	Whole body	Ethane, propane, pentane	Gas chromatography	10^{-2}–10^{-4}
Glutathione release	Liver (bile)	Glutathione disulfide	Biochemical analysis	10^{-1}
Chemiluminescence	Liver, brain	hv	Photon counting	10^{-10}–10^{-12}

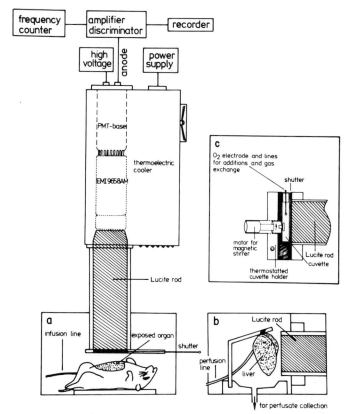

Fig. 1. Scheme of a single-photon-counting apparatus for the measurement of organ chemiluminescence. The Lucite rod used as optical coupler in placed in front of the exposed liver *in situ* (a) or
the perfused liver (b). Thermostatted cuvette holder (c); PMT, photomultiplier tube. From Boveris *et
al.* [26] and Cadenas and Sies [35].

information on chemiluminescence reactions that may contribute to the *in vivo*
light emission. In this chapter we review the data of photoemission by *in situ* and
perfused organs, isolated cells, and subcellular fractions and of some model
reactions.

II. ORGAN CHEMILUMINESCENCE

In general, oxidative stress in the intact organ can be studied with chemiluminescence [26, 27], glutathione release [28, 29], and hydrocarbon expiration
[30–32]. Table I summarizes the characteristics of these noninvasive methods in
terms of their applicability and efficiency. We have reported a parallel increase

TABLE II   Organ Chemiluminescence[a]

Organ	Chemiluminescence (CPS cm^{-2})
Liver	
In situ	10 ± 2
In situ (under HPO[b])	83 ± 12
In situ (*t*-BuOOH-infused)	135 ± 15
Perfused	7 ± 3
Perfused (*t*-BuOOH-infused)	245 ± 18
Brain	
In situ	72 ± 5
In situ (under HPO[b])	140 ± 11
In situ (*t*-BuOOH-infused)	1500 ± 144
Lung	
Perfused (*t*-BuOOH-infused)	160 ± 25
Perfused (H$_2$O$_2$-infused)	75 ± 5

[a]Values of chemiluminescence are expressed over background level and in function of the area exposed. The experimental conditions are described elsewhere [27, 33, 35, 36].
[b]Hyperbaric oxygen conditions at 0.6 mPa.

of chemiluminescence and glutathione release in the perfused rat liver upon hydroperoxide infusion [26], although at low infusion rates glutathione release seems a more sensitive indicator.

Chemiluminescence of the intact organ provides a noninvasive method for estimating the rate of oxidative radical reactions [26, 27, 33]. Chemiluminescence measurements may give a more sensitive estimation of lipoperoxidative processes than the thiobarbituric acid assay [21, 22, 34].

A. Spontaneous Chemiluminescence of Organs

Mammalian liver shows a weak light emission. Tarusov and colleagues [3] were the first to observe it. The level of the detected signal depends largely on the equipment used; in our measurements [26, 27, 35], photoemission was about 200-fold higher than that reported by Tarusov *et al.* [3]. Figure 1 is a scheme of a single-photon-counting apparatus used to measure organ chemiluminescence [35]. *In situ* and perfused liver show comparable spontaneous light emission, the value of which is about 7-fold lower than that of brain when equal photoemissive areas are compared (Table II). Spontaneous light emission of *in situ* brain and liver was found to be enhanced in tumor-bearing animals (Fig. 2) [37], an observation that is reminiscent of the decrease in catalase activity found in most tumors [38].

B. Induced Chemiluminescence of Organs

Spontaneous chemiluminescence can be enhanced by different experimental conditions that involve a challenge to the cellular antioxidant defenses and make them insufficient to control the radical reactions associated to varyious degrees with aerobic metabolism. Examples of conditions capable of triggering oxidative reactions are hyperoxia, oxidative metabolism of organic hydroperoxides, redox cycling of exogenous compounds (e.g., paraquat and antineoplastic agents), and toxic chemicals (carbon tetrachloride and hydrazine).

The values of organ spontaneous chemiluminescence are dramatically increased during hyperoxia or during infusion of *tert*-butyl hydroperoxide (Table II). Organic hydroperoxide-induced light emission of brain is 10-fold higher than that of liver. This higher sensitivity of brain is in accord with its lower glutathione peroxidase and catalase activities along with a relative lower [vitamin E]/[polyunsaturated fatty acid] ratio.

Organic hydroperoxides increase light emission effectively in both *in situ* and perfused liver (Fig. 3); infusion of *tert*-butyl hydroperoxide into either the portal vein of an anesthetized animal or the perfusion fluid increased light emission up to ~30 times. A detail of the initial burst of photoemission, characteristic of *in situ* liver chemiluminescence, is shown in the inset of Fig. 3a. Hydroperoxide-

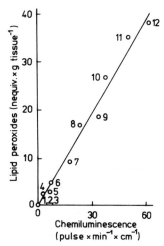

Fig. 2. Relationship between low-level chemiluminescence and content of endogenous peroxides in organs and tissues form normal animals and tumor-bearing animals. 1, Rat skeletal muscles (hindlimbs); 2, mouse sarcoma 180 (days 8–18); 3, Walker rat carcinoma (days 6–20); 4, rat brain (normal); 5, mouse omentum (normal); 6, rat liver (normal); 7, mouse brain with Ehrlich carcinoma (day 11); 8, liver of rat with Walker carcinoma (day 10); 9, 10, brain and liver of mouse with sarcoma 180 (day 18); 11, 12, liver and brain of mouse with sarcoma 180 (day 38). From Neifakh [*37*].

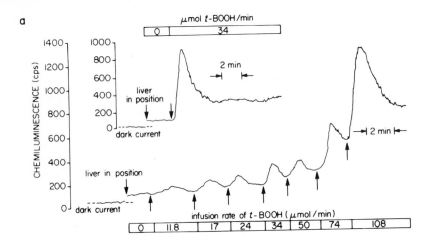

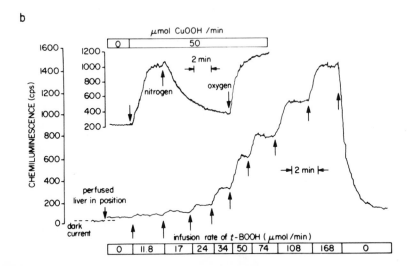

Fig. 3. Organ chemiluminescence. (a) Chemiluminescence of *in situ* liver. *tert*-Butyl hydroperoxide was infused in an intestinal branch of the portal vein at rates indicated by the numbers in the blocks. Inset shows detail of the light emission overshoot observed upon changing hydroperoxide infusion rates. An EMI 9658 photomultiplier was used. (b) Chemiluminescence of the perfused liver. *tert*-Butyl hydroperoxide and cumene hydroperoxide were infused at rates indicated by the numbers in the blocks. For traces in main figure and in inset, red-sensitive EMI 9658 and green-sensitive RCA 8850 phototubes were used, respectively. From Boveris *et al.* [26].

dependent light emission of *in situ* and perfused liver shows similar photoemission at high hydroperoxide infusion rates. However, at low hydroperoxide infusion rates chemiluminescence is higher in the *in situ* liver [26]. Temperature and light absorption of the tissue differentiate the two cases. Chemiluminescence of red blood cell suspensions supplemented with hydroperoxide depends on both hydroperoxide and hemoglobin content. At physiological hemoglobin level (10 mM heme), a significant contribution of blood to *in vivo* organ chemiluminescence is ruled out because light emission is extremely low.

Perfused lung is also a source of light emission when supplemented with organic hydroperoxides [27, 36]. Chemiluminescence of the lung is different in several respects from that of the perfused liver:

1. The concentration of hydroperoxide needed to yield half-maximal chemiluminescence is much higher in the lung.
2. The tolerance of lung to hydroperoxide is greater than that of liver, which shows rapid and severe damage with levels that do not appear to damage the lung. This might be caused by the lower content of polyunsaturated fatty acids in the lung.
3. The chemiluminescence yields with various hydroperoxides are different in lung and liver. Lung shows maximal chemiluminescence with ethyl hydroperoxide, whereas liver shows greater photoemission with cumene hydroperoxide.

Some chemiluminescence reactions involve molecular oxygen or other compounds (peroxides or ozonides) that release molecular oxygen [39]. The oxygen dependence of organ chemiluminescence is shown in the inset of Fig. 3b for the extrinsic hydroperoxide-dependent light emission. The incomplete inhibition of light emission produced by nitrogen-saturated perfusion fluid could be accounted for by an incomplete exclusion of oxygen due to gas diffusion into the perfusion systems or by a persistence of radical reactions in the absence of oxygen.

Chemiluminescence arising from the decomposition of hydroperoxides by the perfused liver seems to be composed mainly of red-light-emitting species [26] with main bands of light emission between 600 and 700 nm; little photoemission seems to be composed of light between 400 and 600 nm. Similar results are observed in the H_2O_2- or *tert*-butyl hydroperoxide-induced photoemission of perfused lung [36].

The molecular source of photoemission is difficult to identify because of the small concentration and rate of production of light-emitting species; 10 cps cm^{-2} corresponds to about 50 cps cm^{-3}, because light could be detected at the most through a 2-mm thickness of the liver lobe. This corresponds to a rate of 8 $\times$ 10^{-20} M^{-1} sec^{-1}. Many side reactions in the complex free-radical reaction sequences may contribute to this minute rate, which is only about 10^{-14} of the overall utilization of *tert*-butyl hydroperoxide or the rate of release of glutathione. Assignment of the prominent red band of chemiluminescence to singlet oxygen dimol emission* is speculative at this stage. Due to the low intensity of this organ chemiluminescence, a higher resolution of this band, as obtained with interference filters, seems difficult to achieve.

The association between oxidative stress exerted by paraquat in perfused organs and lipid peroxidation is a controversial topic. Changes in the [NADPH]/[NADP$^+$] cellular ratio as well as the glutathione status of the cells [42] could be considered to be an index of oxidative stress during the intracellular redox cycling of paraquat, although these changes are not reflected as an increase in spontaneous low-level chemiluminescence in the perfused lung [27] or liver [43]. However, the infusion of organic hydroperoxides into lung of paraquat-treated rats shows higher photoemission than controls [27]. This could be due to the exhaustion of antioxidant defenses upon paraquat treatment.

Pretreatment of animals with hepatotoxic agents such as a carbon tetrachloride and hydrazine enhances the time-dependent spontaneous chemiluminescence of rat liver homogenates [44]. This might reflect *in vivo* peroxidative stress. Photoemission of homogenates is diminished markedly by the *in vitro* addition of antioxidants (e.g., glutathione and vitamin E) or during anaerobiosis [44]. The greater peroxidative capacity of the brain generates a time-dependent spontaneous chemiluminescence 10-fold higher than that of the normal liver [45, 46]. Spectral analysis of oxygen-induced light emission of brain and liver homogenates [46] suggests that singlet molecular oxygen might be formed as a consequence of the peroxidation process (the latter monitored by the formation of thiobarbituric acid-reactive material).

III. CELLULAR CHEMILUMINESCENCE

A. Blood Cells

1. Neutrophils

Reports on cellular chemiluminescence emphasize polymorphonuclear leukocytes, perhaps because of the implications of singlet molecular oxygen in bactericidal action.

The cyanide-insensitive respiratory burst that accompanies phagocytosis has been associated with increased pentose phosphate shunt activity and H_2O_2 production [47–49]. Phagocytosis apparently involves the activation of a membrane-bound NADPH oxidase that primarily generates O_2^- from H_2O_2, and

*Molecular pair interaction of two singlet oxygen molecules ($^1O_2$) emits maximally at 634 and 703 nm with very little light at 668 nm [40, 41].

eventually HO· radicals are produced. The report of Steele and colleagues [*11*] in 1972, which conferred a direct biological role on singlet oxygen in leukocyte microbicidal action, generated much controversy. Later, another possible source of chemiluminescence, through the production of excited carbonyl groups triggered by singlet oxygen-mediated reactions, was suggested [*50*]. Although the evidence for the participation of singlet oxygen was not conclusive [*11, 50*], the pioneering work on chemiluminescence in this laboratory brought a renovating view to the field of microbicidal activity of leukocytes.

Singlet oxygen has been implicated in the phagocytosis process that accompanies the initial oxidative burst [*11, 51, 52*] and in the activity of the isolated enzyme myeloperoxidase [*52*]. Evidence for the latter was based on the effect of singlet oxygen quenchers and enhancement by the presence of D_2O. Conversely there are indications that light emission of intact polymorphonuclear leukocytes is related to the oxygen uptake of the respiratory burst as well as to O_2^{-} production, but it is apparently independent of myeloperoxidase activity [*12*]. This is favored by the insensitivity to cyanide of the chemiluminescence observed (cyanide is an effective inhibitor of myeloperoxidase). Moreover, the myeloperoxidase content of leukocytes ($0.08 \ \mu M$) [*53*] cannot account for the total chemiluminescence yield of the polymorphonuclear leukocytes because fivefold lower chemiluminescence yields were observed with two to five times more enzyme [*54, 55*].

A singlet oxygen contribution to the bactericidal action of leukocytes was suggested by Krinsky [*56*] on the basis of the fact that β-carotene is an effective physical quencher of singlet oxygen [*57*]. No significant killing of a carotenoid-containing wild type of *Sarcina lutea* by polymorphonuclear leukocytes was observed during the first 90 min, whereas the colorless, carotenoid-lacking mutant was rapidly killed [*56*].

2. Macrophages

The oxygen-dependent biochemical reactions of alveolar macrophages are not well known, and it appears that the oxygen species are formed in lower amounts than those described for polymorphonuclear leukocytes [*58*]. The intensity of chemiluminescence of alveolar macrophages has been shown to be lower than that of neutrophils, and luminol was used to amplify the chemiluminescence signal. Little research has been carried out in the absence of luminol. High-sensitivity photon counters showed an ultraweak light emission that has been reported to be sensitive to superoxide dismutase, catalase, and benzoate [*59–61*], which probably reflects a link between the formation of excited chemiluminescence species and the product and metabolism of O_2^{-} and H_2O_2 [*62, 63*]. Spectral analysis of this chemiluminescence was carried out with cut-off filters [*13*], thereby lacking the resolution to identify unequivocally the photoemissive species involved; the red band of chemiluminescence, however, accounted for most of the total light emission observed. Spontaneous chemiluminescence of alveolar macrophages (in the absence of activator) is of the order of 12 to 15 cps per 10^6 cells and increases about 10-fold (induced chemiluminescence) on addition of activators.

3. Eosinophils

Eosinophils are also capable of phagocytosing particles and can ingest and kill bacteria, although both processes occur more slowly than with neutrophils [64–66]. In spite of this, oxygen uptake, hexose monophosphate shunt activity, and H_2O_2 production are higher in the phagocytosing eosinophil than in the neutrophil [66]. The generation of O_2^{-} by the eosinophil has been reported to be equal to or higher than that of the neutrophil [65]. Zymosan-induced light emission of human eosinophils and the more studied chemiluminescence of neutrophils are similar [66], but a precise link of the former chemiluminescence to O_2^{-} generation remains to be established.

4. Erythrocytes

Human erythrocyte stroma shows chemiluminescence upon the addition of iron and immediately after UV irradiation [67]. Preirradiation, which destroys internal antioxidants of the membrane, is required to observe this photoemission. In agreement with this view, vitamin E added to irradiated erythrocytes decreases light emission intensity [67].

A system with the capacity to generate O_2^{-} and H_2O_2 can cause lysis of erythrocytes and peroxidation of liposomes [68]; it was suggested that lysis involves HO· and singlet oxygen, both formed by the reactions between O_2^{-} and H_2O_2. Moreover, preloading erythrocyte membranes with lipid hydroperoxides (by exposing them to a photochemical flux of singlet oxygen) sensitizes them to the lytic attack of enzymatically generated O_2^{-} [69].

Erythrocyte chemiluminescence is also induced by the addition of formalin, which interacts with protein components of the membrane to expose membrane lipids and cause their oxidation [70]. Although the intensity of light emission sharply rises at the initial stage of lipid peroxidation, the accumulation of end products of peroxidation is accompanied by the disappearance of photoemission.

Blood plasma in normal conditions shows a small chemiluminescent signal which is attributed to the content of certain peroxides, probably lipoperoxides [71]. Upon the addition of bivalent iron, blood plasma shows the typical pattern of light emission of lipid-containing systems (see Section IV,A).

Chemiluminescence can by observed in red blood cells only under conditions that increase the fragility of the membrane by the depletion of antioxidants [67,

70] or by a massive attack of oxidizing species [68, 69]. Moreover, the iron-induced chemiluminescence and lipid peroxidation described in mitochondria and microsomes are absent in red blood cells. Although it is difficult to evaluate the contribution of antioxidants and membrane structural arrangements to erythrocyte resistance to lipid peroxidation, it seems that red blood cells have a remarkable resistance to the lipid peroxidative process.

B. Isolated Eukaryotic Cells

1. Hepatocytes

Oxygenation of isolated hepatocytes leads to an increase in low-level chemiluminescence and to an accumulation of malondialdehyde [14], both occurring after a lag phase of 20 to 40 min, in a manner similar to that described for liver homogenates [46]. Malondialdehyde accumulation and chemiluminescence are threefold higher in hepatocytes depleted of glutathione, thus establishing a link between the glutathione status of the cell and chemiluminescence, the latter reflecting a peroxidative state. Moreover, malondialdehyde accumulation and chemiluminescence can be dissociated upon addition of SH-reducing agents (e.g., dithioerythritol and glutathione) [14].

2. Acanthamoeba castellanii

The dependence of low-level photoemission of A. castellanii on oxygen concentration suggests the presence of two components [8]. The intensity of light emission is similar in organisms that have developed different oxidase activities and proportions of mitochondrial respiration to total oxygen uptake. Thus, light emission is not related directly to the reduction of oxygen at respiratory sites other than cytochrome c-oxidase. As observed with polymorphonuclear leukocytes [12], cyanide exerts a stimulatory effect on the chemiluminescence of A. castellanii cells. Photoemission stimulation is dependent on cyanide concentration, resembling closely the inhibitor sensitivity to superoxide dismutase, an enzyme present in many subcellular fractions. Cyanide-stimulated chemiluminescence was interpreted via an increased intracellular concentration of O_2^-, which might, through a reaction with H_2O_2, increase the generation rate of HO· radicals [72, 73]. The formation of oxygen free radicals would lead to lipid peroxide formation and chemiluminescence.

3. Saccharomyces cerevisiae

The weak chemiluminescence observed in the yeast S. cerevisiae [10] seems to lend support to the so-called mitogenic radiation [17, 18]. This chemiluminescence is related to cell division (a yield of one photon per cell division was reported [10] and should be distinguished from the photoemission due to incidental by-products of biological reactions that happen to be chemiluminescent. This phenomenon could be very difficult to detect in asynchronous cultures.

IV. CHEMILUMINESCENCE OF SUBCELLULAR FRACTIONS

A. Mitochondria

1. Iron-Induced Chemiluminescence of Mitochondria

Since 1960 the Russian school, led primarily by Tarusov, Vassil'ev, and Vladimirov, has attributed the main chemiluminescence of organs and cellular and subcellular fractions to the nonenzymatic oxidation of lipids.

The formation of lipid peroxides in mitochondria [74] is associated in certain conditions with the enzymatic system of electron transport and is dependent on the presence of iron. Adrenal cortex mitochondria, which contain the iron–sulfur ferredoxin in a 10 M excess over cytochrome oxidase, are subject to physiologically significant iron-mediated lipid peroxidation [75]. Thus, a lipofuscin-like Schiff base compound in testis mitochondria (which have an electron transfer system that is similar to that of the adrenal cortex) accumulated as the age of rats increased [76].

Vladimirov and colleages have reported extensively on the chemiluminescence of isolated mitochondria [77]. Spontaneous chemiluminescence of mitochondria is almost negligible, but it can be enhanced substantially upon the addition of iron. Figure 4 illustrates the phases of iron-induced chemiluminescence: a latent period (1–2 min) followed by a rapid burst, depression of photoemission, and a slower increase. The nature of the reactions in the first burst was unnoticed for a relatively long time, because chemiluminescence measurements were initiated after termination of the rapid burst [78]. The first flash of light emission is accompanied by a parallel accumulation of the products of lipid peroxidation. Malondialdehyde accumulation ceases 9 min after the addition of iron, and chemiluminescence declines only very slowly.

The relationship between accumulation of malondialdehyde and chemiluminescence seems to depend on the presence of iron in the medium. However, it is not the catalytic activity of iron on lipid peroxides that emits light, but the self-reaction of the lipid peroxyl radicals that are formed [79, 80]. The addition of iron to uv-radiated mitochondria containing lipid peroxides produces an increase in the decomposition of lipid hydroperoxides, oxidation of iron, oxygen uptake, and chemiluminescence [81].

The polyphasic iron-induced chemiluminescence of isolated mitochondria (Fig. 4) has the following characteristics:

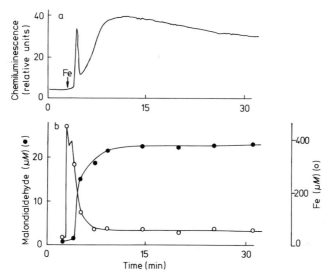

Fig. 4. Development of chemiluminescence, change in the concentration of iron, and malondialdehyde accumulation in suspensions of mitochondria after the addition of FeCl$_2$. (a) Intensity of low-level chemiluminescence in relative units; (b) concentration of iron (○) and malondialdehyde accumulation (•). From Suslova *et al.* [77].

1. Chemiluminescence is dependent on oxygen concentration because an interruption of oxygen supply markedly weakens or even abolishes the chemiluminescence signal. Certain processes in the system, however, are not arrested, because a restoration of oxygen supply causes more intense chemiluminescence. Apparently, in the absence of oxygen, R·-dependent free radical reactions still proceed with concomitant disintegration and solubilization of the membrane [82].

2. Chemiluminescence is dependent on the concentration and redox state of iron. Although mitochondria in suspension accumulate and partially reduce trivalent added iron, an active transport of iron is absent in mitochondria [83]. At high concentration, iron acts as an antioxidant [80], whereas at low concentration iron accelerates lipid peroxidation and chemiluminescence markedly. Processes that reduce and oxidize iron regulate the reactions of lipid peroxidation and chemiluminescence [84]. The effects of ascorbate, glutathione, and cysteine on iron-dependent mitochondrial photoemission could be pro- or antioxidative depending on their reduction of Fe^{3+} to Fe^{2+} [85].

3. The metabolic state of mitochondria does not affect the kinetics of lipid peroxidation and chemiluminescence [83].

4. Carcinogenic hydrocarbons enhance two- to threefold the iron-induced chemiluminescence of mitochondria [86], but they affect neither the kinetics of chemiluminescence nor the amount of malondialdehyde accumulated. Therefore, the carcinogenic hydrocarbons serve as physical activators of chemilumines-

cence (as a result of energy migration into the excited state of the hydrocarbon) and do not influence the initiation of the free-radical chain reactions of lipid peroxidation.

5. The chemiluminescence of iron-supplemented rat liver mitochondria varies with stages of the estral cycle [87]. Antioxidant activity increases toward the estrus stage. The light emission latent period (Fig. 4) increases when the blood content of estrogens increases. This indicates not only antioxidant activity of the estrogens [88], but also the influence *in vivo* of endogenous estrogens on lipid peroxidation of the mitochondrial membrane.

2. Organic Hydroperoxide-Induced Chemiluminescence of Mitochondria

Mitochondrial spontaneous chemiluminescence is increased during the metabolism of organic hydroperoxides [89, 90]. It seems that this light emission is dependent on mitochondrial oxidized cytochrome *c*. This is based on the fact that chemiluminescence intensity is lower than mitochondria are depleted by cytochrome *c*, whereas the cytochrome *c*-containing supernatant retains chemiluminescence activity upon reaction with organic hydroperoxides. Furthermore, the light emission of intact mitochondria and submitochondrial particles is largely enhanced by externally added ferricytochrome *c* [89]. In its oxidized state, mitochondrial cytochrome *c* is more effective in yielding chemiluminescence upon interaction with organic hydroperoxides. Inhibitors of the mitochondrial respiratory chain that block electron flow to cytochrome *c* (antimycin and rotenone) ensure a larger decomposition of hydroperoxide and an enhanced chemiluminescence. Cyanide, however, which keeps cytochrome *c* in a reduced state, exerts an inhibitory effect on light emission [89].

It is known that heavy-metal complexes (e.g., heme) and oxidized hemoproteins can initiate lipid peroxidation reactions [91, 92]. Presumably, a homolytic scission of the peroxide by hemoproteins would trigger the formation of radical species and subsequently attack polyunsaturated fatty acids. In this regard the proximity of hemoproteins such as cytochrome *c* to the unsaturated lipid environment of mitochondria [76] would favor chemiluminescence and lipid peroxidation.

Hydroperoxide-induced chemiluminescence of mitochondria [89, 90] is inhibited by β-carotene and is enhanced by DABCO (an enhancer of singlet oxygen dimol emission in aqueous media [93]). This suggests the formation of singlet oxygen during the oxidative metabolism of hydroperoxide in mitochondria. This does not exclude other possible photoemission sources, as it occurs in systems undergoing lipid peroxidation [24, 94, 95].

Partial inhibition caused by superoxide dismutase on hydroperoxide-dependent chemiluminescence of submitochondrial particles [89] and the competitive effect between superoxide dismutase and *tert*-butyl hydroperoxide [90] indicate a likely contribution of O_2^{-} to the reactions leading to chemiluminescence under certain conditions. This is supported by the enhancing effect on chemiluminescence of antimycin and rotenone, because both inhibitors have been reported to increase the mitochondrial generation of O_2^{-} [96]. Superoxide anion has been claimed to participate in lipid peroxidation reactions [68, 69] either by a self-reaction with polyunsaturated lipids or invoking a metal-catalyzed Haber–Weiss reaction [72, 73] (thus leading to the formation of more effective oxidizing species). In spite of the reported indirect contribution of O_2^{-} to lipid peroxidation and chemiluminescence reactions, its relative participation has not yet been established.

It is unlikely that the observed inhibitory effect of superoxide dismutase on chemiluminescence is due to a direct molecular collision of the enzyme with singlet oxygen [97] because of the short lifetime of singlet oxygen and the low concentration of superoxide dismutase [98]. The quenching rate of singlet oxygen by water is such that its reaction rate with superoxide dismutase would have to be about 10^{15} M^{-1} sec^{-1} to allow the enzyme to compete with water. Furthermore, superoxide dismutase-inhibitable chemiluminescence [97] has not been proved to be associated with singlet oxygen, and it has been demonstrated [99, 100] that singlet oxygen generated by either chemical or photochemical sources is not quenched by superoxide dismutase. Koppenol [101, 102] has shown on thermodynamic grounds that the superoxide dismutase-catalyzed disproportionation of O_2^{-} is an unlikely source of $^1\Delta g$ type of molecular oxygen, because the potential of the couple superoxide dismutase–Cu^{2+}/superoxide dismutase–Cu^+ is too low to oxidize O_2^{-} to $^1\Delta g$ O_2.

B. Microsomes

1. Iron-Induced Chemiluminescence of Microsomes

Howes and Steele [21, 22] and Belova *et al.* [103] opened a field in microsomal chemiluminescence that was investigated further by Nakano and colleages [23, 24]. Both groups related their observations in chemiluminescence to the level of lipid peroxide, and for optimal photoemission they used conditions that produced maximal microsomal lipid peroxidation [104].

The formation of lipid peroxides in microsomes associated with chemiluminescence can proceed either nonenzymatically or with the enzymatic oxidation of NADPH [105]. In both cases the presence of iron and its reduction by NADPH or ascorbic acid are necessary. The NADPH- or ascorbate-induced light emission in microsomes is related directly to lipid peroxidation, with correlation coefficients greater than 0.99 [106]. This chemiluminescence proceeds in the absence of microsomal hydroxylatable substrates [21]. In their presence chemiluminescence and lipid peroxidation are inhibited [22], presumably because electron transfer is then diverted to the oxidation of the substrate rather than to the reduction of iron by NADPH-cytochrome P-450 reductase activity. Microsomes show an "innate" light emission as a function of time [21], which resembles the photoemission from tissue preparations observed by the Russian workers and which might provide a sensitive and direct measurement of lipid hydroperoxide interactions.

Peroxidizing microsomes activated by either NADPH or ascorbate produce a cyanide-sensitive and chloroform-extractable compound that resembles the hydroxyalkyl peroxides. The role of cyanide as reductant, substrate, or catalyst has not been established [107]. Chemiluminescence without lipid peroxidation is obtained by the addition of cyanide to microsomes, the light emission and lipid peroxidation of which had been suppressed previously by pyrophosphate [108]. This establishes a divergence in the microsomal redox chain between NADPH-induced photoemission and lipid peroxidation (as measured by the thiobarbituric acid assay).

Detailed studies on microsomal lipid peroxidation linked to chemiluminescence suggest the formation of singlet oxygen through a characteristic emission spectrum and its behaviour with β-carotene [23, 24]. The location of emission peaks at or near 520 and 635 nm, along with a shoulder at 580 nm, coincides with the $HClO-H_2O_2-NH_3$ aqueous system [40, 41] known to produce singlet oxygen. β-Carotene quenching of chemiluminescence of this system follows the Stern–Volmer equation with a rate constant of $4 \times 10^9 \, M^{-1} \, sec^{-1}$. The participation of NADPH-cytochrome P-450 reductase and free radicals in microsomal chemiluminescence is substantiated by the effect of known inhibitors and scavengers [23, 24]. These characteristics, along with the quadratic dependence of chemiluminescence on hydroperoxide concentration at early stages of lipid peroxidation, suggest a mechanism for light emission. Singlet oxygen and a compound in the triplet state [109] (possibly a carbonyl compound) could be generated by the self-reaction of peroxyl radicals, according to Russell's mechanism [95, 110]. Superoxide dismutase catalase are ineffective, making unlikely the involvement of O_2^- and H_2O_2 in the formation of singlet oxygen.

A flux of singlet oxygen (generated in an aqueous solution of potassium peroxychromate) leads to the peroxidation of microsomal phospholipids and a rapid decline in NADPH-cytochrome c reductase activity [111, 112]. Also, these effects are not inhibited by superoxide dismutase and $HO\cdot$ scavengers, thus precluding the participation of O_2^- and $HO\cdot$.

Another aspect of microsomal lipid peroxidation is the release from the membrane of complex mixture of products including aldehydes, 4-hydroxyalkenals, alk-2-enals, etc. [113], which can exert cytotoxic effects at a distance from the lipid peroxidation loci. The high affinity of these alkenals for thiol groups [114] explains the rapid depletion of intracellular glutathione in the form of a glutathione-S-conjugate, when hepatocytes are incubated with 4-hydroxynonenal

[*115*]. In addition, 4-hydroxynonenal enhances the O_2-induced chemilumines-cence of isolated hepatocytes in a concentration-dependent manner; the alkenal-driven glutathione depletion of hepatocytes seems not to be related to this en-hanced chemiluminescence, which, in turn, is not associated with an extra mal-ondialdehyde or diene conjugate formation [*115*].

2. Hydroperoxide-Induced Chemiluminescence of Microsomes

The oxidative metabolism of organic hydroperoxides by microsomes is ac-companied by photoemission, oxygen uptake, malondialdehyde accumulation, and degradation of cytochrome P-450 molecule [*116*] (Fig. 5). Cumene hydro-peroxide is more effective than *tert*-butyl hydroperoxide and H_2O_2. The organic peroxide-induced chemiluminescence differs from the NADPH–Fe-induced pho-toemission in that there is no addition of NADPH or other electron donors, and therefore the typical initiation phase of peroxidation is omitted.

NADPH and NADH inhibit hydroperoxide-dependent light emission; in this case hydroperoxides do not undergo free-radical decomposition, but rather are

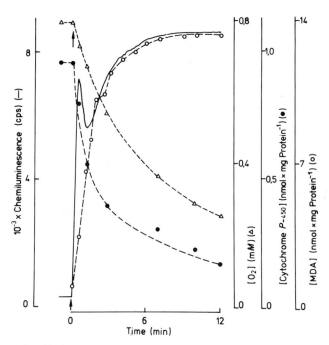

Fig. 5. *tert*-Butyl hydroperoxide-induced chemiluminescence, oxygen uptake, malondialdehyde (MDA) accumulation, and cytochrome P-450 degradation of microsomal fractions. From Cadenas and Sies [*113*].

reduced to the corresponding alcohol over cytochrome P-450 acting in a peroxidase mode [117]. Similarly, organic hydroperoxide-induced lipid peroxidation is inhibited by NADH and NADPH without implying an antioxidant activity of the latter [118].

Monooxygenase substrates also inhibit hydroperoxide-dependent photoemission; in these conditions hydroperoxides could be used to support the monooxygenation of substrates, as already reported to occur in the absence of oxygen and electron donors [119].

In summary, optimal conditions for observing hydroperoxide-induced chemiluminescence as microsomal fractions rely on the oxidized state of cytochrome P-450, as well as on the absence of reducing equivalents and hydroxylatable substrates. The decomposition of hydroperoxides by oxidized cytochrome P-450 would be the first step required for observing this chemiluminescence, followed by a free-radical attack of membrane lipids. In this case, a homolytic scission of the hydroperoxide by cytochrome P-450 would be feasible [120]. An alternative explanation for the mechanism of chemiluminescence, involving cytochrome P-450 activity, takes into account the oxene transferase activity of the cytochrome, which allows the transfer of an oxygen atom from oxene donors like hydroperoxides to substrates of the monooxygenase system [120]. For this reaction, it is postulated a heterolytic cleavage of the O—O bond of the hydroperoxide under formation of a transient $[FeO]^{3+}$ species [121]. This "active oxygen complex" could either hydroxylate or epoxidize a substrate or may react with a second hydroperoxide molecule and yield some active form of oxygen, probably singlet molecular oxygen. Light emission arising from microsomes or isolated cytochrome P-450 supplemented with different oxene donors shows similar photoemission intensities at 634 and 703 nm and a low intensity at 668 nm [122]; this red-band photoemission does not account for the total chemiluminescence observed, therefore, additional photoemissive sources in the 350–600 nm region would be also present.

3. Microsomal Metabolism of Drugs Associated with Chemiluminescence

The NADPH-dependent microsomal metabolism of a proximate carcinogen, 7,8-dihydroxybenzo[a]pyrene, is also associated with chemiluminescence [123, 124]. The emission spectrum of this photoemission shows a broad, structureless band. This is identical to the fluorescence of the eximer of 7,8-dihydroxybenzo[a]pyrene, which the authors [123] conclude is an exiplex chemiluminescence that proceeds through a dioxetane intermediate.

Microsomal NADPH-cytochrome P-450 reductase activity participates in the activation of several compounds, which undergo redox cycling. Herbicides (paraquat), antineoplastic agents, aromatic nitro compounds, quinones, etc., are reduced via a one-electron step to a radical form. Subsequent reaction with oxygen yields $O_2^{\cdot-}$ and other oxygen free-radical species. The activation of paraquat by microsomes and its aerobic metabolism are linked to chemiluminescence and a rapid oxygen uptake [43]. The reaction proceeds only under aerobic conditions. In anaerobiosis, paraquat radical accumulates and no photoemission is observed; of course, a rapid burst in chemiluminescence is observed when oxygen is allowed to react with paraquat radical. This light emission shows similar intensities at 634 and 703 nm and little light emitted at 668 nm, a criterion for the identification of singlet oxygen dimol emission.

Because during paraquat-induced chemiluminescence of microsomes no malon dialdehyde is formed, mechanisms other than the self-reaction of lipid peroxides should be operative in this system. Paraquat chemiluminescence is totally abolished by superoxide dismutase, giving further support to the hypothesis that the enzymatic dismutation of $O_2^{\cdot-}$ is not a source of singlet oxygen. Although the nonenzymatic dismutation of $O_2^{\cdot-}$ ($HO_2^{\cdot-}$) is considered to yield singlet oxygen [101, 125–128], evidence against this has been brought forward [129]. In summary, electron flow through the NADPH-cytochrome P-450 reductase has two alternative pathways. In the absence of paraquat, this electron transport leads to lipid peroxidation and chemiluminescence. In the presence of paraquat, electrons are channeled to paraquat reduction to form paraquat radical and support further redox cycling and chemiluminescence.

It is to be noted that paraquat infused into the perfused liver or lung does not modify the spontaneous chemiluminescence of the organ (see Section II) [27, 43]. However, an absence of photoemission in the intact organ does not indicate that the sources of light emission observed in microsomes do not occur, but rather suggest that they are quenched by existing cellular components (superoxide dismutase, glutathione peroxidase, catalase, etc.) absent in isolated microsomal fractions.

V. CHEMILUMINESCENCE OF BIOLOGICALLY RELATED MODEL SYSTEMS

A. Effect of Catalysts on Oxidation of Fatty Acids

Molecular oxygen rapidly oxidizes unsaturated fatty acids at elevated temperatures. Chemiluminescence accompanies the peroxidation of films of unsaturated fatty acids exposed to oxygen at 85°C [130]. The chemiluminescence spectrum changes during the autoxidation reaction; a red-enhanced photoemission is characteristic of the final stages of the free-radical reaction (Fig. 6). Several chemical species of unknown structure fluorescing in the visible and in the UV region are formed during autoxidation of linolenic acid; however, a contribution of these

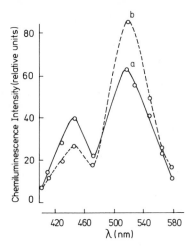

Fig. 6.   Chemiluminescence spectra of linoleic acid. (a) Forty-minute oxidation; (b) 5-h oxidation. The spectrum was evaluated with a set of interference light filters. From Orlov *et al.* [*130*].

substances to photoemission is unlikely because their concentration is extremely low and the overlapping of chemiluminescence emission and fluorescence excitation is poor [*130*]. A spectral distribution of ultraweak chemiluminescence accompanying autoxidation of oleic acid at 77°C has been obtained with a modern photon-counting spectral analyzing system; the spectrum shows five emission maxima at approximately 440, 480, 530, 575, and 620 nm; emission at these wavelengths was assigned to simultaneous transitions of singlet oxygen pairs [*131*]. Another report on light emission accompanying the autoxidation of linoleic acid attributes the emission to singlet oxygen and excited ketone groups [*132*].

Oxidation of unsaturated fatty acids and the accompanying chemiluminescence are accelerated by iron, ceric anion, hemoproteins, or lipoxygenase [*131, 133, 134*]. Iron-catalyzed chemiluminescence of unsaturated fatty acids proceeds autocatalytically for low iron concentrations and develops a pulselike burst of photoemission at high iron levels. The pulse of maximal photoemission is proportional to $[ROOH]^2$ [*134*]. Moreover, chemiluminescence in the presence of iron was proposed as a sensitive assay for endogenous hydroperoxides.

Chemiluminescence associated with unsaturated fatty acid oxidation, either by autoxidation or in the presence of catalysts, depends on the hydroperoxide concentration in the system. Basically, quantum yields and spectral distribution of photoemission are similar in these systems, but kinetic differences are observed.

By using photosensitized luminescence of singlet oxygen, the rate constants for quenching of singlet oxygen by saturated and unsaturated fatty acids were recently determined [*135*]. It was suggested that quenching proceeded by a physical mechanism in the case of saturated fatty acids and by a chemical mechanism in the case of unsaturated fatty acids.

B. Hydroperoxide-Supplemented Hemoproteins

The oxidation of unsaturated fatty acids in the presence of hemoproteins (hemoglobin) leads to light emission with a kinetic pattern that depends on hydroperoxide content in the fatty acids. A drop in chemiluminescence is related to the decline in concentration of hemoprotein destroyed by reaction products. The spectral composition of sodium linoleate chemiluminescence in the presence of hemoglobin or cytochrome c does not differ from that observed in their absence [130], indicating that hemoprotein increases the oxidation rate of the unsaturated fatty acid without changing the reaction mechanism [134].

The decomposition of linoleic acid hydroperoxide by hematin, methemoglobin, and ceric anion is also accompanied by chemiluminescence [136]. The sensitivity of this system to singlet oxygen traps and quenchers, along with an enhancement of light emission by D_2O, led the authors to postulate singlet oxygen as the source of light emission.

Ferricytochrome c supplemented with organic hydroperoxides also yields light emission. This is extended to other hemoproteins such as methemoglobin (MetHb), horseradish peroxidase (HRP), and myoglobin (Mb). When the intensity of the hydroperoxide-induced light emission with different hemoproteins is compared on a heme basis, the following order is found:

$$\text{cyt } c < \text{Mb} = \text{MetHb} < \text{HRP}$$

Light emission intensity is maximal at a [tert-butyl hydroperoxide]/[ferricytochrome c] ratio of about 160. Neither light emission or oxygen uptake is observed when this ratio is equal to 1.0, presumably due to the antioxygenic effect of cytochrome c at high concentrations [137].

Homolytic scission of hydroperoxide by cytochrome c and its catalysis of lipoperoxidation reactions involves free-radical-mediated irreversible damage to the hemoprotein [138, 139]. This is illustrated by a decrease in its absorbance [140] along with an increase in photoemission. Protein damage relies presumably on either cross-linking of the protein by lipid peroxyl radical addition or a process not involving incorporation of lipid radicals [138].

Participation of O_2^- in this chemiluminescent system is unlikely because it does not involve a superoxide dismutase-sensitive reduction of cytochrome c. This is at variance when NADH, as a reductant, is present in the system [141]. The capacity of cytochrome c to induce lipid peroxidation and to yield alkoxyl and HO· radicals as well as its involvement in hydroperoxide breakdown, probably through a unimolecular bond scission, has been described [142]. The formation of peroxyl free radicals by hematin/hydroperoxide systems has been described on the basis of direct electron spin resonance and spin trapping evidence [143].

It could be postulated that the photoemission arising from a hydroperoxide and cytochrome c reaction is part of a process that involves (a) a cytochrome c-catalyzed homolytic scission of the hydroperoxide, (b) a free-radical oxidation process, and (c) the generation of photoemissive species.

The homolytic scission of the hydroperoxide by cytochrome c [Eq. (1)] requires the binding of the hydroperoxide to the hemoprotein molecule. This is

$$\text{cyt } c^{3+} + t\text{-BuOOH} \rightarrow \text{cyt } c^{3+} \text{---OH·} + t\text{-BuO·} \tag{1}$$

supported by the displacement of hydroperoxide by cyanide and subsequent inhibition of chemiluminescence by cyanide [141, 142]. Primary [140, 141] secondary [137, 142], and tertiary [140, 144] hydroperoxides seem capable of binding cytochrome c and giving rise to light emission. Iron-bound HO· may react with either $tert$-butyl hydroperoxide [Eq. (2)] or with residues of the cytochrome c molecule (RH) [Eq. (3)].

$$\text{cyt } c^{3+}\text{---OH·} + t\text{-BuOOH} \rightarrow \text{cyt } c^{3+} + t\text{-BuOO·} + H_2O \tag{2}$$

$$\text{cyt } c^{3+}\text{---OH·} + RH \rightarrow \text{cyt } c^{3+} + H_2O + R· \tag{3}$$

A propagation of free-radical reactions seems feasible, and their participation in chemiluminescence is supported by the inhibitory effect of radical scavengers. Oxygen consumption accompanies chemiluminescence [140, 144], probably through the incorporation of oxygen molecules into alkyl radicals. The formation of hydroperoxy radicals in this system can also be explained by mechanisms other than the primary homolytic scission of the hydroperoxide by the hemoprotein. One of them implies a valency change of the hemoprotein by the hydroperoxide ($Fe^{3+} \rightarrow Fe^{2+}$); oxidation of this intermediate would be also associated with oxygen consumption.

Equations (4) and (5) represent radical termination reactions that can be interpreted on the basis of Russell's mechanism [110] for self-reaction of primary or secondary peroxyl radicals and on the basis proposed by Pryor [145] for tertiary peroxyl radicals. [Eqs. (4) and (5)].

$$\text{ROO·} + \text{ROO·} \rightarrow RO + ROH + {}^1O_2 \tag{4}$$

$$t\text{-BuOO·} + t\text{-BuOO·} \rightarrow t\text{-BuOOBu-}t + {}^1O_2 \tag{5}$$

The observed light emission could take place partly through Eq. (6).

$$ {}^1O_2 + {}^1O_2 \rightarrow 2\ {}^3O_2 + h\nu \tag{6}$$

This photoemission is sensitive to β-carotene and DABCO [140], and spectral analysis carried out with a wedge interference filter shows maximal light emission between 618 and 650 nm [146, 147]. It seems that wavelength depends on the nature of the hydroperoxide that is decomposed rather than on the nature of the hemoprotein itself, because cytochrome c- or myoglobin-containing systems peak at the same wavelength when they are supplemented with the same hydroperoxide.

The generation of excited species occurs at a rate of ~ 1 to 2×10^4 photons per second under described conditions [140, 146, 147]. Assuming that part of the photoemission is due to singlet oxygen, it would correspond to the formation of one $[^1O_2-^1O_2]$ pair per 10^9 molecules of singlet oxygen generated and per $\sim 10^{12}$ molecules of oxygen utilized in the free-radical chain; the steady state of singlet oxygen would be $\sim 10^{-15}$ M.

The addition of lipid vesicles to this system offers a better extrapolation to the membrane-bound electron transport chain [148]. Fatty acid unsaturation is essential to chemiluminescence enhancement, and oxygen uptake rates parallel chemiluminescence yields. Both parameters behave in agreement with the dependence on double bonds and oxygen of the lipid peroxidation free-radical process [76, 145]. When lipids and cytochrome c are part of beef heart submitochondrial particles, chemiluminescence yields and oxygen uptake are threefold higher than the corresponding chemiluminescence and oxygen uptake of a phospholipid–cytochrome c mixture. Considering that lipid unsaturation is similar in both beef heart submitochondrial particles and soybean phosphatidylcholine vesicles [140], it seems that the proximity of the heme groups to lipids [91] in a specific molecular arrangement gives enhanced efficiency of the reactions leading to lipid peroxidation and chemiluminescence.

C. Chemiluminescence Associated with Enzymatic Activity

The xanthine oxidase reaction [150] shows chemiluminescence while operating on its substrates [151, 152]. Stauff and co-workers [125, 151] suggested that light emission is due to excited states of molecular oxygen produced by recombination of oxygen-containing radicals. They stated that emitted light was in the region of 500 nm. The xanthine–xanthine oxidase chemiluminescent reaction affords a quantum yield of 10^{-11}, which is typical of light emission reactions [39].

Photoemission is augmented markedly by H_2O_2 addition and inhibited by superoxide dismutase and catalase [152]. Because the oxidation of xanthine by xanthine oxidase generates O_2^-, the formation of oxygen in an excited state is explained in terms of a nonenzymatic dismutation of O_2^- [101, 125, 128] [Eq. (7)], followed by Eq. (6).

$$O_2^- + O_2^- + 2\,H^+ \rightarrow H_2O_2 + {}^1O_2 \qquad (7)$$

Spectral distribution arising from the acetaldehyde–xanthine oxidase chemiluminescent system has emission maxima at approximately 470, 540, 585, and

640 nm [131]. By using linoleate as a probe molecule for singlet oxygen, it was shown that singlet oxygen is formed only in minimal amounts during the oxidation of xanthine by xanthine oxidase [153].

The weak chemiluminescence obtained from the xanthine oxidase activity was considered to be entirely dependent on the presence of carbonate radicals [154]. Hydroxyl radical generated by the system reacts with $CO_3{}^{2-}$ to yield the carbonate radical, and the emitting species results from the dimerization of carbonate radicals [126, 154]. This is a complication that should be considered in buffer systems containing bicarbonate. However, the validity of this mechanism should be confirmed by spectral analysis of the light emission.

Oxygen free radicals generated by the xanthine oxidase reaction promote the peroxidation of linoleic acid or microsomal lipids [68, 155, 156]. The identity of the species that are responsible for this peroxidation remains obscure, and the participation of singlet oxygen is generally suggested on the basis of effects of scavengers that are unspecific, that is, that can react with a wide variety of oxidants [157]. It was also proposed that $O_2{}^-$ and H_2O_2 could give rise directly to singlet oxygen through a Haber–Weiss reaction [156].

Another chemiluminescence system dependent on enzymatic activity is the lipoxygenase reaction. The chemiluminescence spectrum of lipoxygenase catalysis of unsaturated fatty acids was reported to have maximal emission at 555 nm and kinetics that correlate with the square of lipid peroxyl radicals concentration [37]. The peroxidation kinetics of linoleate catalysis by lipoxygenase involves an increase in the intensity of ultraweak chemiluminescence, accumulation of conjugated hydroperoxides and secondary keto acids, and accumulation of peroxyl radicals. The maximal rate of accumulation of hydroperoxides corresponds to maximal chemiluminescence, whereas accumulation of the subsequently formed keto acids causes a repeated burst of light emission [37]. Chemiluminescence of the lipoxygenase reaction was reported, with spectral distribution peaking at about 450, 550, and 640 nm [158]. Using a cholesterol probe for singlet oxygen, it was concluded that this species is not released into solution during the lipoxygenase reaction [159]. Further evidence that free singlet oxygen does not participate in peroxidation of linoleate by lipoxygenase has been brought forward [153].

The evidence that the bactericidal action of leukocytes resting on myeloperoxidase activity is due to singlet oxygen [52] was lent further support by the fact that chemiluminescence from a lactoperoxidase–H_2O_2–NH_3 system is sensitive to singlet oxygen scavengers [160] and shows 1270-nm chemiluminescence (indicative of singlet oxygen monomol emission), the intensity of which is over 10-fold enhanced in D_2O medium [161].

Chemiluminescence due to the reaction between superoxide dismutase and H_2O_2 [162] is a function of H_2O_2 concentration, and its duration is coincident with inactivation of the enzyme by H_2O_2. Light emission intensity is reduced by urate and formate, which protect the dismutase against inactivation by H_2O_2, and decreased by xanthine and imidazole. Both protection mechanisms and the decrease in chemiluminescence involve competition between the sensitive group of superoxide dismutase and the substrate; in this case the peroxidized substrate produced light with lower efficiency than the group in the enzyme. Thus, for this system the mechanism of chemiluminescence seems to involve some group of the dismutase that is oxidized to an excited state and decays to emit light. This observation emphasizes the risk of using dismutase in chemiluminescence systems producing H_2O_2.

The oxidation of arachidonic acid by prostaglandin synthase is associated with the formation of a strong oxidizing equivalent and chemiluminescence [163–166]. This oxidizing equivalent is formed during the prostaglandin hydroperoxidase-catalyzed reduction of prostaglandin G_2 to prostaglandin H_2, and is involved in the cooxygenation or cooxidation of various substrates [165–167]. The possible relation of chemiluminescence to the formation of the oxidizing equivalent, and its identification as singlet molecular oxygen has failed along with the lack of formation of singlet oxygen oxidation products of cholesterol [163]. However, the spectral distribution of this strong chemiluminescence signal was obtained by the use of a sensitive photon-counting technique [168]. The prominent bands at 634 and 703 nm, along with their enhancement by DABCO, suggest that singlet oxygen is formed during this reaction, probably involving the intermediate generation of an oxo-ferryl complex similar to that described for cytochrome P-450 [122]. This similarity is strongly supported by the fact that prostaglandin synthase has the characteristics of cytochrome P-450 and can catalyze the heterolytic cleavage of the 9,11-endoperoxide of 15-hydroxy-arachidonic acid [169]. Equations (8) and (9) illustrate the hypothetical mechanism for the formation of the excited species.

$$Fe^{3+} + PGG_2 \rightarrow [FeO]^{3+} + PGH_2 \tag{8}$$

$$[FeO]^{3+} + PGG_2 \rightarrow Fe^{3+} + PGH_2 + {}^1O_2 \tag{9}$$

As in other chemiluminescence systems, this red photoemission (beyond 600 nm) does not account for the total light emission detected in the system. Chemiluminescence intensity decreases or is even abolished in the presence of cooxidation substrates, such as quinol, GSH, and phenol [168]. Neither a radical quenching capacity or an electron donor activity of these substrates can be elucidated on their unique effect on low-level chemiluminescence. Cooxidation substrates also exert a quenching effect on epr signals arising from the prostaglandin synthesis [165]. Epr studies on these systems have tentatively identified the oxidizing equivalent as a hydroxyl or peroxyl radical [165] and, more recently, as a hemoprotein-derived radical [170].

Another aspect of chemiluminescence associated with enzymatic activity has been developed by Cilento and colleages in the last 10 years (see [*171*]), who have been devoted to the identification of electronically excited species during biochemical reactions. Special consideration were given to those reactions that proceed through a dioxetane mechanism; cleavage of dioxetanes is known to generate a carbonyl in an electronically excited state [Eq. (10)].

$$\underset{0-0}{\square} \rightarrow >C = 0 + >C = 0^* \tag{10}$$

Among others, the systems investigated by this group involve the peroxidase-catalyzed oxidation of several aldehydes (propanal, isobutanal, etc.) with formation of excited carbonyl compounds [*172–174*] over the above-described mechanism. The peroxidase-catalyzed oxidation of malondialdehyde was reported to yield singlet molecular oxygen [*175*], and the disproportionation of formed peroxy radicals [*95, 110*] would be the operative mechanism for its generation.

VI. CONCLUSIONS

Low-level chemiluminescence is apparently an attribute of life; it has been observed in a variety of animal and plant systems. Chemiluminescence may provide quantitative measurements in a wide range of experimental systems from solution to organs *in vivo*.

On the one hand, chemiluminescence results from a series of oxidative radical reactions related to lipid and organic peroxide formation, and it can be used to assess oxidative stress in physiological and pathological situations. Some chemiluminescence reactions might involve, directly or indirectly, the participation of singlet oxygen (Fig. 7), which can be generated either (*a*) by self-reaction of primary or secondary peroxyl radicals via Russell's mechanism [*110*] or (*b*) by the termination reactions of tertiary peroxyl radicals [*145*]. Singlet oxygen might contribute to light emission (*a*) by singlet oxygen dimol emission at 634 and 703 nm or (*b*) by reacting with a double bond, yielding excited carbonyl groups that can decay and emit light. Figure 7 represents the possible relationships between chemiluminescence and lipid peroxidation.

On the other hand, singlet oxygen may be formed in biological systems that are not related to lipid peroxidation and that produce oxygen free radicals, such as O_2^{-} and H_2O_2 (Fig. 8). Examples of such systems are the bactericidal activity of leukocytes, redox cycling of different xenobiotics, and the xanthine oxidase reaction, among others. Although O_2^{-} and H_2O_2 in those systems can be measured directly, the participation of singlet oxygen can only be suggested by the use of chemical traps or quenchers of singlet oxygen. However, there is some evidence that singlet oxygen could be produced when reactants such as O_2^{-} and H_2O_2 are present in the system [*101, 125–128, 156*]. The scheme of Fig. 8

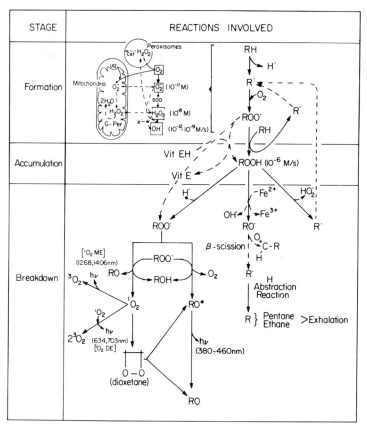

Fig. 7. Scheme showing the generation of photoemissive species as a side reaction of the lipoperoxidation process. The first and second stages are the classical reactions for the formation and accumulation of lipoperoxides. The third stage, the breakdown of lipoperoxides, focuses particularly on radical reactions leading either to chemiluminescence or to the formation of pentane or ethane. Vitamin E (vit EH) stops the chain reaction of lipid peroxidation by reacting with lipid peroxyl radicals (ROO·). The numbers in parentheses express the estimated steady-state concentrations or formation rates of oxygen metabolites. Abbreviations, G-Per, glutathione peroxidase; Cat, catalase; SOD, superoxide dismutase; ROOH, lipid hydroperoxide; RO·, alkoxyl radical; HO·, hydroxyl radical; O_2^-, superoxide anion; $^1O_2$ ME, singlet oxygen monomol emission; $^1O_2$, singlet oxygen dimol emission. From Cadenas *et al.* [*14, 35,* and references therein].

summarizes these possibilities; a detailed discussion on the occurrence of these reactions in biological systems was presented by Krinsky [*176*].

A multitude of light-emitting species are feasible, and both schemes cannot cover all these possibilities. It seems that only careful observation of photoemission in selected bandwidths combined with the effect of known quenchers or enhancers will allow the identification of photoemissive species.

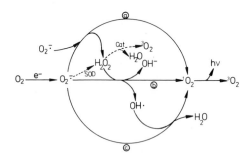

Fig. 8. Scheme showing the possible interrelations between O_2^{-}, H_2O_2, and HO· in the generation of singlet molecular oxygen. (a) Self-disproportionation of O_2^{-}; (b) reaction between O_2^{-} and H_2O_2; (c) reaction between O_2^{-} and HO·. Scheme summarizes experimental and theoretical evidence from Koppenol. [101], Stauff et al. [125, 126], Khan [127, 128], Kellog and Fridovich [156], and Krinsky [176]. A detailed discussion of these reactions was presented by Krinsky [176].

REFERENCES

1. E. N. Harvey, "Living Light." Princeton Univ. Press, Princeton, New Jersey, 1940.
2. E. N. Harvey, "A History of Luminescence." Am. Philos. Soc., Philadelphia, Pennsylvania, 1957.
3. B. N. Tarusov, A. I. Polidova, and A. I. Zhuravlev, *Radiobiologiya* **1**, 150 (1961).
4. N. A. Troitskii, S. V. Konev, and M. A. Katibnikov, *Biofizika* **6**, 238 (1961).
5. B. N. Tarusov, A. I. Polidova, A. I. Zhuralev, and E. N. Sekamova, *Tsitologiya* **4**, 696 (1962).
6. Yu. A. Vladimirov, F. F. Litvin, and Tan'Man'-tsi, *Biofizika* **7**, 675 (1962).
7. G. M. Baremboim, A. N. Domannskii, and K. K. Taroverov, "Luminescence of Biopolymers and Cells," Chapter 4. Plenum, New York, 1969.
8. D. Lloyd, A. Boveris, R. Reiter, M. Filipkowski, and B. Chance, *Biochem. J.* **184**, 149 (1979).
9. J. Stauff and G. Reske, *Naturwissenschaften* **51**, 39 (1964).
10. T. I. Quickenden and S. S. Quee Hee, *Biochem. Biophys. Res. Commun.* **60**, 674 (1974).
11. R. C. Allen, R. L. Stjernholm, and R. H. Steele, *Biochem. Biophys. Res. Commun.* **47**, 679 (1972).
12. K. Kakinuma, E. Cadenas, A. Boveris, and B. Chance, *FEBS Lett.* **102**, 38 (1979).
13. E. Cadenas, R. P. Daniele, and B. Chance, *FEBS Lett.* **123**, 225 (1981).
14. E. Cadenas, H. Wefers, and H. Sies, *Eur. J. Biochem.* **119**, 531 (1981).
15. H. H. Seliger, *Photochem. Photobiol.* **21**, 355 (1975).
16. M. H. Pirenne and E. J. Denton, *Nature (London)* **170**, 1039 (1952).
17. A. A. Gurvitsch, V. F. Eremew, and Yu. A. Karabchievskii, *Nature (London)* **206**, 20 (1965).
18. A. G. Gurvitsch and L. D. Gurvitsch, "Die mitogenitsche Strahlung." Fischer, Jena, 1959.
19. W. S. Metcalf and T. I. Quickenden, *Nature (London)* **216**, 169 (1967).
20. J. Stauff and J. Ostrowski, *Z. Naturforsch., B: Anorg. Chem., Org. Chem., Biochem., Biophys., Biol.* **22B**, 743 (1967).
21. R. M. Howes and R. H. Steele, *Res. Commun. Chem. Pathol. Pharmacol.* **2**, 619 (1971).
22. R. M. Howes and R. H. Steele, *Res. Commun. Chem. Pathol. Pharmacol.* **3**, 349 (1972).
23. M. Nakano, T. Noguchi, K. Sugioka, H. Fukuyama, M. Sato, Y. Shimizu, Y. Tsuji, and H. Inaba, *J. Biol. Chem.* **250**, 2404 (1975).
24. K. Sugioka and M. Nakano, *Biochim. Biophys. Acta* **423**, 203 (1976).

25. J. P. Hamman, D. R. Gorby, and H. H. Seliger, *Biochem. Biophys. Res. Commun.* **75,** 793 (1977).
26. A. Boveris, E. Cadenas, R. Reiter, M. Filipkowski, Y. Nakase, and B. Chance, *Proc. Natl. Acad. Sci. U.S.A.* **77,** 347 (1980).
27. E. Cadenas, I. D. Arad, A. B. Fisher, A. Boveris, and B. Chance, *Biochem. J.* **192,** 303 (1980).
28. N. Oshino and B. Chance, *Biochem. J.* **162,** 509 (1977).
29. G. M. Bartoli and H. Sies, *FEBS Lett.* **86,** 89 (1978).
30. A. L. Tappel and C. I. Dillard, *Fed. Proc., Fed. Am. Soc. Exp. Biol.* **40,** 174 (1981).
31. C. A. Riely and G. Cohen, *Science* **183,** 208 (1974).
32. H. Kappus, U. Koster, D. Koster-Albrecht, H. Kieczka, and H. Remmer, *in* "Functions of Glutathione in Liver and Kidney" (H. Sies and A. Wendel, eds.), p. 176. Springer-Verlag, Berlin and New York, 1978.
33. A. Boveris, B. Chance, M. Filipkowski, Y. Nakase, and K. G. Paul, *in* "Frontiers of Biological Energetics" (P. L. Dutton, J. S. Leigh, and A. Scarpa, eds.), Vol. 2, p. 9750, Academic Press, New York, 1978.
34. T. Noguchi and M. Nakano, *Biochim. Biophys. Acta* **368,** 446 (1974).
35. A. Boveris, E. Cadenas, and B. Chance, *Fed. Proc., Fed. Am. Soc. Exp. Biol.* **40,** 295 (1981). E. Cadenas and H. Sies, *Methods Enzymol.* **105,** 221 (1984).
36. E. Cadenas, I. D. Arad, A. Boveris, A. B. Fisher, and B. Chance, *FEBS Lett.* **111,** 413 (1980).
37. Ye. A. Neifakh, *Biofizika* **16,** 584 (1971).
38. J. P. Greenstein, "Biochemistry of Cancer," p. 202. Academic Press, New York, 1947.
39. R. F. Vassil'ev and A. A. Vichutinskii, *Nature (London)* **194,** 1276 (1962).
40. A. V. Khan and M. Kasha, *J. Chem. Phys.* **39,** 2105 (1963).
41. H. H. Seliger, *J. Chem. Phys.* **4O,** 3133 (1964).
42. R. Brigelius, R. Lenzen, and H. Sies, *Biochem. Pharmacol.* **31,** 1637 (1982).
43. E. Cadenas, R. Brigelius, and H. Sies, *Biochem. Pharmacol.* **32,** 147 (1983).
44. N. R. DiLuzio and T. E. Stege, *Life Sci.* **21,** 1457 (1977).
45. N. R. Di Luzio and A. D. Hartman, *Fed. Proc., Fed. Am. Soc. Exp. Biol.* **26,** 1436 (1967).
46. E. Cadenas, A. I. Varsavsky, A. Boveris, and B. Chance, *Biochem. J.* **198,** 645 (1981).
47. B. Paul and A. J. Sbarra, *Biochim. Biophys. Acta* **156,** 168 (1968).
48. B. M. Babior, R. S. Kipnes, and J. T. Curnutte, *J. Clin. Invest.* **52,** 741 (1973).
49. S. J. Klebanoff, *J. Biol. Chem.* **249,** 3724 (1974).
50. R. C. Allen, S. J. Yerich, R. W. Orth, and R. H. Steele, *Biochem. Biophys. Res. Commun.* **60,** 909 (1973).
51. L. S. Webb, B. B. Keele, Jr., and R. B. Johnston, Jr., *Infect. Immun.* **9,** 1051 (1974).
52. H. Rosen and S. J. Klebanoff, *J. Biol. Chem.* **252,** 4803 (1977).
53. K. Kakinuma and B. Chance, *Biochim. Biophys. Acta* **480,** 96 (1977).
54. J. E. Harrison, B. D. Watson, and J. Schultz, *FEBS Lett.* **92,** 327 (1978).
55. R. C. Allen, *Biochem. Biophys. Res. Commun.* **63,** 684 (1975).
56. N. I. Krinsky, *Science* **186,** 363 (1974).
57. C. S. Foote and R. W. Denny, *J. Am. Chem. Soc.* **90,** 6233 (1968).
58. R. B. Johnston, Jr., *Fed. Proc., Fed. Am. Soc. Exp. Biol.* **37,** 2759 (1978).
59. G. D. Beall, J. E. Repine, J. R. Hoidal, and F. L. Rasp, *Infect. Immun.* **17,** 117 (1977).
60. P. R. Miles, P. Lee, M. A. Trush, and K. Van Dyke, *Life Sci.* **20,** 165 (1977).
61. P. R. Miles, V. Castranova, and P. Lee, *Am. J. Physiol.* **235,** 103 (1978).
62. A. Holian and R. P. Daniele, *FEBS Lett.* **108,** 47 (1979).
63. A. Holian and R. P. Daniele, *J. Appl. Physiol.* **50,** 736 (1981).
64. R. L. Baechner and R. B. Johnston, Jr., *Br. J. Haematol.* **2O,** 277 (1971).

65. A. I. Tauber, E. J. Goetzl, and B. M. Babior, *Blood* **48**, 968 (1976).
66. I. D. Mickenberg, R. K. Roote, and S. M. Wolff, *Blood* **39**, 67 (1972).
67. M. V. Korchagina and Yu. A. Vladimirov, *Biofizika* **17**, 1089 (1972).
68. E. W. Kellog and I. Fridovich, *J. Biol. Chem.* **252**, 6721 (1977).
69. R. E. Lynch and I. Fridovich, *J. Biol. Chem.* **253**, 1838 (1978).
70. N. V. Levina, V. E Nobikov, T. D. Yesakova, Yu. M. Petrusevich, and B. N. Tarusov, *Biofizika* **20**, 963 (1975).
71. Yu. A. Vladimirov, A. P. Sharov, and E. F. Malyugin, *Biofizika* **18**, 157 (1973).
72. J. M. McCord and E. D. Day, *FEBS Lett.* **86**, 139 (1978).
73. B. Halliwell, *FEBS Lett.* **92**, 312 (1978).
74. Yu. A. Vladimirov, O. F. L'Ova, and Z. P. Cheremesina, *Biokhimiya (Moscow)* **31**, 442 (1966).
75. H. P. Wang and T. Kimura, *Biochim. Biophys. Acta* **423**, 374 (1976).
76. A. L. Tappel, *Fed. Proc., Fed. Am. Soc. Exp. Biol.* **32**, 1870 (1973).
77. T. B. Suslova, V. I. Olenev, and Yu. A. Vladimirov, *Biofizika* **14**, 540 (1969).
78. Yu. A. Vladimirov, T. B. Suslova, and Z. P. Cheremesina, *Biokhimiya* **33**, 587 (1968).
79. Yu. A. Vladimirov, M. V. Korchagina, and V. I. Olenev, *Biofizika* **16**, 994 (1971).
80. Yu. A. Vladimirov, T. B. Suslova, and V. I. Olenev, *Biofizika* **14**, 880 (1969).
81. Z. P. Cheremesina, V. I. Olenev, and Yu. A. Vladimirov, *Biofizika* **17**, 631 (1972).
82. R. C. Knight and F. E. Hunter, Jr., *J. Biol. Chem.* **241**, 2757 (1966).
83. V. I. Olenev, T. B. Suslova, and Yu. A. Vladimirov, *Biofizika* **19**, 104 (1974).
84. T. B. Suslova, V. I. Olenev, M. V. Korchagina, and Yu. A. Vladimirov, *Biofizika (Moscow)* **15**, 650 (1970).
85. M. V. Korchagina and Yu. A. Vladimirov, *Biofizika* **19**, 276 (1974).
86. E. Ye. Tafel'shtein, O. B. Utkin, and Yu. A. Vladimirov, *Biofizika* **16**, 991 (1971).
87. V. M. Gukasov, Yu. A. Vladimirov, P. V. Serguyev, and R. D. Seifulla, *Biofizika* **19**, 783 (1974).
88. Yu. A. Vladimirov, P. V. Sergeev, R. D. Seifulla, and Yu. N. Rudnev, *Mol. Biol. (Engl. Transl.)* **7**, 247 (1973).
89. E. Cadenas, A. Boveris, and B. Chance, *Biochem. J.* **186**, 659 (1980).
90. E. Cadenas, A. Boveris, and B. Chance, *in* "Chemical and Biochemical Aspects of Superoxide Anion and Superoxide Dismutase" (J. V. Bannister and H. A. O. Hill, eds.), Vol. 11A, p. 92. Elsevier/North-Holland, Biomedical Press, Amsterdam, 1980.
91. H. E. Demopoulus, *Fed. Proc., Fed. Am. Soc. Exp. Biol.* **32**, 1859 (1973).
92. R. M. Kashnitz and Y. Hatefi, *Arch. Biochem. Biophys.* **171**, 292 (1975).
93. C. F. Deneke and N. I. Krinsky, *Photochem. Photobiol.* **25**, 299 (1977).
94. C. S. Foote, *in* "Free Radicals in Biology" (W. A. Pryor, ed.), Vol. 2, p. 85. Academic Press, New York, 1976.
95. J. A. Howard and K. U. Ingold, *J. Am. Chem. Soc.* **90**, 1056 (1968).
96. A. Boveris and E. Cadenas, *in* "Superoxide Dismutase" (L. W. Oberley, ed.), Vol. 2, p. 75, CRC Press, Boca Raton, Florida, 1982.
97. A. Finazzi-Agro, C. Giovagnoli, P. de Soli, L. Calabrese, G. Rotilio, and V. Mondovi, *FEBS Lett.* **21**, 183 (1972): U. Weser and W. Paschen, *ibid.* **27**, 248 (1972).
98. I. Fridovich, *Annu. Rev. Biochem.* **44**, 147 (1975).
99. H. E. A. Kramer and A. Maute, *Photochem. Photobiol.* **17**, 413 (1973).
100. E. Mayeda and A. J. Bard, *J. Am. Chem. Soc.* **96**, 4023 (1974).
101. W. H. Koppenol, *Nature (London)* **262**, 420 (1976).
102. W. H. Koppenol and J. Butler, *FEBS Lett.* **83**, 1 (1977).
103. V. S. Belova, M. R. Borukaeva, and L. M. Raikhamau, *Biokhimiya* **36**, 383 (1971).

104. P. Hochstein and L. Ernster, *Biochem. Biophys. Res. Commun.* **12**, 388 (1963).
105. T. A. Aleksandrova, A. I. Archakov, Yu. A. Vladimirov, V. I. Olenev, and L. F. Pachenco, *Biofizika* **16**, 987 (1971).
106. J. R. Wright, L. Bleigh, R. C. Rumbaugh, H. D. Colby, and P. R. Miles, *Fed. Proc., Fed. Am. Soc. Exp. Biol.* **37**, 768 (1978).
107. A. R. Schoaf and R. H. Steele, *Biochem. Biophys. Res. Commun.* **61**, 1363 (1974).
108. S. Guthans, R. H. Steele, and W. H. Baricos, *Fed. Proc., Fed. Am. Soc. Exp. Biol.* **38**, 732 (1979).
109. R. F. Vassil'ev, *Opt. Spectrosc. (Engl. Transl.)* **18**, 131 (1964).
110. G. A. Russell, *J. Am. Chem. Soc.* **79**, 3871 (1957).
111. M. B. Baird, H. R. Massie, and M. J. Piekielniak, *Chem. Biol. Interact.* **16**, 145 (1977).
112. J. W. Peters, N. J. Pitts, Jr., I. Rosenthal, and H. Fuhr, *J. Am. Chem. Soc.* **94**, 4348 (1972).
113. H. Esterbauer, K. H. Cheeseman, M. U. Dianzani, G. Poli, and T. F. Slater, *Biochem. J.* **208**, 129 (1982).
114. H. Esterbauer, *in* "Free Radicals, Lipid Peroxidation, and Cancer (D. C. H. McBrien and T. F. Slater, eds.) p. 101, Academic Press, New York, 1982.
115. E. Cadenas, A. Müller, R. Brigelius, H. Esterbauer, and H. Sies, *Biochem. J.* **214**, 479 (1983).
116. E. Cadenas and H. Sies, *Eur. J. Biochem.* **124**, 349 (1982).
117. W. R. Bidlack, *Biochem. Pharmacol.* **29**, 1605 (1980).
118. L. Cavallini, M. Valente, and A. Bindoli, *Biochim. Biophys. Acta* **752**, 339 (1983).
119. F. L. Ashley and B. W. Griffin, *Mol. Pharmacol.* **19**, 146 (1981).
120. R. E. White and M. J. Coon, *Ann. Rev. Biochem.* **49**, 315 (1980).
121. F. Lichtenberger, W. Nastainczyk, and V. Ullrich, *Biochem. Biophys. Res. Commun.* **70**, 939 (1976).
122. E. Cadenas, H. Sies, H. Graf, and V. Ullrich, *Eur. J. Biochem.* **130**, 117 (1983).
123. J. P. Hamman, W. H. Biggley, and H. H. Seliger, *Photochem. Photobiol.* **30**, 519 (1979).
124. H. H. Seliger and J. P. Hamman, *J. Phys. Chem.* **89**, 2296 (1976).
125. J. Stauff, H. Schmidkunz, and G. Hartmann, *Nature (London)* **198**, 281 (1963).
126. J. Stauff, U. Sanders, and W. Jaeschke, *in* "Chemiluminescence and Bioluminescence" (M. J. Cormier, D. H. Hercules, and J. Lee, eds.), p. 131. Plenum, New York, 1973.
127. A. V. Khan, *J. Phys. Chem.* **80**, 2219 (1976).
128. A. V. Khan, *Science* **168**, 476 (1970).
129. C. S. Foote, F. C. Shook, and R. B. Akaberli, *J. Am. Chem. Soc.* **102**, 1503 (1980).
130. S. N. Orlov, V. S. Danilov, A. A. Khruschnhev, and Yu. N. Shvetsov, *Biofizika* **20**, 637 (1975).
131. I. Inaba, Y. Shimizu, Y. Tsuji, and A. Yamagishi, *Photochem. Photobiol.* **30**, 169 (1979).
132. V. Slawson and A. W. Adamson, *Lipids* **11**, 472 (1976).
133. T. G. Mamedov, V. V. Konev, and G. A. Popov, *Biofizika* **18**, 685 (1973).
134. Yu. M. Petrenko, D. I. Roshchupkin, and Yu. A. Vladimirov, *Biofizika* **20**, 617 (1975).
135. A. A. Krasnovsky, Jr., V. E. Kagan, and A. A. Minin, *FEBS Lett.* **155**, 233 (1983).
136. R. J. Hawco, C. R. O'Brien, and P. J. O'Brien, *Biochem. Biophys. Res. Commun.* **76**, 354 (1977).
137. A. Banks, E. Eddie, and J. G. M. Smith, *Nature (London)* **190**, 908 (1961).
138. W. T. Roubal and A. L. Tappel, *Arch. Biochem. Biophys.* **113**, 5 (1966).
139. W. R. Roubal and A. L. Tappel, *Arch. Biochem. Biophys.* **113**, 50 (1966).
140. E. Cadenas, A. Boveris, and B. Chance, *Biochem. J.* **187**, 131 (1980).
141. H. P. Misra and I. Fridovich, *Biochim. Biophys. Acta* **292**, 815 (1973).
142. I. D. Desai and A. L. Tappel, *Lipid Res.* **4**, 204 (1963).

143. B. Kalyanaraman, C. Mottley, and R. P. Mason, *J. Biol. Chem.* **258**, 3855 (1983).
144. E. Cadenas, A. Boveris, R. Reiter, and B. Chance, *Int. Congr. Biochem. 11th 1979*, Abstract, p. 382 (1979).
145. W. A. Pryor, *Photochem. Photobiol.* **28**, 787 (1978).
146. E. Cadenas, A. Boveris, and B. Chance, *FEBS Lett.* **112**, 285 (1980).
147. E. Cadenas, A. I. Varsavsky, A. Boveris, and B.Chance, *FEBS Lett.* **113**, 141 (1980).
148. E. Cadenas, A. Boveris, and B. Chance, *Biochem. J.* **188**, 577 (1980).
149. T. W. Reenan, Y. C. Awasthi, and F. L. Crane, *Biochem. Biophys. Res. Commun.* **40**, 1102 (1970).
150. J. M. McCord and I. Fridovich, *J. Biol. Chem.* **243**, 5753 (1968).
151. J. Stauff and H. Wolf, *Z. Naturforsch., B: Anorg. Chem., Org. Chem., Biochem., Biophys., Biol.* **198**, 97 (1964).
152. R. M. Arneson, *Arch. Biochem. Biophys.* **136**, 352 (1970).
153. M. J. Thomas and W. A. Pryor, *in* "Oxygen and Oxy-radicals in Chemistry and Biology" (M. A. J. Rodgers and E. L. Powers, eds.), p. 761. Academic Press, New York, 1981.
154. E. K. Hodgson and I. Fridovich, *Arch. Biochem. Biophys.* **172**, 292 (1976).
155. T. C. Pederson and S. D. Aust, *Biochem. Biophys. Res. Commun.* **52**, 1071 (1973).
156. E. W. Kellog and I. Fridovich, *J. Biol. Chem.* **250**, 8812 (1975).
157. C. S. Foote, *in* "Biochemical and Clinical Aspects of Oxygen" (W. S. Caughey, ed.), p. 603. Academic Press, New York, 1979.
158. A. Boveris, E. Cadenas, and B. Chance, *Photobiochem. Photobiophys.* **1**, 175 (1980).
159. J. I. Teng and L. L. Smith, *J. Am. Chem. Soc.* **95**, 4060 (1973).
160. J. F. Piatt, A. S. Cheema, and P. J. O'Brien, *FEBS Lett.* **74**, 241 (1977).
161. J. R. Kanofsky, *J. Biol. Chem.* **258**, 5991 (1983).
162. E. K. Hodgson and I. Fridovich, *Biochemistry* **14**, 5299 (1975).
163. L. J. Marnett, P. Wlodawer, and B. Samuelsson, *Biochem. Biophys. Res. Commun.* **60**, 1286 (1974).
164. L. J. Marnett, P. Wlodawer, and B. Samuelsson, *J. Biol. Chem.* **250**, 8510 (1975).
165. F. A. Kuehl, Jr., J. L. Humes, E. A. Ham, R. W. Egan, and H. W. Dougherty, *in* "Prostaglandin and Thromboxane Research" (B. Samuelsson, P. W. Ramwell, and R. Paoletti, eds.) Vol. 6, p. 77, Raven Press, New York, 1980.
166. R. W. Egan, P. G. Gale, and F. A. Kuehl, Jr., *J. Biol. Chem.* **254**, 3295 (1979).
167. L. J. Marnett, M. J. Bienkowsky, and W. R. Pagels, *J. Biol. Chem.* **254**, 5077 (1979).
168. E. Cadenas, H. Sies, W. Nastainczyk, and V. Ullrich, *Hoppe-Seyler's Z. Physiol. Chem.* **364**, 519 (1983).
169. V. Ullrich, L. Castle, and P. Weber, *Biochem. Pharmacol.* **30**, 2033 (1981).
170. B. Kalyanamaran, R. P. Mason, B. Tainer, and T. E. Eling, *J. Biol. Chem.* **257**, 4764 (1982).
171. G. Cilento, *in* "Chemical and Biological Generation of Excited Species" (W. Adam and G. Cilento, eds.), p. 278, Academic Press, New York, 1982.
172. O. M. M. Faria Oliveira, M. Haun, N. Durán, P. J. O'Brien, C. R. O'Brien, E. J. H. Bechara, and G. Cilento, *J. Biol. Chem.* **253**, 4707 (1978).
173. E. J. H. Bechara, O. M. M. Faria Oliveira, N. Durán, R. Casadei de Baptista, and G. Cilento, *Photochem. Photobiol.* **30**, 101 (1979).
174. M. Haun, N. Durán, O. Augusto, and G. Cilento, *Arch. Biochem. Biophys.* **200**, 245 (1980).
175. C. Vidigal-Martinelli, K. Zinner, B. Kachar, N. Durán, and G. Cilento, *FEBS Lett.* **108**, 266 (1979).
176. N. I. Krinsky, *in* "Singlet Oxygen" (H. H. Wasserman and W. A. Murray, eds.) p. 597, Academic Press, New York, 1979.

CHAPTER **8**

Free-Radical Intermediates in the Antiparasitic Action of Drugs and Phagocytic Cells

Roberto Docampo and Silvia N. J. Moreno

Laboratory of Molecular Biophysics
National Institute of Environmental Health Sciences
Research Triangle Park, North Carolina

FREE RADICALS IN BIOLOGY, VOL. VI

243

I. INTRODUCTION

The study of parasitology has led to important developments in free-radical biology. A case in point is the phenomenon of oxygen toxicity. In 1925 Cleveland [1] reported his now famous studies on the oxygen sensitivity and host–parasite relationships of termite flagellates. The 1961 observation by Ritter [2] that the intestine of *Cryptocercus*, which harbors a similar fauna, is virtually oxygen free suggests that anaerobic environmental conditions are needed for the survival of parasites and that oxygen is toxic to those flagellates.

Another instructive example concerns the dye-induced light sensitivity of *Paramecium caudatum* (a free-living protozoan). Some 80 years ago Raab [3] showed that low concentrations of acridines and certain other dyes, which had no deleterious effects in the dark, rapidly killed the protozoa on illumination. These investigations were the cornerstones on which the vast edifice of contemporary photobiology research was erected, a discussion of which lies outside the scope of this chapter [4].

Parasitology-related research also retarded later developments in free-radical biology. The well-known British biochemist and parasitologist David Keilin began to study *Gastrophilus* (an endoparasitic insect) in 1919 because he was interested in tracing the fate of hemoglobin, which is a characteristic constituent of the larvae but which does not occur even in traces in the adult flies. This study had a tremendous influence on general biochemistry, because it led Keilin directly to the rediscovery of the cytochromes, as he described in his book on the history of cell respiration and cytochromes [5]. Paradoxically, this landmark discovery, which favored the hypothesis of the exclusive tetravalent reduction of oxygen to water, eclipsed for several decades the investigation of univalent oxygen reduction leading to the superoxide anion ($O_2{}^-$) or divalent reduction leading to H_2O_2 [6].

Since the early 1970s significant progress has been made in acquiring knowledge of the action of free radicals on parasites. Studies on intermediary metabolism, on the mode of action of chemotherapeutic agents, and on phagocytosis have contributed to a better understanding of free-radical toxicity to protozoa and helminths and are the subject of this chapter.

II. PROTOZOA AND HELMINTHS OF MEDICAL IMPORTANCE

In a restricted sense the term *parasitology* applies only to animal parasites and does not include plants such as bacteria and fungi and borderline forms such as spirochetes and viruses. Animal parasites of humans and most vertebrates are contained in three major groups: (*a*) protozoa, which are unicellular organisms;

TABLE I   Protozoa of Medical Importance

Protozoan	Habitat	Disease
Balantidium coli	Large intestine	Balantidiasis
Entamoeba histolytica	Large intestine, liver, brain, lung, spleen	Amebiasis
Giardia lamblia	Small intestine, gallbladder	Giardiasis
Leishmania brasiliensis	Skin, mucous membranes of nose and mouth	Leishmaniasis (mucocutaneous)
Leishmania donovani	Liver, spleen, skin	Leishmaniasis (Kala azar)
Leishmania tropica	Skin	Leishmaniasis (cutaneous)
Plasmodium malarie	Blood, liver, spleen, brain	Malaria
Plasmodium falciparum	Blood, liver, spleen, brain	Malaria
Plasmodium ovale	Blood, liver, spleen, brain	Malaria
Plasmodium vivax	Blood, liver, spleen, brain	Malaria
Pneumocystis carinii	Lung	Interstitial pneumonitis
Toxoplasma gondii	Liver, spleen	Toxoplasmosis
Trichomonas vaginalis	Vagina	Trichomoniasis
Trypanosoma cruzi	Blood, heart, brain, muscles	Chagas' disease
Trypanosoma gambiense	Blood, lymphatics, brain	Sleeping sickness (chronic form)
Trypanosoma rhodesiense	Blood, lymphatics, brain	Sleeping sickness (acute form)

(*b*) helminths, which are multicellular animals that include cestodes (or tapeworms), nematodes (or roundworms), and trematodes (or flukes); and (*c*) arthropods, which are mainly ectoparasites.

Information about the protozoa and helminths of medical importance is summarized in Tables I and II. In addition to the parasites mentioned in these tables we discuss several species of parasites that are important to man because they cause disease in domestic animals, for example, *Trypanosoma brucei*, which causes nagana in cattle, and *Tritrichomonas foetus*, which causes trichomoniasis in cattle. We also discuss other species that are nonpathogenic to man or domestic animals but that are used as models for the pathogenic species, such as *Crithidia fasciculata*, an insect parasite used as a model for trypanosomatids, and *Plasmodium berghei*, a parasite of rodents used as a model for malaria parasites. For a complete discussion of these pathogenic and nonpathogenic parasites the reader is referred to previous reviews and books [7, 8].

III. INTERMEDIATES OF OXYGEN REDUCTION IN PARASITES

Several oxidative enzymes produce substantial amounts of O_2^- and/or H_2O_2 in mammalian tissues [6]. Some of these enzymes have been identified in several

TABLE II   Helminths of Medical Importance

Helminth	Habitat	Main lesions
Cestodes		
Diphyllobothrium latum	Small intestine	Enteritis
Dipylidium caninum	Intestine	Enteritis
Echinococus granulosus	Liver, other viscera	Hydatid cyst
Hymenolepis nana	Small intestine	Enteritis
Taenia saginata	Small intestine, tissues	Enteritis
Taenia solium	Small intestine	Enteritis, cysticerci
Nematodes		
Ancylostoma brazilienze	Small intestine, skin	Creeping eruption, enteritis
Ancylostoma duodenale	Small intestine	Pneumonitis, enteritis
Ascaris lumbricoides	Small intestine	Enteritis, pneumonitis
Dracunuculus medinensis	Subcutaneous tissues	Cutaneous ulcers, abscesses
Enterobius vermicularis	Intestine	Enteritis
Loa loa	Subcutaneous tissues, blood, eye	"Calabar" swellings
Necator americanus	Small intestine	Pneumonitis, enteritis
Onchocerca volvulus	Skin, subcutaneous tissues	Cutaneous nodules, iridocyclitis
Strongyloides stercolaris	Small intestine	Enteritis, pneumonitis
Trichinella spiralis	Small intestine, muscles	Enteritis, myositis
Trichuris trichiura	Intestine	Enteritis
Wuchereria bancrofti	Lymphatics, blood	Lymphagitis, elephantiasis
Trematodes		
Clonorchis sinensis	Liver, bile ducts	Hepatic cirrhosis, biliary fibrosis
Fasciola hepatica	Liver, bile ducts	Hepatitis, hepatic cirrhosis
Fasciolopsis buski	Small intestine	Enteritis
Heterophyes heterophyes	Small intestine	Enteritis
Metagonimus yokogawai	Small intestine	Enteritis
Opisthorchis felineus	Liver, bile ducts	Biliary fibrosis
Paragonimus westermani	Lung, other tissues	Ulcerative tubercles
Schistosoma hematobium	Vesical and pelvic plexures	Hyperplasia and fibrosis of bladder
Schistosoma japonicum	Mesenteric veins	Splenic fibrosis, hepatic cirrhosis
Schistosoma mansoni	Mesenteric veins	Fibrosis

parasites and are listed in Table III. However, a complete and detailed identification of enzymes and substrates that form $O_2{}^-$ and H_2O_2 in parasites is not yet possible.

Better known is the succinate oxidase of large helminths such as *Ascaris lumbricoides*, *Fasciola hepatica*, and *Monieza expansa*. This mitochondrial system, in contrast to the mammalian succinate oxidase, does not form water but forms H_2O_2 from oxygen and is not inhibited by cyanide, azide, or antimycin A [22, 23]. The *Ascaris* system works more efficiently at high than at low oxygen tensions, and it has been postulated to be mediated by a flavin-containing oxidase, a view strengthened by the observation that purified preparations contain FAD as a major component [24]. However, there is evidence [25–27, 30, 31] that the mitochondria of these large helminths contain a functional branched respiratory chain system with two terminal oxidases: cytochromes o and a_3. The major pathway with cytochrome o as its terminal oxidase is apparently linked with H_2O_2 formation [27]:

$$\text{Succinate} \rightarrow [\text{rhodoquinone } b \text{ complex}] \rightarrow c_1 \rightarrow c \rightarrow a \rightarrow a_3 \rightarrow H_2O$$
$$\phantom{\text{Succinate} \rightarrow [\text{rhodoquinone } b \text{ complex}]} \rightarrow o \rightarrow H_2O_2$$

However, it has been postulated that the generation of H_2O_2 by *Ascaris* muscle mitochondria may be an artifact due to the effect of an increase in the pressure of oxygen in the assay systems [27]. In this connection it has been stated that, when a helminth is transferred from its oxygen-poor environment to air, the sudden increase in the partial pressure of oxygen may result in metabolic changes analogous to those that occur when aerobic organisms are placed in hyperbaric oxygen [27]. The utilization of NAD(P)H and succinate by *Hymenolepis diminuta* mitochondrial membranes also results in H_2O_2 formation [32], although the presence of a branched respiratory chain in this cestode has not been established.

TABLE III   Parasite Enzymes Producing Hydrogen Peroxide or Superoxide Anion

Enzyme	Parasite	References
L-Aminoacyl oxidase	*Plasmodium berghei*	9
	Hymenolepis diminuta	10
	Ascaris lumbricoides	11
Monoamine oxidase	*Balantidium coli*	12
	Schistosoma mansoni	13
Dihydroorotate oxidase	*Crithidia fasciculata*	14
	Trypanosoma brucei	14
	Leishmania mexicana	15
Dihydroorotate dehydrogenase	*Plasmodium knowlesi*	16
	Plasmodium berghei	16
	Plasmodium gallinaceum	16
NADPH-cytochrome c reductase	*Trypanosoma cruzi*	17, 18
NADH-cytochrome c reductase	*Trypanosoma cruzi*	18, 19
	Ascaris lumbricoides	20, 21
Succinate oxidase	*Ascaris lumbricoides*	22–27
	Fasciola hepatica	28, 29
	Monieza expansa	30, 31
	Hymenolepis diminuta	32

Branched cytochrome chains are not unique to parasitic helminths but occur widely in microorganisms, parasitic protozoa, and plants. However, no H_2O_2 production has been implicated in the function of cytochrome o as the terminal oxidase of different trypanosomatids [17, 33–38].

In contrast to the *Ascaris* muscle mitochondria, H_2O_2 formation by the protozoan *C. fasciculata* mitochondrial preparations is qualitatively the same as that by mitochondria from vertebrate tissues [39]. The quantity of H_2O_2 produced is also of the same order of magnitude as that reported with vertebrate mitochondria [6].

It has been reported [40] that *Tritrichomonas foetus* aerobically excretes H_2O_2. In contrast, respiring *Trypanosoma cruzi* does not release either O_2^{-} or H_2O_2 to the suspending medium [41–53], although *T. cruzi* homogenates supplemented with reduced nicotinamide adenine dinucleotides are effective sources of O_2^{-} and H_2O_2 (Table IV). The mitochondrial fraction of *T. cruzi* also generates O_2^{-} and H_2O_2 in the presence of NADH or NADPH [41, 44, 45, 47, 48, 50, 52]. The use of reduced nicotinamide adenine dinucleotides as electron donors is compulsory because, at variance with *C. fasciculata* [39] and other mitochondrial preparations [6], succinate plus antimycin A does not modify the rate of H_2O_2 production with *T. cruzi* mitochondrial fractions [41]. *Tryponosoma cruzi* microsomes also generate O_2^{-} and H_2O_2, NADPH being more effective than NADH as electron donor [41, 44, 45, 48, 52, 53]. Comparable results are observed with *T. cruzi* supernatants [41].

The intracellular concentration of H_2O_2 in *Trypanosoma brucei* bloodstream forms has been estimated to be 70 μM, or 0.8 nmol/mg protein, approximately 30 times as high as in a rat liver preparation that contained 0.03 nmol/mg, or ~2.6 μM [54]. These values are apparently overestimated, because other reports set H_2O_2 in rat liver at 0.1 to 0.001 μM [55]. Mitochondrial preparations of the *T. brucei* bloodstream form produce 0.2–0.7 nmol H_2O_2 per milligram protein each minute, which corresponds to 1 to 3% of the oxygen consumed [56].

TABLE IV Generation of Superoxide Anion and Hydrogen Peroxide by *Trypanosoma cruzi* Homogenates[a]

Stage	Substrate	O_2^{-} generation (nmol/min $\times$ mg protein)[b]	H_2O_2 production (nmol/min + mg protein)
Epimastigote (culture	NADH	6.7	23.3 ± 5.2
form)	NADPH	0	14.0 ± 0.9
Amastigote (intracellu-	NADH	7.9	12.7 ± 2.9
lar form)	NADPH	ND	14.7 ± 5.7
Trypomastigote (blood-	NADH	ND	6.3 ± 2.1
stream form)	NADPH	ND	4.2 ± 0.6

[a]From [46] and [51].
[b]ND, Not determined.

Although no direct demonstration of $O_2{}^-$ and/or H_2O_2 formation by *Plasmodium* has been reported, a *Plasmodium*-induced erythrocyte oxidant sensitivity has been postulated [57–70a] on the basis of the following evidence:

1. *Plasmodium berghei*-infected mouse red blood cells accumulate methemoglobin *in vivo* and are more sensitive to exogenous oxidants *in vitro*. In the absence of added oxidants, catalase within infected cells is readily inactivated by 3-amino-1,2,4-triazole, a reaction that requires the presence of H_2O_2. These phenomena have been attributed to parasite-mediated NADH (and probably NADPH [59]) oxidation with the concomitant formation of H_2O_2 [57]. However, whether H_2O_2 originates from the parasites or from the red blood cells remains to be established. A similar sensitivity to exogenous oxidants such as alloxan, phenylhydrazine, and H_2O_2 is observed in *Plasmodium vinckei*-infected mice. These oxidants cause a rapid reduction in parasitemia and a transient hemolysis in malaria-infected mice but not in controls [70a].

2. *Plasmodium berghei*-infected mouse red blood cells show an abrupt lowering in the activity of superoxide dismutase and a rise in the concentration of the lipid peroxidation product, malondialdehyde, with the growth of parasitemia [61].

3. Reduced glutathione and glutathione reductase activity are elevated in *P. vinckei-* [62] and *P. berghei*-infected mouse erythrocytes [59]. Isolated *P. berghei* cells contain most of the reduced glutathione found in whole infected red blood cells [59]. Because the pentose shunt metabolism is accelerated in parasitized erythrocytes [63, 64], the oxidation of erythrocyte NADPH during reduction of parasite-oxidized glutathione has been postulated to explain both the increased oxidant sensitivity of infected erythrocytes and the acceleration of pentose shunt activity in infected cells [59].

4. The vitamin E concentration in plasma of *P. berghei*-infected mice is approximately half that in uninfected animals [58], and it has been suggested that oxidants of parasitic origin might increase vitamin E consumption [58]. Accordingly, malnourished African herding people, whose diet is normally low in vitamin E, are relatively resistant to malaria infections. Feeding these people cereal grains (an excellent source of vitamin E) causes recrudescence of previously undetectable *P. falciparum* infections [58, 65].

5. One inherited alteration of human red blood cell metabolism, quantitative deficiency of the X-linked enzyme glucose-6-phosphate dehydrogenase, has attained high frequency in many areas of endemic malaria and may protect against fulminant *P. falciparum [66–68]*. Evidence has been presented that glucose 6-phosphate-deficient cells as well as α- and β-thalasemia red blood cells are refractory to parasite development because of decreased protection against oxidant damage [60]. It is interesting that malaria induced by *P. berghei* in mice causes a decrease in superoxide dismutase and drug-metabolizing enzyme ac-

tivities as well as enhanced lipid peroxidation damage in the liver [69, 70]. However, it has been reported [71] that glutathione stability is enhanced in P. falciparum-infected red cells and, although there is an increase in glucose utilization and in the pentose pathway in infected red blood cells, this may be interpreted as being necessary for protein and nucleic acid synthesis and cell division and not as a response to oxidative stress [71].

Finally, as occurs with mammalian systems [6], nonenzymatic sources of O_2^{-} and H_2O_2 such as the autoxidation of ferredoxins [72] (present in parasites such as Trichomonas [73, 73a]), tetrahydropteridines [74], thiols [75], or other soluble parasite constituents are of unknown importance under physiological conditions.

IV. ENZYMES UTILIZING OXYGEN REDUCTION INTERMEDIATES IN PARASITES

A. Catalase

Catalase is present in different cell types. In most cases the enzyme is localized in subcellular organelles such as the peroxisomes (microbodies) of liver and kidney or in much smaller aggregates such as the microperoxisomes found in a variety of other cells [76, 77].

There are few reports of peroxisomes or catalase-containing microbodies in parasites. Incubation of aldehyde-fixed cells in alkaline 3,3'-diaminobenzidine medium is a frequently used procedure for demonstrating the presence of peroxisomes, because peroxidation of 3,3'-diaminobenzidine determines the formation of electron-opaque material easily visualized by electron microscopy [78]. However, this reaction is also positive with other hemoproteins such as peroxidases [79] and cytochrome oxidase [80], both of which can act as peroxidases, and in some cases there is no agreement between the cytochemical and biochemical findings. For example, Trypanosoma cruzi epimastigotes [81] contain no catalase activity but show 3,3'-diaminobenzidine positivity; this has been attributed to the presence of a peroxidase in their microbodies [81].

The microbodies of higher trypanosomatids, such as T. cruzi and the T. brucei group, and of several species of Trichomonas, such as Tritrichomonas foetus, are entities biochemically distinct from other known redox organelles. They are termed glycosomes [82] and hydrogenosomes [83], respectively. The glycosomes of trypanosomatids contain the early enzymes of the glycolytic pathway [82, 84]. The hydrogenosomes of Trichomonas contain the enzymes of H_2 production [83]. Catalase activity is absent in both organelles [85–87], although a 3,3'-diaminobenzidine reaction, apparently nonenzymatic [88], has been reported in the denser area (core) of hydrogenosomes [88a]. Catalase is found in the cytoplasm of *Tritrichomonas foetus* [89] but is absent in several other *Trichomonas* species [90, 91]. Catalase is also present in the cytoplasm of *Entamoeba invadens* [8] and *Toxoplasma gondii* [92] but is absent from *Entamoeba histolytica* [93].

Although catalase is completely absent in the bloodstream form of African pathogenic trypanosomes [94–96] and in *Trypanosoma cruzi* [81], it is present in other mammalian trypanosomes [96] as well as avian [96], amphibian [96], and insect trypanosomes [97]. 3,3'-Diaminobenzidine-positive microbodies of typical structure are present in these lower trypanosomatids [98, 99]. In *Crithidia fasciculata* these microbodies occupy ~2.5% of the total cell volume [98]. In contrast to higher trypanosomatids, *Crithidia* species possess an unusually high activity of catalase [86, 97, 100].

A high content of catalase has been reported in the intracellular (amastigote) forms of *Leishmania donovani* [101]. Because amastigotes were obtained from hamster spleen homogenates, it is difficult to exclude the presence of contaminating catalase from the host.

Catalase activity is low in cestodes [102, 103] and nematodes [104–106] and is undetectable in newborn larvae, adult worms, and muscle larvae of *Trichinella spiralis* [107], as well as in *Fasciola hepatica* [108] (see Table V).

B. Glutathione Peroxidase and Other Peroxidase Activities

Glutathione peroxidase catalyzes the reaction of hydroperoxides with reduced glutathione (GSH) to form oxidized glutathione disulfide (GSSG) and the reduction product of the hydroperoxides:

$$ROOH + 2\ GSH \rightarrow ROH + GSSH + H_2O$$

Glutathione peroxidase is specific for its hydrogen donor, reduced glutathione, and nonspecific for the hydroperoxide, which may be H_2O_2 or organic hydroperoxides [6, 109].

No H_2O_2-utilizing glutathione peroxidase activity is present in *Trypanosoma cruzi* [110] and *Trypanosoma brucei* [56]. A very low glutathione peroxidase activity measured with organic hydroperoxides (*tert*-butyl hydroperoxide and cumene hydroperoxide) in *T. cruzi* extracts [110] may reflect a relatively nonspecific reaction such as that catalyzed by glutathione *S*-transferase B, in which oxidation of reduced glutathione by hydroperoxides occurs at hydrophobic areas of the protein [111]. In this connection a glutathione *S*-transferase similar to glutathione *S*-transferase B has been purified from *T. cruzi* [112].

Glutathione peroxidase is present in *Toxoplasma gondii* [93], *Entamoeba his-*

TABLE V Catalase, Glutathione Peroxidase, and Superoxide Dismutase Activities in Parasites[a]

Organism	Stage	Catalase (mU/mg)	Glutathione peroxidase (nmol/min × mg)	Superoxide dismutase (U/mg)
Tritrichomonas foetus	Culture form	32 ± 6 [128]	ND	5.5 ± 1 [128]
Toxoplasma gondii	Trophozoites	48 ± 4 [93][b]	117 ± 8 [93]	6.1 ± 2.2 [93]
Entamoeba histolytica				
HM-1 strain	Culture form	0 [93][b]	11.7 ± 1.7 [93]	0.45 ± 0.31 [93]
H-303 strain	Culture form	0 [93][b]	14.2 ± 3.1 [93]	0.61 ± 0.12 [93]
Crithidia luciliae	Culture form	160 × 10³ [97]	ND	ND
Crithidia fasciculata	Culture form	328 × 10³ [97]	ND	14.2 ± 2.6 [130]; 9.16 [129]
Leishmania tropica	Culture form	<50 [113]	14.4 [113]	23.0 ± 2.3 [113]; 6.4 ± 0.5 [93]
Leishmania donovani	Culture form	30 ± 20 [93][b]	0.2 ± 0.2 [93]	4.1 ± 0.9 [101]
	Intracellular	50 ± 20 [93][b]	0.5 ± 0.5 [93]	12.8 ± 1.3 [101]
Tripanosoma brucei	Culture form	140 ± 20 [101]	7.2 ± 2.4 [101]	10.3 [130]
	Culture form	0 [130]	ND	13.7 ± 1.6 [85]
	Bloodstream form	0 [54]	0 [54]	4.8 ± 0.5 [130]
Trypanosoma cruzi				
Colombia strain	Culture form	ND	ND	11.0 [130]
Y strain	Culture form	ND	ND	11.8 [130]
Tulahuen strain	Culture form	0 [81]	0 ± 1.7 [110]	0.06 ± 0.007 [110]
Hymenolepis diminuta	Adult worm	0 [103]	<2 [103]	7.9 ± 1.6 [103]
Moniezia expansa	Adult worm	0 [103]	<2 [103]	9.5 ± 1.8 [103]
Fasciola hepatica	Adult worm	0 [108]	0 [108]	7.2 ± 2.5 [108]
Trichinella spiralis	Newborn larvae	<50 [107]	3 [107]	6.8 ± 1.2 [107]
	Adult worm	<250 [107]	16 ± 8 [107]	19.2 [107]
	Muscle larvae	<50 [107]	36 ± 3 [107]	30.8 ± 4.0 [107]

[a] ND, Not determined. References are in brackets.
[b] Baudhin units [93].

tolytica [*93*], and *Leishmania tropica* promastigotes [*113*]. Although a high content of glutathione peroxidase has been reported in *Leishmania donovani* intracellular forms [*101*], it is difficult to rule out contamination from the host.

Newborn larvae of *Trichinella spiralis* lack detectable activity of glutathione peroxidase, whereas adult worms and muscle larvae contain a high activity of this enzyme [*107*]. *Fasciola hepatica* also lacks glutathione peroxidase [*108*].

A peroxidase activity has been detected by cytochemical methods in *Trypanosoma cruzi* [*81, 114*], *Trypanosoma gambiense* [*115*], and *Trypanosoma brucei* [*116*]. A very weak ascorbate peroxidase activity (7.5 ± 1.2 mU per 10^8 cells) has also been demonstrated in *Trypanosoma cruzi* homogenates [*81, 110*].

A cytochrome *c* peroxidase activity has been detected in mitochondrial preparations of *Crithidia fasciculata* [*39*]. The distribution of activity after sonic disruption of mitochondrial preparations is that expected for a soluble enzyme and, although the activity in mitochondrial fractions isolated from *C. fasciculata* was calculated to be 0.3% that of isolated yeast mitochondria, it has been suggested that the *in vivo* activity may be considerably higher than this estimate [*39*]. A cytochrome *c* peroxidase activity has also been detected in *Ascaris lumbricoides* [*117*] and *Fasciola hepatica* [*108*].

Other peroxidases have been demonstrated in *Fasciola hepatica* [*118*], *Hymenolepis diminuta* [*119*], *Monieza expansa* [*120*], the hemocele fluid of *Ascaris lumbricoides* [*121*] and the miracidium of *Schistosoma mansoni* [*122*].

C. Superoxide Dismutases

Superoxide dismutase is a widely distributed enzyme that exists in a variety of forms [*123*]. The copper–zinc enzyme is found in the cytosol and in the mitochondrial intermembrane space of eukaryotic cells and is sensitive to high concentrations of cyanide [*124*]. Mitochondria also contain, in the matrix space, a distinctive cyanide-insensitive manganese enzyme similar to that found in prokaryotes [*125*]. In addition, a ferrienzyme has been identified in bacteria [*126*] that is also insensitive to cyanide.

In general, protozoa, as eukaryotic algae, lack the copper–zinc enzyme [*127*]. However, a cyanide-sensitive superoxide dismutase activity has been reported in *Trypanosoma cruzi* [*41*]. *Plasmodium berghei* isolated from mouse red blood cells also contains a cyanide-sensitive superoxide dismutase activity [*127a, 127b*]. Plasmodial and mouse enzymes are indistinguishable electrophoretically [*127a, 127b*]. These results suggest that the malarial superoxide dismutase may be entirely of host origin [*127a, 127b*]. Accordingly, plasmodia isolated from mouse red blood cells contain mouse superoxide dismutase, whereas rat-derived parasites contain the rat enzyme [*127b*].

Superoxide dismutase activity is present in the facultative anaerobic flagellate *Tritrichomonas foetus* [*128*]. About five-sixths of the activity of *T. foetus* is in the nonsedimentable portion of the cytoplasm. The rest is connected with the hydrogenosome and exhibits structure-bound latency, which can be abolished by freezing and thawing or with the detergent Triton X-100. The enzymes from both locations differ in their electrophoretic mobility on polyacrylamide gels. They are similar in their molecular weight (38,000) and in their sensitivity to heat and organic solvents. Neither of them is inhibited by cyanide [128].

Cyanide-insensitive superoxide dismutases are also present in trypanosomatids such as Crithidia fasciculata [127] and Trypanosoma brucei [85]. The bimodal distribution of superoxide dismutase in T. brucei [85] is similar to that described for other eukaryotes, the particle-bound activity showing a complex distribution after isopycnic centrifugation [85]. Both superoxide dismutase activities in T. brucei are insensitive to 1 mM cyanide [85]. The superoxide dismutase activities of T. brucei, Leishmania tropica, and T. cruzi obtained after four cycles of freezing and thawing, Triton X-100 treatment, and centrifugation are also cyanide insensitive but peroxide and azide sensitive [129, 130]. Studies on the superoxide dismutase of C. fasciculata have revealed that this enzyme is located in the cytosol and exists in three forms, which may represent three distinct isozymes [129, 130]. The major superoxide dismutase isozyme has a molecular weight of 43,000 and consists of two equally sized subunits, each of which contains 1.4 atoms of iron. Comparisons of the amino acid content of this superoxide dismutase with those of superoxide dismutases from other sources suggests that the crithidial system is closely related to bacterial ferrienzyme and only distantly related to human manganese and copper–zinc superoxide dismutase and to the ferrienzyme of the alga Euglena gracilis [129, 130].

Superoxide dismutase activities have also been detected in Leishmania donovani [101], Leishmania tropica [113, 130], Toxoplasma gondii [93], and Entamoeba histolytica [93], although they were not identified.

Superoxide dismutase activity has also been detected in newborn larvae, adult worms, and muscle larvae of Trichinella spiralis, the activity of newborn larvae being significantly lower than that of the other stages [107]. A copper–zinc, cyanide-sensitive superoxide dismutase has been isolated and purified from muscle larvae T. spiralis [107a] that is apparently secreted extracellularly [107a]. Extracts of Fasciola hepatica show appreciable superoxide dismutase activity [108]. A superoxide dismutase is also present in Hymenolepys diminuta extracts [103], and in the Ascaris lumbricoides [108a].

V. TOXICITY OF OXYGEN REDUCTION INTERMEDIATES IN PARASITES

Parasitic habitats can be divided into oxygen-free (or oxygen-poor) environments and aerobic habitats [7]. The best known anaerobic, or at least nearly anaerobic, habitats are the bile duct [131] and the intestines of larger vertebrates

[*132*]. However, most tissue parasites live in environments with moderate to high oxygen tension.

As to other respiring organisms, oxygen is toxic to them, and their sensitivity to oxygen poisoning varies from species to species. Obligate anaerobes, which lack defenses against oxygen toxicity, are killed readily by oxygen even at low tensions. Only two groups of protozoa, rumen ciliates [*133, 134*] and termite flagellates [*135, 136*], belong to this category. Other protozoa are less sensitive to oxygen damage and consume oxygen when available; they can be cultivated for indefinite periods under completely or nearly anaerobic conditions. Examples are *Entamoeba* spp. [*137, 138*] and *Trichomonas* spp. [*139, 140*]. Trypanosomatids are also able to survive under anaerobic conditions but, in contrast to the aforementioned forms, they cannot multiply in anaerobiosis [*141–143*].

All parasitic helminths studied thus far are facultative aerobes, showing various degrees of tolerance to a lack of oxygen [7]. Because large intestinal helminths (*Ascaris, Monieza,* and others) and *Fasciola hepatica* live in oxygen-poor environments, their oxygen consumption is strictly dependent on the oxygen tensions, and most authors assume that they lead a predominantly anaerobic life *in vivo* [7].

The phenomenon of oxygen toxicity to parasites was first noted by Cleveland [*1*]. He found that, after exposing the termite *Termopsis nevadensis* to 1 atm oxygen for 24 h, the flagellate *Trichomonas termopsidis* disappeared completely from its intestine. Other parasitic protozoa were removed after the termite had been exposed to oxygen for 72 h. Termites did not suffer direct ill effects from the oxygen but died of starvation within 3 to 4 weeks after deparasitization [*1*]. These studies provided the initial impetus for the use of oxygen as an antiparasitic agent. For instance, intrarectal oxygen has been used to treat acute ascaridiasis [*144*].

It is interesting that, for experimental purposes, parasites are frequently tested in media with greater oxygen tensions than their oxygen-poor natural environments. This sudden increase in the partial pressure of oxygen may result in metabolic changes analogous to those that occur when aerobic organisms are placed in hyperbaric oxygen [*145*]. In fact, any gas mixture with a P_{O_2} greater than that to which organisms are adapted may be considered hyperbaric oxygenation [*145*].

It is known that oxygen pressurization increases the generation of O_2^- and H_2O_2, leading to an increased steady-state concentration of these intermediates in different tissues [6]. Accordingly, it is possible to demonstrate appreciable levels of H_2O_2 accumulation by mitochondrial preparations of some helminths at P_{O_2} greater than those to which they are adapted [*32, 117*].

Certain O_2^--generating systems such as thiols [*95*], xanthine–xanthine oxidase [*92, 146, 147*], and dihydroorotate–dihydroorotate oxidase [*146*] are toxic to parasites. In some instances the addition of superoxide dismutase is protective, implicating O_2^- in the toxicity. However, it should be emphasized that inhibition by superoxide dismutase does not necessarily indicate a direct effect of O_2^-. An agent dependent on O_2^- for its formation may be responsible. Thus, it is probable that O_2^- exerts its parasiticidal action largely through the products that it forms, namely, H_2O_2, OH·, and possibly $^1O_2$ [123, 148]. Accordingly catalase (which destroys H_2O_2), mannitol and benzoate (scavengers of OH·), and histidine and diazabicyclooctane (scavengers of $^1O_2$) reverse *Toxoplasma gondii* killing by xanthine–xanthine oxidase [92].

Hydrogen peroxide is also toxic to parasites [149]. Evidence regarding the toxicity of H_2O_2 in parasites was obtained in earlier studies in which the trypanocidal activity of certain thiols and ascorbate could be reversed by the addition of catalase [149, 150]. Different thiols cause inhibition of the respiration and mobility of *Trypanosoma rhodesiense in vitro* and accumulation of H_2O_2 in the medium. The addition of catalase prevents this inhibitory effect of thiols [95]. Although these trypanosomes are devoid of catalase, the pyruvate (which scavenges H_2O_2) generated by their metabolism partially destroys the H_2O_2 formed in the system [95].

To determine the toxic effect of H_2O_2 on parasites, several groups of investigators have studied the sensitivity of different parasites to either reagent H_2O_2 or enzymatically generated H_2O_2 (glucose–glucose oxidase system). The following parasites were found to be sensitive to these systems: epimastigotes, amastigotes, and trypomastigotes of *Trypanosoma cruzi* [151, 151a], promastigotes [152, 153, 153a] and amastigotes [101, 153, 153a] of *Leishmania donovani*, amastigotes of *Leishmania enrietti* [154, 155], epimastigotes and trypomastigotes of *Trypanosoma dionisii* [156, 157], trypomastigotes of *Trypanosoma equiperdum* [148], *Entamoeba hystolitica* [93], *Plasmodium yoelii* [157a, 157b], *Plasmodium berghei* [157a], and *Toxoplasma gondii* [92], newborn larvae of *Trichinella spiralis* [158], and schistosomula of *Schistosoma mansoni* [159].

Injection of H_2O_2 [157a] or *tert*-butyl hydroperoxide [159a, 159b] *in vivo* also reduces *P. yoelii* [157a, 159a] and *P. vinckei* [159a] parasitemias but have less effect on *P. berghei* [157a]. In addition, *tert*-butyl hydroperoxide inhibits the development of *P. falciparum* in culture [159a].

In circulating polymorphonuclear leukocytes and monocytes, peroxidases serve an important role in the peroxidation of ingested microbes by employing H_2O_2 and a halide as substrates [160]. The importance of these peroxidases in microbial killing by phagocytes has been widely acknowledged [161]. Accordingly, the *in vitro* addition of a halide and lactoperoxidase [92, 152, 153a], horseradish peroxidase [154, 161, 162], myeloperoxidase [154, 157–159], or eosinophil peroxidase [158, 159, 163] enhances parasite killing by H_2O_2.

Hydrogen peroxide is also involved in the toxic action of purines and xanthine oxidase present in calf serum on epimastigotes of *Trypanosoma cruzi*, because its trypanocidal action is reversed by catalase [164].

Thus, there is no doubt that H_2O_2 is toxic to parasites. However, it seems possible that H_2O_2 is damaging not by virtue of direct attack upon parasite components, but rather because it can, by reaction with O_2^{-} or with Fe^{2+}, give rise to the extremely reactive hydroxyl radical [123, 148].

VI. FREE-RADICAL INTERMEDIATES IN THE METABOLISM OF ANTIPARASITIC COMPOUNDS

Antiparasitic drugs range widely in structural complexity and action at the subcellular and molecular levels. However, a number of antiparasitic drugs are thought to exert their action by generating free radicals. In this section we describe the role of free radicals in the antiparasitic action of these agents.

A. Quinones and Quinoneimines

Perhaps the earliest unambiguous demonstration of the enzymatic formation of a drug-derived free radical in intact eukaryotic organisms resulted from studies with the trypanocidal agent β-lapachone (3,4-dihydro-2,2′-dimethyl-2H-naphtho[1,2b]pyran-5,6-dione) [43].

β-LAPACHONE

β-Lapachone was originally isolated as a contaminant of lapachol preparations from *Tabebuia* sp. [165]. The incubation of *Trypanosoma cruzi* intact epimastigotes with β-lapachone generates a five-line ESR spectrum identified as the semiquinone free radical [43]. This ESR spectrum is time dependent and progressively changes from a symmetric, motionally narrowed spectrum to an immobilized spectrum [43]. Similar progressive immobilization of the semiquinone on the ESR time scale has been observed in rat liver microsomal incubations of daunorubicin in the presence of NADPH [166].

The addition of β-lapachone to *T. cruzi* epimastigotes increases the intracellular rate of O_2^{-} and H_2O_2 generation and releases the oxygen reduction intermediates to the suspending medium. The immediate response indicates a rapid permeation of the quinone through the epimastigote cell membrane [42–45]. The chemical reactions that would explain the trypanocidal action of β-lapachone (β-L) could be written as follows:

$$\text{NAD(P)H} + \text{H}^+ + 2\ \beta\text{-L} \rightarrow \text{NAD(P)}^+ + 2\ \beta\text{-LH}\cdot \tag{1}$$

$$2\ \beta\text{-LH}\cdot + \text{O}_2 \rightarrow 2\ \beta\text{-L} + 2\ \text{O}_2^{\cdot-} + 2\ \text{H}^+ \tag{2}$$

$$\text{O}_2^{\cdot-} + \text{O}_2^{\cdot-} + 2\ \text{H}^+ \overset{\text{SOD}}{\rightarrow} \text{H}_2\text{O}_2 + \text{O}_2 \tag{3}$$

$$\text{O}_2^{\cdot-} + \text{H}_2\text{O}_2 \overset{\text{Fe}^{3+}}{\rightarrow} \text{OH}\cdot + \text{OH}^- + \text{O}_2 \tag{4}$$

Quinones such as β-lapachone can be reduced enzymatically by flavoproteins in either a one-electron or two-electron transfer process. In mitochondrial and microsomal membranes, β-lapahone reduction to the semiquinone form (β-LH·) [Eq. (1)] is favored over bivalent reduction [43, 167–169]. Semiquinone radicals (β-LH·) usually reduce O_2 to $\text{O}_2^{\cdot-}$ [Eq. (2)]. The quinone (β-lapachone) is regenerated in the process, and a redox cycle is initiated, forming a large amount of $\text{O}_2^{\cdot-}$. Superoxide anion subsequently dismutates to form H_2O_2 and $\text{O}_2^{\cdot-}$ in a reaction catalyzed by superoxide dismutase (SOD) [Eq. (3)] [124]. Superoxide anion and H_2O_2 may also take part in metal-catalyzed reactions to form more toxic species of active oxygen, such as hydroxyl radical (OH·) [148, 170–172] [Eq. (4)] and singlet oxygen [173]. These reactions seem to be the most reasonable mechanisms underlying the complex chain reaction that leads to extensive lipid and organic peroxide formation and to biological damage [43, 174]. In this connection it seems pertinent to recall that T. cruzi epimastigotes supplemented with β-lapachone show (a) increased lipid peroxidation [43]; (b) severe ultrastructural alterations: damage of the nuclear, mitochondrial, and cytoplasmic membranes, formation of blebs on the cell membrane and disappearance of subpellicular microtubules where the blebs form, and condensation of the nuclear chromatin in patches (Fig. 1) [174]; (c) severe metabolic alterations: initial stimulation (probably related to $\text{O}_2^{\cdot-}$ formation) and later inhibition of cell respiration (related to cell death), correlated with inhibition of glucose and pyruvate oxidation [174]; and (d) lysis of the cells after prolonged incubations [43]. The metabolic and ultrastructural alterations observed could be easily accounted for by $\text{O}_2^{\cdot-}$ and other reactive oxygen-derived species that cause DNA strand breaks [175, 176], enzyme inhibition [177], lipid peroxidation [178], and oxidation of thiol groups of proteins [179]. It is interesting that ultrastructural alterations similar to those observed after treatment of T. cruzi with β-lapachone (swelling of the mitochondria and condensation of the nuclear chromatin in patches [46, 174]) are also observed after treatment of Trypanosoma dionisii with H_2O_2 and myeloperoxidase [157]. Moreover, in the presence of cytotoxic concentrations of

Fig. 1.   Trypanosoma cruzi epimastigotes treated with β-lapachone (10 μg/ml for 180 min). (a) Note the mitochondrion (M) swelling and alterations of the mitochondrial membrane. Kinetoplast (K) DNA seems not to be affected. ×30,400. (b) Bleb on the cell membrane showing vesicles. ×45,900. (c) Section of a β-lapachone-treated epimastigote showing swelling of the mitochondrion, alterations of the nuclear, mitochondrial, endoplasmic reticulum, and cell membranes, and chromatin arranged in patches. B, Bleb. ×20,250. From Docampo et al. [174] with permission.

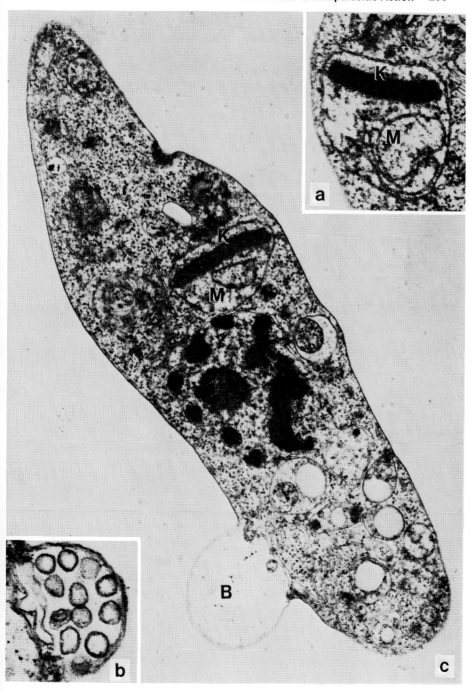

menadione (2-methyl-1,4-naphthoquinone) isolated hepatocytes show similar ultrastructural alterations (formation of surface blebs) related to a redox cycle similar to that exerted by β-lapachone in *T. cruzi* [*180*]. Similar alterations in the surface structure of isolated hepatocytes also occur during the metabolism of *tert*-butyl hydroperoxide, and it has been reported [*181*] that these changes are associated with alterations in intracellular thiol and Ca^{2+} homeostasis [*180*]. These alterations in Ca^{2+} homeostasis might explain the disappearance of subpellicular microtubules in β-lapachone-treated *T. cruzi* [*174*].

Results similar to those obtained with *T. cruzi* epimastigotes are observed with β-lapachone-supplemented *T. cruzi* amastigotes (the intracellular forms) and trypomastigotes (the bloodstream infective forms) [*46*]. In addition, when 10 β-lapachone-related naphthoquinones are compared for O_2^- generation and toxic action on *T. cruzi* [*45*], the correlation coefficient *r* is .88 (*p* < .01), suggesting a toxic action mediated by this radical [*45*].

A complex mechanism of action involving a sequence of reactions among β-lapachone, a reducing agent (dithiothreitol), and the target has been proposed as a factor in the inhibition of *Oncornavirus* reverse transcriptase and eukaryotic DNA polymerase α [*182*]. It is possible that in the aforementioned observations the reduction of β-lapachone, followed by spontaneous autoxidation generating O_2^- and H_2O_2, contributes to the inhibition of the activity of those enzymes. This may be in some way related to the ultrastructural alterations observed in the nuclear DNA of *T. cruzi* [*174*].

The *in vivo* activity of β-lapachone in infected mice does not parallel its *in vitro* effects on *T. cruzi,* suggesting an inactivation of the drug by the host [*183*]. Moreover, fetal calf serum or hemoglobin solution protects the epimastigotes against the motility inhibition exerted by β-lapachone [*183*]. A similar protective effect of blood and serum is observed against the toxicity of menadione to *T. cruzi* epimastigotes [*184*]. In addition, like menadione [*185*], β-lapachone increases methemoglobin formation from oxyhemoglobin [*183*]. A β-lapachone derivative, 3-allyl-β-lapachone, is effective against *T. cruzi* trypomastigotes, is not inactivated in blood, does not increase the formation of methemoglobin, and is therefore potentially useful in preventing blood transmission of Chagas' disease [*49*].

β-Lapachone is also endowed with antitumor properties for sarcoma 180 cells, induces the formation of the semiquinone radical, stimulates O_2^- and H_2O_2 formation by the mitochondrial and microsomal fractions, increases lipid peroxidation in intact cells, and decreases the viability of these cancer cells [*186*]. Vitamin E maintains the transplantability of these cells in the presence of β-lapachone [*186*]. β-Lapachone toxicity in bacteria is also mediated by the formation of O_2^- and H_2O_2 and is not observed under anaerobic conditions [*187*].

A variety of other naphthoquinones are toxic against parasites. For example, menadione is toxic *in vitro* against *T. cruzi* [*184*], *Leishmania mexicana ama-*

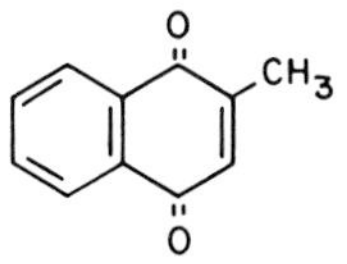

MENADIONE

zonensis [*188*], and *Trypanosoma brucei* [*56*]. It also increases oxygen consumption and H_2O_2 production by *T. brucei* mitochondrial preparations [*56*]. The addition of sublytic concentrations of menadione and heme leads to a synergistic lysis of *T. brucei in vitro* [*56*]. Menoctone [2-hydroxy-3-(8-cyclohexyloctyl)-1,4-naphthoquinone] and other 2-alkyl-3-hydroxy-1,4-naphthoquinones are

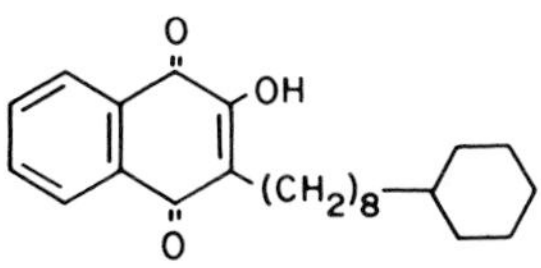

MENOCTONE

effective against *Theileria parva* both *in vivo* and *in vitro* [*189, 190*]. Menoctone is also active against *Plasmodium* spp. [*16*].

Anthracycline antibiotics have both quinone and hydroquinone structures. The anthracycline daunorubycin (daunomycin) is a potent trypanocidal agent *in vitro* but not *in vivo* [*191, 192*]. It permanently abolishes the infectivity of *Try-*

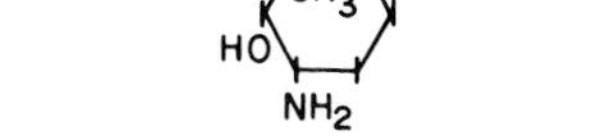

DAUNORUBICIN

panosoma rhodesiense at less than nanomolar concentrations [*192*]. In contrast, other anthracyclines such as doxorubicin (adriamycin) and nogalamycin must be present in millimolar concentrations to achieve the same effect [*192*].

Mitomycin C (a quinone) and actinomycin D (a quinoneimine) are also toxic against *T. cruzi in vitro* [*193*]. Both of these drugs, as well as anthracycline

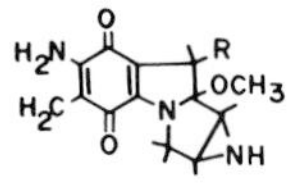

MITOMYCIN C

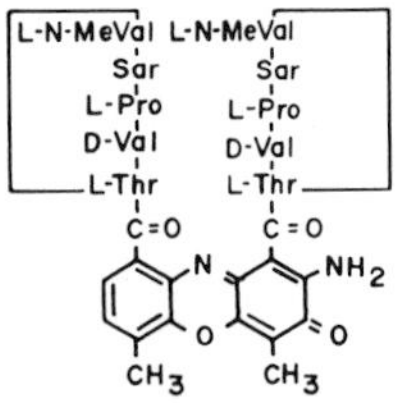

ACTINOMYCIN D

antibiotics, have been proposed to produce strand breaks in DNA and cytotoxic effects through oxidoreduction processes and reductive activation of molecular oxygen in mammalian cells [166, 194–196]. It is not known whether they have the same effect in parasites.

Some N-substituted 2-amino-5-hydroxy-4-methylnaphtho[1,2-d]thiazoles have substantial rapid trypanolitic activity *in vitro* on *T. brucei,* but their mechanism of action is unknown [197].

The reactivity toward DNA of some pyridocarbazole derivatives has been demonstrated [198], and some ellipticine (5,11-dimethyl-6H-pyrido[4,3-b]carbazole) derivatives have been tested *in vitro* against *Trypanosoma cruzi* [199, 200]. Also, the trypanocidal activity in infected blood of several substituted 1,2,3,4-tetrahydrocarbazoles and dihydro[c]benzocarbazoles has been reported [201]. Olivacine (1,5-dimethyl-1,6H-pyrido[4,3-b]carbazole), an alkaloid isolated from *Aspidosperma nigricans,* is also toxic *in vitro* against *T. cruzi* [202]. The phenolic derivative 9-hydroxyellipticine has been identified as a main product of the oxidative metabolism of ellipticine in rats [203]. It has been recognized

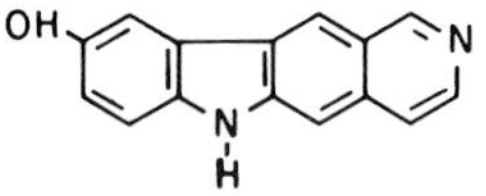

9-HYDROXYELLIPTICINE

for several years that the addition of a hydroxyl group on the ellipticine ring in position 9 results in a strong increase in both cytotoxic and antitumor activities [204]. Two main hypotheses have been proposed concerning the mechanism of action of the antitumor drugs derived from ellipticine: (a) the possible interaction of these drugs with DNA through an intercalation process [198] and (b) the oxidation of these drugs to chemically reactive products with potential toxicity [204]. In agreement with the latter hypothesis, using a myeloperoxidase–H_2O_2 system as a model of bioactivation, it was demonstrated that 9-hydroxyellipticine derivatives may undergo oxidation through a one-electron process yielding quinoneimine radical intermediates, the autoxidation of which may generate O_2^{-} and H_2O_2 [204]. It is interesting that 9-hydroxyellipticine has the most potent inhibitory effect on *T. cruzi* polymerase activity [199], but the possible involvement of a free-radical mechanism in its trypanocidal action was not investigated.

Gossypol, a phenolic compound isolated from the cotton plant, is a powerful inhibitor of NAD-linked enzymes of *T. cruzi* and is toxic *in vitro* against *T. cruzi* epimastigotes [204a] and *Plasmodium* [204b]. Gossypol is being tested as a male contraceptive in China [204c], and its mode of action as spermicide may be through the production of oxygen radicals [204d]. This mechanism has not been investigated in parasites.

B. 5-Methylphenazinium Methyl Sulfate

5-Methylphenazinium methyl sulfate, often referred to as phenazine methosulfate, has been widely used in the study of biological oxidation and reduction reactions, particularly in the respiratory chain [205] and in the study of pho-

$$\left[\underset{\underset{CH_3}{\overset{H}{N}}}{\bigcirc\bigcirc} \right]^{+} \left[CH_3OSO_3 \right]^{-}$$

5-METHYLPHENAZINIUM METHYL
SULFATE

tophosphorylation [206]. Zaugg [207] was the first to carry out a comprehensive optical study of the various oxidation states of the 5-methylphenazinium ion (MP^+) and also reported a broad nine-line ESR spectrum of the $MPH^{+\cdot}$ cation-radical. There followed a series of papers in which $MPH^{+\cdot}$ was shown by ESR to interact with certain bacteria [208], with DNA [209], and with adenosine 5'-monophosphate [210]. It was shown that air oxidation of reduced phenazine methosulfate produces hydrogen peroxide [211] and superoxide anion [212].

When phenazine methosulfate is added to live suspensions of *Trypanosoma cruzi* epimastigotes, the corresponding $MPH^{+\cdot}$ cation-radical is enzymatically generated intracellularly, and the viability of the cells is affected [47]. Phenazine methosulfate also significantly increases the rate of O_2^{-} formation in intact cells when ethanol is the electron donor and, in mitochondrial and microsomal fractions, when NADH is the electron donor. No significant O_2^{-} formation is detected under assay conditions in the absence of the fractions, and the rates of the reactions are directly proportional to the amount of cell protein, thus excluding a significant nonenzymatic reduction of phenazine methosulfate by NADH [47].

Exposure of macrophages infected with *Leishmania mexicana amozonensis,*

Leishmania enriettii, or *Trypanosoma cruzi* to phenazine methosulfate results in rapid damage and disappearance of the intracellular amastigotes without obvious ill effects on the host cells [*188, 212, 212b*]. Phenazine methosulfate also reduces the growth of *Leishmania* promastigotes in culture. These results are compatible with a direct effect of the drug on the intracellular amastigotes, involving only a permissive participation of the macrophage [*188*].

C. Nitro Heterocyclic Compounds

Nitro compounds such as nitrofurans and nitroimidazoles have been widely used in the treatment of several parasitic diseases [*213, 214*]. These compounds have also been used in cancer therapy, both as radiosensitizers and as cytotoxic agents [*215, 216*]. Radiosensitization, hypoxic cell toxicity, and chronic aerobic toxicity correlate with the electron affinity of the nitro compounds [*217*]. This similarity suggests that redox processes are involved in each phenomenon but does not necessarily indicate a common mechanism [*217*].

The first step in the reduction of nitro heterocyclic compounds is the formation of the nitro anion-radical $(R—NO_2^{\,-})$ [*218*]. The intermediate $R—NO_2^{\,-}$ is the first in a sequence of reduction products, any of which may be responsible for the cytotoxicity:

$$R—NO_2 \xrightarrow{e^-} R—NO^{\,-}_2 \xrightarrow[2\,H^+]{R—NO_2^{\,-}} R—NO \xrightarrow[2\,H^+]{2\,e^-} R—NHOH \xrightarrow[2\,H^+]{2\,e^-} R—NH_2$$

In addition, under aerobic conditions this radical $(R—NO_2^{\,-})$ is rapidly oxidized by molecular oxygen, with regeneration of the nitro group and production of superoxide anion [*219*]:

$$R—NO_2^{\,-} + O_2 \rightarrow R—NO_2 + O_2^{\,-}$$

It seems likely that one or more of the nitro reduction products and superoxide and hydrogen peroxide may participate in the mechanism of toxicity of nitro compounds in the types of cells in which this toxic action occurs. Hence, depending on the presence of appropriate nitroreductases, the presence and relative activities of biochemical and enzyme defense mechanisms, the capacity to repair nitro compound-mediated damage, and the aerobic or anaerobic habitat of each parasite, one process or the other may be more important.

Nifurtimox [4-(5-nitrofurfurylidene)amino]-β-methylthiomorpholine 1,1-dioxide), a nitrofuran derivative, is one of the most effective drugs used in the

NIFURTIMOX

treatment of acute Chagas' disease [214]. The addition of nifurtimox to NAD(P)H-containing homogenates of *Trypanosoma cruzi* (epi-, trypo- or amastigote forms) determines the ESR spectrum corresponding to the nitro anion-radical. The anion-radical signal is observed after an induction period that depends on the oxygen concentration and the pyridine nucleotide in the incubation medium [51]. Incubation of intact *T. cruzi* forms with nifurtimox also leads to the appearance of the anion-radical signal [51]. The nitro anion-radical, which is assumed to be the first product of nitroreductase activity, reacts with oxygen under aerobic conditions as demonstrated by (*a*) the increase in the cyanide-insensitive respiration of and the release of H_2O_2 from the whole cells to the suspending medium [48], (*b*) the increase in the rate of O_2^{-} and H_2O_2 production after the addition of nifurtimox to homogenates of *T. cruzi* in the presence of NAD(P)H [51], (*c*) the stimulation of O_2^{-} production by the *T. cruzi* mitochondrial fraction supplemented with NADH (or succinate) [48, 50], and (*d*) the enhancement of O_2^{-} generation by the microsomal fraction with NADPH as reductant [48, 50]. It is noteworthy that the serum concentration of nifurtimox after the administration of a single oral dose of 15 mg/kg in man is ~ 10–20 μM, which is as high as that which produces maximal stimulation of O_2^{-} production by the *T. cruzi* mitochondrial fraction, initiates diffusion of H_2O_2 outside of the cells, and completely inhibits growth [48, 50].

Nifurtimox is also reduced by rat liver, testes, and brain microsomes to a nitro anion-radical, as indicated by ESR spectroscopy [220–222]. It also produces, in rat liver microsomes, an increase in (*a*) electron flow from NADPH to molecular oxygen; (*b*) generation of O_2^{-}, H_2O_2, and hydroxyl radical; and (*c*) lipid peroxidation [223–225]. In addition, treatment of rats with nifurtimox induces a time- and dose-dependent depletion of liver glutathione, a release of both oxidized and reduced glutathione into bile [226, 226a], and ultrastructural alterations in the testes similar to those observed after whole-body X-irradiation of animals [224, 227]. These results confirm that nifurtimox generates active oxygen radicals *in vivo* as observed *in vitro* with liver microsomal incubations [223]. This free-radical generation may be the basis of nifurtimox toxicity in mammals [214].

Benznidazole (*N*-benzyl-2-nitro-1-imidazoleacetamide) is a nitroimidazole de-

$$CH_2\text{-}CONH\text{-}CH_2 \text{—} \bigcirc$$

BENZNIDAZOLE

rivative that is also effective in the treatment of acute Chagas' disease [214]. This drug is reduced to the nitro anion-radical and induces the generation of both O_2^{-} and H_2O_2 by rat liver micrsomes [53]. In spite of the many similarities in the mode of action of nitrofurans and nitroimidazoles in several biological systems

[218], benznidazole increases O_2^{-} and H_2O_2 production by rat liver microsomes at a concentration that is one order of magnitude higher than that of nifurtimox [53]. This difference fits in well with the fact that the electronegativity of the 2-nitroimidazoles is lower than that of the nitrofurans, which should decrease the rate of enzymatic radical formation [217]. These results may imply that the effect of benznidazole on O_2^{-} and H_2O_2 formation in mammalian tissues is small, hardly exceeding the basal levels. The difference noted between the capacity of nifurtimox and benznidazole to generate oxygen free radicals in mammalian cells may be significant in the treatment of Chagas' disease when these drugs are compared for their toxicity to the host.

Addition of benznidazole to NADH-containing homogenates of *T. cruzi* epimastigotes determines the appearance of a weak signal corresponding to the nitro anion-radical [228]. However, no nitro anion-radical formation is observed after incubation of benznidazole with NADPH and *T. cruzi* microsomal preparations, and O_2^{-} and H_2O_2 generation is not stimulated by benznidazole under the same experimental conditions [53]. In addition, benznidazole decreases the respiration rate of the whole cells [228]. Moreover, benznidazole inhibits growth of *T. cruzi* at concentrations that do not stimulate O_2^{-} and H_2O_2 generation, indicating that the trypanocidal effect does not depend on the effect of oxygen radicals [53]. The difference between the mechanisms of benznidazole and nifurtimox toxicity on *T. cruzi*, therefore, is significant, and stresses the caution that should be exercised before considering a common mechanism of action for all nitro heterocyclic drugs or extrapolating results obtained with cells that are not related to *T. cruzi* [229].

Metronidazole [Flagyl, 1-(2-hydroxyethyl)-2-methyl-5-nitroimidazole] and other nitroimidazoles are used extensively to treat infections due to anaerobic

METRONIDAZOLE   SECNIDAZOLE

RONIDAZOLE   MISONIDAZOLE

protozoa such as *Trichomonas vaginalis, Entamoeba hystolitica*, and *Giardia lamblia*, as well as bacterial and helminth (*Schistosoma mansoni*) infections [213, 230]. Although the mechanism of the cytotoxic action of nitroimidazoles on these organisms is not well understood, derivatives arising from the reduction of the nitro group may be responsible for the observed activity [231–234]. The nitro anion-radical and the hydroxylamine derivative of nitroimidazole reduction were suggested as the main candidates for the toxic intermediates [235, 236]. *Tritrichomonas foetus* intact cells are able to reduce metronidazole (Fig. 2), ronidazole (1-methyl-5-nitroimidazole-2-methanol carbamate), secnidazole

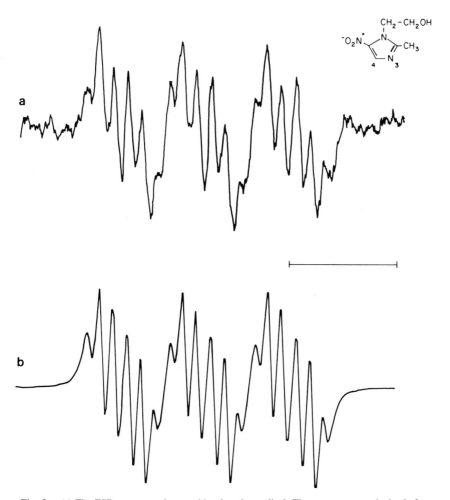

Fig. 2. (a) The ESR spectrum of metronidazole anion-radical. The spectrum was obtained after anaerobic incubation of 12.5 mM metronidazole with 10 mM glucose and 10^9 *Tritrichomonas foetus* intact cells per milliliter in 0.1 M potassium phosphate buffer (pH 7.5). The spectrum was obtained with a nominal microwave power of 20 mW and a modulation amplitude of 1 G. Bar. 20G. (b) Computer simulation of the metronidazole anion-radical spectrum. The hyperfine coupling constants were $a_N^{NO_2} = 15.656$; $a_H^{(4)} = 5.42$ G; $a_H^{CH_3} = 2.29$G. From Moreno *et al.* [237] with permission.

[1-(2-hydroxypropyl)-2-methyl-5-nitroimidazole], benznidazole, and misonidazole [1-(2-nitro-1-imidazolyl)-3-methoxy-2-propanol] to their respective anion-radicals [237]. This activity appears to be related to the cellular content of reducing substrates because the formation of nitro anion-radical is stimulated in the presence of glucose or pyruvate, the two main sources of *T. foetus* reducing equivalents [237]. Anaerobic homogenates of *T. foetus* also reduce metronidazole to the nitro anion-radical when pyruvate is added as electron donor [237]. Trichomonad homogenates contain a pyruvate : ferredoxin oxidoreductase [238], which reduces nitroimidazoles via systems containing ferredoxin- or flavodoxin-type electron transport proteins [213]. These proteins are reduced enzymatically by the substrates, and it is supposed that the proteins, in turn, reduce the nitroimidazoles [213]. Nitroimidazole anion-radical formation cannot be detected under aerobic conditions; this can be attributed either to the reaction between oxygen and the nitro anion-radicals, as has been postulated for rat liver microsome incubations of metronidazole [239], or to a competition for electrons between the nitroimidazoles and molecular oxygen [213]. The competitive nature of this inhibition is suggested by the observation that metronidazole does not stimulate oxygen consumption in *T. foetus* intact cells or homogenates even in the presence of pyruvate [237]. The metronidazole anion-radical signal is also observed in the homogenate in the presence of an NADPH- or an NADH-generating system [237]. These results demonstrate that pyruvate synthase is not the sole system in trichomonads capable of reducing metronidazole [237].

Two carcinogenic nitroquinoline derivatives, 4-nitroquinoline 1-oxide and 4-hydroxylaminoquinoline 1-oxide, are toxic to *Leishmania tarentolae* [240],

4-NITROQUINOLINE 1-OXIDE   4-HYDROXYAMINOQUINOLINE 1-OXIDE

Leishmania donovani [241], and *Trypanosoma cruzi* [50]. 4-Nitroquinoline 1-oxide is also toxic to *Crithidia fasciculata* [242], and a *C. fasciulata*-resistant strain that shows an increase in the number of peroxisomes parallel to an increase in catalase activity of those microorganisms has been isolated [98]. 4-Nitroquinoline 1-oxide can be enzymatically reduced to 4-hydroxylaminoquinoline [242, 243], and the air oxidation of this compound can generate H_2O_2 [244]. Therefore, air oxidation of both the 4-nitroquinoline 1-oxide anion free radical and the corresponding hydroxylamine (nitroxide) radical [245] is likely to reduce oxygen to superoxide anion, the dismutation of which produces H_2O_2. Accordingly, the addition of 4-nitroquinoline 1-oxide or 4-hydroxylaminoquinoline 1-oxide to *T. cruzi* induces the release of H_2O_2 from the whole cells to the suspension medium [50]. The addition of both compounds stimulates O_2^- production by *T. cruzi* mitochondrial fractions in the presence of NADH and by microsomal fractions in the presence of NADPH. 4-Nitroquinoline 1-oxide and 4-hydroxylaminoquinoline 1-oxide concentrations that increase O_2^- and H_2O_2 production by *T. cruzi* and cell fractions are effective in inhibiting parasite growth [50].

D. Porphyrins

Hematoporphyrin has photodynamic properties. The exposure of biomolecules or cells to hematoporphyrin and light can result in damage [246] and can even be lethal [247]. There is evidence that the cytotoxic action of hematoporphyrin and light is the result of the photosensitized production of singlet oxygen [248]; however, a radical mechanism is also possible [249] because hematoporphyrin can produce O_2^- in aqueous solution upon illumination [250].

Hematoporphyrin and several other related porphyrins have a lytic effect on *Trypanosoma brucei in vitro* in the presence of light [251]. Hematoporphyrin is also effective *in vivo* [251]. This *in vivo* effect is apparently unrelated to the photodynamic action observed *in vitro* [251] and is related to the formation of zinc–hematoporphyrin in infected mice [251]. Both zinc–hematoporphyrin and ferriprotoporphyrin IX chloride (hemin) have a lytic effect on *T. brucei in vitro* even in the absence of light [54, 251]. A free-radical mechanism of T. brucei lysis by these metalloporphyrins in the dark has been postulated [54] on the basis of the following findings: (*a*) ultrastructural alterations observed after incubation of *T. brucei* in the presence of hemin consisting of swelling and disruption of endoplasmic reticulum membranes, (*b*) protection from lysis by riboflavin, (*c*) potentiation of hemin lysis by depletion of cellular glutathione by pretreatment of trypanosomes with diamide, and (*d*) greater susceptibility to hemin lysis in trypanosomes from animals maintained on diets enriched with unsaturated fatty acids rather than saturated fatty acids [54]. However, other mechanisms of lysis cannot be dismissed. In this connection hemin also lyses *Plasmodium berghei* [252] and isolated normal erythrocytes [253]. In this case hemin impairs the capacity of erythrocytes to maintain cation gradients and induces hemolysis by a colloid osmotic mechanism [253].

E. Tryarylmethane Dyes

Crystal violet, a tryarylmethane dye, has been used for the control of intestinal parasites in humans [254–256] and is widely used in blood banks in attempts to eliminate the blood transmission of Chagas' disease [257–260].

When crystal violet is metabolized in the dark under nitrogen by *T. cruzi* intact

$$(CH_3)_2N-\!\!\!\bigcirc\!\!\!-\overset{||}{C}-\!\!\!\bigcirc\!\!\!-N(CH_3)_2$$

CRYSTAL VIOLET

epimastigote or trypomastigote stages (Fig. 3), a single-line ESR spectrum is obtained [261]. This ESR spectrum is that of the tri(p-dimethylamino-phenyl)methyl radical, which is the one-electron reduction product of crystal violet [262, 263]. The formation of this carbon-centered free radical is completely inhibited by oxygen. After exposure to room light or the light of a slide

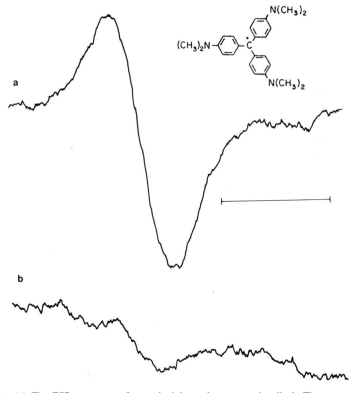

Fig. 3.   (a) The ESR spectrum of crystal violet carbon-centered radical. The spectrum was obtained after anaerobic incubation of 5 mM crystal violet and *Trypanosoma cruzi* epimastigotes (25 mg protein per milliliter) in 0.1 M potassium phosphate buffer (pH 7.4). The nominal microwave power was 20 mW, and the modulation amplitude was 3.5 G. Bar, 20G. (b) Identical with a, but without crystal violet. From Docampo *et al.* [261] with permission.

projector, the signal grows in intensity. The incubation of *T. cruzi* homogenates in the presence of an NADH-generating system also generates a single-line ESR spectrum in the dark [*261*]. Although NADPH can replace NADH in these incubations, its activity is about 80% lower [*261*]. Growth inhibition of *T. cruzi* by crystal violet is also enhanced by light [*261*].

It is interesting that crystal violet-resistant mutants of *Crithidia fasciculata* show cross-resistance with another free-radical-generating drug, the quinone-imine actinomycin D [*264*]. In addition, exposure of macrophages infected with *Leishmania mexicana amazonensis* or *Trypanosoma cruzi* to crystal violet results in rapid damage and disappearance of the intracellular amastigotes without ob-vious damage to the host cells [*188, 212a*]. Crystal violet also reduces the growth of *Leishmania* promastigotes in culture [*188*].

F. Phenothiazines

Phenothiazines have been used as biological dyes, as antihelminthic agents, and as antipsychotic agents [*265*]. Chlorpromazine was the first and is the best known of these compounds [*265*]. The antihelminthic activity correlates with the

$$CH_2-CH_2-CH_2N(CH_3)_2$$

CHLORPROMAZINE

oxidation potential of the phenothiazines, which is related to the formation of their cation free radicals [*266, 267*]. In addition, phenothiazines are lethal to both the extracellular and the intracellular stage of *Leishmania donovani* [*267a*] and the culture form of *Trypanosoma brucei* [*267b*], although their mode of action in these parasites has not been investigated.

G. Bleomycins

The bleomycins, a family of glycopeptide antibiotics, are effective in the chemotherapy of several solid tumors and malignant lymphomas in man [*268*]. Their cytostatic activity is apparently due to DNA strand breaks, which have been shown to correlate with the reduction of cell proliferation [*269*]. Several reports suggest that DNA breakage is produced via a tertiary oxygen–ferrous ion–bleomycin complex [*270*] that is similar to heme-containing oxygenases [*271*]. Whether this complex produces DNA breakage through the formation of O_2^- and OH· [*272–274*] or whether a ferryl ion intermediate is involved [*169*] is not known. However, in either case free radicals are involved in its mechanism of action [*169*].

Bleomycin inhibits nuclear division and causes malformation of the nuclear microtubules in *Trypanosoma gambiense* [275]. Administration of bleomycin alone [276] or in combination with a DL-α-difluoromethylornithine, an irreversible inhibitor of polyamine biosynthesis, cures mice infected with *Trypanosoma brucei* [277, 278]. Polyamines antagonize the curative effects of this drug combination in mice [278].

H. Aminoquinolines

The formation of H_2O_2 by the reaction of primaquine with oxyhemoglobin is thought to be a major reaction leading to primaquine toxicity in red blood cells [279]. A primaquine-mediated increase of flux through the hexose monophos-

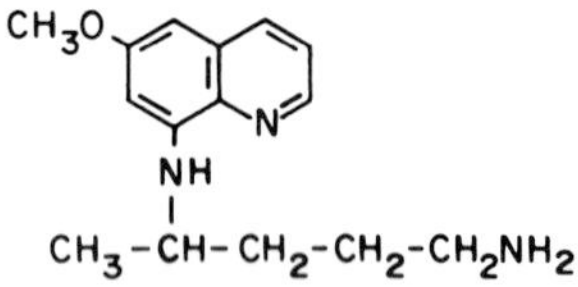

CH$_3$O

NH

CH$_3$-CH-CH$_2$-CH$_2$-CH$_2$NH$_2$

PRIMAQUINE

phate shunt [280] is assumed to result from the removal of H_2O_2 by glutathione peroxidase [279] or from the oxidation of NADPH by the drug [281]. Primaquine in aqueous solution autoxidizes at a slow rate, generating H_2O_2, O_2^{-}, and OH· [282, 282a]. Whether these radicals are involved in the hemolytic properties of primaquine [283] or in the antimalarial activity [284] of primaquine or other 4-amino- or 8-aminoquinolines (quinine, quinacrine, chloroquine, hydroxychloroquine, and amodiaquine) is, as yet, unknown.

VII. REDUCED OXYGEN BY-PRODUCTS IN THE ANTIPARASITIC ACTION OF PHAGOCYTES

Phagocytosis of microorganisms by polymorphonuclear leukocytes and mononuclear phagocytes initiates a sequence of events that generally results in the death of the ingested organisms [160]. A variety of antimicrobial systems is responsible, some dependent on oxygen and others operative in its absence [160]. Oxygen-dependent antimicrobial systems have been the subject of considerable interest [160, 161, 285, 286].

All phagocytic cells have in common a mechanism or mechanisms by which perturbation of the plasma membrane results in a burst of oxygen consumption, and much, if not all, the extra oxygen consumed is converted to O_2^{-} [160]. Hydrogen peroxide is generated by dismutation of O_2^{-}. Hydroxyl radical and perhaps singlet oxygen also may be formed. These potentially toxic products of oxygen reduction or excitation released into the phagosome or extracellular fluid appear to contribute significantly to the toxic activity of phagocytes [160].

Several research groups have clearly demonstrated that polymorphonuclear leukocytes as well as mononuclear phagocytes produce reduced oxygen by-products in the presence of different parasites. These studies have involved two areas of phagocyte effector function: intracellular killing of certain protozoan parasites and extracellular lysis of helminths.

A. Intracellular Killing of Parasitic Protozoa

Polymorphonuclear leukocytes and mononuclear phagocytes are important defenses against parasites. However, mononuclear phagocytes interact with parasites not only as effector cells of immune response that participate in the control of the infection, but also as host cells in which several intracellular protozoa such as *Toxoplasma gondii* [287], *Leishmania* spp. [288], and *Trypanosoma cruzi* [289] multiply and differentiate. The focus of this section is on the evidence implicating the role of reduced oxygen by-products in the destruction of these and other protozoa within phagocytic cells.

1. Polymorphonuclear leukocytes

It has been clearly established that oxygen-requiring microbicidal mechanisms play a role in the destruction of pathogens by neutrophils and eosinophils [160]. One such system involves myeloperoxidase, an enzyme abundant in neutrophils, which mediates microbial killing in the presence of H_2O_2 and a halide ion. The H_2O_2 used by this system arises from the dismutation of O_2^{-}, the product of an enzyme-catalyzed reaction [160]. Like myeloperoxidase, eosinophil peroxidase combines with H_2O_2 and a halide to form a cytotoxic system effective against microbes [160]. In addition, it has become apparent that polymorphonuclear leukocytes employ other oxygen-requiring microbicidal mechanisms that do not depend on peroxidases, such as the generation of hydroxyl radical and singlet oxygen [160]. Some of these oxygen-requiring microbicidal mechanisms of polymorphonuclear leukocytes have been tested against parasites.

Contact of human neutrophils with anitbody-coated *Trypanosoma cruzi* epimastigotes or trypomastigotes triggers the respiratory burst [290–292]. Oxygen consumption and O_2^{-} and H_2O_2 release are stimulated under conditions of neutrophil-mediated killing. This stimulation does not occur under nonkilling conditions when antibody is omitted [290–292]. When neutrophils are incubated with the myeloperoxidase inhibitors aminotriazole [290], cyanide [151a, 290], or azide [151a], an inhibition of cytotoxicity [151a, 290] without any effect on phagocytosis or on the respiratory burst is observed [290], indicating the *in vivo* involvement of myeloperoxidase in the destruction of *T. cruzi* by normal human neutrophils [*151a, 290*]. Incubation of neutrophils with antibody-coated *T. cruzi* epimastigotes or trypomastigotes and the spin trap 5,5-dimethyl-1-pyrroline *N*-oxide (DMPO) results in an ESR signal with splitting constants of $a_N = a_H =$ 15.0 ± 0.2 G and the 1 : 2 : 2 : 1 intensity distribution characteristic of the DMPO–hydroxyl radical adduct. Because the DMPO–hydroxyl radical adduct formation is inhibited by superoxide dismutase but not by catalase, the generation of the DMPO–hydroxyl radical adduct by *T. cruzi*-stimulated neutrophils can be interpreted as being due to the generation of O_2^{-} [*292*].

Hydrogen peroxide generation by human neutrophils in the presence of *T. cruzi* epimastigotes [*291*] and amastigotes [*151a*] or by human neutrophils and eosinophils in the presence of *Leishmania donovani* amastigotes [*154*] has also been detected by cytochemical techniques. In addition, it has been shown that neutrophils or eosinophils from patients with chronic granulomatous disease (which lack the respiratory burst) show limited leishmanicidal activity [*154*].

Human neutrophils incubated with opsonized *Leishmania donovani* promastigotes also produce O_2^{-}, as detected by ferricytochrome *c* reduction [*152*]. This stimulation does not occur under nonkilling conditions with unopsonized promastigotes [*152*].

The mechanism of killing of trypomastigotes of *Trypanosoma dionisii* by human neutrophils also appears to involve myeloperoxidase and H_2O_2 [*156*], trypomastigotes being as effective as epimastigotes in inducing the formation of H_2O_2 by these neutrophils [*156*].

2. Mononuclear Phagocytes

The mononuclear phagocytic system consists of circulating monocytes and tissue macrophages. Cells of the mononuclear phagocytic series are concentrated in organs of the reticuloendothelial system, such as spleen, liver, and bone marrow. Freshly isolated human monocytes respond to soluble and particulate surface-active stimuli by the secretion of O_2^{-} and H_2O_2 in amounts similar to those released by polymorphonuclear leukocytes [*292a*]. However, as monocytes mature into macrophages, their capacity to release O_2^{-} and H_2O_2 declines precipitously, probably reflecting diminished capacity to produce them [*292a*].

In the latter stages of infection or after the host recovers, lymphocytes encountering antigens of the infecting organism confer upon macrophages an enhanced capacity to kill the same or unrelated pathogens [*292b*]. This process, as well as that resulted from exposure of macrophages to inflammatory agents [*292a*], is termed *macrophage activation* [*292b*]. The soluble products of sensitized lymphocytes are known as lymphokines [*292c*], and it has been demonstrated that some of these lymphokines increase the capacity of mononuclear phagocytes to secrete O_2^{-} and H_2O_2 in the presence of several parasites [*292c*].

a. Monocytes.   Although evidence is accumulating for the generation of reduced oxygen by-products by macrophages in the presence of parasites, few reports have been presented on the destructive potential of monocyte-derived oxygen metabolites.

Human monocytes, which generate O_2^- and H_2O_2 during phagocytosis of unsensitized or antibody-coated *Toxoplasma gondii*, kill this organism and inhibit intracellular replication of those few organisms that are not killed [293]. Killing of *Toxoplasma* is impaired in monocytes from patients with chronic granulomatosus disease or from heterozygotes for this defect [294].

The survival of *Leishmania donovani* in human monocytes is dependent on the parasite stage [153a, 294a]. Human monocytes cultivated for 1 day readily generate H_2O_2 in response to *L. donovani* promastigotes ingestion and kill 90% of them by 24 h. In contrast, the same cells release little H_2O_2 during amastigotes ingestion, kill only <30% of them by 24 h, and support intracellular amastigotes replication. The same cells from patients with chronic granulomatous disease or normal cells whose oxidative activity is impaired by catalase pretreatment or glucose deprivation exert considerably less or no antileishmanial activity during the early postphagocytic period. A longer incubation period or the addition of lymphokines result in near-normal levels of promastigocidal activity [294a]. These results indicate that human monocytes utilize both oxygen-dependent and -independent mechanisms to achieve activity against ingested *Leishmania* [294a].

Unsensitized or antibody-coated epimastigotes of *Trypanosoma dionisii* are killed by the peroxidase–H_2O_2 system of human monocytes [157], the formation of H_2O_2 being induced during the interaction between the parasites and the effector cells. Monocytes are unable to kill unsensitized or antibody-coated trypomastigotes (the infective forms) of *T. dionisii*. This resistance does not appear to occur because of a lack of H_2O_2 formation, because antibody-coated trypomastigotes are as effective as antibody-coated epimastigotes as stimulators of H_2O_2 generation. The reason for this resistance is apparently that trypomastigotes of *T. dionisii* are less sensitive than epimastigotes to the damaging effects of peroxidase and H_2O_2 [157].

b. Macrophages. A positive correlation between the capacity of *in vivo* or *in vitro* activated mouse peritoneal macrophages to secrete H_2O_2 and their activity against *Trypanosoma cruzi* [151] and *Toxoplasma gondii* [295, 296] has been observed. Similarly, *in vitro* activated macrophages display parallel increases in hexose monophosphate shunt activity and O_2^- production and destruction of *Leishmania enriettii* amastigotes [155].

More recent studies correlating the secretion of oxygen reduction by-products with antiparasitic activity have used the parasite itself as the trigger of the macrophage respiratory burst.

The promastigote form of *Leishmania donovani* and *Leishmania tropica* parasitizes nonactivated J774G8 macrophage-like cells, whereas 80–95% of the same promastigotes are killed within resident macrophages (also nonactivated) from normal mice [297, 298]. This striking difference in intracellular anti-leishmanial activity correlates closely with the capacity of these macrophages to generate oxygen reduction by-products. Thus, the ingestion of these cells readily triggers the resident macrophage oxidative burst resulting in the generation of considerable amounts of O_2^- and H_2O_2, whereas triggering of the J774G8 cells releases 5- to 10-fold less O_2^- and H_2O_2. However, activation of the latter macrophages [298], or of human monocyte-derived macrophages [294a] by lymphokines effectively enhances their oxidative response and intracellular anti-leishmanial activity [298]. Inhibiting macrophage H_2O_2 production by glucose deprivation of macrophages or addition of catalase [294a, 297, 298] consistently decreases the killing of Leishmania by activated or resident macrophages. Comparative results are obtained using Leishmania donovani amastigotes [101, 294a, 299, 300]. Upon phagocytosis by mouse peritoneal macrophages, amastigotes elicit a significantly weaker respiratory burst than promastigotes [299]. Immunologically activated macrophages with acquired leishmanicidal capacity can be triggered to release a substantial amount of H_2O_2 in the presence of amastigotes [101, 299]. In contrast, macrophages with a decreased capacity to release a significant amount of H_2O_2 after several days in culture cannot eliminate their parasite burden. Calalase markedly inhibits the elimination of amastigotes by lymphokine- activated macrophages [299]. In addition, lymphokine treatment of human monocyte-derived macrophages induces an oxygen-independent promastigocidal and amastigocidal activity [294a].

A similar correlation between oxidative metabolism and killing of Trypanosoma cruzi epimastigotes [301] or trypomastigotes [147a] is observed when macrophage cell lines with different deficiences in the capacity to produce O_2^- and H_2O_2 are used. When a T. cruzi-infected deficient macrophage clone is reconstituted with an H_2O_2-generating system (glucose–glucose oxidase), intracellular killing of parasites occurs [147a, 301].

Phagocytosis of Trypanosoma cruzi trypomastigotes by normal mouse and human macrophages induces the release of H_2O_2 in a quantity that is sufficient to kill the organisms if they are coated by eosinophil peroxidase [163]. The intracellular killing of eosinophil peroxidase-coated trypomastigotes can be inhibited by catalase and azide, suggesting that toxicity is mediated through the small amount of hydrogen peroxide generated by the phagocytic event in normal macrophages and the peroxidatic activity of eosinophil peroxidase [163]. Eosinophil peroxidase-coated organisms can also be killed extracellularly when exposed to normal macrophages at high parasite-to-cell ratios or when a high phagocytic load of another particle is given simultaneously. This effect can be inhibited by both azide and catalase but not by superoxide dismutase. This suggests that enough H_2O_2 is released by phagocytosis of a large number of organisms to generate toxic concentrations of H_2O_2 outside the confines of the vacuolar system [163].

Activated mouse macrophages kill or inhibit intracellular replication of *Toxoplasma gondii* that are not antibody-coated, whereas normal human and mouse macrophages do not [293]. Each of these types of macrophages can kill antibody-coated *Toxoplasma* [293]. Phagocytosis of antibody-coated *Toxoplasma* stimulates the respiratory burst by each of these types of macrophage, whereas phagocytosis of organisms that are not antibody-coated stimulates the respiratory burst only by activated mouse macrophages. Phagocytosis of *Toxoplasma* does not inhibit the production of reactive oxygen metabolites by normal macrophages; rather, it fails to stimulate their production [293]. However, these cells acquire significant toxoplasmicidal activity when *Toxoplasma* organisms are coated with eosinophil peroxidase before ingestion [294], an effect that is decreased by the hemoprotein inhibitors aminotriazole and azide [294].

When mouse peritoneal macrophages are exposed to superoxide dismutase, catalase, benzoate, histidine, and diazabicyclooctane, both the inhibition of intracellular multiplication and the killing of antibody-coated *Toxoplasma gondii* are reversed [295]. Depriving cells of glucose, which markedly reduces H_2O_2 release, also results in a similar reversal of macrophage antitoxoplasma activity [295]. These results suggest a role for the products of the interaction of $O_2{}^-$ and H_2O_2 such as the hydroxyl radical and perhaps singlet oxygen. Aminotriazole, an inhibitor of catalase, renders macrophages from normal mice more toxoplasmastatic; immunologically activated macrophages from mice genetically deficient in catalase are more toxoplasmastatic than those from wild-type mice [92]. Moreover, lymphokines alone render macrophages from untreated, acatalasemic mice toxoplasmicidal in contrast to macrophages from untreated, wild-type mice, which require a combination of lymphokines and other agents such as heart infusion broth [92]. Interferon-γ has been identified as the lymphokine that enhances both the production of H_2O_2 by human macrophages and their ability to kill *T. gondii* [301a]. In addition, lymphokines enhance the oxygen-independent toxoplasmicidal activity in human macrophages obtained from patients with chronic granulomatous disease [301b]. This indicates that lymphokines can regulate both oxygen-dependent and oxygen-independent antiparasitic responses.

The antiprotozoal effects of macrophage-derived $O_2{}^-$ and H_2O_2 are further illustrated by the marked impairment of the capacity of normal, lymphokine-stimulated, and *in vivo*-activated mouse peritoneal macrophages to kill *Toxoplasma gondii, Leishmania donovani,* and *Trypanosoma cruzi* by phorbol myristate acetate pretreatment [302]. This agent destroys the capacity of the macrophage to generate $O_2{}^-$ and H_2O_2 [302].

Phagocytosis of free parasites or parasitized erythrocytes by macrophages has been recognized as playing a crucial role in the defense of the host during malaria infection, particularly when associated with acquired immunity [302a]. Accordingly, mouse peritoneal macrophages activated *in vitro* by lymphokines show higher *Plasmodium berghei*-induced chemiluminiscence response than does non-activated macrophages, indicating the possible involvement of oxygen reduction derivatives in their cytotoxic capacity [302b]. Phagocytosis-associated oxidative mechanisms also mediate the destruction of *P. yoelii* by lymphokines-activated mouse peritoneal macrophages [157b].

B. Extracellular Killing of Helminths

The striking association between peripheral blood and tissue hypereosinophilia and diverse invasive parasitic infestations in animals and humans was documented in the late nineteenth century [303]. In certain helminth infections, blood eosinophil counts reach levels 10 to 100 times higher than normal, and massive eosinophil accumulation occurs in the tissues around the invading parasites.

Few protozoan infections have been shown to trigger an eosinophilic response. Eosinophils may be seen in the small intestine of patients infected with *Balantidium coli* [303], in the hepatic abscess of amebiasis [304], in the intestinal inflammatory reaction following *Isospora* infections [304], in the cardiac inflammatory reaction in patients with Chagas' disease [303], and in cutaneous lesions of mice infected with *Leishmania mexicana* [303].

In contrast to protozoa, helminth parasites represent large, nonphagocytosable targets and exhibit surface features of extreme diversity. The demonstration that eosinophils produce large quantities of O_2^- for a longer period relative to neutrophils may provide some insight into the mechanisms by which they are equipped to damage nonphagocytosable parasites such as large helminths [305].

The mechanism whereby eosinophils adhere to and kill helminths has been documented in experimental systems using schistosomula of *Schistosoma mansoni* as a target and has been reviewed elsewhere [303].

Eosinophils generate H_2O_2 when incubated with opsonized schistosomula (with antibody and/or complement). Hydrogen peroxide is produced only by the cells attached to the parasites, and catalase inhibits both H_2O_2 production and schistosomula killing, as detected by methylene blue dye exclusion [306]. Eosinophils from patients of chronic granulomatous disease show a significant impairment in parasite cytotoxicity [306]. Peroxidase is required for H_2O_2-mediated killing [306]. The addition of the peroxidase inhibitors azide, cyanide, and aminotriazole to eosinophil–schistosomula mixtures significantly reduces cytotoxicity. Iodination of parasites inhibited by azide also occurs [306].

Superoxide anion production is also stimulated in eosinophils by the presence of antibody-coated schistosomula [158]. However, under completely anaerobic conditions eosinophils are still capable of efficiently killing antibody-coated schistosomula, as detected by ^{51}Cr release [158]. The addition of catalase or superoxide dismutase does not inhibit the damage to schistosomula in this system. These studies suggest that eosinophils also possess a potent nonoxidative killing mechanism against schistosomula [158].

Neutrophils also generate H_2O_2 when incubated with opsonized schistosomula of *Schistosoma mansoni* [306]. Neutrophil-mediated parasite killing correlates with the rate of H_2O_2 generation, and both processes are inhibited by catalase [305]. Accordingly, neutrophils from patients with chronic granulomatous disease show a significant reduction of neutrophil-mediated parasite mortality [306]. The addition of the peroxidase inhibitor azide, cyanide, or aminotriazole to neutrophil–schistosomula mixtures significantly reduces parasite cytotoxicity [306]. Contact with neutrophils also results in iodination of schistosomula [306].

Contact of human basophils with opsonized schistosomula of *Schistosoma mansoni* increases the rate of H_2O_2 production by these cells. Basophil-mediated parasite killing correlates with the rate of H_2O_2 generation [306].

Neutrophils are effective in the killing of newborn larvae of *Trichinella spiralis*. The killing also appears to be mediated by the oxidative metabolic burst with its generation of H_2O_2 because (*a*) killing is inhibited by catalase but is unchanged by superoxide dismutase and is enhanced by cyanide and azide, and (*b*) neutrophils from patients with chronic granulomatous disease demonstrate a markedly suppressed larvicidal effect [307].

VIII. CONCLUSIONS

There are a number of biological oxidations in parasites that generate O_2^{-} and/or H_2O_2. In addition, intracellular reduction followed by autoxidation yielding O_2^{-} and H_2O_2 has been suggested as the mode of action of several antiparasitic agents. Thus, naphthoquinones, phenazine methosulfate, and nitro heterocyclic drugs have been shown to act in intact parasites or extracts as electron carriers between NADH or NADPH and oxygen, with concomitant production of either O_2^{-} or H_2O_2. Moreover, oxygen reduction products have been implicated in the mechanism of cytotoxicity of phagocytic cells against parasites. The ability of organisms to prevent the lethal action of oxygen reduction intermediates depends on their content of superoxide dismutase, catalase, and peroxidases, and many parasites have been shown to be deficient in these enzymes. This deficiency has been correlated with their sensitivity to both intracellular generators of oxygen reduction intermediates and cell-derived oxygen metabolites. Furthermore, free-radical intermediates not related to oxygen reduction products have also been found in the metabolic pathways of a wide variety of antiparasitic drugs. The possibility that the mechanism of antiparasitic action of a number of other compounds might be mediated by free radicals has yet to be explored.

ACKNOWLEDGMENTS

The authors wish to thank Ronald P. Mason for his encouragement and stimulating discussions and Mary J. Mason for her help with the manuscript. We are indebted to S. R. Meshnick, N. Le Trang, A. Cerami, M. Müller, N. Nogueira, and C. F. Nathan for sending us several preprints.

Work from the authors received financial support from the UNDP/World Bank/Work Health Organization Special Program for Research and Training in Tropical Diseases. R. Docampo is Career Investigator, Consejo Nacional de Investigaciones Cientificas y Tecnicas, Republica Argentina, and Silvia N. J. Moreno is Visiting Fellow, U.S. National Institutes of Health.

REFERENCES

1. L. R. Cleveland, *Biol. Bull. (Woods Hole, Mass.)* **48,** 309 (1925).
2. H. Ritter, *Biol. Bull. (Woods Hole, Mass.)* **121,** 330 (1961).
3. O. Raab, *Z. Biol. (Munich)* **39,** 524 (1900).
4. C. S. Foote, in "Free Radicals in Biology" (W. A. Pryor, ed.), Vol. 2, p. 85. Academic Press, New York, 1976.
5. D. Keilin, "The History of Cell Respiration and Cytochrome." Cambridge Univ. Press, London and New York, 1966.
6. B. Chance, H. Sies, and A. Boveris, *Physiol Rev.* **59,** 527 (1979).
7. T. Von Brand, "Biochemistry of Parasites," 2nd ed. Academic Press, New York, 1973.
8. W. E. Gutteridge and G. H. Coombs, "Biochemistry of Parasitic Protozoa." The Macmillan, New York, 1977.
9. B. W. Langer and P. Phisphumvidhi, *J. Parasitol.* **57,** 677 (1971).
10. J. W. Daugherty, *Exp. Parasitol.* **4,** 455 (1955).
11. R. Cavier and J. Savel, *C. R. Hebd. Seances Acad. Sci.* **238,** 2448 (1954).
12. N. N. Sharma and G. H. Bourne, *Acta Histochem.* **17,** 293 (1964).
13. R. H. Nimmo-Smith and C. G. Raison, *Comp. Biochem. Physiol.* **24,** 403 (1968).
14. R. A. Pascal, Jr., N. Le Trang, A. Cerami, and C. Walsh, *Biochemistry* **22,** 171 (1983).
15. A. M. Gero and G. H. Coombs, *Exp. Parasitol.* **54,** 185 (1982).
16. W. E. Gutteridge, D. Dave, and W. H. G. Richards, *Biochim. Biophys. Acta* **582,** 390 (1979).
17. M. Agosin, C. Náquira, J. Capdevila, and J. Paulin, *Int. J. Biochem.* **7,** 585 (1976).
18. M. Agosin, C. Náquira, J. Paulin, and J. Capdevila, *Science* **194,** 195 (1976).
19. J. F. de Boiso and A. O. M. Stoppani, *Proc. Soc. Exp. Biol. Med.* **136,** 215 (1971).
20. S. Matuda, *J. Biochem. (Tokyo)* **85,** 343 (1979).
21. S. Matuda, *J. Biochem. (Tokyo)* **85,** 351 (1979).
22. H. Laser, *Biochem. J.* **38,** 333 (1944).
23. E. Bueding, N. Entner, and F. Farber, *Biochim. Biophys. Acta* **18,** 305 (1955).
24. E. Bueding, in "Control Mechanisms in Respiration and Fermentation" (B. Wright, ed.), p. 167. Ronald Press, New York, 1963.
25. K. S. Cheah and B. Chance, *Biochim. Biophys. Acta* **223,** 55 (1970).
26. E. Kmetec and E. Bueding, *J. Biol. Chem.* **236,** 584 (1961).
27. K. S. Cheah, in "Biochemistry of Parasites and Host-Parasite Relationships" (H. Van den Bossche, ed.), p. 133. Elsevier/North-Holland Bromedical Press, Amsterdam, (1976).
28. L. W. De Zoeten and J. Tipker, *Hoppe-Seyler's Z. Physiol. Chem.* **350,** 691 (1969).
29. R. K. Prichard and P. J. Schoefield, *Exp. Parasitol.* **29,** 215 (1971).
30. K. S. Cheah and C. Bryant, *Comp. Biochem. Physiol.* **19,** 197 (1966).
31. K. S. Cheah, *Biochim. Biophys. Acta* **153,** 718 (1968).

32. C. F. Fioravanti, *Comp. Biochem. Physiol. B.* **72B**, 591 (1982).
33. S. K. Ray and G. A. M. Cross, *Nature (London), New Biol.* **237**, 174 (1972).
34. G. C. Hill and G. A. M. Cross, *Biochim. Biophys. Acta* **305**, 590 (1973).
35. W. E. Gutteridge and G. W. Rogerson, *in* "Biology of the Kinetoplastida" (W. H. R. Lumsden and A. Evans, eds.), Vol. 2, p. 619. Academic Press, New York, 1979.
36. G. C. Hill and D. C. White, *J. Bacteriol.* **95**, 2151 (1968).
37. C. Edwards and B. Chance, *J. Gen. Microbiol.* **128**, 1409 (1982).
38. P. Kronick and G. C. Hill, *Biochim. Biophys. Acta* **368**, 173 (1974).
39. J. P. Kusel, A. Boveris, and B. T. Storey, *Arch. Biochem. Biophys.* **158**, 799 (1973).
40. M. Ninomiya and Z. Suzuoki, *J. Biochem. (Tokyo)* **39**, 321 (1952).
41. A. Boveris and A. O. M. Stoppani, Experientia **33**, 1306 (1977).
42. F. S. Cruz, R. Docampo, and W. De Souza, *Acta Trop.* **35**, 35 (1978).
43. R. Docampo, F. S. Cruz, A. Boveris, R. P. A. Muniz, and D. M. S. Esquivel, *Arch. Biochem. Biophys.* **186**, 292 (1978).
44. A. Boveris, R. Docampo, J. F. Turrens, and A. O. M. Stoppani, *Biochem. J.* **175**, 431 (1978).
45. A. Boveris, A. O. M. Stoppani, R. Docampo, and F. S. Cruz, *Comp. Biochem. Physiol. C.* **61C**, 327 (1978).
46. R. Docampo, W. De Souza, F. S. Cruz, I. Roitman, B. Cover, and W. E. Gutteridge, *Z. Parasitenkd.* **57**, 189 (1978).
47. R. Docampo, F. S. Cruz, R. P. A. Muniz, D. M. S. Esquivel, and M. E. L. Vasconcellos, *Acta Trop.* **35**, 221 (1978).
48. R. Docampo and A. O. M. Stoppani, *Arch. Biochem. Biophys.* **197**, 317 (1979).
49. A. M. Conçalves, M. E. L. Vasconcellos, R. Docampo, F. S. Cruz, W. De Souza, and W. Leon, *Mol. Biochem. Parasitol.* **1**, 167 (1980).
50. R. Docampo and A. O. M. Stoppani, *Medicina (Buenos Aires)* **40** (Suppl. 1), 10 (1980).
51. R. Docampo, S. N. J. Moreno, A. O. M. Stoppani, W. Leon, F. S. Cruz, F. Villalta, and R. P. A. Muniz, *Biochem. Pharmacol.* **30**, 1947 (1981).
52. R. Docampo, S. N. J. Moreno, J. F. Turrens, A. M. Katzin, S. M. Gonzalez-Cappa, and A. O. M. Stoppani, *Mol. Biochem. Parasitol.* **3**, 169 (1981).
53. S. N. J. Moreno, R. Docampo, R. P. Mason, W. Leon, and A. O. M. Stoppani, *Arch. Biochem. Biophys.* **218**, 585 (1982).
54. S. R. Meshnick, K. P. Chang, and A. Cerami, *Biochem. Pharmacol.* **26**, 1923 (1977).
55. N. Oshino, B. Chance, H. Sies, and T. Bucher, *Arch. Biochem. Biophys.* **154**, 106 (1973).
56. S. R. Meshnick, S. H. Blobstein, R. W. Grady, and A. Cerami, *J. Exp. Med.* **148**, 569 (1978).
57. N. L. Etkin and J. W. Eaton, *in* "Erythrocyte Structure and Function" (G. J. Brewer, ed.), p. 219. Alan R. Liss, Inc., New York, 1975.
58. J. W. Eaton, J. R. Eckman, E. Berger, and H. S. Jacob, *Nature (London)* **264**, 758 (1976).
59. J. R. Eckman and J. W. Eaton, *Nature (London)* **278**, 754 (1979).
60. M. J. Friedman, *Nature (London)* **280**, 245 (1979).
61. V. D. Pomoinetsky, V. V. Pokrovsky, L. S. Kladukhina, and V. A. Yurkiv, *Byull. Eksp. Biol. Med.* **92**, 307 (1981).
62. A. Picard-Maureau, E. Hempelmann, G. Krammer, G. Jackisch, and A. Jung, *Tropenmed. Parasitenkd.* **26**, 405 (1975).
63. P. G. Shakespeare and P. I. Trigg, *Nature (London)* **241**, 538 (1973).
64. M. G. Barnes, H. Polet, T. R. Denison, and C. F. Barr, *J. Lab. Clin. Med.* **74**, 1 (1969).
65. M. J. Murray, A. B. Murray, M. B. Murray, and C. J. Murray, *Lancet i*, 1283 (1976).
66. F. B. Livingstone, *Annu. Rev. Genet.* **5**, 33 (1971).
67. M. Krvatrachue, K. Kiongkumnuanhara, and C. Harinasuta, *Lancet* **1**, 404 (1966).
68. L. Luzzato, E. A. Usanga, and S. Reddy, *Science* **164**, 839 (1969).
69. O. P. Sharma, R. P. Shukla, C. Singh, and A. B. Sen, *Indian J. Med. Res.* **69**, 944 (1979).

70. C. R. Nair, P. H. Gupta, D. P. Chauhan, A. Bhatia, and V. K. Vinayak, *Indian J. Med. Res.* **74**, 829 (1981).
70a. I. A. Clark and H. Hunt, *Infect. Immun.* **39**, 1 (1983).
71. E. F. Roth Jr., C. Raventos-Suarez, M. Perkins, and R. L. Nagel, *Biochem. Biophys. Res. Commun.* **109**, 335 (1982).
72. H. P. Misra and I. Fridovich, *J. Biol. Chem.* **246**, 6886 (1971).
73. D. G. Lindmark and M. Müller, *Antimicrob. Agents Chemother.* **10**, 476 (1976).
73a. R. Marczak, T. E. Gorrell, and M. Müller, *J. Biol. Chem.* **258**, 12427 (1983).
74. D. B. Fisher and S. Kaufman, *J. Biol. Chem.* **248**, 4300 (1973).
75. H. P. Misra, *J. Biol. Chem.* **249**, 2151 (1974).
76. C. De Duve and P. Baudhin, *Physiol. Rev.* **46**, 323 (1966).
77. A. B. Novikoff, P. M. Novikoff, N. Quintana, and C. Davis, *J. Histochem. Cytochem.* **21**, 1010 (1973).
78. A. B. Novikoff and S. Goldfischer, *J. Histochem. Cytochem.* **17**, 675 (1969).
79. K. I. Hirai, *J. Histochem. Cytochem.* **19**, 434 (1971).
80. S. Goldfischer, *J. Cell Biol.* **34**, 398 (1967).
81. R. Docampo, J. R. de Boiso, A. Boveris, and A. O. M. Stoppani, *Experientia* **32**, 972 (1976).
82. F. R. Opperdoes and P. Borst, *FEBS Lett.* **80**, 360 (1977).
83. M. Müller, in "The Eukariotic Microbial Cell" (G. W. Gooday, D. Lloyd, and A. P. J. Trinci, eds.), p. 127. Cambridge Univ. Press, London and New York, 1980.
84. J. J. B. Cannata, E. Valle, R. Docampo, and J. J. Cazzulo, *Mol. Biochem. Parasitol.* **6**, 151 (1982).
85. F. R. Opperdoes, P. Borst, and H. Spits, *Eur. J. Biochem.* **76**, 21 (1977).
86. F. R. Opperdoes, P. Borst, S. Bakker, and W. Leene, *Eur. J. Biochem.* **76**, 29 (1977).
87. M. Müller, *Annu. Rev. Microbiol.* **29**, 467 (1975).
88. J. Kulda, J. Králová, and J. Vávra, *J. Protozool.* **24**, 51 (1977).
88a. J. Kulda, J. Králová, and J. Vávra, *J. Protozool.* **21**, 456 (1974).
89. M. Müller, *J. Cell Biol.* **57**, 453 (1978).
90. D. J. Doran, *J. Protozool.* **4**, 182 (1957).
91. D. J. Doran, *J. Protozool.* **5**, 89 (1958).
92. H. W. Murray, C. F. Nathan, and Z. A. Cohn, *J. Exp. Med.* **152**, 1610 (1980).
93. H. W. Murray, S. B. Aley, and W. A. Scott, *Mol. Biochem. Parasitol.* **3**, 381 (1981).
94. S. C. Harvey, *J. Biol. Chem.* **179**, 435 (1949).
95. J. D. Fulton and D. F. Spooner, *Biochem. J.* **63**, 475 (1956).
96. M. Tasaka, *Fukuoka Acta Med.* **28**, 127 (1935).
97. D. M. Wertlieb and H. N. Guttman, *J. Protozool.* **10**, 109 (1963).
98. L. A. Cohen, Some ultrastructural, biochemical and physiological changes induced by the carcinogen 4-nitroquinoline-N-oxide in the flagellate, *Crithidia fasciculata*. Ph.D. Thesis, City University, New York, 1972.
99. K. E. Muse and J. F. Roberts, *Protoplasma* **78**, 343 (1973).
100. Y. Eeckhout, in "Comparative Biochemistry of Parasites" (H. Van den Bossche, ed.), p. 297. Academic Press, New York, 1972.
101. H. W. Murray, *J. Immunol.* **129**, 351 (1982).
102. E. Pennoit-De Cooman and G. Van Grembergen, *Verh. K. Vlaam. Acad. Wet., Lett. Schone Kunsten Belg., Kl. Wet.* **4**(6) 7 (1942).
103. J. M. Paul and J. Barrett, *Int. J. Parasitol.* **10**, 121 (1980).
104. V. C. Glocklin and D. Fairbairn, *J. Cell. Comp. Physiol.* **39**, 341 (1952).
105. M. Monteoliva, *Rev. Iber. Parasitol.* **21**, 339 (1961).
106. W. J. Vaatstra, *Hoppe-Seyler's Z. Physiol. Chem.* **350**, 701 (1969).
107. J. W. Kazura and S. R. Meshnick, *Mol. Biochem. Parasitol.* **10**, 1 (1984).
107a. M. L. Rhoads, *Exp. Parasitol.* **56**, 41 (1983).

108. J. Barrett, *J. Parasitol.* **66,** 697 (1980).
108a. A. M. Gerasimov, N. V. Kasatkina, and E. N. Darmova, *J. Evol. Biochem. Physiol.* **15,** 126 (1979).
109. G. C. Mills, *J. Biol. Chem.* **229,** 189 (1957).
110. A. Boveris, H. Sies, E. E. Martino, R. Docampo, J. F. Turrens, and A. O. M. Stoppani, *Biochem. J.* **188,** 643 (1980).
111. R. F. Burk, Jr., K. Nishiki, R. A. Lawrence, and B. Chance, *J. Biol. Chem.* **253,** 43 (1978).
112. A. Yawetz and M. Agosin, *Comp. Biochem. Physiol. B.* **68B,** 237 (1981).
113. S. R. Meshnick and J. W. Eaton, *Biochem. Biophys. Res. Commun.* **102,** 970 (1981).
114. V. D. Kallinikova, *Acta Protozool.* **5,** 395 (1968).
115. R. Michel, *Z. Tropenmed. Parasitol.* **15,** 400 (1964).
116. R. F. Steiger, *Acta Trop.* **30,** 64 (1973).
117. T. Hayashi and D. Terada, *Jpn. J. Parasitol.* **22,** 1 (1973).
118. L. T. Threadgold and C. P. Read, *Exp. Parasitol.* **23,** 221 (1968).
119. L. T. Threadgold, C. Arme, and C. P. Read, *J. Parasitol.* **54,** 802 (1968).
120. K. S. Cheah, *Comp. Biochem. Physiol.* **21,** 351 (1967).
121. M. Monteoliva, *Rev. Iber. Parastol.* **24,** 43 (1964).
122. B. J. Bogitsh, *J. Parasitol.* **61,** 621 (1975).
123. I. Fridovich, *in* "Free Radicals in Biology" (W. A. Pryor, ed.), Vol. 1, p. 239. Academic Press, New York, 1976.
124. J. M. McCord and I. Fridovich, *J. Biol. Chem.* **244,** 6049 (1969).
125. I. Fridovich, *Horiz. Biochem. Biophys.* **1,** 1 (1974).
126. F. J. Yost, Jr. and I. Fridovich, *J. Biol. Chem.* **248,** 4905 (1973).
127. K. Asada, S. Kanematsu, and K. Uchida, *Arch. Biochem. Biophys.* **179,** 243 (1977).
127a. U. Suthipark, J. Krunkai, A. Jearpipatkul, Y. Yuthavong, and B. Panijpan, *J. Parasitol.* **68,** 337 (1983).
127b. A. S. Fairfield, S. R. Meshnick, and J. W. Eaton, *Science* **221,** 764 (1983).
128. D. G. Lindmark and M. Müller, *J. Biol. Chem.* **249,** 4634 (1974).
129. N. Le Trang, S. R. Meshnick, K. Kitchener, J. W. Eaton, and A. Cerami, *J. Biol. Chem.* **258,** 125 (1983).
130. S. R. Meshnick, N. Le Trang, K. Kitchener, A. Cerami, and J. W. Eaton, *in* "Oxy Radicals and their Scavenger Systems: Molecular Aspects" (R. A. Greenwald and G. Cohen, eds.) Vol. 1, p. 348. Elsevier/North-Holland Biomedical Press, Amsterdam, 1983.
131. T. von Brand and W. Weise, *Z. Vergl. Physiol.* **18,** 339 (1932).
132. M. Chaigneau and G. Charlet-Lery, *C. R. Hebd. Seances Acad. Sci.* **245,** 2536 (1957).
133. R. E. Hungate, *Biochem. Physiol. Protozoa* **2,** 159 (1955).
134. B. Sugden, *J. Gen. Microbiol.* **9,** 44 (1953).
135. W. Trager, *Biol. Bull. (Woods Hole, Mass.)* **66,** 182 (1934).
136. R. E. Hungate, *Ecology* **20,** 230 (1939).
137. C. Dobell and P. P. Laidlaw, *Parasitology* **18,** 283 (1926).
138. G. J. Jackson and N. R. Stoll, *Am. J. Trop. Med. Hyg.* **13,** 520 (1964).
139. L. R. Cleveland, *Am. J. Hyg.* **8,** 256 (1928).
140. R. Cavier, P. Georges, and J. Savel, *Exp. Parasitol.* **15,** 556 (1964).
141. M. H. Soule, *J. Infect. Dis.* **36,** 245 (1925).
142. S. Adler and O. Theodor, *Proc. R. Soc. London, Ser. B* **108,** 453 (1931).
143. I. C. Ray, *Indian J. Med. Res.* **20,** 355 (1932).
144. N. Islam and R. I. Chowdhury, *J. Trop. Med. Hyg.* **66,** 285 (1963).
145. S. F. Gottlieb, *Annu. Rev. Microbiol.* **25,** 111, (1971).
146. J. L. Avila, A. Bretaña, P. J. Jacques, M. A. Casanova, and J. Convit, *Curr. Chemother., Proc. Int. Congr. Chemother., 10th, 1977* p. 153 (1978).

147. Y. Tanaka, H. Tanowitz, and B. B. Bloom, *Infect. Immun.* **41,** 1322 (1983).
148. S. D. Aust and A. B. Svingen, in "Free Radicals in Biology" (W. A. Pryor, ed.), Vol. 5, p. 1. Academic Press, New York, 1982.
149. L. Reiner and C. S. Leonard, *Proc. Soc. Exp. Biol. Med.* **29,** 951 (1932).
150. W. I. Strangeways, *Ann. Trop. Med. Parasitol.* **31,** 405 (1937).
151. C. F. Nathan, N. Nogueira, C. Juangbhanich, J. Ellis, and Z. A. Cohn, *J. Exp. Med.* **149,** 1056 (1979).
151a. F. Villalta and F. Kierszenbaum, *J. Immunol.* **131,** 1504 (1983).
152. R. D. Pearson and R. T. Steigbigel, *J. Immunol.* **127,** 1438 (1981).
153. N. E. Reiner and J. W. Kazura, *Infect. Immun.* **36,** 1023 (1982).
153a. R. D. Pearson, J. L. Harcus, D. Roberts, and G. R. Donowitz, *J. Immunol.* **131,** 1994 (1983).
154. K. P. Chang, *Am. J. Trop. Med. Hyg.* **30,** 322 (1981).
155. Y. Buchmüller and J. Mauel, *J. Reticuloendothel. Soc.* **29,** 181 (1981).
156. K. J. I. Thorne, R. J. Svvennsen, and D. Franks, *Infect. Immun.* **21,** 798 (1978).
157. K. J. I. Thorne, A. M. Glauert, R. J. Svvennsen, H. Thomas, J. Morris, and D. Franks, *Parasitology* **83,** 115 (1981).
157a. H. M. Dockrell and J. H. L. Playfair, *Infect. Immun.* **39,** 456 (1983).
157b. C. F. Ockenhouse and H. L. Shear, *J. Immunol.* **132,** 424 (1984).
158. D. A. Bass and P. Szejda, *J. Clin. Invest.* **64,** 1558 (1979).
159. E. C. Jong, A. A. F. Mahmoud, and S. J. Klebanoff, *J. Immunol.* **126,** 468 (1981).
159a. I. Clark, W. B. Cowden, and G. A. Butcher, *Lancet* **i,** 234 (1983).
159b. A. C. Allison and E. M. Eugui, *Lancet* **ii,** 1431 (1983).
160. S. J. Klebanoff, *Adv. Host Def. Mech.* **1,** 111 (1982).
161. R. L. Baehner, L. A. Boxer, and L. M. Ingraham, in "Free Radicals in Biology" (W. A. Pryor, ed.), Vol. 5, p. 91. Academic Press, New York, 1982.
162. J. L. Avila, J. Convit, M. E. Pinardi, and P. J. Jacques, *Biochem. Soc. Trans.* **4,** 680 (1976).
163. N. Nogueira, S. J. Klebanoff, and Z. A. Cohn, *J. Immunol.* **128,** 1705 (1982).
164. F. S. Cruz, R. L. Berens, and J. J. Marr, *J. Parasitol.* **69,** 237 (1982).
165. O. G. Lima, I. C. D'Alburquerque, C. F. Lima, and M. H. D. Maia, *Rev. Inst. Antibiot.* **4,** 3 (1962).
166. B. Kalyanaraman, E. Perez-Reyes, and R. P. Mason, *Biochim. Biophys. Acta* **630,** 119 (1980).
167. T. Iyanagi and I. Yamazaki, *Biochim. Biophys. Acta* **216,** 282 (1970).
168. G. Powis and P. L. Appel, *Biochem. Pharmacol.* **29,** 2567 (1980).
169. V. Favaudon, *Biochimie* **64,** 457 (1982).
170. K. L. Fong, P. B. McCay, J. L. Poyer, B. B. Keele, and H. Misra, *J. Biol. Chem.* **248,** 7792 (1973).
171. R. Zimmerman, L. Flohé, U. Weser, and H. J. Hartmann, *FEBS Lett.* **29,** 117 (1973).
172. J. M. McCord and E. D. Day, Jr., *FEBS Lett.* **86,** 139 (1978).
173. E. W. Kellog III and I. Fridovich, *J. Biol. Chem.* **250,** 8812 (1975).
174. R. Docampo, J. N. Lopes, F. S. Cruz, and W. De Souza, *Exp. Parasitol.* **42,** 142 (1977).
175. S. A. Lesko, R. J. Lorentzen, and P. O. P. Ts'o, *Biochemistry* **19,** 3023 (1980).
176. K. Brawn and I. Fridovich, *Arch. Biochem. Biophys.* **206,** 414 (1981).
177. H. Sies and E. Cadenas, in "Biological Basis of Detoxication" (J. Caldwell and W. B. Jakoby, eds.), p. 181. Academic Press, New York, 1983.
178. J. S. Bus and J. E. Gibson, *Rev. Biochem. Toxicol.* **1,** 125 (1979).
179. J. M. McCord, *Rev. Biochem. Toxicol.* **1,** 109 (1979).
180. H. Thor, M. T. Smith, P. Hartzell, G. Bellomo, S. A. Jewell, and S. Orrenius, *J. Biol. Chem.* **257,** 12419 (1982).
181. G. Bellomo, S. A. Jewell, H. Thor, and S. Orrenius, *Proc. Natl. Acad. Sci. U.S.A.* **79,** 6842 (1982).
182. A. R. Schuerch and W. Wehrli, *Eur. J. Biochem.* **84,** 197 (1978).

183. J. N. Lopes, F. S. Cruz, R. Docampo, M. E. L. Vasconcellos, M. C. R. Sampaio, A. V. Pinto, and B. Gilbert, *Ann. Trop. Med. Parasitol.* **72**, 523 (1978).
184. R. Lopetegui and C. Sosa Miatello, *Rev. Soc. Argent. Biol.* **37**, 134 (1961).
185. B. Goldberg and A. Stern, *Biochim. Biophys. Acta* **437**, 628 (1976).
186. R. Docampo, F. S. Cruz, A. Boveris, R. P. A. Muniz, and D. M. S. Esquivel, *Biochem. Pharmacol.* **28**, 723 (1979).
187. F. S. Cruz, R. Docampo, and A. Boveris, *Antimicrob. Agents Chemother.* **14**, 630 (1978).
188. M. Rabinovitch, J. P. Dedet, A. Ryter, R. Robineaux, G. Topper, and E. Brunet, *J. Exp. Med.* **155**, 415 (1982).
189. N. McHardy, A. J. B. Haigh, and T. T. Dolan, *Nature (London)* **261**, 698 (1976).
190. P. Boehm, K. Cooper, A. T. Hudson, J. P. Elphick, and N. McHardy, *J. Med. Chem.* **24**, 295 (1981).
191. J. Williamson, T. J. Scott-Finnigan, M. A. Hardman, and J. R. Brown, *Nature (London)* **292**, 466 (1981).
192. J. Williamson and T. J. Scott-Finnigan, *Antimicrob. Agents Chemother.* **13**, 735 (1978).
193. J. F. Fernandes, M. Halsman, and O. Castellani, *Exp. Parasitol.* **18**, 203 (1966).
194. K. Handa and S. Sato, *Gann* **66**, 43 (1975).
195. N. R. Bachur, S. L. Gordon, and M. V. Gee, *Cancer Res.* **38**, 1745 (1978).
196. N. R. Bachur, M. V. Gee, and S. L. Gordon, *Proc. Am. Assoc. Cancer Res.* **19**, 75 (1978).
197. P. Ulrich and A. Cerami, *J. Med. Chem.* **25**, 654 (1982).
198. J. B. Le Pecq, N. Dat-Xuong, C. Gosse, and C. Paoletti, *Proc. Natl. Acad. Sci. U.S.A.* **71**, 5078 (1974).
199. J. Bénard and G. Riou, in "Biochemistry of Parasites and Host-Parasite Relationships" (H. Van den Bossche, ed.), p. 477. Elsevier/North-Holland Biomedical Press, Amsterdam, 1976.
200. J. Bénard and G. Riou, *Biochem. Biophys. Res. Commun.* **77**, 1189 (1977).
201. N. Poliakoff, S. M. Albonico, M. Alvarez, J. Gallo Pecca, and M. J. Vernengo, *J. Med. Chem.* **16**, 1411 (1973).
202. L. Leon, M. E. L. Vasconcellos, W. Leon, F. S. Cruz, R. Docampo, and W. De Souza, *Exp. Parasitol.* **45**, 151 (1978).
203. V. Rheinhold, L. Bittman, R. Bruni, K. Thrun, and D. Silveria, *Proc. Am. Assoc. Cancer Res.* **16**, 135 (1975).
204. C. Auclair and C. Paoletti, *J. Med. Chem.* **24**, 289 (1981).
204a. E. E. Montamat, C. Burgos, N. M. Gerez de Burgos, L. Rovai, A. Blanco, and E. L. Segura, *Science* **218**, 288 (1982).
204b. J. E. Heidrich, L. A. Hunsaker, and D. L. Van der Jagt, *IRCS Med. Sci. Biochem.* **11**, 304 (1983).
204c. B. N. Ames, *Science* **221**, 1256 (1983).
204d. M. Coburn, P. Sinsheimer, S. Segal, M. Burgos, and W. Troli, *Biol. Bull. (Woods Hole, Mass.)* **159**, 468 (1980).
205. E. B. Kearney and T. B. Singer, *J. Biol. Chem.* **219**, 963 (1956).
206. J. W. Newton and M. D. Kamen, *Biochim. Biophys. Acta* **25**, 462 (1957).
207. W. S. Zaugg, *J. Biol. Chem.* **239**, 3964 (1964).
208. J. R. White and H. H. Dearman, *Proc. Natl. Acad. Sci. U.S.A.* **54**, 887 (1965).
209. K. Ishizu, H. H. Dearman, M. T. Huang, and J. R. White, *Biochemistry* **8**, 1238 (1969).
210. Y. Nosaka, K. Akasaka, and H. Hatano, *J. Am. Chem. Soc.* **100**, 706 (1978).
211. H. L. White and J. R. White, *Mol. Pharmacol.* **4**, 549 (1968).
212. N. Nishikimi, N. A. Rao, and K. Yagi, *Biochem. Biophys. Res. Commun.* **46**, 849 (1972).
212a. M. J. M. Alves and M. Rabinovitch, *Infect. Immun.* **39**, 435 (1983).
212b. J. Mauel, C. Bonnard, J. Schnyder, and M. Baggiolini, *Experientia* **37**, 628 (1981).
213. M. Müller, *Scand. J. Infect. Dis., Suppl.* **26**, 31 (1981).

214. H. Van den Bossche, *Nature (London)* **173**, 626 (1978).
215. G. E. Adams, *in* "Treatment of Radioresistant Cancers" (M. Abe, K. Sakamoto, and T. L. Phillips, eds.), p. 3. Elsevier/North-Holland Biomedical Press, Amsterdam, New York, 1979.
216. J. F. Fowler, G. E. Adams, and J. Denekamp, *Cancer Treat. Rev.* **3**, 227 (1977).
217. G. E. Adams, I. J. Stratford, R. G. Wallace, P. Wardman, and M. E. Watts, *JNCI, J. Natl. Cancer Inst.* **64**, 555 (1980).
218. R. P. Mason, *in* "Free Radicals in Biology" (W. A. Pryor, ed.), Vol. 5, p. 161. Academic Press, New York, 1982.
219. R. P. Mason and J. L. Holtzman, *Biochem. Biophys. Res. Commun.* **67**, 1267 (1975).
220. R. Docampo, *Rev. Soc. Argent. Biol.* **56**, 15 (1980).
221. R. Docampo, *in* "The Host-Invader Interplay" (H. Van den Bossche, ed.), p. 677. Elsevier/North-Holland Biomedical Press, Amsterdam, 1980.
222. R. Docampo, R. P. Mason, C. Mottley, and R. P. A. Muniz, *J. Biol. Chem.* **256**, 10930 (1981).
223. R. Docampo, S. N. J. Moreno, and A. O. M. Stoppani, *Arch. Biochem. Biophys.* **207**, 316 (1981).
224. S. N. J. Moreno, D. J. Palmero, K. E. de Palmero, R. Docampo, and A. O. M. Stoppani, *Medicina (Buenos Aires)* **40**, 553 (1980).
225. S. N. J. Moreno, R. Docampo, and A. O. M. Stoppani, *Rev. Soc. Argent. Biol.* **56**, 28 (1980).
226. R. Docampo, M. Dubin, E. E. Martino, S. N. J. Moreno, and A. O. M. Stoppani, *Medicina (Buenos Aires)* **43**, 33 (1983).
226a. M. Dubin, S. N. J. Moreno, E. E. Martino, R. Docampo, and A. O. M. Stoppani, *Biochem. Pharmacol.* **32**, 483 (1983).
227. J. Hugon and M. Borgers, *Anat. Rec.* **155**, 15 (1966).
228. R. Docampo and S. N. J. Moreno, *in* "Oxygen Radicals in Chemistry and Biology" (W. Bors, M. Saran, and D. Tait, eds.) de Gruyter, Berlin, 1984 (in press).
229. W. E. Gutteridge, J. Ross, M. R. J. Hargadon, and J. E. Hudson, *Trans. R. Soc. Trop. Med. Hyg.* **76**, 493 (1982).
230. E. J. Baines and J. A. McFadzean, *Adv. Pharmacol. Chemother.* **18**, 223 (1981).
231. G. H. Coombs, *J. Protozool.* **20**, 524 (1973).
232. D. I. Edwards, M. Dye, and H. Carne, *J. Gen. Microbiol* **76**, 135 (1973).
233. R. M. J. Ings, J. A. McFadzean, and W. E. Ormerod, *Biochem. Pharmacol.* **23**, 1421 (1974).
234. B. B. Bealieu, Jr., M. A. McLafferty, R. L. Koch, and P. Goldman, *Antimicrob. Agents Chemother.* **20**, 410 (1981).
235. D. I. Edwards, *J. Antimicrob. Chemother.* **5**, 499 (1979).
236. R. L. Willson, W. A. Cramp, and R. M. J. Ings, *Int. J. Radiat. Biol.* **26**, 557 (1974).
237. S. N. J. Moreno, R. P. Mason, R. P. A. Muniz, F. S. Cruz, and R. Docampo, *J. Biol. Chem.* **258**, 4051 (1983).
238. D. G. Lindmark and M. Müller, *J. Biol. Chem.* **248**, 7724 (1973).
239. E. Perez-Reyes, B. Kalyanaraman, and R. P. Mason, *Mol. Pharmacol.* **17**, 239 (1979).
240. W. Leon and S. M. Krassner, *J. Parasitol* **62**, 115 (1976).
241. G. Crabtree, W. Leon, and S. M. Krassner, *Comp. Biochem. Physiol. C* **57C**, 143 (1977).
242. W. Leon, F. S. Cruz, and M. E. L. Vasconcellos, *J. Protozool.* **22**, 227 (1975).
243. T. Sugimura, K. Okabe, and M. Nagao, *Cancer Res.* **26**, 1717 (1966).
244. M. Hozumi, *Gann* **60**, 83 (1969).
245. N. Kataoka, A. Imamura, Y. Kawazoe, G. Chihara, and C. Nagata, *Bull. Chem. Soc. Jpn.* **40**, 62 (1967).
246. W. L. Fowlks, *J. Invest. Dermatol.* **32**, 233 (1959).
247. T. W. Sery, *Cancer Res.* **39**, 96 (1979).

248. K. R. Weishaupt, C. J. Gomer, and T. J. Dougherty, *Cancer Res.* **36,** 2326 (1976).
249. S. Cannistraro and A. Van de Vorst, *Biochem. Biophys. Res. Commun.* **74,** 1177 (1977).
250. G. R. Buettner and L. W. Oberley, *FEBS Lett.* **121,** 161 (1980).
251. S. R. Meshnick, R. W. Grady, S. H. Blobstein, and A. Cerami, *J. Pharmacol. Exp. Ther.* **207,** 1041 (1978).
252. A. U. Orjih, H. S. Banyal, R. Chevli, and C. D. Fitch, *Science* **214,** 667 (1981).
253. A. C. Chou and C. D. Fitch, *J. Clin. Invest.* **68,** 672 (1981).
254. B. H. Kean and D. W. Haskins, *in* "Drugs of Choice" (W. Modell, ed.), p. 371. Mosby, St. Louis, Missouri, 1978.
255. J. J. Procknow, *Lab. Invest.* **11,** 1217 (1962).
256. W. N. Wright, F. J. Brady, and J. Bozicevick, *Proc. Helminthol. Soc. Wash.* **5,** 5 (1938).
257. V. Nussensweig, R. Sonntag, A. Biancalana, J. L. Pedreira de Freitas, V. Amato Neto, and J. Koletzel, *Hospital (Rio de Janeiro)* **44,** 731 (1953).
258. V. Amato Neto and O. Mellone, *Hospital (Rio de Janeiro)* **55,** 343 (1959).
259. J. M. Rezende, W. Zupeli, and M. G. Bufuto, *Rev. Goiana Med.* **11,** 35 (1965).
260. Z. Brener, *Adv. Pharmacol. Ther.* **7,** 71 (1979).
261. R. Docampo, S. N. J. Moreno, R. P. A. Muniz, F. S. Cruz, and R. P. Mason, *Science* **220,** 1292 (1983).
262. I. H. Leaver, *Photochem. Photobiol.* **16,** 189 (1972).
263. W. G. Harrelson and R. P. Mason, *Mol. Pharmacol.* **22,** 239 (1982).
264. D. E. Hughes, C. A. Schneider, and L. Simpson, *J. Parasitol.* **68,** 642 (1982).
265. D. C. Borg, *in* "Biological Applications of Electron Spin Resonance" (H. M. Swartz. J. R. Bolton, and D. C. Borg, eds.), p. 265. Wiley (Interscience), New York, 1972.
266. J. C. Craig, M. E. Tate, G. P. Warwick, and W. P. Roberts, *J. Med. Chem.* **2,** 659 (1960).
267. T. N. Tozer, L. D. Tuck, and J. C. Craig, *J. Med. Chem.* **12,** 294 (1969).
267a. R. D. Pearson, A. A. Manian, J. L. Harcus, D. Hall, and E. L. Hewlett, *Science* **217,** 369 (1982).
267b. T. Seebeck and P. Gehr, *Mol. Biochem. Parasitol.* **9,** 197 (1983).
268. S. T. Crooke and W. T. Bradner, *J. Med.* **7,** 333 (1977).
269. K. W. Kohn and R. A. G. Ewig, *Cancer Res.* **36,** 3839 (1976).
270. R. M. Burger, J. Peisach, and S. B. Horwitz, *Life Sci.* **28,** 715 (1981).
271. R. M. Burger, J. Peisach, E. W. Blumberg, and S. B. Horwitz, *J. Biol. Chem.* **254,** 10906 (1979).
272. L. W. Oberley and G. R. Buettner, *FEBS Lett.* **97,** 47 (1979).
273. Y. Sugiura and T. Kikuchi, *J. Antibiot.* **31,** 1310 (1978).
274. J. M. C. Gutteridge and X. C. Fu, *FEBS Lett.* **123,** 71 (1981).
275. T. Ono and T. Nakabayashi, *Biken J.* **23,** 143 (1980).
276. H. C. Nathan, C. J. Bacchi, T. T. Sakai, D. Rescigno, D. Stumpf, and S. H. Hutner, *Trans. R. Soc. Trop. Med. Hyg.* **75,** 394 (1981).
277. P. P. McCann, C. J. Bacchi, A. B. Clarkson, Jr., J. R. Seed, H. C. Nathan, B. O. Amole, S. H. Hutner, and A. Sjoerdsma, *Med. Biol.* **59,** 434 (1981).
278. C. J. Bacchi, H. C. Nathan, S. H. Hutner, P. P. McCann, and A. Sjoerdsma, *Biochem. Pharmacol.* **31,** 2833 (1982).
279. G. Cohen and P. Hochstein, *Biochemistry* **3,** 895 (1964).
280. A. Szeinberg and P. A. Marks, *J. Clin. Invest.* **40,** 914 (1961).
281. S. N. Kelman, S. G. Sullivan, and A. Stern, *Biochem. Pharmacol.* **31,** 2409 (1982).
282. M. Summerfield and G. R. Tudhope, *Br. J. Clin. Pharmacol.* **6,** 319 (1978).
282a. P. J. Thornalley, A. Stern, and J. V. Bannister, *Biochem. Pharmacol.* **32,** 3571 (1983).
283. R. J. Dern, E. Beutler, and A. S. Alving, *J. Lab. Clin. Med.* **44,** 171 (1954).

284. B. B. Gaitonde, *in* "Recent Advances in Malaria Research" (Organization of Pharmaceuticals Producers of India), p. 41. Indian Council of Medical Research, 1977.

285. C. F. Nathan and A. Nakagawara, *in* "Self Defense Mechanisms, Role of Macrophages" (D. Mizuno, Z. A. Cohn, K. Takeya, and N. Ishida, eds.), p. 279. Univ. of Tokyo Press, Tokyo, 1982.

286. N. Nogueira, *in* "Contemporary Topics in Immunobiology: Immunoparasitology" (J. Marchalonis, ed.) Vol. 12, p. 53. Plenum Press, New York, 1984.

287. T. Jones and J. Hirsch, *J. Exp. Med.* **136**, 1173 (1972).

288. F. Ebert, G. L. Enriquez, and H. Mühlpfordt, *Behring Inst. Mitt.* **60**, 65 (1976).

289. N. Nogueira and Z. A. Cohn, *J. Exp. Med.* **148**, 288 (1978).

290. R. L. Cardoni, R. Docampo, and A. M. Casellas, *J. Parasitol.* **68**, 547 (1982).

291. R. Docampo, R. L. Cardoni, A. M. Casellas, and W. De Souza, *in* "Nuclear Techniques in the Study of Parasitic Infections," p. 61. IAEA, Vienna, 1982.

292. R. Docampo, A. M. Casellas, E. D. Madeira, R. L. Cardoni, S. N. J. Moreno, and R. P. Mason, *FEBS Lett.* **155**, 25 (1983).

292a. A. Nakagawara, C. F. Nathan, and Z. A. Cohn, *J. Clin. Invest.* **68**, 1243 (1981).

292b. G. B. Mackaness, *in* "Infectious Agents and Host Reactions" (S. Mudd, ed.) P. 61. Saunders, Philadelphia, Pennsylvania, 1970.

292c. A. Nakagawara, N. M. De Santis, N. Nogueira, and C. F. Nathan, *J. Clin. Invest.* **70**, 1042 (1982).

293. C. B. Wilson, V. Tsai, and J. S. Remington, *J. Exp. Med.* **151**, 328 (1980).

294. R. M. Locksley, C. B. Wilson, and S. J. Klebanoff, *J. Clin. Invest.* **69**, 1099 (1982).

294a. H. W. Murray and D. M. Cartelli, *J. Clin. Invest.* **72**, 32 (1983).

295. H. W. Murray, C. W. Juangbhanich, C. F. Nathan, and Z. A. Cohn, *J. Exp. Med.* **150**, 950 (1979).

296. H. W. Murray and Z. A. Cohn, *J. Exp. Med.* **152**, 1596 (1980).

297. H. W. Murray, *J. Exp. Med.* **153**, 1302 (1981).

298. H. W. Murray, *J. Exp. Med.* **153**, 1690 (1981).

299. C. G. Haidaris and P. F. Bonventre, *J. Immunol.* **129**, 850 (1982).

300. R. D. Pearson, J. L. Harcus, P. H. Symes, R. Romito, and G. R. Donowitz, *J. Immunol.* **129**, 1282 (1982).

301. Y. Tanaka, C. Kiyotaki, H. Tanowitz, and B. R. Bloom, *Proc. Natl. Acad. Sci. U.S.A.* **79**, 2584 (1982).

301a. C. F. Nathan, H. W. Murray, M. E. Wiebe, and B. Y. Rubin, *J. Exp. Med.* **158**, 670 (1983).

301b. H. W. Murray, G. I. Byrne, C. D. Rothermel, and D. M. Cartelli, *J. Exp. Med.* **158**, 234 (1983).

302. H. W. Murray, *J. Reticuloendothel. Soc.* **31**, 479 (1982).

302a. P. R. Cannon and W. H. Taliaferro, *J. Prev. Med.* **5**, 37 (1931).

302b. S. Makimura, V. Brinkmann, H. Mossman, and H. Fischer, *Infect. Immun.* **37**, 800 (1982).

303. A. J. Dessein and J. R. David, *Adv. Host Def. Mech.* **1**, 243 (1982).

304. P. B. Beeson and D. A. Bass, "The Eosinophil." Saunders, Philadelphia, Pennsylvania, 1977.

305. E. J. Goetzl and K. F. Austen, *Am. J. Trop. Med. Hyg.* **26**, 142 (1977).

306. J. W. Kazura, M. M. Fanning, J. L. Blumer, and A. A. F. Mahmoud, *J. Clin. Invest.* **67**, 93 (1981).

307. D. A. Bass and P. Szejda, *J. Clin. Invest.* **64**, 1415 (1979).

CHAPTER 9

Multiple Mechanisms of Metabolic Activation of Aromatic Amine Carcinogens

Peter J. O'Brien

Department of Biochemistry, Memorial University of Newfoundland, St. John's, Newfoundland, Canada

I. INTRODUCTION

The most widely accepted theory for the initiation of processes leading to carcinogenesis is that carcinogens elicit their activity after conversion to electrophiles that interact with a nucleophilic region of a macromolecule [1]. The covalent interaction with DNA must result in a minimal change in the physical structure of the DNA so as to avoid setting into motion the highly effective *in vivo* repair mechanisms. This modified DNA must persist long enough *in vivo* to replicate during cell division. Subsequent proliferation of these cells leads to the

FREE RADICALS IN BIOLOGY, VOL. VI

development of the neoplasmic state. Some investigators suggest that RNA and protein adducts should be considered [2]. Evidence against genomic DNA as the target in chemical carcinogenesis has been presented [3].

The metabolic activation of the carcinogen is believed to occur within the endoplasmic reticulum and the nucleus of mammalian cells. A family of cytochrome P-450 isoenzymes is implicated in the initial oxygenation as part of a mixed-function oxidase activity. In the activation of polycyclic aromatic hydrocarbons, epoxide hydrase is also required whereas, in the activation of arylamides, various cytosolic or endoplasmic reticular transferases are involved in the formation of the electrophile. The activation by the mixed-function oxidase is considered to be a two-electron oxidation of the polycyclic aromatic hydrocarbon to an epoxide or of the arylamine to an N-hydroxyarylamine [1].

However, it is clear that other mechanisms of activation also occur in the target tissues of various animals. Thus, peroxidases and prostaglandin synthase readily catalyze the covalent binding of various carcinogens to DNA or the formation of mutagens. The mechanism is believed to involve a one-electron oxidation to a radical-cation that can then directly interact with DNA or be oxidized to an electrophile. Unlike the two-electron oxidation, activation by no other enzymes is required. Free radicals formed by ionizing radiation and UV light clearly result in carcinogenesis, so that it is unlikely that the complexity of chemical carcinogenesis will be understood unless free-radical pathways are considered. Furthermore, some cytochrome P-450 isoenzymes carry out a one-electron oxidation of substrates. Some of the highly reactive radicals formed have been trapped and identified and shown to form adducts with heme, protein, and DNA.

As described in Section VII, the activation by different target tissues may reflect the different activating systems present. Thus, liver hepatocytes have a very high level of cytochrome P-450, but the Kupffer cells of the liver contain a peroxidase. The skin, colon, rectum, and bone marrow also contain cytochrome P-450 and peroxidase. The bladder and stomach contain cytochrome P-450 and prostaglandin synthase. Target tissues containing all three systems include the lungs, mammary gland, brain, and bone marrow. However, the thyroid, salivary gland, Zymbal's gland, and the Harderian gland contain active peroxidases and little cytochrome P-450. Kidney medulla contains an active prostaglandin synthase but no cytochrome P-450. The uterus contains prostaglandin synthase and peroxidase but no cytochrome P-450.

In this chapter we present DNA covalent binding and mutagenicity evidence that multiple mechanisms of activation exist in the cell for various carcinogenic arylamine compounds. In this series the metabolism of 2-acetylaminofluorene (AAF) with particular emphasis on the free-radical aspects has been reviewed [4] as has the ESR evidence for free radicals formed in xenobiotic metabolism [5].

II. FREE-RADICAL FORMATION DURING CYTOCHROME *P*-450 FUNCTION

A. Reductive Activation

Carbon tetrachloride is reductively metabolized to the trichloromethyl free radical ($CCl_3\cdot$), which by abstracting a hydrogen atom is capable of initiating lipid peroxidation and binding covalently to microsomal protein and lipid. It can also react with oxygen, yielding a peroxyl free radical ($CCl_3O_2\cdot$) that can interact with tryptophan and tryosine [6]. The levels of cytochrome *P*-450 are markedly decreased as a result of heme catabolism. Oxygen is not required, and carbon monoxide is not an end product.

Reduced cytochrome *P*-450 is also probably involved in the reactive activation of a triarylmethane dye, gentian violet, to a carbon-centered free radical [7]. Reductive activation of the herbicide diquat to bipyridilium cation free radical is inhibited by carbon monoxide, suggesting the involvement of reduced cytochrome *P*-450 [8].

B. Superoxide-Catalyzed Oxidation

The mixed-function oxidase can result in superoxide formation, probably as a result of dissociation of the oxy–cytochrome *P*-450 complex [9]. The released superoxide has been shown to oxidize the hydroxylamine TEMPO (2,2,6,6-tetranethyl-4-piperidone-*N*-oxyl) to a nitroxyl radical. The mixed-function oxidase-catalyzed oxidation of catechols [10], methyl-dopa [10], 2-hydroxyestradiol [11], or epinephrine [10] is mediated by superoxide and results in protein binding. 2,2′-Dichlorobiphenyl binding to microsomes also seems to be mediated by superoxide [12]. Microsomal mixed-function oxidase-catalyzed alcohol oxidation is partly mediated by superoxide [13]. A hydroxyethyl radical is formed by the oxidation of the alcohol [14].

C. Peroxidase Activity

The peroxidase activity of cytochrome *P*-450 was first discovered when it was found to be responsible for the decomposition of fatty acid hydroperoxides to a wide range of lipid free-radical products [15]. Phenylenediamines readily act as a hydrogen donor, but cytochrome *P*-450 drug substrates are oxidized poorly [16]. Hydrogen peroxide is a poor substrate for cytochrome *P*-450. The much higher preference for organic hydroperoxides is presumably due to the hydrophobic substrate binding site in the heme pocket. However, tertiary hydroperoxides, particularly cumene hydroperoxide, can substitute for NADPH, NADPH-cytochrome *P*-450 reductase, and molecular oxygen in the mixed-function oxidase reaction for all cytochrome *P*-450 drug substrates [reviewed in reference *17*]. This hydroperoxide system catalyzes epoxidation, aromatic ring hydroxylation, side-chain oxidation, *N*-hydroxylation, and O-dealkylation by a two-electron oxidation [reviewed in reference *18*]. However, a one-electron oxidation to free radicals occurs with phenylenediamines and aminophenols. Thus, *N,N,N',N'*-tetramethylphenylenediamine is oxidized to the stable Wurster's blue cation-radical. We have shown that most unacetylated benzidines are oxidized by a one-electron oxidation pathway to diimines. No N-hydroxylation could be detected with either the mixed-function oxidase activity or peroxidase activity. However, N-hydroxylation has been demonstrated with aminoazobenzene [*19*], 2-naphthylamine (2-NA) [*20*], and AAF [*18*]. *N*-Alkyl compounds readily N-demethylate via a one-electron oxidation with peroxidase–peroxide systems. Because of its hydrophobic binding site, cytochrome *P*-450 is unlike other hemes and peroxidases in catalyzing the N-demethylation of hexobarbital, benzphetamine, and ethylmorphine. Thus, mixed-function oxidase and cytochrome *P*-450–cumene hydroperoxide readily carry out an N-demethylation of these compounds. Some violene cation-radical is formed in the N-demethylation of aminopyrine by cytochrome *P*-450–cumene hydroperoxide. The authors suggested that this indicates that cytochrome *P*-450 carries out N-demethylation via a one-electron oxidation mechanism [*21*]. However, the amount of cation-radical formed is very small and is not formed in the presence of reducing agents that do not affect the N-demethylation. It is more likely that a two-electron oxidation to a carbinolamine is involved.

Benzopyrene is also oxidized by phenobarbital-induced microsomes–cumene hydroperoxide to a much higher level of quinones [*22*] as a result of an enhanced one-electron oxidation reaction to the radical-cation and further oxidation of the phenols (formed from the epoxides) to quinones. Oxidation by microsomes–H_2O_2 forms products more similar to those found with the mixed-function oxidase activity [*23*] because the phenols are not further oxidized in this system.

In summary, the peroxidase activity of cytochrome *P*-450 normally carries out the same two-electron oxidation pathway as the mixed-function oxidase. However, in the case of phenylenediamines, aminophenols, unacetylated benzidines, hydroxylamines, and alcohols, a one-electron oxidation pathway is important. The carcinogenic consequences of this is the greatly enhanced reactivity of the peroxidase activity in forming DNA adducts (discussed later) when compared to that with the mixed-function oxidase activity. The addition of spin traps to trap the small amounts of methyl or alkyl radicals formed from the cumene hydroperoxide during the peroxidase activity has no effect on these reactions. It is therefore concluded that the higher oxidation states of cytochrome *P*-450 catalyze these reactions [*24*].

The relevance of the peroxidase activity is that it makes the highly reactive one-electron peroxidation pathway of cytochrome P-450 more apparent. During the *in vitro* mixed-function oxidase assay the high levels of NADPH would be expected to decrease this pathway by reducing the radicals involved. Physiological hydroperoxides such as those of steroids and unsaturated fatty acids could play a role in cytochrome P-450-catalyzed activation of phenylenediamines, aminophenols, and benzidines by a one-electron oxidation.

D. Cytochrome *P*-450 Adduct Formation and Inactivation

Cytochrome P-450 is believed to be destroyed by alkylation of one of the nitrogen atoms of the prosthetic heme during catalytic transfer of activated oxygen to the unsaturated bond (π-bond) in certain substrates and forming epoxides. Thus, vinyl fluoride, vinyl bromide, and fluorene alkylate the prosthetic heme group of cytochrome P-450 enzymes that catalyze their metabolism. Major losses of cytochrome P-450 occur with green pigment formation. The identity of the heme adduct indicates that oxygen is introduced at the trifluoroethoxy- or halide-substituted terminus of the bond, and the unsubstituted terminus reacts with a heme nitrogen atom. This reaction orientation is consistent with a radical intermediate, possibly formed by way of an initial π-bond radical-cation. The consequence of the adduct formation is an inactivation of the cytochrome P-45O [25]. Other investigators have also concluded that the destruction of cytochrome P-450 by inhalation of vinyl chloride is not mediated by the corresponding epoxide metabolite [26]. The suicidal nature of cytochrome P-450 inactivation by olefins [27] implies that a common catalytic pathway results in both heme alkylation and epoxide formation. This would suggest that the epoxidation is also mediated by free radicals. The regiospecificity of the reaction of substituted olefins with prosthetic heme argues for a radical rather than a cationic intermediate in olefin oxidation [25].

A number of alkyl-containing xenobiotics are suicide substrates of cytochrome P-450. Thus, 2-alkyl-2-isopropylacetamide (AIA) produces a green discoloration of the liver due to heme degradation products. A free-radical intermediate preceding epoxide formation is probably responsible for the heme alkylation [28]. Approximately 250 molecules of AIA are turned over for each cytochrome P-450 molecule that is inactivated. Acetylenes and ethylene are like olefins in that the entire substrate plus an oxygen atom is attached to a nitrogen in the prosthetic heme group [29, 30].

Cytochrome P-450 is also inactivated during the catalytic turnover of 3,5-bis(ethoxycarbonyl)-4-ethyl-2,6-dimethyl-1,4-dihydropyridine (DDEP). Spin trapping evidence indicates that the oxidation of the nitrogen in DDEP by cytochrome P-450 proceeds by a one-electron oxidation, resulting in the ejection of an ethyl radical that binds to the nitrogen of the prosthetic heme [28]. This ethylation results in destruction of the cytochrome P-450; H_2O_2 could replace NADPH in this reaction.

Hydrazines (e.g., ethylhydrazine, acetylhydrazine, and phenylhydrazine [32] cause a rapid inactivation of cytochrome P-450. Isoniazid, iproniazid [32], and 1,1-disubstituted hydrazines [33] are particularly effective. Radicals have been trapped during the microsomal oxidation of ethylhydrazine [34]. The inactivation of hemoglobin during phenylhydrazine oxidation results in an N-arylheme derivative. A phenyl radical is formed and explains the heme adduct formation [35]. These results provide direct evidence of carbon radical formation in the microsome- or hemoglobin-catalyzed metabolism of alkyl- and acylhydrazines. The mixed-function oxidase-catalyzed oxidation of phenylhydrazine has been shown to form products similar to those with ferrihemoglobin. The oxygen-sensitive phenyldiazene produced forms a ferrihemochrome with the cytochrome P-450 and therefore behaves as a suicide substrate [36]. The 1,1-disubstituted hydrazines readily form complexes with cytochrome P-450 when added to a suspension of rat liver microsomes in the presence of NADPH and oxygen. This abortive complex has been attributed to nitrene, formed by an amine oxidase or a cytochrome P-450-catalyzed two-electron oxidation of the hydrazine [37]. These reactions would be in resonance equilibrium with the corresponding diazene. Nitrosamine formation was ruled out in this complex formation. The trapping of ethyl radicals suggests that the diazene is formed by direct dehydrogenation of the hydrazine rather than by dehydration of an N-hydroxy metabolite.

E. Other Substrates

The rate of sulfur compound oxidation by microsomal mixed-function oxidase correlates well with the one-electron oxidation potentials of the sulfur atoms [38]. The oxidation of N-hydroxycocaine to nitroxy radicals catalyzed by microsomal mixed-function oxidase was shown not to be mediated by superoxide, and thus a direct one-electron nitrogen oxidation by cytochrome P-450 occurs [39]. N-Hydroxy-2-naphthylamine and N-hydroxyacetylaminofluorene are also oxidized to nitroxyl radicals by microsomal mixed-function oxidase [40, 41]. Because of its hydrophobic binding site, cytochrome P-450 is unlike other hemes and peroxidases in catalyzing alkane hydroxylation. Seminal experiments on carbon hydroxylation using deuterium scrambling as a probe have provided evidence that alkane hydroxylation by mixed-function oxidase proceeds via a free-radical oxidative mechanism [42, 43]. A free-radical mechanism for cyclohexane hydroxylation to cyclohexanol by the cytochrome P-450–iodosobenzene system has been suggested from ^{18}O experiments [44].

Microsomal mixed-function oxidase would be expected to carry out a two-electron oxidation of the substituted aminophenol paracetamol (acetaminophen) to an N-hydroxy derivative. However, paracetamol is readily oxidized to a quinoneimine, which forms an adduct with glutathione or binds to protein, and N-hydroxylation did not occur [45].

III. NITROFURANS

Many 5-nitrofurans possess antibacterial, mutagenic, and carcinogenic activity. Metabolic activation is required because the nitrofurans induce tumors at sites remote from the point of administration. Cytosolic or microsomal NADPH nitroreductases are involved in the anaerobic reductive activation of 5-nitrofurans to reactive nitroso or hydroxylamine intermediates. The cytosolic nitroreductase activity is associated with xanthine oxidase and aldehyde oxidase. The microsomal reductase activity is associated with the NADPH-cytochrome P-450 reductase. Thus, bacterial mutants insensitive to these chemicals are also deficient in nitroreductases. 2-Amino-4-(5-nitro-2-furyl)arylamide (ANFT) and nitrofurantous are activated by nitroreductase to species that bind to DNA, RNA, and protein [47, 49]. Xanthine oxidase has also been shown to catalyze the binding of nitropyrene or nitrofurazone to DNA [46, 48].

With some nitrofurans an oxidative pathway is also available. Thus, ANFT binding to proteins and nucleic acids is catalyzed by prostaglandin hydroperoxidase activity of rabbit bladder translational epithelial and ram seminal vesicle microsomes [50, 51, 57]. The prostaglandin synthase activity of rabbit kidney inner medullary microsomes is one-third as active as the nitroreductase in catalyzing protein binding [52]. N-[4-(5-Nitro-2-furyl)-2-thiazolyl]formamide (FANFT) is a very potent urinary bladder carcinogen that can also be metabolized by prostaglandin synthase [58]. Allopurinol, an inhibitor of xanthine oxidase, increases the incidence of FANFT-induced bladder cancer [56], whereas aspirin, an inhibitor of prostaglandin synthase, can completely prevent the development of morphological bladder lesions induced by FANFT [55]. Deformylation of FANFT to ANFT is probably an important step in FANFT-induced carcinogenesis. Peroxidase does not metabolize ANFT [62].

The microsomal nitroreductases are distributed in both kidney cortex and medulla, whereas the prostaglandin synthase oxidative pathway predominates in the kidney inner medulla. Immunohistochemical fluorescence techniques have localized prostaglandin synthase in the renal tubular epithelium, a common site of carcinogenicity. Both enzyme systems are also associated with the endoplasmic reticulum and nuclear membrane [53]. Bladder epithelium has also been shown to contain prostaglandin synthase [54].

Under aerobic conditions the nitroaromatic anion free radical formed by the nitroreductases reduces molecular oxygen to superoxide and results in hydroxyl radical and H_2O_2 formation [60] and can lead to red blood cell hemolysis [59].

IV. HYDRAZINES

Phenylhydrazine once found clinical use in the lysing of erythrocytes to lower the red blood cell count in polycythemia vera [65]. Phenylhydrazine also causes hemoglobin degradation and a precipitation of Heinz bodies in erythrocytes. The erythrocyte lysis is believed to be the result of the reaction between phenylhydrazine and hemoglobin. The reaction of phenylhydrazine with oxyhemoglobin leads to the formation of phenyl radicals that are preferentially trapped in the erythrocyte membrane and could be responsible for initiating the oxidation of unsaturated fatty acid residues in phospholipids, the oxidation of membrane sulfhydryl groups, and the aggregation of membrane proteins that result in erythrocyte lysis [66]. Erythrocytes deficient in glutathione (GSH) are remarkably sensitive [64], and hemolytic anemia is readily induced in individuals following accidental hydrazine exposure if they have a glucose-6-phosphate dehydrogenase deficiency. The phenyl radicals also react with the heme to form a green N-phenyl adduct of protoporphyrin [61, 63]. The initial step in the oxidation of phenylhydrazine is a two-electron oxidation to phenyldiazene that can autoxidize and form superoxide, phenyl radicals, and nitrogen [32]. The spontaneous dismutation of superoxide and/or the reaction with oxyhemoglobin produce H_2O_2. Ferrous complexes could then result in hydroxyl radical formation, which could also be responsible for the membrane lipid peroxidation. The overall stoichiometry [62] seems to be as follows:

$$6 \text{ PhNHNH}_2 + 6 \text{ O}_2 + 1 \text{ heme (active)} \rightarrow 5 \text{ benzene} + 1 \text{ } N\text{-phenylheme} + 6 \text{ H}_2\text{O}_2 + 6 \text{ N}_2$$

Phenyldiazene can also form a ferrihemochrome with methemoglobin [32] or cytochrome P-450 [33]. Both N-phenyl and C-phenyl derivatives of the heme of hemoglobin are formed [32]. The ferrihemochrome of both hemoproteins is unstable, and the heme is destroyed. The ferrihemochrome with 1,1-disubstituted hydrazines in contrast decomposes to regenerate cytochrome P-450 [34]. Whether phenyl radicals result in phenylation and cause the suicide destruction of these hemoproteins remains to be determined. Phenylhydrazine, iproniazid, and isopropylhydrazine bind covalently and inactivate other heme compounds such as lactoperoxidase, horseradish peroxidase, myoglobin, and cytochrome c as well as enzymes with essential thiol groups [71].

Other derivatives of hydrazines have gained wide use as tranquilizers, antitubercular agents, and antitumor agents. Acetylhydrazine has been identified as a metabolite of the widely used antitubercular drug isonizid in humans and has been shown to be hepatotoxic. A dose-dependent decline in hepatic cytochrome P-450 in the phenobarbital-pretreated rat occurs [68]. The microsomal metabolism of isopropylhydrazine and acetylhydrazine results in covalent binding of their metabolites to microsomal protein [72]. Ethyl and acetyl radicals have been trapped during the microsomal-catalyzed oxidative metabolism of ethylhydrazine and acetylhydrazine [31].

1,2-Dimethylhydrazine induces colon adenocarcinomas, liver hemangioen-dothelioma, and kidney tumors. 1,2-Dimethylhydrazine is believed to be activated *in vivo* via azomethane, azoxymethane, and methylazoxymethanol (MAM) to yield a reactive diazonium ion and methylcarbonium ions. Methylazoxymethanol is also degraded by alcohol dehydrogenase via MAM aldehyde [67]. Methylation of nucleic acids [69] and histones [70] of liver and colon as indicated by the formation of 7-methylguanine, O^6-methylguanine, and 3-methyladenine are observed after treatment of rats and mice with 1,2-dimethylhydrazine, a colon carcinogen [74]. A much smaller methylation is observed with 1-methyl-hydrazine [69]. Wild mushrooms contain several carcinogenic hydrazine compounds such as N-methyl-N-formylhydrazine and 1-methylhydrazine. Although no *in vitro* system has demonstrated adduct formation, DNA fragmentation, presumably as a result of activated oxygen species formed during the autoxidation of the hydrazine, has been reported [73]. The carcinogenic potency of 16 hydrazine derivatives tested correlated with their potency in inducing DNA fragmentation *in vivo* and not their mutagenicity in the *Salmonella* cell microsome test [82]. Unscheduled DNA synthesis is also observed in hepatocytes after exposure to 1,2-dimethylhydrazine [83]. Unscheduled DNA synthesis is observed in cultured human fibroblasts after exposure to isoniazid, iproniazid, and hydrazine. The effect is dependent on transition metals, and activated oxygen species are implicated [84]. Chromosome aberrations are also induced in Chinese hamster ovary cells [85].

Hydrazine, carbon tetrachloride, and ethanol at nearly toxic doses for long periods also induce 7-methylguanine and O^6-methylguanine formation in liver DNA. This may be the result of a change in the specificity of S-adenosylmethionine-dependent DNA methyltransferase or a direct chemical methylation by S-adenosylmethionine or inhibition of the enzymes involved in demethylation [86–88].

V. 2-ACETYLAMINOFLUORENE

The multiplicity of electrophilic metabolites of precarcinogens has been exemplified by studies on AAF. This compound is a hepatocarcinogen in rats, hamsters, and mice but not guinea pigs. N-Hydroxylation of AAF by a cytochrome *P*-448-dependent monooxygenase followed by activation by cytosolic sulfotransferases or seryltransferases is believed to form AAF–DNA *in vivo* [1]. The deacetylated aminofluorene–DNA adducts, which form 80% of the adducts *in vivo,* may involve either N-hydroxyaminofluorene (N-OH-AF) formed by deacetylases [76, 77]. The guanosine adduct isolated from the DNA adduct is shown in Fig. 1. Space-filling molecular models of the DNA adduct indicate that the N-OH-AAF could fit within the major groove of the DNA helix and escape detection by excision repair systems.

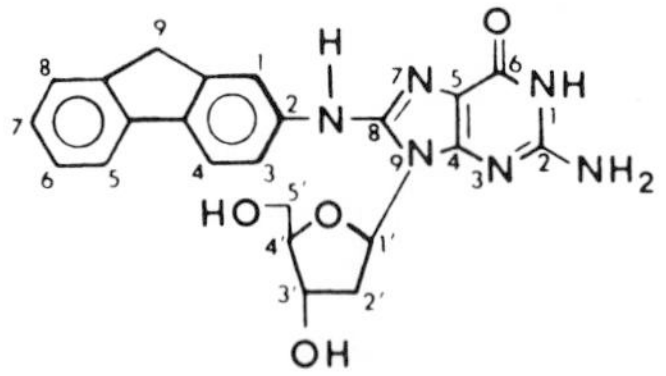

Fig. 1.   Aminofluorene–guanosine adduct.

Male rats are also much more susceptible to carcinogenesis by AAF than female rats and have a four- to fivefold higher sulfotransferase activity [89]. However, it is absent in nonhepatic tissues, which are also target tissues [81]. It is also absent in rat liver nuclei [80], and the N-OSO_3-AAF has a half-life in water of less than 1 min [79], which makes it unlikely that it can travel from the cytosol into the nucleus. The sulfate ester, although binding to DNA, is not mutagenic [90] and has not yet been shown to be carcinogenic. The specific sulfotransferase inhibitor pentachlorophenol decreases the *in vivo* acetylated adducts from 40 to 13% of total, which could indicate that 70% of the acetylated DNA adduct involves sulfotransferase. The level of hepatic sulfotransferase correlates with susceptibility to hepatic tumor formation [79], and the hepatotoxicity and hepatocarcinogenicity of N-OH-AAF for the rat are decreased when a sulfate-deficient diet is given [80]. However, the nonacetylated adducts are increased [84]. Sulfotransferase seems to be involved in aminofluorene-evoked repair but not mutagen formation in isolated rat liver cells [91].

Stout *et al.* [92] first suggested the biological importance of the deacetylated DNA adducts by proposing that N-OH-AF is the principal mutagen formed from N-OH-AAF. A paraoxon-insensitive cytosol activation of N-OH-AAF to a mutagen was formed in the Ames test [78]. In the absence of S9 activation, the activation can be carried out by a paraoxon-sensitive bacterial deacetylase. The mutagenicity correlates with the formation of the N-OH-AF–DNA adduct in the bacteria [78]. Microsomes and nuclei also have a paraoxon-sensitive deacetylase. Deacetylase inhibitors decrease the covalent binding of N-OH-AAF to microsomal protein [93], its mutagenic activation [90], and its binding to nuclear DNA [94]. The N-OH-AF formed probably reacts via the nitrenium ion with nucleophilic sites on nucleic acids and proteins at acidic pH [77].

Ascorbate markedly enhances the mutagenesis of the sulfate ester of N-OH-AAF 12-fold while decreasing DNA adduct formation by 80% [95, 96]. The effect on DNA adduct formation was explained by the ascorbate reduction of the highly reactive electrophilic arylnitrenium ion to a free-radical intermediate that may be highly mutagenic as a result of abstracting an electron from DNA [97]. Ascorbate, however, decreases the mutagenic activity of N-hydroxyphenacetin [98].

Bartsch and Hecker first demonstrated that nitroxyl free radicals were formed when N-OH-AAF was oxidized by peroxidases and H_2O_2 [*100*] and that they dismutate at a rate of $2.7 \times 10^5 \ M^{-1} \ sec^{-1}$ to N-acetoxy-AAF and 2-nitrosofluorene. The former readily binds covalently to DNA, tRNA, and guanosine [*101*] to form acetylated aminofluorene adducts. Nitrosofluorene is one of the most mutagenic compounds known [*102*] and binds covalently to proteins. Since then, Floyd et al. have demonstrated nitroxyl free-radical formation when N-OH-AAF is oxidized by linoleic acid hydroperoxide with methemoglobin [*103*] or high-spin cytochrome P-420 [*104*]. Similar results are obtained with arachidonate and platelets, mammary gland parenchymal cell homogenates, or ram seminal vesicular prostaglandin synthase [*105*]. Deacetylase inhibitors did not affect nitrosofluorene formation. This suggests that prostaglandin synthase or lipoxygenase catalyzes the activation of N-OH-AAF in AAF-induced mammary gland carcinogenesis. Ascorbate, propyl gallate, or GSH prevents the formation of nitroxyl radicals [*106*]. However, the only DNA adduct isolated from the target organ rat mammary gland, after rats were treated with N-hydroxy-2-AAF, has been identified as N-(deoxy-guanosin-8-yl)-2-aminofluorene, indicating that sulfotransferase, peroxidase, lipoxygenase, or prostaglandin synthase are not involved in the activation [*99a,b*]. The researchers favor activation by mammary cytosolic N,O-acetyl-transferase rather than microsomal deacetylase. In this case, the reactive electrophile is probably N-acetoxy-2-aminofluorene. With Zymbal's gland, another target organ, the microsomal deacetylase activity is very high and is believed to be responsible for the activation [*99c*]. Deacetylated adducts were also isolated from dog bladder in vivo [*99d*].

We have discovered another one-electron oxidation pathway that results in the formation of deacetylated DNA adducts from AAF [*99e, 99f*]. This involves a peroxidase-catalyzed oxidation of aminofluorene formed by the microsomal deacetylase. Prostaglandin synthase and arachidonate also oxidize aminofluorene to nitrofluorene and azoaminofluorene [*108*] and products that are mutagenic [*209*] and bind to DNA [*109*]. A direct one-electron oxidation to cation-radicals is probably involved, and an ortho-semidine type of dimer formed by ortho C—C coupling could also be a product [*110*]. A similar AAF dimer is a product of the sulfation of N-OH-AAF and may bind to protein [*111*]. The authors suggested a reaction between AAF and its nitrenium ions [*111*], but a radical mechanism may be involved.

Stier et al. [*112*] obtained very stable nitroxyl free radicals in chloroform–methanol extracts of liver microsomes (from phenobarbital-induced rabbits) metabolizing arylamine carcinogens. They were identified as lipid arylamine nitroxide adducts and were stable for days in organic solvents and hours in the incubation mixture. A strong correlation among 28 arylamines exists between carcinogenicity and free-radical formation after incubation with liver microsomes and NADPH [*113*]. Similar adducts are formed with N-OH-AAF and rat liver microsomes [*115*]. A pseudo-Diels–Alder or ene reaction of 2-nitrofluorene with the unsaturated bonds of fatty acids or cholesterol was proposed. A similar ESR signal is observed in the liver 90 min after injecting N-OH-AAF intraperitoneally into rats [105]. In the case of 2-NA, 2-aminoanthracene, 4-aminoazobenzene, and N-methylaniline, the arylamine nitroxide radical precedes the adduct radical, presumably because the nitroxide radicals are less reactive than those formed from the more carcinogenic arylamines. Only 0.2% of the phospholipids may be involved in the adduct formation. The adduct radical is readily reduced by NADPH to N-hydroxy adducts and regained by ferricyanide oxidation [105]. Stier suggested that autoxidation of this adduct results in superoxide formation and DNA strand breaks [113]. The adduct may be mutagenic [105].

VI. NAPHTHYLAMINES

2-Naphthylamine was one of the first human carcinogens to be discovered. Although industrial exposure has been severely restricted, nanogram quantities of 2-NA and other arylamines in cigarette smoke could be responsible for the increased risk of smokers to urinary bladder cancer [114]. It also induces bladder cancer in dogs (100%), guinea pigs (50%), rabbits (50%), hamsters (50%), and rats (10%) but induces liver cancer in mice. Two days after 2-NA administration the binding to dog urothelium DNA is fourfold higher than DNA. In both cases the adducts are 1-(deoxyguanosin-N^6-y1)-2-naphthylamine, an imidazole ring-opened derivative [formed by hydrolysis of N-(deoxyguanosin-8-yl)-2-naphthylamine] and 1-(deoxyguanosin-N^2-yl)-2-naphthylamine [115]. At 7 days the difference is eight-fold due to the greater persistence of these adducts, particularly the ring-opened derivative, in the urothelium as compared with the liver. In contrast, 1-NA produces much lower levels of adducts and thus correlates with the apparent lack of carcinogenic activity in the dog. Presumably, the repair mechanisms are less effective in the bladder than in the liver. It is believed that the liver cytochrome P-448 [116] or microsomal amine oxidase [117] is responsible for the N-oxidation of 2-NA. However, in the dog the N-oxidation is induced by phenobarbitone [118]. The fact that the levels of N-OH-2-NA in the urine can be either increased relative to its N-glucuronide conjugate by lowering urine pH or decreased by elevating the pH [119] suggests that acidic hydrolysis of the glucuronide (formed by the liver and excreted in the bile) occurs in urine retained in the bladder lumen. No detectable sulfotransferase, seryltransferase, O-acetyltransferase, or transacetylase activity that could act on N-hydroxyarylamines are found in the dog urothelial cytosol or microsomes [119]. Thus, extracellularly generated arylnitrenium ion–carbocation species, which can be formed in a slightly acidic urine, may be capable of entering an intact cell and causing a malignant transformation. The major metabolite (90%) of 2-NA, 2-amino-1-naphthol, is excreted in the urine.

2-Naphthylamine is also rapidly oxidized by prostaglandin synthase [109] or peroxidases to free radicals and mutagenic products that bind readily to DNA and protein [107, 209]. The ESR radicals observed in benzene extracts of a reaction mixture of liver microsomes [from rats induced by 3-methylcholanthrene (MC)], NADPH, and 2-NA includes a radical from 2-Amino-1-naphthol [120] and the nitroxide radical [116]. 1-Naphthylamine forms a much weaker signal attributed to 1-Amino-2-naphthol, but no nitroxide radical was found. Stier [113] also obtained lipid 2-NA nitroxide adducts with phenobarbital-induced rabbit liver microsomes, NADPH, and 2-NA. No studies on the liver or bladder microsomal-catalyzed binding of 2-NA to DNA have been reported.

VII. N-ALKYL COMPOUNDS

The primary target tissue for the carcinogenic effect of *trans*-4-dimethyl-aminostilbene in the rat is the Zymbal's gland, a sebaceous gland located in the external ear duct [121]. This organotropic effect is best explained by activation by the peroxidase present in the gland [122]. Peroxidases also catalyze the binding to the DNA of derivatives of 4-acetylaminostilbene and 4-aminostilbene [122]. There is little mixed-function oxidase N-hydroxylase, sulfotransferase, or acyltransferase activity in this gland.

Although N,N-dimethyl-4-aminoazobenzene (DAB) is well known as a liver carcinogen in rats, it can also induce bladder tumors in dogs but is noncarcinogenic in guinea pigs and hamsters. An analysis of the DNA adduct isolated from rat liver 1 or 14 days after a stomach intubation of methyl-4-aminoazobenzene (MAB) showed that N-(deoxyguanosin-N^2-yl)-MAB persists for at least 14 days [123]. Similar adducts are formed when DNA is reacted with N-benzoyloxy-N-methyl-4-aminoazobenzene [123]. Space-filling models show that N^2-yl-MAB may be situated in the minor groove of the DNA helix. Similar results are obtained with mice [124]. It has been suggested that MAB is N-oxidized by microsomal cytochrome P-448 [125, 127] and amine oxidase [126] and is esterified to an ultimate carcinogen by sulfotransferase. Dietary supplementation with sulfate also enhances the carcinogenicity of $3'$-CH_3-DAB.

Extensive DNA binding of MAB is observed when the MAB is oxidized to free radicals by a peroxidase–H_2O_2 system [128]. In this system the possibility of MAB binding to the protein, which then binds to the DNA, as suggested by others [109] is unlikely because the DNA is in excess of the protein by a factor of 10^5 (by weight) and the DNA is treated with phenol as well as proteases in the DNA isolation procedure. N-Methyl-4-aminoazobenzene readily binds to DNA in intact leukocytes activated by a tumor promoter [128]. This binding is prevented by cyanide and is presumably catalyzed by myeloperoxidase. Prostaglandin synthase can also oxidize MAB to radical species that bind to DNA and protein [129]. The peroxidase activity of cytochrome P-450 (using cumene hydroperoxide) also results in the binding to DNA and protein [130]. Best results are obtained with mouse, hamster, or rabbit liver microsomes. Binding to DNA in liver nuclei can also be demonstrated. With microsomes and cumene hydroperoxide, no nitroxyl radicals are formed [17], but N-dealkylation occurs and N-hydroxyaminoazobenzene is formed [126]. With microsomes and NADPH, N-OH-MAB is readily formed [126], and nitroxyl radicals are formed with MC-induced microsomes [125]. The rate of nitroxyl radical formation of various MAB and DAB derivatives also correlates with their carcinogenic activity [125]. The binding to DNA or protein could therefore involve an imine or a benzidinediimine following dimerization of a cation-radical. We have shown that other N-methyl compounds, for example, N-methylaniline, and N-methyl-o-toluidine, also bind covalently to DNA and protein during oxidative demthylation by a peroxidase–H_2O_2 system [107].

Hematin and nearly all hemoproteins with peroxidase activity can carry out N-demethylation of N-alkyl compounds with most hydroperoxides. Not surprisingly, prostaglandin synthase–arachidonate can also carry out N-demethylation to formaldehyde (HCHO), and the intermediate cation-radicals are readily detected [131, 134]. The mechanism of aminopyrine N-demethylation is believed to involve a one-electron oxidation to a radical-cation and then disproportionates to the methiminium cation. The latter is probably hydrolyzed to the demethylated amine and HCHO. The aminopyrine radical-cation steady-state concentration is linear to the square root of enxyme concentration, indicating that the radical is formed enzymatically but decays by a dimerization or disproportionation [131] rather than being oxidized by the peroxidase to the iminium cation as suggested by others [132]. Oxygen is not required for the N-dealkylation, and the oxygen for the HCHO is derived from H_2O rather than the hydroperoxide [132]. We have also found that GSH totally inhibits the N-dealkylation, presumably by reducing the radical-cation and by forming an adduct with the imine. GSH was oxidized in the process. Ascorbate also inhibits, presumably by reducing the intermediate back to the N-methyl compounds. A similar mechanism may exist for the microsomal mixed-function oxidase activity [18], but it is unlikely that this involves the release of free radicals or imines involved in the N-demethylation reaction because the N-demethylation was not affected by GSH or ascorbate. The N-dealkylation mechanism catalyzed by cytochrome P-450 could also act by forming an unstable carbinolamine compound with the N-alkyl group that breaks down spontaneously to aldehyde and dealkylated amine [135]. Normally, GSH has no effect on microsomal N-demethylation and does not form GSH adducts. In MAB N-demethylation, however, a GSH adduct is formed and is clearly an important detoxification route for MAB via the bile [177]. Presumably, GSH reacts with an imine or the N-methylol intermediate, and some inhibition of N-demethylation is found.

The peroxidase activity of microsomes is very effective at N-dealkylation [136]. Studies with cytochrome P-450 show that N,N-dimethylaniline is converted to stoichiometric quantities of N-methylaniline and HCHO [137]. However, our studies show that the peroxidase–H_2O_2 reaction results in $N,N,N'N'$-tetramethylbenzidine as a major product. The ESR signal formed during the oxidation has also been identified as the N,N,N',N'-tetramethylbenzidine cation-radical [138]. This indicates that a free-radical pathway results in dimerization as the principal consequence, confirming that the peroxidase activity of cytochrome P-450 in N-dealkylation may not involve the release of free radicals. Other investigators, however, demonstrated the formation of the aminopyrine cation-radical (0.25%) during the hydroperoxide-dependent N-dealkylation by cytochrome P-450 [18], so that some leakage of radical may occur in the peroxidase activity. We have also shown that GSH and ascorbate prevent cation-radical formation but stimulate HCHO formation two- to fivefold, indicating that the N-dealkylation does not involve the release of radical-cation or iminium cation intermediate. Presumably, the hydroperoxide and/or the imine inactivate cytochrome P-450, and GSH or ascorbate prevents this. Clearly, if the N-dealkylation of the peroxidase activity of cytochrome P-450 involves a one-electron oxidation, the free radicals are transient and formed within a "cage."

VIII. BENZIDINES

Benzidine and benzidine congener-based dyes are widely used in dye manufacturing, textile dyeing, paper printing, and leather industries [139]. The incidence of urinary bladder cancer is also elevated among industrial workers exposed to benzidine-based dyes. The incidence of tumors of the stomach, rectum, lung, and prostate is also elevated. Metabolic azoreduction of these dyes in animals or exposed humans results in the appearance of monoacetylbenzidine derivatives in the urine or bile [140–142]. Mutagens also appear in the urine of rats treated with benzidine but not 3,3',5,5'-tetramethylbenzidine [143]. Anaerobic intestinal bacteria are probably largely responsible for the azoreduction [144]; however, liver azoreductases [140] may be involved in the liver carcinogenicity demonstrated when certain azo dyes are fed to rats [140]. Bacterial azoreductases are a thousand times more active than mammalian enzymes [144].

Although benzidine has been shown to be a human urinary bladder carcinogen, it is not carcinogenic in monkeys and is a poor bladder carcinogen in dogs or rabbits [139]. The liver is the major target organ in mice, hamsters, and rats. The Harderian gland and lymphoreticular system of mice and the Zymbal's gland, mammary gland, and intestine of rats also develop tumors [145]. Unlike other arylamines, benzidine does not cause bladder cancer in mice and rats. In acute poisoning benzidine attacks the kidney of humans, dogs, and rabbits, causing hematuria and chronic cystitis, and the liver of rats, dogs, and rabbits, causing cirrhosis [139].

Ring hydroxylation seems to be the major metabolite pathway for aromatic amines. In the case of benzidine, 3-hydroxybenzidine is the major urinary product in man [146]. However, the discovery of a metabolite of N-acetyl-benzidine in human urine [147] capable of forming a chloroform-soluble copper chelate, a known property of arylhydroxamic acids, lends some credence to the suggestion that the initial steps in the metabolic activation of benzidine in the rat liver involves a microsomal-catalyzed N-hydroxylation following a cytosolic-catalyzed acetylation [148]. Experiments have suggested that the N-hydroxy derivative of N,N'-diacetylbenzidine is further activated by an N,O-acyltransferase [148] or sulfotransferase [149] to an electrophile that reacts with methionine to form 3-methylmercapto-N,N'-diacetylbenzidine [149] or binds to DNA or RNA and is mutagenic [148, 149]. However, Martin et al. [151] have identified a persistent DNA adduct isolated from mouse or rat liver after the administration of benzidine or N-acetylbenzidine as N-(deoxy-guanosin-8-yl)-N'-acetylbenzidine. Neither the nonacetylated or diacetylated derivative is found. No DNA binding is observed after the administration of N,N'-diacetylbenzidine. The same adduct is found when DNA is reacted with N-hydroxy-N'-acetylbenzidine at pH 5 or when deoxyguanosine is reacted with N-acetoxy-N,N'-diacetylbenzidine. It is suggested that N-hydroxy-N'-acetylbenzidine is formed from N'-acetylbenzidine or N-hydroxydiacetylbenzidine (formed from N-hydroxy-N'-acetylbenzidine). Whether N-hydroxy-N'-acetylbenzidine can react directly with DNA or requires prior activation by esterification remains to be determined.

Space-filling models of this DNA adduct indicate that the molecule is unhindered in the major groove, causing no apparent perturbation of conformation of the DNA, and explains its persistence as a result of nonrecognition by repair enzymes. These results clearly go a long way in explaining the mechanism of liver cancer induction by benzidine in rats or mice. However, in nonacetylating species such as dogs, the bladder is the target [152]. 2-Naphthylamine has been suggested to cause adduct formation in the urothelium as a result of the urine acidity releasing N-hydroxyarylamine from the N-glucuronide originally formed by the liver [119]. The difference between man and dogs, on the one hand, and rodents, on the other, is ascribed to facilitated hydrolysis at the lower urinary pH of the former species. However, the mechanism of targeting to tissues such as rat ear duct or mammary gland would remain unexplained.

A more economical hypothesis is that the activated species is formed in the target organ. The bladder urothelium contains a prostaglandin synthase [129] and low levels of cytochrome P-450 [153]. There is no evidence, however, that cytochrome P-450 can carry out the N-hydroxylation of unacetylated benzidine. A one-electron oxidation of the benzidine catalyzed by prostaglandin synthase [154] of the urothelium results in extensive DNA binding. No attempts to induce bladder or liver cancer with any oxidized benzidine derivatives has yet been reported, although mammary gland and Zymbal's gland tumors are induced in rats injected intraperitoneally with benzidine or N-hydroxy-N,N'-diacetylbenzidine [150]. Nor have DNA adducts been isolated from the urothelium to ascertain whether they are acetylated.

Isolated rat liver cells are more active in forming mutagens from benzidine than the 9000 g supernatant from these cells [156]. Cytosol also enhances the microsomal-catalyzed mutagenic activation of benzidine. This indicates that N-hydroxylation of benzidine requires prior N-acetylation. However, deacetylation may then be required for mutagenicity because paraoxon, an inhibitor of the deacetylation reaction, decreases the formation of mutagens by liver cells considerably. Studies of DNA excision repair, however, suggest the involvement of sulfotransferase in DNA adduct formation [156].

The peroxidase activity of hemoglobin catalyzes the oxidation of benzidine to an intensely blue charge-transfer complex, and this has formed the basis of a forensic test for hemoglobin in urine and stools. 3,5,3'5'-Tetramethylbenzidine is a noncarcinogen [157] and forms a similar complex in a peroxidase reaction [157, 158]. It is also four times more sensitive than benzidine and should make an excellent substitute. The peroxidase activity of liver cytochrome P-450 and cytochrome P-420 using benzidine as a substrate have been used to stain SDS-polyacrylamide gels so as to locate cytochrome P-450 [159]. Tetramethylbenzidine oxidation at pH 5 to a blue charge-transfer complex is probably a preferred method [169].

The peroxidase-catalyzed oxidation of benzidine derivatives to brown polymers has also formed the basis of a cytochemical method for determining the location of peroxidases and catalase. The principal intracellular sites for peroxidases appear to be the cisternae of endoplasmic reticulum and the nuclear envelope in the cells studied. It is interesting that most of the peroxidase-containing cells are target organs for carcinogenesis. The cells studied include leukocytes, eosinophils, peritoneal macrophages, bone marrow cells [160], liver Kupffer cells [161], thyroid follicular epithelial cells [162], salivary gland [162], uterus endometrium [163], and colon crypts of Lieberkuln mucous secretory cells [164] and Zymbal's gland [165]. Lactoperoxidase is also active in the submaxillary, sublingual, Harderian [166], and mammary gland [167]. It is unlikely that tetramethylbenzidine will be a useful noncarcinogenic substitute for cytochemists because it does not polymerize and therefore diffuses freely. Polymerization presumably involves the ortho ring position. Thus, tetramethylbenzidine is unsatisfactory for determining hemoglobin on polyacrylamide gels at acidic pH [168]. Diaminobenzidine may remain the benzidine derivative of most use to cytochemists because of its ready polymerization on oxidation. It can be readily used to locate endogenous and exogenous peroxidase activity in a variety of cell types. Its phenylenediamine character also makes it readily reduced by cytochrome c, and it will thus stain cytochrome oxidase [170]. Mitochondria should therefore be important intracellularly in the metabolic oxidation of di-aminobenzidine [171] as an alternative to the peroxidase pathway.

Prostaglandin synthase was shown to have peroxidase activity [172] and catalyze xenobiotic cooxidation [173]. Zenser et al. [154] later demonstrated that the prostaglandin synthase activity of rabbit renal medullary preparations with arachidonate as substrate catalyzes the metabolism of benzidine to products that are not organic extractable and bind covalently to DNA or RNA [174]. Kadlubar et al. [109] found that the prostaglandin synthase activity of ram seminal vesicles catalyzes the binding to DNA and protein of benzidine 300 times more effectively than the potent carcinogen 2-aminofluorene. We have obtained similar results with pig or dog bladder microsomes, so that the macromolecular binding could be important in explaining nephrotoxicity (tubule damage) or bladder cancer in dogs and rabbits. In the case of the cooxidation of 3,5,3′,5′-tetramethylbenzidine by prostaglandin synthase, ESR has shown the formation of a cation-radical, presumably in equilibrium with the blue-colored charge-transfer complex with the diimine [175]. However, in the case of benzidine no such radical is observed. The microsomal lipids and arachidonic acid facilitate the charge-transfer complex formation and stabilize it, so that there is no cation-radical ESR signal. Up to 25% of the product is diimine (trapped with phenol as a pink indoaniline), and the major organic soluble product is azobenzidine (1%) [176]. Similar products are probably formed with tetramethylbenzidine [175]. However, benzidinediimine at neutral pH undergoes polymerization to a brown polymer (benzidine-derived *melanin*). Further oxidation of the diimine may be involved in this process. However, the narrow-line ESR signal seen during the oxidation of benzidine by prostaglandin synthase–arachidonate at pH 7 is believed to be the polymer before precipitating from the aqueous solution.

Monoacetylbenzidine is oxidized by the peroxidase–H_2O_2 system but not to colored products [175]. However, benzidine is oxidized much more rapidly. The high acetylating activity of liver cytosol and the lack of peroxidase activity in liver cells other than Kupffer cells favors the N-hydroxylation pathway of N-acetylbenzidine for DNA adduct formation in the liver. It is also clear that 5–15% of aromatic amine carcinogens are excreted unchanged in the urine [140–142] and could pose a carcinogenic hazard to the bladder as well as other organs as a result of reabsorption of aromatic amines by the bladder [119, 178]. N-Acetylation may be involved in carcinogenesis of the liver but is probably not involved in carcinogenesis of the urinary bladder because arylamines, for example, aminobiphenyl and 2-NA, usually cause bladder, but not liver cancer in dogs that are incapable of N-acetylation. Furthermore, some acetylarylamines (e.g., acetylnaphthylamine) do not induce bladder tumors in dog, presumably because of the very low deacetylase activity of dog liver microsomes, and N-hydroxyacetyl-2-naphthylamine is excreted [152]. The dog, however, has a good ability to deacetylate acetylarylamines, and AAF and acetylaminobiphenyl can cause bladder and liver cancer in dogs. The rate of N-acetylation in humans and rabbits reflects genetic polymorphism. Individuals are distributed bimodally according to their capacity for N-acetylation as either rapid or slow acetylators of these drugs [179]. Rapid acetylators may therefore be more susceptible to liver tumors, whereas slow acetylators may be more susceptible to bladder tumors [179]. A survey of 20 inbred mouse strains also revealed a 20-fold variation in the N-acetylation of benzidine by liver N-acetyltransferase, with three strains being slow acetylators [180].

Bovine bladder mucosa microsomes have been shown to contain cytochrome P-450 (0.13 nmol/mg protein) and catalyze the N-hydroxylation of the bladder carcinogen 4-aminobiphenyl [153]; direct activation of this carcinogen in the target bladder tissue seems a more likely mechanism than the N-hydroxylation and glucuronidation by the liver followed by release of the free N-hydroxyarylamine in the mildly acidic environment of urine.

We have shown that benzidinediimine readily reacts with double-stranded DNA to form a covalently bound brown product. 3,3,5,5'-Tetramethylbenzidinediimine did not bind to DNA, so that it is likely that the benzidine–DNA adduct involves the initial binding of guanine or adenine to the o-position of the benzidine. In Fig. 2 the arylamine–DNA adducts that have been identified are outlined. The o-position of arylamines binds to the 2-NH_2 group of guanine or the 6-NH_2 group of adenine and the adducts are usually more persistent than the adducts formed between the amine group of the arylamine and the C-8 position of guanine. It is possible that the peroxidase-catalyzed benzidine–DNA adducts will be shown to be the same as the *in vivo* adducts.

We have also shown [107] that the peroxidase activity of liver cytochrome P-450 and cytochrome P-420 catalyzes extensive covalent binding to DNA and microsomal protein. The product responsible was benzidinediimine so that extensive one-electron oxidation is involved. NADPH and microsomes catalyze some microsomal protein binding but no DNA binding.

The most plausible theory is that in the liver the benzidine is N-acetylated and is readily N-hydroxylated. In nonhepatic tissues the unacetylated benzidine is readily oxidized by peroxidase or prostaglandin synthase to a diimine, which forms persistent DNA adducts. Thus, N-acetyl-benzidine is more mutagenic than benzidine in liver microsome-mediated mutagenicity assays [210], but the opposite is true with a prostaglandin synthase–arachidonate-mediated activation [209].

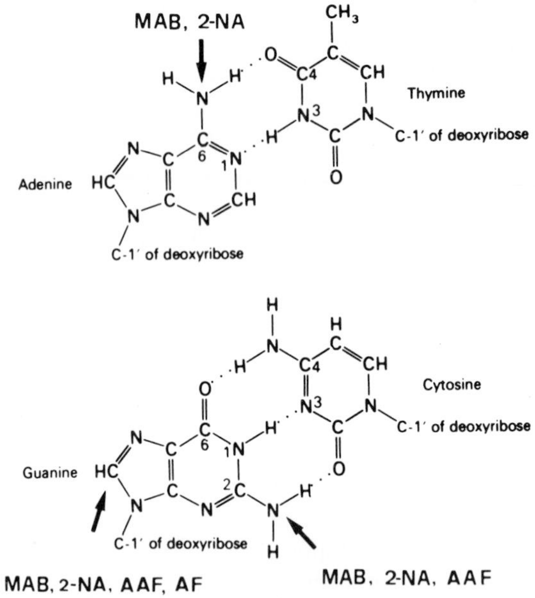

Fig. 2. Nucleic acid adduct-forming sites with carcinogenic arylamines (AF, aminofluorene).

IX. PHENYLENEDIAMINES

Phenylenediamines are oxidized by peroxidases or ceruloplasmin via a one-electron oxidation to a cation-radical [181], which is further oxidized or disproportionates to form the very reactive diamagnetic benzoquinonediimine. The latter is then oxidized to trimers known as Brandrowski's base.

By far the largest of the domestically produced phenylenediamines is 2,4-diaminotoluene, which is used to make toluene diisocyanate, a cross-linking agent used in the synthesis of polyurethanes. They are also used in the dyestuff industry. 2,4-Diaminotoluene induces liver tumors in rats and mice and mammary tumors in rats but not bladder tumors at doses of 0.01% of the diet administered for 80 weeks [182]. It also readily induces methemoglobin, possibly due to 5-hydroxy and 6-hydroxy metabolites [183]. N-Acetylation readily occurs with liver cytosol and in vivo]184]. 2,4-Diaminotoluene in vivo results in covalent binding to rat liver nuclear RNA and proteins [185]. No evidence of N-hydroxylation has yet been found. Binding to microsomal protein and RNA was found after in vitro incubation, but no binding to DNA was found in vivo [186]. We have found, however, that peroxidase–H_2O_2 readily catalyzes the binding of 2,4-diaminotoluene to DNA and protein to form colored adducts. Mammary tumor development could therefore involve mammary peroxidase. Liver tumor formation presumably involves the microsomal mixed-function oxidase. However, rather than an N-hydroxylation activation mechanism, a one-electron oxidation to a diimine seems more likely because binding is greatly enhanced by substituting NADPH with cumene hydroperoxide. Indeed, with the latter system exogenous DNA became colored [107]. NADPH reduces the diimine and cation-radicals and could explain the lack of binding to exogenous DNA by the mixed-function oxidase when NADPH is used. Prostaglandin synthase–arachidonate also activates 2,4-diaminotoluene, 2,5-diaminotoluene, 4-methoxy-m-phenylenediamine, and 4-chloro-o-phenylenediamine to products that cause DNA strand breakage in cultured human fibroblasts [187].

Permanent and semipermanent hair dye formulations and synthetic rinses also consist of a mixture of phenylenediamines, aromatic nitro derivatives, and phenols that are mixed with H_2O_2 just before use. Over 50% of 25 hair dye compounds are mutagenic in bacteria [188]. Some of these compounds are mutagenic in yeast and *Drosophilia* and induce chromosomal damage in cultured Chinese hamster cells [188]. Epidemiological studies are still not clear as to whether there is a positive association between hair dye use and bladder cancer [189]. However, it is clear that dyeing the hair can result in the skin absorption of the dye and excretion in the urine [190]. Several derivatives are also carcinogenic in rats and mice fed at 0.2 to 1% in the diet. Thus, 4-chloro-o-phenylenediamine induces rat and mouse bladder cancer, whereas 4-chloro-m-phenylenediamine induces liver carcinoma and adenoma in mice and adrenal pheochromocytoma in male rats [191]. O-Phenylenediamine induces hepatic tumors in rats and mice [192], which suggests that the electron-attracting chlorine may hinder acetylation as well as increase the level and stability of the ultimate or proximate carcinogen. In support of this theory both m- and p-phenylenediamines are inactive in rats and mice [191].

2,4-Diaminoanisole (2,4-DAA), a black hair dye component, is unusual in inducing goiter in rats. It also induces neoplasms of the thyroid (follicular cell carcinomas), skin, Zymbal's gland, and mammary gland of rats at doses of 0.5% fed for 80 weeks [182, 183]. A dose-related deposition of brown pigment occurs within the follicular cells of the thyroid glands [193]. It has been proposed that oxidative o-demethylation with subsequent semiquinone formation and N-hydroxylation followed by nitrenium ion formation could explain the bioactivation of 2,4-DAA [186]. However, although o-demethylation to 2,4-diacetylaminophenol, ring hydroxylation to 5-hydroxy, and ω-oxidation of the N-acetyl have been demonstrated with the rat *in vivo* [194], there is no evidence of N-hydroxylation. No unchanged compound is found in the urine. Phenobarbital increases mutagenicity and binding of 2,4-DAA to liver and kidney proteins *in vitro* and *in vivo*, apparently as a result of increased monoacetylation or o-demethylation but not ring hydroxylation or ω-oxidation [195]. Ring-labeled 2,4-DAA bound three- to fourfold more than methyl-labeled 2,4-DAA, indicating that o-demethylation is important in metabolic activation. Phenobarbital and naphthoflavone induction increases binding up to 70%. Methylcholanthrene pre-treatment of mice also causes an increase in 2,4-DAA mutagenicity with the S9 fractions as a result of cytochrome P-450-catalyzed metabolism [196, 197]. Similarly, nuclei isolated from α-naphthoflavone- induced rats are more effective in mutagenic activation of 2,4-DAA [198] and in the binding of 2,4-DAA to nuclear proteins. It is suggested that N-hydroxylation is induced and is responsible for the activation. However, no binding to nuclear DNA occurs [198]. Furthermore, no binding to DNA occurs *in vivo* [186] or with a microsomal activation system [196, 199]. It is therefore possible that the bacteria used in the mutagenesis assay carry out the final activation inside the bacteria. Binding to microsomal RNA occurs [199], but there is no binding to liver RNA (186).

It is likely that the phenylenediamines are not mutagenic but become so after oxidation. These oxidation products are also present as impurities in the phenylenediamines used. Mutagenic products of *p*-phenylenediamine include Brandrowski's base [200] or the resorcinol–phenylenediamine adduct 2-hydroxyindoaniline [188]. The urines of rats treated topically with mutagenic hair dyes are also mutagenic [188]. The most mutagenic phenylenediamines in order of mutagenicity are 2,4-DAA, 4-nitro-*o*-phenylenediamine, and 2-nitro-*p*-phenylenediamine. Weaker mutagens include *m*-phenylenediamine, 2,5-DAA, and 2,4-diaminotoluene. Hydrogen peroxide [200], light [201], or aging in dimethyl sulfoxide [202] greatly enhances the mutagenicity of several phenylenediamines, presumably as a result of oxidation or enhanced autoxidation. The mutagenicity has been attributed to the trimer Brandrowski's base. It is interesting, however, that the S9 fraction is required for the mutagenicity of the mixtures with 2,4-DAA, 2,5-diaminotoluene, 2,5-DAA [200], *o*-phenylenediamine, 2,4-diaminotoluene, *p*-phenylenediamine, or *p*-phenylenediamine–resorcinol [203], which suggests that S9 activates the aforementioned trimer products. However, in the case of 2-nitro-*p*-phenylenediamine and 4-nitro-*o*-phenylenediamine, the S9 fraction is not required for mutagenicity so that bacterial nitroreductases may be involved [202]. Nor does H_2O_2 increase their mutagenicity. The nitro-phenylenediamine derivatives are also teratogenic.

We have shown that the peroxidase–H_2O_2 system at pH 4 oxidizes *p*-phenylenediamines to products that readily bind to DNA to form purple DNA adducts, suggesting that indamines bind to DNA [107]. The *p*-phenylenediamine products in Fig. 3 have been synthesized [203–206] and shown to form a colored DNA adduct after further peroxidase–H_2O_2 activation. This suggests that it is the imine derivatives of these trimers that bind to form a purple DNA adduct. Oxidized 2,4-Diaminoanisole binds covalently to DNA at pH 5.0, forming a purple-gray DNA adduct. 2,4-Diaminoanisole and 2,5-DAA are also oxidized to mutagenic products by prostaglandin synthase–arachidonate [209].

The intracellular oxidation of phenylenediamines in the liver may be due to an oxidase in the outer membrane of the mitochondria as well as in the endoplasmic reticulum [207]. Our research indicates that cytochrome P-450 has some phenyl-

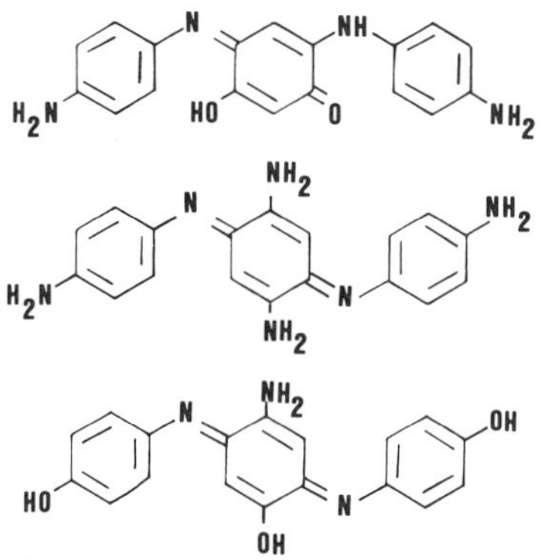

Fig. 3.   *p*-Phenylenediamine oxidation products.

enediamine oxidase activity, and the peroxidase activity of cytochrome *P*-450 rapidly catalyzes phenylenediamine oxidation to the cation-radical [*208*].

X. AMINOPHENOL DERIVATIVES

The nephrotoxicity resulting in renal tubular necrosis caused by the analgesic phenacetin or its metabolite paracetamol has been attributed to *p*-aminophenol. It is also a potent teratogen in the hamster [*211*]. Methemoglobinemia induced by aniline has been attributed to *p*-aminophenol. *p*-Aminophenol administered to rats results in decreased cytochrome and GSH levels in the kidney and binding to kidney proteins. This binding is not affected by inducers or inhibitors of the microsomal mixed-function oxidase systems. Similar results were obtained with microsomes *in vitro*. The binding is enhanced by prior autoxidation of *p*-aminophenol [*212*]. The mechanism could involve prostaglandin synthase located in the renal medulla, which can oxidize *p*-aminophenol [*213*]. By means of ESR spectroscopy, peroxidase has been shown to oxidize *p*-aminophenol readily to a phenoxyl free radical [*214*], which is stable only at alkaline pH [*181*]. A transient purple compound, a red-brown polymeric precipitate, and an organically soluble orange indophenol (2% yield) are also formed. It was suggested that *p*-benzo-quinoneimine is responsible for the protein binding [*214*] (Fig. 4). We have also found extensive covalent binding to DNA resulting in a purple adduct at pH 4 and a pale brown adduct at pH 7, which suggests that the indoaniline binds to DNA. Peroxidase-catalyzed oxidation of the Brandrowski's base formed from *p*-aminophenol also results in DNA binding, suggesting that the imine derivative of these trimers reacts with DNA.

p-Hydroxyacetanilide (paracetamol, acetaminophen) is an analgesic and anti-pyretic drug that is widely used as an aspirin substitute. However, it produces a centrilobular liver necrosis in man when taken in overdoses, probably as a result

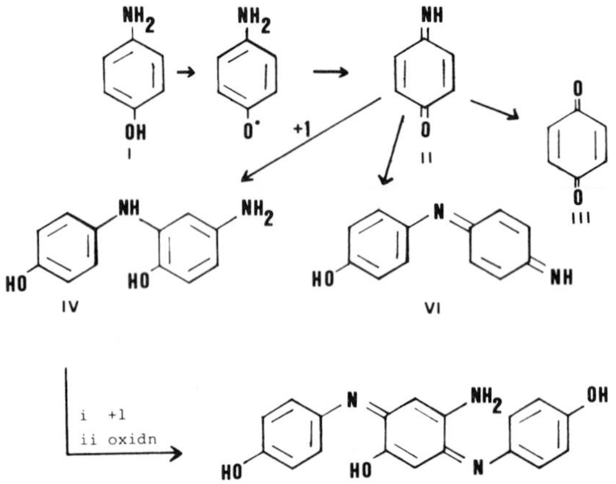

Fig. 4.   p-Aminophenol oxidation products.

of GSH depletion, which allows extensive covalent binding to proteins. The severity of the necrosis caused by paracetamol correlates with the amount of covalent binding [215]. Indeed, N-acetylcysteine has proved to be an effective antidote to the hepatotoxic effect of a paracetamol overdose in humans, presumably by increasing hepatic GSH levels [216]. GSH-Filled liposomes also protect against paracetamol-induced liver necrosis in mice [217]. The metabolic activation of paracetamol was originally postulated to occur via N-oxidation to the N-hydroxy derivative, which dehydrated to N-acetyl-p-benzoquinoneimine. However, microsomes catalyze the N-hydroxylation of phenacetin but not of paracetamol [44]. Furthermore, the N-hydroxyparacetamol derivative did not react with GSH [218]. The quinoneimine synthesized by lead tetraacetate oxidation of paracetamol, however, did react with GSH to form the adduct [219]. Instead, paracetamol may behave like p-aminophenol in which the quinoneimine is formed by a one-electron oxidation [220] (Fig. 5). It is not clear whether the GSH can also react with the phenoxyl radical to form the GSH adduct. It has also been shown that kidney medulla prostaglandin synthase–arachidonate catalyzes the formation of the GSH adduct and could explain the necrosis of the kidney medulla and renal papillary necrosis after chronic paracetamol use. However, the rates of adduct formation are low [221, 222]. Peroxidase–H_2O_2 also readily oxidizes paracetamol to give an ESR signal tentatively identified as the phenoxyl radical [155] but more likely a polymer signal [214]. The mechanism for quinoneimine formation in the liver could involve the one-electron oxidation pathway for cytochrome P-450 in the mixed-function oxidase activity. Evidence for this is that, with liver microsomes, a fivefold increase in GSH adduct formation occurs with cumene hydroperoxide compared with that occurring with NADPH [221]. The mixed-function oxidase seems to be involved in the liver because phenobarbital or MC dramatically increases the sensitivity up to 28-fold

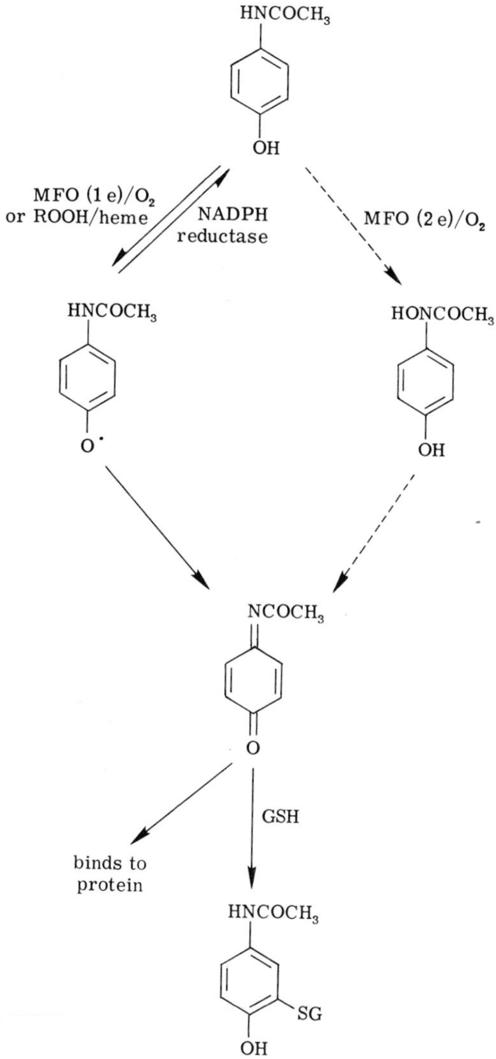

Fig. 5. Mechanism for the metabolic formation of paracetamol–GSH adduct (MFO, mixed-function oxidose).

for the hepatotoxic effects of paracetamol [44]. Inhibitors of drug metabolism such as piperonyl butoxide [44] or cobaltous chloride decrease the toxicity [223]. The quinoneimine can be reduced back to paracetamol with ascorbate, and ascorbate prevents the protein binding [224].

p-Ethoxyacetanilide (phenacetin) is another analgesic and antipyretic found in a number of nonprescription compounds. Long-term consumption leads to renal papillary necrosis and tumors of the renal pelvis and bladder in humans [225]. It induces tumors in bladder, mammary gland, and ear duct in rats [225]. As described earlier the necrosis may be due to a prostaglandin synthase activation of the metabolite paracetamol. However, the carcinogenesis may involve N-hydroxylation to N-hydroxyphenacetin [229], which can become mutagenic and bind to tRNA following activation by a cytosolic aryltransferase [226]. The microsomal-catalyzed mutagenicity and tRNA binding involve a deacetylase to form N-hydroxyphenetidine, which is converted to p-nitrosophenetole, the responsible reactive species [226]. Sulfate conjugation also leads to tRNA and protein binding, and N-acetylbenzoquinoneimine has been suggested to be responsible for protein binding [227]. Microsomes also catalyze the binding of labeled phenacetin to DNA, RNA, and protein [228]. Acetyl[^{3}H]phenacetin binds, indicating that deacetylation to phenetidine is not involved. Another major metabolite of phenacetin in man is p-phenetidine [230]. Prostaglandin synthase and peroxidases have been shown to cause protein binding by paracetamol [230] and p-phenetidine [231]. In addition, DNA strand breaks are found in cultured human skin fibroblasts incubated with vesicular gland microsomes containing high prostaglandin synthase activity [213]. Our results have shown that most of the extensive binding to DNA by p-phenetidine, p-toluidine, or p-anisidine is noncovalent with peroxidase activation [107]. Prostaglandin synthase–arachidonate also does not catalyze p-phenetidine binding to DNA [109]. However, liver microsomes catalyze the binding of phenetidine to DNA [228], which suggests that p-nitrophenetole formed from N-hydroxyphenetidine is responsible.

o-Aminophenols may play a role in carcinogenesis. They are formed by the metabolic ring hydroxylation of carcinogenic arylamines. At one time the o-hydroxylation hypothesis was thought to be important in chemical carcinogenesis, but the two o-aminophenols of 2-acetamidofluorene (2-amino-1-fluorenol and 2-amino-3-fluorenol) are not carcinogenic on bladder implantation. However, the 2-amino-1-naphthol metabolite of 2-NA and 3-hydroxy-4-aminobiphenyl metabolite of 4-aminobiphenyl are as carcinogenic as the N-hydroxy derivative on implantation into mouse bladder [232]. Furthermore, the dog excretes 3.5–13% of ingested 2-NA as 2- amino-1-naphthyl-glucuronide [233]. o-Aminophenols readily combine with protein after their oxidation to o-quinoneimines [234]. Autoxidized 2-amino-1-naphthol readily binds to proteins of the rat bladder, forming a purple 2-naphthoquinone–protein derivative. Either naphthoquinoneimine or naphthoquinone could be responsible for the protein binding [235]. The autoxidation is catalyzed by cytochrome c–cytochrome oxidase or mitochondria. No adduct with yeast RNA is obtained, however, with autoxidizing 2-amino-1-naphthol, although binding to bovine serum albumin readily occurs [236]. However, a decrease in the priming activity of DNA with RNA polymerase is found after incubation with 2-amino-1-naphthol [237]. The formation of active oxygen species could explain the latter effect. Furthermore, we have found that the peroxidase–H_2O_2 catalyzes a small amount of irreversible binding of 2-amino-1-naphthol to DNA [107].

NOTE ADDED IN PROOF

Recently, intramolecular isotope effects have been determined for the N-demethylation of N-methyl-N-trideuteriomethylaniline catalyzed by cytochrome P-450 and several peroxidases [238]. It has also been confirmed that ^{16}O from the oxidant rather than ^{18}O from H_2O is incorporated into the product of the cytochrome P-450-catalyzed N-demethylation of N-methylcarbazole when the reaction is supported by hydroperoxides or by NADPH, reductase, and O_2 [239]. These results could indicate that peroxidases catalyze the N-demethylation of N,N-dimethylaniline via the hydrolysis of a methimine formed by α-carbon hydrogen atom abstraction, while cytochrome P-450 proceeds via α-carbon deprotonation from an aminium cation radical to a neutral carbon radical. The latter recombines with the enzyme-bound oxygen moiety to form a carbinolamine which readily N-demethylates. At low peroxidase concentrations (<0.15 μg/ml), N-demethylation has been reported to result in stoichiometic quantities of N-methylaniline [240]. We have found however that N,N,N^1,N^1-tetramethylbenzidine becomes the major product at slightly higher peroxidase concentration and suggests that the dimerization of the radicals occurs if they are formed fast enough [109]. GSH has also been readily shown to be oxidized during the peroxidase-catalyzed N-demethylation while inhibiting the rate of HCHO formation [241]. The lack of effect of GSH on cytochrome P-450 catalyzed N-demethylation suggests that the methimine is not involved in the N-demethylation. The methimine, however, may be involved in the GSH oxidation during peroxidase-catalyzed N-demethylation. Evidence for a free radical mechanism for cytochrome P-450 catalyzed o-demethylation of anisole derivative has also been reported [242]. A large intramolecular kinetic isotope effect was observed for p-methoxy-d_3-anisole. The o-demethylation may proceed by an initial hydrogen abstraction to an alkyl radical with subsequent hemiacetal formation.

REFERENCES

1. E. C. Miller and J. A. Miller, *Cancer* **47**, 1055 (1982).
2. M. Gronow, *Chem.-Biol. Interact.* **29**, 1 (1980).
3. J. Cairns, *Nature (London)* **289**, 353 (1981).
4. R. A. Floyd, in "Free Radicals in Biology" (W. Pryor, ed.), Vol. 4, p. 187. Academic Press, New York, 1980.
5. R. Mason, in "Free Radicals in Biology" (W. Pryor, ed.), Vol. 5, p. 161. Academic Press, New York, 1982.
6. J. E. Packer, T. F. Slater, and R. L. Wilson, *Life Sci.* **23**, 2617 (1978).
7. W. G. Harrelson and R. P. Mason, *Mol. Pharmacol.* **22**, 239 (1982).
8. R. C. Baldwin, A. Pasi, J. T. MacGregor, and C. H. Hine, *Toxicol. Appl. Pharmacol.* **32**, 298 (1975).
9. H. Kuthan, V. Ullrich, and R. W. Estabrook, *Biochem. J.* **203**, 551 (1982).
10. E. Dybing, S. D. Nelson, J. R. Mitchell, H. A. Sasame, and J. R. Gillette, *Mol. Pharmacol.* **12**, 911 (1976).
11. H. M. Bolt and H. Kappus, *Biochem. Biophys. Res. Commun.* **70**, 1157 (1976).
12. S. Hesse, M. Mezger, and T. Wolff, *Chem.-Biol. Interact.* **20**, 355 (1978).
13. A. I. Cederbaum and G. Cohen, *Arch. Biochem. Biophys.* **204**, 397 (1980).

14. E. G. Janzen, in "Free Radicals in Biology" (W. Pryor, ed.), Vol. 4, p. 115. Academic Press, New York, 1980.
15. C. Little and P. J. O'Brien, *Can. J. Biochem.* **47**, 493 (1969).
16. E. G. Hrycay and P. J. O'Brien, *Arch. Biochem. Biophys.* **147**, 14 (1971).
17. P. J. O'Brien, *Pharmacol. Ther., Part A* **2**, 517 (1978).
18. P. J. O'Brien, in "Lipid Peroxides in Biology and Medicine" (K. Yagi, ed.), p. 137. Academic Press, New York, 1983.
19. F. F. Kadlubar, J. A. Miller, and E. C. Miller, *Cancer Res.* **36**, 1196 (1976).
20. P. J. O'Brien and C. Nagata, unpublished.
21. B. W. Griffin, C. Marth, Y. Yasukochi, and B. S. S. Masters, *Arch. Biochem. Biophys.* **175**, 543 (1980).
22. J. Capdevila, R. W. Estabrook, and R. A. Prough, *Arch. Biochem. Biophys.* **200**, 186 (1980).
23. R. Renneberg, J. Capdevila, N. Chacos, R. W. Estabrook, and R. A. Prough, *Biochem. Pharmacol.* **30**, 843 (1981).
24. P. J. O'Brien and A. D. Rahimtula, in "Microsomes, Drug Oxidations, and Chemical Carcinogenesis" (A. H. Conney, M. J. Coon, R. W. Estabrook, H. V. Gelboin, J. R. Gillette, and P. J. O'Brien, eds.), p. 253. Academic Press, New York, 1980.
25. P. R. Ortiz de Montellano, K. L. Kunze, H. S. Beilan, and C. Wheeler, *Biochemistry* **21**, 1331 (1982).
26. F. P. Guengerich and T. W. Strickland, *Mol. Pharmacol.* **13**, 993 (1977).
27. P. R. Ortiz de Montellano and K. L. Kunze, *J. Biol. Chem.* **255**, 5578 (1980).
28. P. R. Ortiz de Montellano and B. A. Mico, *Arch. Biochem. Biophys.* **206**, 43 (1981).
29. P. R. Ortiz de Montellano and K. L. Kunze, *Arch. Biochem. Biophys.* **209**, 710 (1981).
30. P. R. Ortiz de Montellano, H. S. Beilan, K. L. Kunze, and B. A. Mico, *J. Biol. Chem.* **256**, 4395 (1981).
31. O. Augusto, K. L. Kunze, and P. R. Ortiz de Montellano, *J. Biol. Chem.* **257**, 11288 (1982).
32. R. Kato, A. Takanaka, and H. Shoji, *Jpn. J. Pharmacol.* **19**, 315 (1969).
33. R. N. Hines and R. A. Prough, *J. Pharmacol. Exp. Ther.* **214**, 80 (1980).
34. O. Augusto, P. R. Ortiz de Montellano, and A. Quintanilha, *Biochem. Biophys. Res. Commun.* **101**, 1324 (1981).
35. O. Augusto, K. L. Kunze, and P. R. Ortiz de Montellano, *J. Biol. Chem.* **257**, 6231 (1982).
36. H. G. Jonen, J. Werringloer, R. A. Prough, and R. W. Estabrook, *J. Biol. Chem.* **257**, 4404 (1982).
37. R. A. Prough, P. C. Freeman, and R. N. Hines, *J. Biol. Chem.* **256**, 4178 (1981).
38. Y. Watanabe, T. Iyanagi, and S. Oae, *Tetrahedron Lett.* p. 3685 (1980).
39. E. J. Rauckman, G. M. Rosen, and J. Cavagnaro, *Mol. Pharmacol.* **21**, 458 (1982).
40. C. Nagata, M. Kodama, Y. Ioki, and T. Nagata, in "Free Radicals and Cancer" (R. A. Floyd, ed.), p. 1. Dekker, New York, 1982.
41. T. Nakayama, T. Kimura, M. Kodama, and C. Nagata, *Gann* **73**, 382 (1982).
42. J. T. Groves, O. F. Akinbote, and G. E. Avaria, in "Microsomes, Drug Oxidations and Chemical Carcinogenesis" (M. J. Coon, A. H. Conney, R. W. Estabrook, H. V. Gelboin, J. R. Gillette, and P. J. O'Brien, eds.), p. 253. Academic Press, New York, 1980.
43. R. F. White and M. J. Coon, *Annu. Rev. Biochem.* **49**, 315 (1980).
44. T. L. Macdonald, L. T. Burka, S. Tracy Wright, and F. Guengerich, *Biochem. Biophys. Res. Commun.* **104**, 620 (1982).
45. J. A. Hinson, L. R. Pohl, and J. R. Gillette, *Life Sci.* **24**, 2133 (1979).
46. P. C. Howard and F. A. Beland, *Biochem. Biophys. Res. Commun.* **104**, 727 (1981).
47. M. R. Boyd, A. W. Stiko, and H. Sasame, *Biochem. Pharmacol.* **28**, 601 (1979).
48. K. Tatsumi, S. Kitamura, and H. Yoshimura, *Chem. Pharm. Bull.* **25**, 2948 (1977).
49. S. Swaminathan, G. M. Lower, and T. T. Bryan, *Cancer Res.* **42**, 4479 (1982).
50. M. B. Mattammal, T. V. Zenser, and B. B. Davis, *Cancer Res.* **41**, 4961 (1981).

51. T. V. Zenser, M. B. Mattammal, and B. B. Davis, *J. Lab. Clin. Med.* **96**, 425 (1980).
52. M. B. Mattammal, T. V. Zenser, and B. B. Davis, *Carcinogenesis* **3**, 1339 (1982).
53. T. E. Rollins and W. L. Smith, *J. Biol. Chem.* **255**, 4872 (1980).
54. W. W. Brown, T. V. Zenser, and B. B. Davis, *Am. J. Physiol.* **239**, 452 (1980).
55. S. N. Cohen, T. V. Zenser, G. Murazaki, S. Fukushawa, M. B. Mattammal, N. S. Rapp, and B. B. Davis, *Cancer Res.* **41**, 3355 (1981).
56. C. Y. Wang, S. Hayashida, A. M. Pamukai, and G. T. Bryan, *Cancer Res.* **36**, 1551 (1976).
57. R. W. Wise, T. V. Zenser, and B. B. Davis, *Cancer Res.* **43** 1518 (1983).
58. T. V. Zenser, M. B. Mattammal, and B. B. Davis, *Cancer Res.* **40**, 114 (1980).
59. M. Dershivitz and R. F. Novak, *J. Biol. Chem.* **257**, 75 (1982).
60. H. A. Sasame and M. R. Boyd, *Life Sci.* **24**, 1091 (1979).
61. S. Saito and H. A. Itano, *Proc. Natl. Acad. Sci. U.S.A.* **78**, 5508 (1981).
62. H. A. Itano and J. L. Matteson, *Biochemistry* **21**, 2421 (1982).
63. P. R. Ortiz de Montellano and L. L. Kunze, *J. Am. Chem. Soc.* **103**, 6534 (1981).
64. H. Loos, D. Roos, R. Weening, and J. Houwerzil, *Blood* **48**, 53 (1976).
65. H. Z. Griffin and E. V. Allen, *Am. J. Med. Sci.* **185**, 1 (1933).
66. H. A. O. Hill and P. J. Thomally, *Can. J. Chem.* **60**, 1528 (1982).
67. S. Wolter and N. Frank, *Chem.-Biol. Interact.* **42**, 335 (1982).
68. A. K. Bahri, C. S. Chiang, and J. A. Timbrell, *Toxicol. Appl. Pharmacol.* **60**, 561 (1981).
69. A. Hawks and P. N. Magee, *Br. J. Cancer* **30**, 440 (1974).
70. L. C. Boffa, R. J. Guoss, and V. G. Allfrey, *Cancer (Amsterdam)* **42**, 382 (1982).
71. W. S. Allison, L. C. Swain, S. M. Tracy, and L. V. Benitez, *Arch. Biochem. Biophys.* **155**, 400 (1973).
72. S. D. Nelson, J. R. Mitchell, J. A. Timbrell, W. R. Snodgrass, and G. B. Corcoran, *Science* **193**, 901 (1976).
73. E. Freese, S. Sklarow, and E. B. Freese, *Mutat. Res.* **5**, 343 (1968).
74. K. J. Rogers and A. E. Peggy, *Cancer Res.* **37**, 4082 (1977).
75. C. M. King and W. T. Allaben, *in* "Enzymatic Basis of Detoxification" (W. B. Jakoby, ed.), Vol. 2, p. 187. Academic Press, New York, 1980.
76. R. A. Cardona and C. M. King, *Biochem. Pharmacol.* **25**, 1051 (1976).
77. C. B. Frederick, J. B. Mays, D. M. Ziegler, F. P. Guengerich, and F. F. Kadlubar, *Cancer Res.* **42**, 2671 (1982).
78. D. T. Beranek, G. L. White, R. H. Heflich, and F. A. Beland, *Proc. Natl. Acad. Sci. U.S.A.* **79**, 5175 (1982).
79. J. A. Miller, *Cancer Res.* **30**, 559 (1970).
80. J. H. Weisburger, R. S. Yamamoto, G. M. Williams, P. H. Grantham, T. Matsushima, and E. K. Weisburger, *Cancer Res.* **32**, 491 (1972).
81. C. C. Irving, D. H. Janss, and L. T. Russell, *Cancer Res.* **31**, 387 (1971).
82. P. Parodi, S. De Flora, M. Cavanna, A. Pino, L. Robbiano, C. Bennicelli, and G. Brambilla, *Cancer Res.* **41**, 1469 (1981).
83. J. G. Lewis and J. A. Swenberg, *Cancer Res.* **42**, 89 (1982).
84. R. F. Whiting, L. Wei, and H. F. Stich, *Mutat. Res.* **62**, 605 (1979).
85. R. F. Whiting, L. Wei, and H. F. Stich, *Biochem. Pharmacol.* **29**, 842 (1980).
6. A. Quintero-Ruiz, L. L. Paz-Neri, and S. Villa-Trevino, *J. Natl. Cancer Inst.* **67**, 613 (1981).
87. R. A. Becker, L. R. Barrows, and R. C. Shank, *Carcinogenesis (London)* **2**, 1181 (1981).
88. L. R. Barrows and R. C. Shank, *Toxicol. Appl. Pharmacol.* **60**, 334 (1981).
89. J. R. DeBaun, E. C. Miller, and J. A. Miller, *Cancer Res.* **30**, 577 (1970).
90. H. A. J. Schut, P. J. Wirth, and S. S. Thorgeirsson, *Mol. Pharmacol.* **14**, 682 (1978).
91. R. M. E. Brouns, R. V. Van Doorn, R. P. Bos, L. J. S. Mulleners, and P. T. Henderson, *Toxicology* **19**, 67 (1981).
92. D. L. Stout, J. N. Baptist, T. S. Matney, and C. R. Shaw, *Cancer Lett.* **1**, 269 (1976).

93. M. A. Kaderbhai, T. K, Bradshaw, and R. B. Freedman, *Chem.-Biol. Interact.* **36**, 211 (1981).
94. S. Sakai, C. E. Reinhold, P. J. Wirth, and S. S. Thorgeirsson, *Cancer Res.* **36**, 2058 (1978).
95. L. S. Andrews, J. A. Hinson, and J. R. Gillette, *Biochem. Pharmacol.* **27**, 2399 (1978).
96. J. A. Hinson, L. S. Andrews, and J. R. Gillette, in "Free Radicals and Cancer" (R. A. Floyd, ed.), p. 423. Dekker, New York, 1982.
97. L. S. Andrews, J. M. Fysh, J. A. Hinson, and J. R. Gillette, *Life Sci.* **24**, 59 (1979).
98. P. J. Wirth, E. Dybing, C. von Bahr, and S. S. Thorgeirsson, *Mol. Pharmacol.* **18**, 117
99a. W. T. Allaben, C. C. Weiss, N. F. Fullerton, and F. A. Beland, *Carcinogenesis* **4**, 1067 (1983).
99b. W. T. Allaben, C. E. Weeks, C. C. Weiss, G. T. Burger, and C. M. King, *Carcinogenesis* **3**, 233 (1982).
99c. W. T. Allaben and C. C. Weiss, *Proc. Am. Assoc. Cancer Res.* **23**, Abstr. 321.
99d. F. A. Beland, D. T. Peranek, K. L. Dooley, R. H. Heflich, and F. F. Kadlubar, *Environ. Health Perspectives* **49**, 125 (1983).
99e. W. Marshall and P. J. O'Brien, in "Icosanoids and Cancer" (D. H. Thaler, ed.). Raven Press, New York (in press).
99f. P. J. O'Brien, *J. Am. Oil Chem. Soc.* (in press).
100. H. Bartsch and E. Hecker, *Biochim. Biophys. Acta* **237**, 567 (1971).
101. H. Bartsch, J. A. Miller, and E. C. Miller, *Biochim. Biophys. Acta* **273**, 40 (1972).
102. M. J. Hampton, R. A. Floyd, J. B. Clark, and J. H. Lancaster, *Mutat. Res.* **69**, 231 (1980).
103. R. A. Floyd, "Free Radicals and Cancer." Dekker, New York, 1982.
104. R. A. Floyd, *Radiat. Res.* **806**, 243 (1981).
105. R. A. Floyd, *Can. J. Chem.* **60**, 1577 (1982).
106. R. A. Floyd, L. M. Soong, and P. L. Culver, *Cancer Res.* **36**, 1510 (1976).
107. P. J. O'Brien, unpublished observations.
108. J. A. Boyd, D. J. Harvan, and T. E. Eling, *J. Biolog Chem.* **258**, 8246.
109. F. F. Kadlubar, C. B. Frederick, C. C. Weiss, and T. V. Zenser, *Biochem. Biophys. Res. Commun.* **108**, 253 (1982).
110. K. Yasukouchi, I. Taniguchi, H. Yamaguchi, K. Miyaguchi, and K. Horie, *Bull. Chem. Soc. Jpn.* **52**, 3208 (1979).
111. L. S. Andrews, L. R. Pohl, J. A. Hinson, C. L. Fisk, and J. R. Gillette, *Drug Metab. Dispos.* **7**, 296 (1979).
112. A. Stier, I. Reitz, and E. Sackmann, *Naunyn-Schmiedeberg's Arch. Pharmacol.* **274**, 189 (1972).
113. A. Stier, R. Class, A. Lucke, and I. Reitz, in "Free Radicals, Lipid Peroxidation and Cancer" (D. C. H. McBrien and T. F. Slater, eds.), p. 329. Academic Press, New York, 1982.
114. C. Patrianakos and D. Hoffman, *J. Anal. Toxicol.* **3**, 150 (1979).
115. F. F. Kadlubar, J. F. Anson, K. L. Dosley, and F. A. Beland, *Carcinogenesis (London)* **2**, 467 (1981).
116. T. Nakayana, T. Kimura, M. Kodama, and C. Nagata, *Gann* **73**, 382 (1982).
117. L. L. Poulson, B. S. S. Masters, and D. M. Ziegler, *Xenobiotica* **6**, 481 (1976).
118. H. Uehleke and E. Brill, *Biochem. Pharmacol.* **17**, 1459 (1968).
119. L. A. Oglesby, T. J. Flammang, D. L. Tullis, and F. F. Kadlubar, *Carcinogenesis* **2**, 15 (1981).
120. T. Kimora, M. Kodama, and C. Nagata, *Biochem. Pharmacol.* **28**, 557 (1979).
121. A. Haddow, R. J. C. Harris, G. A. R. Kon, and E. M. F. Roe, *Philos. Trans. R. Soc. London, Ser. A* **241**, 147 (1948).
122. J. C. Osborne, M. Metzler, and H. G. Neumann, *Cancer Lett.* **8**, 221 (1980).
123. F. A. Beland, D. L. Tullis, F. F. Kadlubar, K. M. Straub, and F. E. Evans, *Chem. Biol. Interact.* **31**, 1 (1980).

124. W. G. Tarpley, J. A. Miller, and E. C. Miller, *Cancer Res.* **40,** 2493 (1980).
125. T. Kimura, M. Kodama, and C. Nagata, *Carcinogenesis (London)* **3,** 1393 (1982).
126. F. F. Kadlubar, J. A. Miller, and E. C. Miller, *Cancer Res.* **36,** 1196 (1976).
127. S. Igarashi, H. Yonekawa, K. Kawajiri, and J. Watanabe, *Biochem. Biophys. Res. Commun.* **106,** 164 (1982).
128. K. Takanaka, P. J. O'Brien, Y. Tsuruta, and A. D. Rahimtula, *Cancer Lett.* **15,** 311 (1982).
129. S. Vasdev, Y. Tsuruta, and P. J. O'Brien, in "Prostaglandin and Cancer" (T. Powles, R. S. Bockman, K. V. Honn, and P. Ramwell, eds.), p. 163. Alan R. Liss, Inc., New York, 1981.
130. S. Vasdev and P. J. O'Brien, *Biochem. Pharmacol.* **31**(4), 607 (1982).
131. J. M. Lasker, K. Sivarajah, R. P. Mason, B. Kalyanaraman, M. B. Abou Donia, and T. E. Eling, *J. Biol. Chem.* **256,** 7764 (1981).
132. B. W. Griffin and P. L. Ting, *Biochemistry* **17,** 2206 (1978).
133. J. P. Shea, G. L. Valentine, and S. D. Nelson, *Biochem. Biophys. Res. Commun.* **109,** 231 (1982).
134. K. Sivarajah, J. M. Lasker, T. E. Eling, and M. B. Abou Donia, *Mol. Pharmacol.* **21,** 133 (1981).
135. B. B. Brodie, J. R. Gillette, and B. N. LaDu, *Annu. Rev. Biochem.* **27,** 427 (1958).
136. F. F. Kadlubar, K. C. Mortons, and D. M. Ziegler, *Biochem. Biophys. Res. Commun.* **54,** 1255 (1973).
137. G. D. Nordblom, R. E. White, and M. J. Coon, *Arch. Biochem. Biophys.* **175,** 524 (1976).
138. B. W. Griffin, D. K. Davis, and G. V. Bruno, *Bioorg. Chem.* **10,** 342 (1981).
139. T. J. Haley, *Clin. Toxicol.* **8,** 13 (1975).
140. J. C. Kennelly, P. J. Herzog, and C. N. Martin, *Carciongenesis (London)* **3,** 947 (1982).
141. J. E. Robens, G. S. Dill, J. M. Ward, J. R. Joiner, R. A. Griesemer, and J. F. Douglas, *Toxicol. Appl. Pharmacol.* **54,** 431 (1980).
142. L. K. Lowry, W. P. Tolos, M. F. Boeniger, C. R. Nony, and M. C. Bowman, *Toxicol. Lett.* **7,** 29 (1980).
143. R. P. Bos, R. M. E. Brouns, R. V. Van Doorn, J. L. G. Theuws, and P. T. Henderson, *Toxicology* **16,** 113 (1980).
144. C. N. Martin and J. C. Kennelly, *Carcinogenesis (London)* **2,** 307 (1981).
145. S. D. Vesslinovitch, K. V. N. Rao, and N. Mihailovich, *Cancer Res.* **35,** 2814 (1975).
146. L. J. Sciarini and J. W. Meigs, *Arch. Environ. Health* **2,** 584 (1961).
147. S. Belman, W. Troll, G. Teebor, and F. Mukai, *Cancer Res.* **28,** 535 (1968).
148. K. C. Morton, C. M. King, and K. P. Baetche, *Cancer Res.* **39,** 3107 (1979).
149. K. C. Morton, F. A. Beland, F. E. Evans, N. F. Fullerton, and F. F. Kadlubar, *Cancer Res.* **40,** 751 (1980).
150. K. C. Morton, C. Y. Wang, C. D. Garner, and T. Shira, *Carcinogenesis (London)* **2,** 747 (1981).
151. C. N. Martin, F. A. Beland, R. W. Roth, and F. F. Kadlubar, *Cancer Res.* **42,** 2678 (1982).
152. G. M. Lower and G. T. Bryan, *J. Toxicol. Environ. Health* **1,** 421 (1976).
153. J. M. Pouplo, J. L. Radomiski, and W. Lee Hearn, *Cancer Res.* **41,** 1306 (1981).
154. T. V. Zenser, M. B. Mattammal, and B. B. Davis, *J. Pharmacol. Exp. Ther.* **211,** 460 (1979).
155. S. D. Nelson, D. C. Dahlin, E. J. Rauckman, and G. M. Rosen, *Mol. Pharmacol.* **20,** 195 (1981).
156. R. M. E. Brouns, R. V. Van Doorn, and P. T. Henderson, *Toxicology* **23,** 235 (1982).
157. V. R. Holland, B. C. Saunders, F. L. Rose, and A. L. Walpole, *Tetrahedron* **30,** 3299 (1974).
158. P. D. Josephy, T. E. Eling, and R. P. Mason, *J. Biol. Chem.* **257,** 3669 (1982).
159. A. F. Welton and S. D. Aust, *Biochem. Biophys. Res. Commun.* **56,** 898 (1984).
160. M. E. Bentfeld, B. A. Nichols, and D. F. Bainton, *Anat. Rec.* **187,** 219 (1977).
161. H. D. Fahimi, *J. Cell Biol.* **47,** 247 (1970).
162. J. M. Strum and M. J. Karnovsky, *J. Cell Biol.* **44,** 655 (1970).

163. J. Brokelman and D. W. Fawcett, *Biol. Reprod.* **1,** 59 (1969).
164. M. A. Venkatachalam, M. H. Soltani, and H. D. Fahimi, *J. Cell Biol.* **46,** 168 (1970).
165. J. C. Osborne, M. Metzler, and H. G. Neumann, *Cancer Lett.,* **8,** 221 (1980).
166. M. Morrison and P. Z. Allen, *Science,* **152,** 1626 (1966).
167. W. A. Anderson, J. Trantalis, and Y. H. Kang, *J. Histochem. Cytochem.* **23,** 295 (1975).
168. R. H. Bryoles, B. M. Pack, S. Berger, and A. R. Dom, *Anal. Biochem.* **94,** 211 (1979).
169. P. E. Thomas, D. Ryan, and W. Levin, *Anal. Biochem.* **75,** 168 (1976).
170. A. M. Seligman, M. J. Karnovsky, H. L. Wasserkrug, and J. S. Hanker, *J. Cell Biol.* **38,** 1 (1968).
171. W. Cammer and C. L. Moore, *Biochemistry* **12,** 2502 (1973).
172. P. J. O'Brien and A. D. Rahimtula, *Biochem. Biophys. Res. Commun.* **70,** 832 (1976).
173. L. J. Marnett, P. Wlodawer, and B. Samuelsson, *J. Biol. Chem.* **250,** 8510 (1975).
174. T. V. Zenser, M. B. Mattammal, H. J. Armbrecht, and B. B. Davis, *Cancer Res.* **40,** 2839 (1980).
175. P. D. Josephy, T. Eling, and R. P. Mason, *J. Biol. Chem.* **258,** 5561 (1983).
176. P. D. Josephy, R. P. Mason, and T. Eling, *Cancer Res.* **42,** 2567 (1982).
177. B. Ketterer, D. Tullis, F. Evans, and F. F. Kadlubar, *Chem.-Biol. Interact.* **38,** 287 (1982).
178. R. G. Rowland and R. Oyasu, *Urol. Res.* **8,** 101 (1980).
179. G. M. Lower, T. Nilsson, C. E. Nelson, H. Wolf, T. E. Gamsky, and G. T. Bryan, *Environ. Health Perspect.* **29,** 71 (1979).
180. I. B. Glowinski and W. E. Weber, *J. Biol. Chem.* **257,** 1424 (1982).
181. S. N. Young and G. Curzon, *Biochem. J.* **129,** 273 (1972).
182. J. M. Sontag, *JNCI, Natl. Cancer Inst.* **66,** 591 (1981).
183. R. H. Waring and A. E. Pheasant, *Xenobiotica* **6,** 257 (1976).
184. T. Glinsukon, T. Benjamin, P. H. Grantham, N. L. Lewis, and E. K. Weisburger, *Biochem. Pharmacol.* **25,** 95 (1976).
185. Y. Hiasi, *J. Nara Med. Assoc.* **21,** 1 (1970).
186. E. Dybing, T. Aune, and S. D. Nelson, *Biochem. Pharmacol.* **28,** 43 (1979).
187. A. Rahimtula, B. Andersson, P. Moldeus, M. Nordenskjold, in ''Protaglandin and Cancer'' (T. Powles, R. S. Bockman, K. V. Honn, and P. Ramwell, eds.), p. 159. Alan R. Liss, New York (1981).
188. G. Albano, A. Carere, R. Crabelli, and R. Zito, *Chem. Toxicol.* **20,** 171 (1982).
189. P. Hartge, R. Hoover, R. Altman, and D. W. West, *Cancer Res.* **42,** 4784 (1982).
190. H. I. Maibach, M. A. Leaffer, and W. A. Skinner, *Arch. Dermatol.* **111,** 1444 (1975).
191. E. K. Weisburger, A. K. Murphy, R. W. Fleischman, and M. Hagopian, *Carciongenesis, (N.Y.)* **1,** 495 (1980).
192. E. K. Weisburger, A. B. Russfield, F. Homburger, J. H. Weisburger, E. Boger, C. G. Van Dongen, and K. C. Chu, *J. Environ. Pathol. Toxicol.* **2,** 325 (1978).
193. A. M. Ward, S. F. Stinson, J. F. Hardisty, B. Y. Cockrell, and D. W. Hayden, *JNCI, J. Natl. Cancer Inst.* **62,** 1067 (1979).
194. E. P. Evarts and C. A. Brown, *JNCI, J. Natl. Cancer Inst.* **65,** 197 (1980).
195. M. Ruchirawat, P. H. Grantham, T. Benjamin, and E. K. Weisburger, *Biochem. Pharmacol.* **30,** 2715 (1981).
196. T. Aune and E. Dybing, *Biochem. Pharmacol.* **28,** 2791 (1979).
197. E. Dybing and S. S. Thorgeirsson, *Biochem. Pharmacol.* **26,** 279 (1977).
198. T. Aune, E. Dybing, and S. D. Nelson, *Chem.-Biol. Interact.* **31,** 35 (1980).
199. E. Dybing, T. Aune, and S. D. Nelson, *Biochem. Pharmacol.* **28,** 51 (1979).
200. B. Ames, H. Kammen, and E. Yamasaki, *Proc. Natl. Acad. Sci. U.S.A.* **72,** 2423 (1975).
201. K. Nishi and H. Nishioka, *Mutat. Res.* **104,** 347 (1982).
202. C. Burnett, C. Fuchs, J. Corbett, and J. Menhart, *Mutat. Res.* **103,** 1 (1982).

203. R. Crebelli, L. Conti, A. Carere, and R. Zito, *Food Cosmet. Toxicol.* **19**, 79 (1981).
204. J. F. Corbett, *J. Soc. Cosmet. Chem.* **24**, 103 (1973).
205. M. Altman and M. M. Rieger, *J. Soc. Cosmet. Chem.* **19**, 141 (1968).
206. K. C. Brown and J. F. Corbett, *J. Chem. Soc., Perkin Trans.* **II**, p. 308 (1979).
207. J. L. Holtzman and A. M. Seligman, *Arch. Biochem. Biophys.* **155**, 237 (1973).
208. E. G. Hrycay and P. J. O'Brien, *Arch. Biochem. Biophys.* **153**, 480 (1972).
209. I. G. C. Robertson, K. Sivarajah, T. E. Eling, and E. Zeigler, *Cancer Res.* **43**, 476 (1983).
210. K. Tanaka, T. Mu, S. Mariu, I. Matsubara, and H. Igaki, *Int. Arch. Occup. Environ. Health* **49**, 177 (1981).
211. J. V. Rutkowski and V. H. Fern. *Toxicol. Appl. Pharmacol.* **63**, 264 (1982).
212. I. C. Calder, A. C. Yong, R. A. Woods, C. A. Crowe, K. N. Ham, and J. D. Tauge, *Chem.-Biol. Interact.* **27**, 245 (1979).
213. B. Andersson, M. Nordenskjold, A. Rahimtula, and P. Moldeus, *Mol. Pharmacol.* **22**, 479 (1982).
214. P. D. Josephy, T. E. Eling, and R. P. Mason, *Mol. Pharmacol.* **23**, 766 (1983).
215. J. R. Mitchell, D. J. Jallow, W. Z. Potter, D. C. Davis, J. R. Gillette, and B. B. Brodie, *J. Pharmacol. Exp. Ther.* **187**, 195 (1973).
216. E. Piperno and D. A. Berssenbuegge, *Lancet* **2**, 738 (1976).
217. A. Wendel, H. Jaeschke, and M. Gloger, *Biochem. Pharmacol.* **31**, 3601 (1982).
218. S. D. Nelson, A. J. Forte, and D. C. Dahlin, *Biochem. Pharmacol.* **29**, 1617 (1980).
219. I. A. Blair, A. R. Boobis, and D. S. Davies, *Tetrahedron Lett.* **21**, 4947 (1980).
220. D. J. Miner and P. T. Kissinger, *Biochem. Pharmacol.* **28**, 3285 (1979).
221. P. Moldeus, B. Andersson, A. Rahimtula, and M. Berggren, *Biochem. Pharmacol.* **31**, 1363 (1982).
222. G. A. Boyd and T. E. Eling, *J. Pharmacol. Exp. Ther.* **249**, 659 (1981).
223. W. Z. Potter, S. S. Thorgeirsson, D. J. Jollow, and J. R. Mitchell, *Mol. Pharmacol.* **12**, 129 (1974).
224. E. B. Corcoran, J. R. Mitchell, Y. N. Vaishnaw, and E. C. Horning, *Mol. Pharmacol.* **18**, 536 (1980).
225. S. L. Johansson, *Int. J. Cancer* **27**, 521 (1981).
226. J. B. Vaught, P. B. McGarvey, M. S. Lee, C. D. Garner, C. Y. Wang, E. M. Linsmaier-Bednay, and C. M. King, *Cancer Res.* **41**, 3424.
227. S. D. Nelson, A. J. Forte, Y. Vaishnaw, J. R. Mitchell, J. R. Gillette, and J. A. Hinson, *Mol. Pharmacol.* **19**, 140 (1981).
228. R. Nery, *Biochem. J.* **122**, 311 (1971).
229. J. A. Hinson and J. R. Mitchell, *Drug Metab. Dispos.* **4**, 430 (1976).
230. J. Raaflaub and U. C. Dubach, *Klin. Wochenchr.* **23**, 1286 (1969).
231. B. Andersson, R. Larsson, A. Rahimtula, and P. Moldeus, *Biochem. Pharmacol.* **32**, 1019 (1983).
232. D. M. Clayson, *Br. Med. Bull.* **20**, 115 (1964).
233. W. Troll, S. Belman, N. Nelson, M. Levitz, and G. H. Trombley, *Proc. Soc. Exp. Biol. Med.* **100**, 75 (1959).
234. H. T. Nagasawa and H. R. Gutman, *J. Biol. Chem.* **233**, 1593 (1958).
235. S. Belman and W. Troll, *J. Biol. Chem.* **237**, 746 (1962).
236. C. M. King and E. Kriek, *Biochim. Biophys. Acta* **111**, 147 (1965).
237. S. Belman, T. Huang, E. Levine, and W. Troll, *Biochem Biophys. Res. Commun.* **14**, 463 (1964).
238. G. T. Miwa, J. S. Walsh, G. L. Kedderis, and P. F. Hollenberg, *J. Biol. Chem.* **258** 14445 (1983).

239. G. L. Kedderis, L. A. Dwyer, D. E. Rickert, and P. F. Hollenberg, *Mol. Pharmacol.* **23,** 758 (1983).
240. G. L. Kedderis and P. F. Hollenberg, *J. Biol. Chem.* **258,** 8129 (1983).
241. P. Moldéus, P. J. O'Brien, H. Thor, M. Berggren, and S. Orrenius, *FEBS Lett.* **162,** 411 (1983).
242. Y. Watanabe, S. Oae, and T. Iyanagi, *Bull. Chem. Soc. Jpn.* **55,** 188 (1982).

CHAPTER **10**

One-Electron and Two-Electron Oxidation in Aromatic Hydrocarbon Carcinogenesis

Ercole L. Cavalieri and Eleanor G. Rogan

Eppley Institute for Research in Cancer and Allied Diseases
University of Nebraska Medical Center
Omaha, Nebraska

FREE RADICALS IN BIOLOGY, VOL. VI
Copyright © 1984 by Academic Press, Inc.
All rights of reproduction in any form reserved.
ISBN 0-12-566506-7

I. INTRODUCTION

Xenobiotics, including chemical carcinogens, undergo metabolism in various animal tissues. Although a small group of these compounds are alkylating or acylating agents that can react directly with cellular macromolecules, most carcinogens require some type of metabolic activation. Overall metabolism of these compounds yields a variety of products, and only a small fraction of the generated intermediates can react with cellular macromolecules to trigger the transformation of normal to neoplastic cells.

Chemical carcinogens encompass a broad variety of compounds which do not have any common feature that can be related to their tumorigenic activities, even though the ultimate carcinogenic metabolites are all electron deficient, or electrophilic [1, 2]. The electrophilicity of the critical intermediates has been an important concept for unifying the variety of structures known as chemical carcinogens [1, 2].

Polycyclic aromatic hydrocarbons (PAH) require one or more activating steps to produce proximate and/or ultimate carcinogenic forms capable of reacting with informational macromolecules. Electrophilic intermediates involved in determining the carcinogenicity of various PAH can be produced by two general mechanisms: two-electron oxidation, or oxygenation, and one-electron oxidation. Because of the strong electron-donating properties of PAH, transfer of one electron from these compounds generates radical-cations that can directly react with various cellular nucleophiles. Two-electron oxidation, or oxygenation, is the transfer of a catalytically activated oxygen, which reacts as an oxene species [3], from the enyzme to the substrate. Some of the oxygenated products can also form covalent bonds with cellular macromolecules.

We discuss in this chapter the reactive intermediates of PAH generated by both one-electron and two-electron oxidation that are involved in carcinogenesis. Studies of these electrophiles also provide information about the enzymes that catalyze carcinogenic activation.

II. ENZYMOLOGY OF ONE-ELECTRON AND TWO-ELECTRON OXIDATION

A. Cytochrome *P*-450-Catalyzed Two-Electron Oxidation

Acting as the terminal oxidase of the monooxygenase enzyme system, cytochrome *P*-450 is involved in the oxygenation of various foreign and endogenous substrates [4, 5]. This reaction depends on NADPH and oxygen,

$$RH + O_2 + NADPH + H^+ \rightarrow ROH + H_2O + NADP^+$$

where RH represents the substrate and ROH the product. Three main types of these reactions can occur: (*a*) insertion of an oxygen atom into the bond between a hydrogen and a carbon or nitrogen atom to yield the corresponding hydroxyl derivative, (*b*) addition of an oxygen atom to a carbon–carbon π-bond to give an epoxide, and (*c*) addition of an oxygen to the electron pair of a nitrogen or sulfur to yield a dipolar oxide.

Some of these oxygenated derivatives of chemical carcinogens can be the proximate or ultimate electrophilic metabolites. In the case of PAH, simple arene oxides [6–8] and more commonly vicinal diol-epoxides have been found to be ultimate carcinogenic metabolites [9–11]. For *meso*-methylated PAH, hydroxylation of the benzylic carbon atom, followed by esterification, could also produce a potential ultimate carcinogen [12–14].

B. Cytochrome *P*-450-Catalyzed One-Electron Oxidation

It has become apparent that cytochrome *P*-450 acting as a monooxygenase with NADPH and oxygen can also carry out one-electron oxidation of a variety of substrates. Autocatalytic destruction of cytochrome *P*-450 by derivatives of cyclopropylamine [15, 16] and dihydropyridine [17] has been proven to occur via aminium radical intermediates.

Cytochrome *P*-450-catalyzed oxygenation of sulfides to sulfoxides and sulfoxides to sulfones proceeds via a one-electron transfer process, which generates a sulfinium radical intermediate [18, 19]. The following mechanism has been postulated for the formation of sulfoxides:

In this case the one-electron acceptor would be the cytochrome P-450–oxygen complex containing $(FeO)^{3+}$. The reduced form, $(FeO)^{2+}$, of this complex would release a nucleophilic oxygen rather than the usual electrophilic oxene species. Sulfoxides are oxidized to sulfones by a similar mechanism. The one-electron oxidation of the substrate in these studies is demonstrated by a correlation between the ionization potential (IP) of the sulfur substrates and the rate of oxygenation by cytochrome P-450.

Similarly, cytochrome P-450 catalyzes one-electron oxidation of N-hydroxynorcocaine, the pathway by which cocaine induces hepatic toxicity [20].

In addition to one-electron oxidation with NADPH and oxygen, cytochrome P-450 catalyzes one-electron oxidation of organic substrates at the expense of peroxy compounds [21–25]. The stoichiometry of this reaction is

$$RH + R'OOH \rightarrow ROH + R'OH$$

where RH and ROH represent the substrate and product, and R'OOH the peroxy compound serving as oxygen donor.

The metabolic profile of benzo[a]pyrene (B[a]P) with cytochrome P-450 acting as a peroxidase is quite different from that obtained with cytochrome P-450 acting as a monooxygenase [26, 27]. Whereas cytochrome P-450 in the presence of NADPH and oxygen metabolizes B[a]P to yield approximately 50% phenols and 25% each of dihydrodiols and quinones, with cumene hydroperoxide as cofactor the products are virtually all quinones with trace amounts of dihydrodiols and phenols.

Griffin *et al.* [28] have suggested that N-demethylation of a tertiary amine, aminopyrine, mediated by the cytochrome P-450–cumene hydroperoxide system proceeds by one-electron oxidation through an aminium radical intermediate. This step is presumably followed by loss of a proton and by a second one-electron oxidation to yield a carbonium ion, which, by hydrolysis, generates formaldehyde and a secondary amine product.

C. Horseradish Peroxidase

Horseradish peroxidase (HRP) catalyzes the one-electron oxidation of a variety of chemicals by the following mechanism:

$$HRP + H_2O_2 \rightleftharpoons HRP-H_2O_2 \longrightarrow E_1$$

$$E_1 + RH \rightleftharpoons E_1-RH \longrightarrow E_2 + RH^{\dot{+}}$$

$$E_2 + RH \rightleftharpoons E_2-RH \longrightarrow HRP + RH^{\dot{+}}$$

Here, HRP is the ferric enzyme, RH is the substrate, $RH^{\dot{+}}$ is the corresponding radical-cation, and E_1 and E_2 are the oxidized forms of the enzyme at two and one higher levels of oxidation of HRP, respectively. Because this purified enzyme is often predictive of oxidation by mammalian peroxidases, it has been used as a model system to study peroxidative activation.

One-electron oxidation of N-hydroxy-2-acetylaminofluorene by HRP–H_2O_2 generates an oxy radical, which dismutates to N-acetoxy-2-acetylaminofluorene and nitrosofluorene [29–32].

A wide variety of other compounds are activated by one-electron oxidation in the HRP–H_2O_2 system. These include the following:

1. Diethylstilbestrol [33]. This compound yields p-semiquinone and p-quinone by stepwise one-electron oxidation. The latter in turn tautomerizes to β-dienestrol.

2. Phenol [*34*]. One-electron oxidation of phenol yields *p,p'*-phenol, which by further oxidation generates the *p*-diphenylquinone, presumably via the intermediate *p*-diphenylsemiquinone.

3. Aminopyrine [*35*] and other tertiary amines [*36–38*]. Horseradish peroxidase–H_2O_2-catalyzed activation of aminopyrine is similar to that mediated by cytochrome *P*-450 (see Section II,B, p. 326). The same type of activation scheme has been suggested for tertiary amines.
4. Benzidine and derivatives [*39, 40*]. Stepwise one-electron oxidation in this case produces an unstable diimine dication and a diimine monocation. The latter couples nonenzymatically with phenolic compounds.

5. Tetramethylhydrazine [*41*]. A stable radical-cation has been detected by ESR when this compound is incubated with HRP–H_2O_2. This pathway of

   activation his been suggested to play a role in the carcinogenicity of hydrazine compounds.
6. Cyclopropanone hydrate [*42*]. This compound autocatalytically inactivates HRP after covalent binding of the proposed radical intermediate to the heme moiety of HRP.

D. Mammalian Peroxidases

Peroxidases present in many organs, including skin, lung, mammary gland, bone marrow, and uterus, activate various compounds by one-electron oxidation. Rat mammary cell peroxidase, mouse uterine peroxidase, and rat bone marrow peroxidase follow this mechanism of activation with N-hydroxy-2-acetylaminofluorene [32, 43], diethylstilbestrol [33], and phenol [34], respectively.

One type of mammalian peroxidase implicated in the oxidation of carcinogens is prostaglandin endoperoxide synthase (PES). The possible role of this enzyme in carcinogenesis is under intensive investigation. Activation of 5-nitrofuran derivatives and benzidine by PES has been proposed in kidney and bladder carcinogenesis [44, 45]. Josephy et al. have reported that metabolic activation of tetramethylhydrazine and benzidine by PES occurs by one-electron oxidation, which is similar to that in the $HRP-H_2O_2$ system [46, 47]. PES-Catalyzed cooxidation of N-hydroxy-2-acetylaminofluorene shows the characteristic N-oxy radical, suggesting one-electron oxidation [32]. Metabolism of the transplacental carcinogen diethylstilbestrol by PES yields β-dienestrol in a manner similar to $HRP-H_2O_2$-catalyzed one-electron oxidation (see Section II,C, p. 327) [48]. Furthermore, PES is implicated in the metabolism of diethylstilbestrol in cultured Syrian hamster embryo fibroblasts [49, 50]. PES-Catalyzed oxidation of B[a]P produces exclusively B[a]P-1,6-,B[a]P-3,6-, and B[a]P-6,12-dione [51, 52], whereas the other common metabolites formed by mammalian microsomes, namely, phenols and dihydrodiols, are not present. There is substantial evidence that the formation of B[a]P quinones involves initial one-electron transfer from B[a]P (see Section V). These experiments clearly point to the extensive role of PES in catalyzing one-electron oxidation of various substrates.

III. ONE-ELECTRON OXIDATION

A. Chemical Properties of Polycyclic Aromatic Hydrocarbon Radical-Cations

The structures of the most common PAH are presented in Fig. 1. Radical-cations are reactive intermediates obtained by the removal of one electron from PAH. These electrophiles can easily be produced by virtue of the strong electron-donating properties of PAH and the presence in biological systems of several oxidants in the form of metal-ion-containing enzymes, which include cytochrome P-450 with coordinated Fe^{3+}. It has been reported that Fe^{3+} oxidizes B[a]P, 7,12-dimethylbenz[a]anthracene (DMBA), and 3-methylcholanthrene (MC) to their respective radical-cations, which then react specifically with various nucleophiles [53–56].

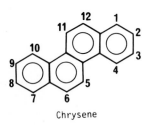

Chrysene

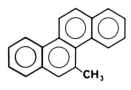

5-Methylchrysene

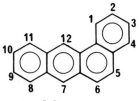

Benz[*a*]anthracene
(B[a]A)

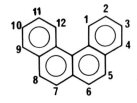

Benzo[*c*]phenanthrene

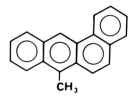

7-Methylbenz[*a*]anthracene
(7-methyl-B[a]A)

CH₃

CH₃

CH₃

7,12-Dimethylbenz[*a*]anthracene
(DMBA)

H₃C

3-Methylcholanthrene
(MC)

Dibenz[*ah*]anthracene
(DB[ah]A)

Fig. 1.   Structures of selected PAH.

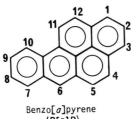

Benzo[*a*]pyrene
(B[a]P)

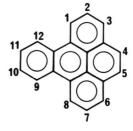

Benzo[*e*]pyrene

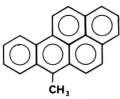

6-Methylbenzo[*a*]pyrene
(6-methyl-B[a]P)

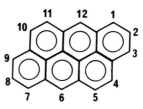

Anthanthrene

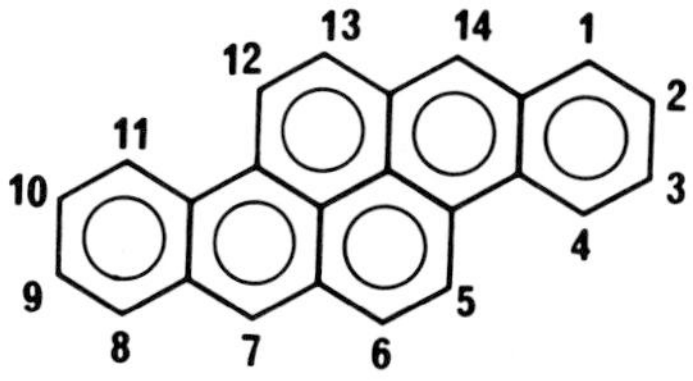

Dibenzo[*ah*]pyrene
(DB[a h]P)

Dibenzo[*ai*]pyrene
(DB[a i]P)

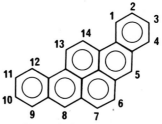

Cyclopenta[*cd*]pyrene

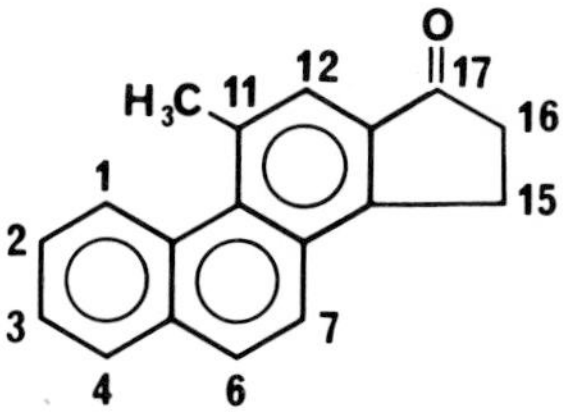

15,16-Dihydro-11-methylcyclo-
penta[*a*]phenanthren-17-one

Iodine has been demonstrated to be a one-electron oxidant of PAH [53, 55, 57–59], and the resulting intermediate PAH radical-cations bind covalently to DNA [60, 61]. To establish some of the basic properties of the radical-cations, we have generated them in two one-electron oxidant systems. The first consists of iodine as oxidant and pyridine as nucleophile and solvent [62, 63], whereas the second is manganic acetate in acetic acid [64].

The results of nucleophilic substitution in the iodine–pyridine system are shown in Table I. Reaction yields with various PAH are presented together with their IP. The compounds 5-methylchrysene and dibenz[a,h]anthracene (DB[ah]A), which have relatively high IP, are not oxidized. Nucleophilic trapping for unsubstituted PAH, namely, benz[*a*]anthracene (B[a]A), anthracene, B[a]P, and anthanthrene, occurs specifically at the position(s) of highest charge density in their respective

TABLE I   One-Electron Oxidation by the Iodine–Pyridine System [62, 63][a]

Compound	Position of pyridine substitution	Yield[b] (%)	Ionization potential[c] (eV)
5-Methylchrysene	No reaction	0	~7.7
Dibenz[*a,h*]anthracene	No reaction	0	7.57
Benz[*a*]anthracene	7	54	7.54
6-Methylbenz[*a*]anthracene	7	48	7.50
	12	14	
11-Methylbenz[*a*]anthracene	7	82	7.48
2-Methylbenz[*a*]anthracene	7	83	7.46
5-Methylbenz[*a*]anthracene	7	85	7.46
8-Methylbenz[*a*]anthracene	7	58	7.46
	12	14	
Anthracene	9	60	7.43
7-Ethylbenz[*a*]anthracene	12	68	7.39
12-Methylbenz[*a*]anthracene	7	78	7.38
7-Methylbenz[*a*]anthracene	7-CH$_3$	32	7.37
	12	11	
Benzo[*a*]pyrene	6	58	7.23
7,12-Dimethylbenz[*a*]anthracene	5	58	7.22
	7-CH$_3$	18	
	12-CH$_3$	15	
3-Methylcholanthrene	1	96	7.12
6-Methylbenzo[*a*]pyrene	6-CH$_3$	74	7.08
Anthanthrene	6	20	6.96

[a]Reaction at 30 to 35°C for 20 h.

[b]The remainder is starting material and/or undetected minor products.

[c]Determined from absorption maximum of the charge-transfer complex of each compound with chloranil [65], with the exception of dibenz[*a,h*]anthracene, determined by polarographic oxidation [66].

radical-cations. For 2-, 5-, 11-, and 12-monomethyl-B[a]A, in which the steric environment of the position of highest charge density, C-7, is unaltered, substitution occurs specifically at this position. When some steric restriction exists at C-7, as in the case of 6- and 8-methyl-B[a]A, reaction occurs competitively at C-12, the position of second highest charge density. When the methyl group occupies the position of highest charge density, as in 7-methyl-B[a]A, nucleophilic substitution proceeds competitively at C-12, as well as at the 7-methyl group. The latter mechanism is exclusive for 6-methyl-B[a]P. When the two positions of highest charge density are blocked by a methyl group, as in DMBA, competitive reactions occur at the unsubstituted 5-position, as well as on the 7- and 12-methyl groups. No reaction at the benzylic methylene group is observed for 7-ethyl-B[a]A radical-cation, because substitution takes place only at C-12. Conversely, MC radical-cation yields exclusive substitution at the 1-methylene group. Because charge distribution in the 7-ethyl-B[a]A radical-cation is not appreciably different from that in the MC radical-cation, the relative rates of proton loss are responsible for the two different reaction pathways [62].

The results obtained by one-electron oxidation of PAH with manganic acetate in acetic acid are illustrated in Table II. In this case the weaker nucleophile, acetate ion, should be more sensitive to differences in charge density. Compounds such as phenanthrene and chrysene, which have relatively high IP, are not oxidized by Mn^{3+} at 40°C. For PAH with lower IP, acetoxy substitution occurs at the position(s) of highest charge density on the aromatic nucleus (compounds 3–6, 8, 9, 12–15) and/or at the methyl group blocking the position of highest charge density (compounds 8, 10–12). The higher selectivity of the weaker nucleophile, acetate ion, can be observed in DMBA, in which substitution takes place only at the 7- and 12-methyl groups and not at C-5, as in the iodine–pyridine system. For anthracene, a second acetate ion trapping occurs before the loss of a proton with the formation of a stable dihydrodiacetoxy derivative.

These studies reveal, as anticipated, that the capacity of PAH to form radical-cations is related to IP, which depends on the π-electron charge distribution of the compound. Other factors influencing the reactivity of radical-cations are the localization of charge on one or a few carbon atoms, the steric environment of centers on which the charge is preferentially localized, and the strength of the nucleophiles. For unsubstituted and methyl-substituted PAH, one-electron oxidation and subsequent binding to a nucleophile occur as outlined in Fig. 2.

The removal of one electron from the π-system generates a radical-cation. The position(s) of highest charge density is the most susceptible to nucleophilic substitution. In path 1, in which the charge is localized mainly at an unsubstituted carbon atom on the ring, nucleophilic attack proceeds at that position. The resulting radical is very easily oxidized to an arenonium ion, with loss of a proton

TABLE II   One-Electron Oxidation by the Manganic Acetate–Acetic Acid System [64][a]

Compound	Time	Position of acetoxy substitution	Yield (%)	Starting material (%)	Ionization potential[b] (eV)
1. Phenanthrene	96 h	No reaction	0	100	8.19
2. Chrysene	96 h	No reaction	0	100	~7.8
3. 5-Methylchrysene	96 h	6	28	72	~7.7
4. Benzo[e]pyrene	96 h	1	14	66	7.62
5. Benz[a]anthracene	48 h	7	90–100	Traces	7.54
6. Pyrene	96 h	1	60	24	7.50
		1,6	16		
7. Anthracene[c]	66 h	9,10-Dihydro-9,10-diacetoxy[d]	100		7.43
8. 7-Methylbenz[a]anthracene	24 h	7-CH$_3$	85	Traces	7.37
		12	10		
9. Benzo[a]pyrene	<10 min	6	95	Traces	7.23
		Quinones	5		
10. 7,12-Dimethylbenz[a]anthracene	<10 min	7-CH$_3$	50–60	Traces	7.22
		12-CH$_3$	40–50		
11. 3-Methylcholanthrene	<10 min	1	75–80	Traces	7.12
12. 6-Methylbenzo[a]pyrene	10 min	6-CH$_3$	75–80	Traces	7.08
		1[e]			
		3[e]	15–20		
13. Perylene[f]	<10 min	1	22	30	7.06
		1,7	26		
		Triacetoxy	11		
14. Anthanthrene[f]	<10 min	6	41	26	6.96
		6,12	12		
15. 6-Methylanthanthrene[f]	5 min	12	51	37	6.85

[a]Reaction at 40°C unless otherwise specified.
[b]Determined from absorption maximum of the charge-transfer complex of each compound with chloranil [65].
[c]Reaction at 22°C.
[d]Cis–trans ratio, 3 : 4.
[e]Isolated by HPLC.
[f]Reaction at 55°C.

to complete the substitution reaction. When the maximum charge density is located at a position adjacent to a methyl group, ejection of a methyl proton generates a benzylic radical, which is rapidly oxidized to a benzylic carbonium ion, with subsequent reaction with a nucleophile. These properties of PAH radical-cations allow us to predict the position(s) involved in the covalent binding of PAH to cellular macromolecules, if activation occurs by one-electron oxidation.

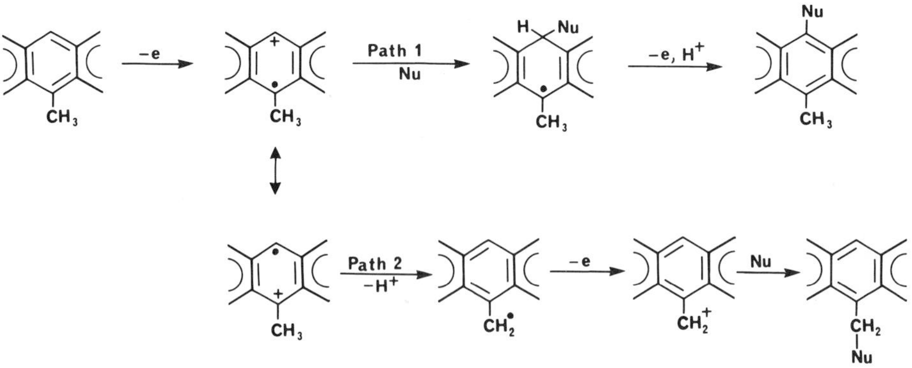

Fig. 2. Mechanisms of trapping in radical-cations of unsubstituted and methyl-substituted PAH by nucleophiles (Nw).

B. Binding of Polycyclic Aromatic Hydrocarbons to Nucleic Acids by One-Electron Oxidation

One-electron oxidation of various chemicals by HRP–H_2O_2 has been demonstrated in several laboratories [29–42]. We have studied the relationship between the IP of various PAH (Table III) and the extent of their HRP–H_2O_2-catalyzed binding to DNA [65] to show that this mechanism is also common to PAH. A sharp cutoff is observed at an IP of ~7.35 eV. With IP > 7.35 eV the level of binding is low, whereas for PAH with IP < 7.35 eV it is high. These results suggest that HRP–H_2O_2 catalyzes the binding of PAH to DNA by one-electron oxidation and that this mechanism is operative only for certain PAH with IP < 7.35 eV.

Binding of PAH to DNA and determination of the structures of adducts constitute an approach for substantiating or eliminating a hypothesized mechanism of PAH activation. If radical-cations are the critical intermediates in this process, reaction with cellular nucleophiles occurs via a substitution reaction with loss of the hydrogen atom from the position of binding. Thus, specific tritiation of PAH at a postulated position of binding should produce loss of tritium if that position participates in a covalent bond between DNA and the PAH. In chemical experiments nucleophilic attack on the B[a]P radical-cation occurs almost exclusively at C-6, the position of highest charge density (Tables I and II) [55, 57–59, 63, 64, 67, 68]. Positions C-1 and C-3 are the sites of next highest charge density. Thus, these three positions are expected to lose tritium if binding proceeds by one-electron oxidation.

Study of the binding of double-labeled [^{3}H,^{14}C]B[a]P to nucleic acids has

TABLE III   Ionization Potentials and HRP–H_2O_2-Catalyzed Binding to DNA [65]

Compound	Ionization potential[a] (eV)	DNA-Bound [^{14}C]- or [^{3}H]PAH[b] (μmol/mol DNA)	DNA-Bound PAH in controls (μmol/mol DNA)
Phenanthrene	8.19	3.8 ± 0.8 (11)	1.2 ± 0.4 (7)
5-Methylchrysene	~7.7	1.4 ± 0.5 (8)	0.6 ± 0.2 (8)
Benzo[e]pyrene	7.62	5.1 ± 0.9 (5)	1.5 ± 0.3 (3)
Dibenz[a,h]anthracene	7.57	4.3 ± 1.0 (10)	1.4 ± 0.5 (8)
Benz[a]anthracene	7.54	4.0 ± 0.5 (12)	0.9 ± 0.2 (8)
Pyrene	7.50	2.8 ± 1.4 (4)	1.5 ± 1.0 (3)
Anthracene	7.43	8.8 ± 1.6 (9)	6.3 ± 0.7 (4)
7-Methylbenz[a]anthracene	7.37	5.6 ± 0.6 (6)	2.6 ± 0.2 (4)
Benzo[a]pyrene	7.23	89.2 ± 5.6 (8)	1.3 ± 0.3 (18)
7,12-Dimethylbenz[a]anthracene	7.22	63.9 ± 4.6 (12)	1.0 ± 0.2 (8)
3-Methylcholanthrene	7.12	60.6 ± 4.1 (10)	3.4 ± 0.2 (4)
6-Methylbenzo[a]pyrene	7.08	39.8 ± 5.3 (9)	2.2 ± 0.2 (7)
Anthanthrene	6.96	27.0 ± 7.1 (8)	2.3 ± 0.8 (6)
6,12-Dimethylanthanthrene	6.68	62.0 ± 13 (5)	4.4 ± 1.0 (3)

[a]The IP were calculated from absorption maximum of the charge-transfer complex of each compound with chloranil [64], with the exception of dibenz[a,h]anthracene, which was determined by polarographic oxidation [66].

[b]Control levels of binding have been subtracted from these levels, which are presented as average ± standard error of measurement. Numbers in parentheses represent number of determinations.

been conducted in several *in vitro* systems, as well as in mouse skin [69, 70]. Because HRP–H_2O_2 provides a good *in vitro* biological model for one-electron oxidation, we have used it along with binding of [^{3}H,^{14}C]B[a]P to DNA catalyzed by liver microsomes from MC-induced rats and binding to endogenous DNA in liver nuclei from the same rats. These *in vitro* systems are compared with the binding occurring in mouse skin treated for 4 h with [^{3}H,^{14}C]B[a]P.

The results summarized in Table IV show that microsome-catalyzed B[a]P binding to poly(G) and denatured DNA does not result in a significant loss of tritium from the 1,3- or 6-position, indicating that activation does not occur by one-electron oxidation. Most of the microsome-catalyzed binding to native DNA, however, is observed at the three positions of highest charge density, with no loss of tritium at C-7. Binding of B[a]P to endogenous DNA in isolated rat liver nuclei occurs mostly at C-6, whereas smaller amounts at C-1 and C-3 are also observed. Tritium is lost almost exclusively at C-6 in HRP–H_2O_2-catalyzed binding to DNA. A small loss occurs at C-7, but none occurs at C-8. The small loss of tritium from C-7 observed in almost all systems may be explained by the proximity of C-7 to position 6, where most binding occurs.

When mouse skin is treated with B[a]P *in vivo*, binding to DNA and RNA appears to occur predominantly at the 6-position, with smaller amounts at the 1-

TABLE IV Loss of Tritium during Binding of B[a]P to DNA and RNA *in Vitro* and *in Vivo* [*69, 70*]

System	Loss of tritium from indicated position in $[^3H, ^{14}C]B[a]P$ (%)			
	6	1,3	7	8
In vitro				
Microsomes + poly(G)	0	16	—	—
Microsomes + denatured DNA	0	3	—	—
Microsomes + native DNA	42	42	0	—
Nuclei	69	37	10	—
HRP + native DNA	94	—	12	1
In vivo				
Skin DNA	77	37	17	—
Skin RNA	79	35	17	—

and 3-positions. This indicates that in the target organ mouse skin B[a]P is bound predominantly at its position of highest charge density and activation may occur by one-electron oxidation. However, other laboratories have observed binding of PAH to mouse skin primarily through diol-epoxide derivatives (see Section IV,A, p. 345) [*71–73*]. Several factors in our experiments have differed from others, including mouse strain and contact time of the PAH and the skin, and further experiments must be conducted to attempt to reconcile the contrasting results.

Identification of DNA adducts in target tissues is one type of evidence that one-electron oxidation is an important mechanism of activation. It is difficult to analyze B[a]P–DNA adducts formed by one-electron oxidation because the elucidation of adducts requires synthesis of model compounds, and thus the synthesis of adducts with B[a]P bound at C-6 to nucleosides must be developed. We have chosen, therefore, analysis of 6-methyl-B[a]P–DNA adducts prepared chemically and enzymatically because one-electron oxidation yields 6-methyl-B[a]P bound to nucleophiles at the methyl group (Tables I and II). When DNA adducts of 6-[^{14}C]methyl-B[a]P bound in an HRP–H_2O_2-catalyzed reaction are analyzed by HPLC (Fig. 3a), a major peak (peak 2) is the DNA adduct with 6-methyl-B[a]P, in which the 6-methyl group is bound to the 2-amino group of dG [*74*]. The structure of the adduct indicates that activation of 6-methyl-B[a]P by the HRP–H_2O_2 system proceeds through the radical-cation formed by one-electron oxidation.

Comparison of DNA adducts formed in mouse skin treated with 6-[^{14}C]methyl-B[a]P (Fig. 3b) with those formed in the HRP–H_2O_2 system (Fig. 3a) demonstrates that the adduct profiles are qualitatively similar. Peaks 3 and 6 are identified as 6-hydroxymethyl-B[a]P and 6-methyl-B[a]P, respectively. Peak 1, which may

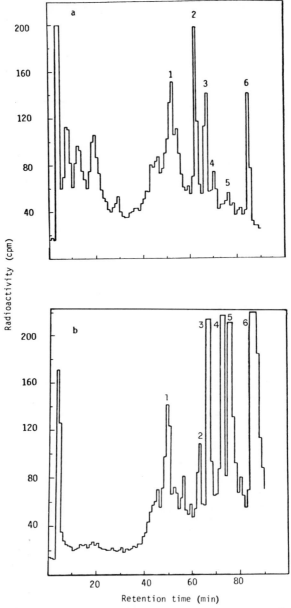

Fig. 3. HPLC profile of 6-methyl-B[a]P–DNA hydrolyzed to deoxyribonucleosides. (a) 6-[^{14}C]Methyl-B[a]P bound to DNA in the HRP–H_2O_2 system, (b) DNA from mouse skin treated *in vivo* in 6-[^{14}C]methyl-B[a]P [*74*].

contain one or more adducts, has not yet been identified. Thus, peak 2 in both the HRP–H_2O_2 *in vitro* system and mouse skin appears to be the major identifiable adduct.

The presence of the peak 2 adduct in mouse skin DNA and the qualitative similarities between the mouse skin and HRP–H_2O_2 profiles suggest that 6-methyl-B[a]P may be activated by one-electron oxidation in mouse skin *in vivo*, although hydroxylation of the methyl group and subsequent esterification cannot be ruled out as an alternative pathway of activation (see Section IV,B, p. 353). These results provide the first biochemical evidence that one-electron oxidation may be a significant mechanism of activation for PAH in a target organ.

C. Comparative Carcinogenicity in Polycyclic Aromatic Hydrocarbon Target Organs

A variety of experimental results suggest that multiple mechanisms of activation occur in mouse skin, a major target organ for PAH carcinogenesis [75]. Studies of PAH binding to mouse skin nucleic acids have supported both diol-epoxides [71–73] and radical-cations [69, 74] as ultimate carcinogenic metabolites. To obtain clearer evidence that activation of PAH by one-electron oxidation plays an important role in PAH carcinogenesis, we have considered other target organs in which to conduct experiments. We have selected the rat mammary gland as an organ in which to test carcinogenicity for two major reasons: (*a*) N-Hydroxy-2-acetylaminofluorene is activated in rat mammary cells by one-electron oxidation [32, 43], and (*b*) the only PAH observed to have carcinogenic activity in mammary glands are those with a relatively low IP, namely, IP < 7.35 eV, a tentative cutoff point derived from DNA-binding studies in the HRP–H_2O_2 system (see Table III).

To limit the activation of PAH to the mammary gland in these studies, compounds have been directly applied to the organ [76–78]. A series of PAH that are or are not expected to be carcinogenic by one-electron oxidation because of their low or high IP, respectively, has been tested, as have PAH in which the diol-epoxide region is blocked but which can be activated by one-electron oxidation. Comparison of the carcinogenic activity of these PAH in mouse skin and rat mammary gland (Tables V and VI) suggests that one-electron oxidation could be the predominant, if not selective, mechanism of activation in the latter tissue. The only PAH observed to be carcinogenic in this organ, namely, DMBA, 7-methyl-B[a]A, MC, 2-methyl-MC, 8-fluoro-MC, 10-fluoro-MC, B[a]P, and 6-methyl-B[a]P (Table V) are favored to be activated by one-electron oxidation because of their low IP and sufficient charge localization in their radical-cation. Although 1-methyl-MC has a low IP, it is inactive in rat mammary gland, presumably because of steric hindrance at C-1, the position of nucleophilic

TABLE V Comparative Carcinogenicity of PAH in Mouse Skin and Rat Mammary Gland: Compounds Active in Mammary Gland

Compound	Structure	Ionization potential[a] (eV)	Carcinogenicity[b] in:	
			Mouse skin	Rat mammary gland
7-Methylbenz[a]anthracene		7.37	++	+
Benzo[a]pyrene		7.23	++++	+++
7,12-Dimethylbenz[a]anthracene		7.22	+++++	+++++
10-Fluoro-3-methylcholanthrene		7.17	NT	++

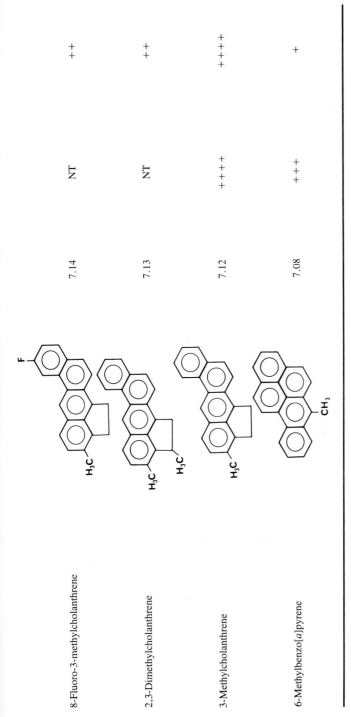

8-Fluoro-3-methylcholanthrene		7.14	NT	++
2,3-Dimethylcholanthrene		7.13	NT	++
3-Methylcholanthrene		7.12	++++	++++
6-Methylbenzo[a]pyrene		7.08	+++	+

[a]Determined from absorption maximum of the charge-transfer complex of each compound with chloranil [65].

[b]Extremely active, ++++; very active, +++; active, ++; moderately active, ++; weakly active, +; very weakly active, +; inactive, ±; not tested, NT.

341

TABLE VI Comparative Carcinogenicity of PAH in Mouse Skin and Rat Mammary Gland: Compounds Inactive in Mammary Gland

Compound	Structure	Ionization potential[a] (eV)	Carcinogenicity[b] in:	
			Mouse skin	Rat mammary gland
Cyclopenta[cd]pyrene		—	+ +	—
Benzo[a]pyrene-7,8-dihydrodiol		—	+ + + +	—
5-Methylchrysene		~7.7	+ + +	—

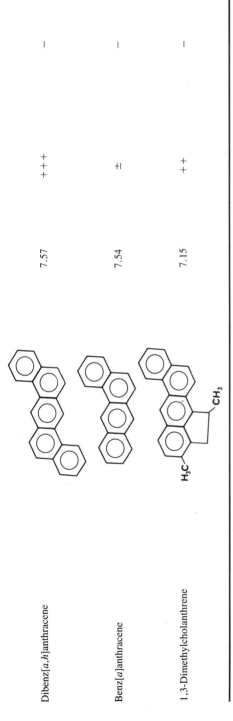

Compound			
Dibenz[a,h]anthracene	7.57	+++	—
Benz[a]anthracene	7.54	±	—
1,3-Dimethylcholanthrene	7.15	++	—

[a]Determined from absorption maximum of the charge-transfer complex of each compound with chloranil [65], with the exception of dibenz[a,h]anthracene determined by polarographic oxidation [66].

[b]Extremely active, +++++; very active, ++++; active, +++; moderately active, ++; weakly active, +; very weakly active, ±; inactive, —.

substitution in the MC radical-cation. This is in contrast to 2-methyl-MC, in which the methyl substituent at C-2 does not prevent nucleophilic substitution at C-1. The carcinogenicity of 8-fluoro-MC and 10-fluoro-MC in spite of the blocking of the bay region by fluoro substitution suggests that the formation of a diolepoxide is not the mechanism of activation. Although the carcinogenicity of 2-methyl-MC, 8-fluoro-MC, and 10-fluoro-MC has not yet been tested in mouse skin, the remaining PAH have all been observed to be active there.

Compounds with relatively high IP, such as 5-methylchrysene, DB[ah]A, and B[a]A, are not carcinogenic when directly applied to the mammary gland (Table VI). 5-Methylchrysene has been demonstrated to be carcinogenic in mouse skin by the diol-epoxide mechanism [79], and the potent carcinogenicity of DB[ah]A in this target organ presumably occurs by the same mechanism [11]. The inactivity of these two skin carcinogens in mammary gland suggests that diol-epoxides are not formed in the latter target organ. Similarly, the skin carcinogens B[a]P-7,8-dihydrodiol and cyclopenta[cd]pyrene, which require a simple epoxidation to become active, do not induce tumors in the rat mammary gland.

The main conclusions that can be reached from these experiments are that (a) oxygenation of PAH by monooxygenase enzymes to form epoxides or diolepoxides does not seem to play a role in the carcinogenicity of these compounds in mammary glands; (b) the results on mammary glands support the hypothesis that one-electron oxidation might be the predominant, if not selective, mechanism of activation in this target organ; and (c) multiple mechanisms of activation appear to occur in mouse skin.

IV. TWO-ELECTRON OXIDATION

A. Bay-Region Vicinal Diol-Epoxides

The formation and significance of bay-region vicinal diol-epoxide derivatives of PAH have been thoroughly reviewed repeatedly [9–11], and our discussion here is not intended to be comprehensive.

The bay-region theory has evolved from studies on the formation and binding of B[a]P metabolites to DNA. Borgen et al. have observed that covalent binding of B[a]P-7,8-dihydrodiol activated by liver microsomes is greater than that of B[a]P and other B[a]P metabolites [80]. Further studies have provided evidence that the B[a]P-7,8-diol-9,10-epoxide may be involved in the binding [81]. This seminal work has led to the bay-region theory, which predicts the greater ease of formation of benzylic carbocations for bay-region epoxides and diol-epoxides [82, 83]. The perturbational molecular orbital calculations by the method of Dewar [84] take into account only the π-electron energy change, $\Delta E_{deloc}/\beta$, which occurs as the oxirane ring breaks at the benzylic carbon atom to form a carbocation. These calculations neglect all effects due to the presence of hydroxyl groups in the saturated ring and their stereochemistry. Large values of $\Delta E_{deloc}/\beta$, which correspond to greater stabilization of the carbocation, depend on the aromatic structure from which the intermediate is formed (Table VII). In general, as can be observed in Table VII, the benzylic carbonium ion formed in the bay-region for a given hydrocarbon has a larger $\Delta E_{deloc}/\beta$ than that formed elsewhere. Thus, these values offer some preliminary guidelines for predicting the PAH that can become carcinogenic through the formation of a bay-region diol-epoxide.

1. Benzo[a]pyrene

Metabolism of B[a]P by liver microsomes and purified hepatic cytochrome P-450 yields phenols (predominantly the 3-, 7-, and 9-hydroxy-B[a]P), quinones (B[a]P-1-,6-,B[a]P-3,6-, and B[a]P-6,12-dione), and B[a]P-dihydrodiols (B[a]P-4,5-,B[a]P-7,8-, and B[a]P-trans-9,10-dihydrodiol) [85–87]. Benzo[a]pyrene is oxidized by cytochrome P-450 to B[a]P-7,8-oxide, which in turn by epoxide hydrase produces B[a]P-7,8-dihydrodiol. This intermediate is readily oxidized by cytochrome P-450 to the diastereomeric B[a]P-7,8-diol-9,10-epoxides (Fig. 4).

Marked stereoselectivity is observed in the metabolism of B[a]P to the 7,8-dihydrodiol, and the major enantiomer with both uninduced and induced liver microsomes is (−)-B[a]P-7,8-dihydrodiol [88, 89]. Stereoselective metabolism is also observed for the biotransformation of the two enantiomeric B[a]P-7,8-dihydrodiols to the optically active diol-epoxides, with (+)-B[a]P-7,8-diol-9,10-epoxide-2 predominating in general [89–91]. These differences in stereoselectivity are the result of different proportions of cytochrome P-450 isozymes in various microsomal preparations [92, 93].

The major nucleic acid adduct of B[a]P observed in various cultured cells is the (+)-enantiomer of B[a]P-7,8-diol-9,10-epoxide-2 bound at C-10 to the 2-amino group of deoxyguanosine [94–100]. The same adduct predominates when the skin of C57B1 mice is treated with B[a]P [71]. Comparative studies of B[a]P and (±)-B[a]P-7,8-dihydrodiol or (−)-B[a]P-7,8-dihydrodiol by repeated application [101–103] and initiation–promotion [76, 104–107] in mouse skin show that their carcinogenicity is similar. The putative proximate metabolite B[a]P-7,8-dihydrodiol or, even better, the (−)-B[a]P-7,8-dihydrodiol would be expected to be more active than the parent B[a]P. This result poses a serious question about this pathway of activation as the exclusive one determining the carcinogenicity of B[a]P.

Both enantiomers of B[a]P-7,8-diol-9,10-epoxide-2 are less active than B[a]P in mouse skin [108, 109]. In this case the low carcinogenicity of the reactive diol-epoxides can be due to decomposition before reaching critical cellular tar-

TABLE VII $\Delta E_{deloc}/\beta$ **Values for Bay-Region and Non-Bay-Region Benzylic Carbonium Ions** [82]

Compound	$\Delta E_{deloc}/\beta$
	0.871
	0.848
	0.794
	0.766
	0.738
	0.714
	0.658
	0.640
	0.628

(*continued*)

TABLE VII (*Continued*)

Compound	$\Delta E_{deloc}/\beta$
	0.604
	0.594
	0.572
	0.572
	0.540
	0.526
	0.526
	0.506
	0.506
	0.488

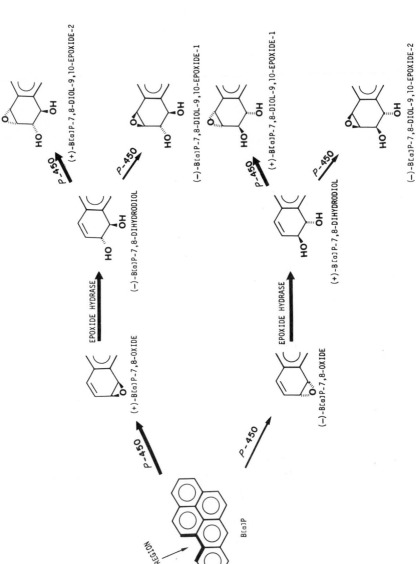

Fig. 4. Absolute stereochemistry of B[a]P metabolites responsible for some of the carcinogenicity of the parent compound.

BAY REGION

B[a]P

P-450

P-450

(+)-B[a]P-7,8-OXIDE

(−)-B[a]P-7,8-OXIDE

EPOXIDE HYDRASE

EPOXIDE HYDRASE

(−)-B[a]P-7,8-DIHYDRODIOL

(+)-B[a]P-7,8-DIHYDRODIOL

P-450

P-450

P-450

P-450

(+)-B[a]P-7,8-DIOL-9,10-EPOXIDE-2

(−)-B[a]P-7,8-DIOL-9,10-EPOXIDE-1

(+)-B[a]P-7,8-DIOL-9,10-EPOXIDE-1

(−)-B[a]P-7,8-DIOL-9,10-EPOXIDE-2

gets. Evidence that (+)-B[a]P-7,8-diol-9,10-epoxide-2 is an ultimate carcinogenic metabolite and that B[a]P-7,8-dihydrodiol is a proximate carcinogenic metabolite has been obtained from studies in newborn mice using intraperitoneal injection [110–113]. In these studies the incidence of pulmonary carcinomas is higher for the proximate and ultimate metabolites when compared with B[a]P.

2. Benz[a]anthracene

This compound displays a very weak carcinogenic activity [114, 115], but shows consistent tumor-initiating activity [116]. Metabolism of B[a]A by rat liver microsomes or purified cytochrome P-450 plus epoxide hydrase affords B[a]A-3,4-dihydrodiol among the various metabolites [117, 118]. The predominant enantiomer of B[a]A-3,4-dihydrodiol is the (3R,4R)-(−)-enantiomer [119].

In turn, metabolism of this enantiomer forms stereoselectively B[a]A-3,4-diol-1,2-epoxide-2, which has the oxirane ring trans to the benzylic hydroxyl group [120].

In initiation–promotion experiments racemic B[a]A-3,4-dihydrodiol causes more tumors than the parent compound B[a]A or any other possible dihydrodiol [121, 122]. Other tumor initiation studies have shown that the (3R,4R)-(−)-enantiomer of B[a]A-3,4-dihydrodiol is more active than the (3S,4S)-(+)-isomer [123]. B[a]A-3,4-Dihydrodiol is more active in inducing lung tumors in newborn mice than B[a]A or other possible dihydrodiols [124]. Racemic B[a]A-3,4-diol-1,2-epoxide-2 is more tumorigenic in mouse skin and newborn mice than the parent compound, other B[a]A diol-epoxides, and in some cases B[a]A-3,4-dihydrodiol [122, 123, 125]. Because the (3R, 4R)-(−)-B[a]A-3,4-dihydrodiol and the racemic B[a]A-3,4-diol-1,2-epoxide-2 are strongly active, the weak carcinogenicity of B[a]A has been attributed to the low conversion of B[a]A to B[a]A-3,4-dihydrodiol [10, 118].

3. 7-Methylbenz[a]anthracene

Substitution of a methyl group at the 7-position of B[a]A leads to substantially enhanced carcinogenicity [126, 127]. Among the methyl-substituted B[a]A, the 7-methyl is the most carcinogenic, followed by the 12-, 6- and 8-substituted compounds, whereas the others are inactive [126, 127]. Mouse skin maintained in short-term organ culture and rat liver microsomes convert 7-methyl-B[a]A to several metabolites, including the 7-methyl-B[a]A-3,4-dihydrodiol [*128–130*]. Adducts formed with DNA in mouse skin treated with 7-methyl-B[a]A *in vitro* [*128*] and *in vivo* [*131*] show an anthracene-like fluorescence spectrum similar to that found when DNA is treated with 7-methyl-B[a]A-3,4-diol-1,2-epoxide, but this result does not rule out a bound form that retains the B[a]A aromaticity.

The tumor-initiating activity of 7-methyl-B[a]A-3,4-dihydrodiol is greater than that of the parent compound and three other 7-methyl-B[a]A dihydrodiols [*132*]. Thus, metabolism, binding, and tumorigenicity studies suggest that the pathway of 7-methyl-B[a]A activation leading to the 3,4-dihydrodiol and, subsequently, to the bay-region diol-epoxide may account at least in part for the carcinogenic activity of this compound. However, the carcinogenicity of 2-fluoro-7-methyl-B[a]A [*133*] suggests that diol-epoxide formation cannot explain all the carcinogenicity, because fluoro substitution blocks the formation of the bay-region diol-epoxide. Various experimental results indicate that other mechanisms of activation can be invoked to explain the carcinogenicity of 7-methyl-B[a]A (see Section V,B).

4. 7,12-Dimethylbenz[a]anthracene

This compound is among the strongest carcinogenic PAH [*115*]. It is metabolized by rat liver microsomes to the 3,4-dihydrodiol, among the various products [*130, 134*]. The same dihydrodiol is also found to arise from the primary metabolite 7-hydroxymethyl-12-methyl-B[a]A [*130, 134*].

The DNA from mouse skin treated with DMBA shows an anthracene-type fluorescence spectrum [*135*], and a 1,2,3,4-ring diol-epoxide adduct is observed [*136*]. The 3,4-dihydrodiols of DMBA are more potent tumor initiators than the parent compound [*137, 138*].

Derivatives of DMBA fluorinated at C-1, C-2, C-5, and C-11 are weakly active or inactive as tumor initiators, with the exception of the 11-fluoro derivative [*139*]. The inactivity of 1-, 2-, and 5-fluoro-DMBA, in which the formation of the bay-region diol-epoxide is hindered, is consistent with this mechanism of activation. However, the 1,2,3,4-tetrahydro-DMBA, although less active than DMBA, displays potent tumor-initiating activity [*140*]. In addition, the induction of sarcomas by 4-fluoro-DMBA subcutaneously injected in rats [*133*] and papillomas in mice treated by initiation–promotion [*140*] clearly indicates that DMBA is activated by mechanisms other than the formation of bay-region diol-epoxides.

5. 3-Methylcholanthrene

After DMBA, MC is one of the most potent carcinogens by various routes of administration [*115*]. The bay-region theory predicts that one or more of the MC-9,10-diol-7,8-epoxides are ultimate carcinogenic forms. In studies on the metabolism of MC by rat liver microsomes, MC is not converted to the presumed proximate metabolite MC-9,10-dihydrodiol, whereas 1-hydroxy-MC, which is one of the major metabolites, is converted to two diastereomeric *trans*-9,10-dihydrodiols [*141, 142*]. This *in vitro* metabolism implies that formation of the ultimate carcinogenic bay-region diol-epoxide from MC requires four distinct steps, namely, hydroxylation at C-1, epoxidation at the 9,10-π-bond, enzymatic hydrolysis to the 9,10-dihydrodiol, and final epoxidation at the 7,8-double bond. Tumor-initiating activity in mouse skin, however, shows that the activity of the two diastereomeric 1-hydroxy-MC-9,10-dihydrodiols is no greater than that of the parent compound MC, whereas 1-hydroxy-MC is less active than the other three compounds [*143*]. The presumed proximate metabolite 1-hydroxy-MC is consistently less tumorigenic than MC [*144, 145*]. In another mouse skin experiment, MC-9,10-dihydrodiol was somewhat less tumorigenic than MC, but more active than the other possible MC-dihydrodiols (MC-4,5-, MC-7,8-, or MC-11,12-dihydrodiol) [*146*]. More lung and hepatic tumors are induced in newborn mice by one of the 1-hydroxy-MC-9,10-dihydrodiol diastereomers than by MC or the marginally active 1-hydroxy-MC [*143*].

The fluorescence of MC bound to mouse skin DNA is consistent with the formation and binding of the 9,10-diol-7,8-epoxide [*135*], but the evidence is not very conclusive. As a result of metabolic, binding, and tumorigenicity studies, it has been proposed that the carcinogenic activity of MC is due to the bay-region diol-epoxide of MC and/or its 1- or 2-hydroxy derivatives, the two primary metabolites of MC [*9,11*]. Although this metabolic pathway can contribute to the carcinogenic activity of MC, the metabolic, binding, and tumorigenicity data discussed here suggest mechanisms of activation for MC other than formation of the bay-region diol-epoxide.

6.   Dibenz[*a,h*]anthracene

The symmetric DB[ah]A, the first pure PAH tested [*147*], is a relatively potent carcinogen in various assays [*114, 115*]. This compound has two equivalent bay regions. DB[ah]A-3,4-Dihydrodiol, the proximate carcinogenic metabolite that yields the ultimate 3,4-diol-1,2-epoxide, is metabolically formed in addition to the 1,2- and 5,6-dihydrodiol [*148, 149*]. However, the mouse skin tumor-initiating activity and lung tumor induction in newborn mice of DB[ah]A-3,4-dihydrodiol are no greater than that of the parent compound, although this dihydrodiol is more potent than the 1,2- and 5,6-dihydrodiol [*150*]. DB[ah]A-3,4-Diol-1,2-epoxide-2 has little or no tumorigenic activity by initiation–promotion [*151*]. These results, similar to those obtained for MC and B[a]P, as discussed earlier, indicate that formation of the DB[ah]A-3,4-dihydrodiol followed by epoxidation at the 1,2-position constitutes a pathway of activation for DB[ah]A, but the total activity of this compound must also be attributed to other electrophilic ultimate metabolites.

7. Chrysene and 5-Methylchrysene

Chrysene is a weak carcinogen [*114, 115*] with moderate tumor-initiating activity [*152, 153*]. The bay-region theory predicts that one or more chrysene-1,2-diol-3,4-epoxides are its ultimate carcinogenic form. The major metabolites of chrysene, which has two equivalent bay regions, are the 1,2- and 3,4-dihydrodiol [*154, 155*]. The major enantiomer of the proximate *trans*-1,2-dihydrodiol is the (1R,2R)-(−)-form. This metabolite in turn is converted predominantly to the 1,2-diol-3,4-epoxide-2, which has the epoxide trans to the benzylic hydroxyl group [*154, 155*]. The racemic *trans*-1,2-dihydrodiol has higher tumor-initiating activity than chrysene, whereas the 3,4- and 5,6-dihydrodiol are inactive [*151, 156*]. The same is also observed in the induction of lung and liver tumors in newborn mice [*156*]. The tumorigenicity on mouse skin and in newborn mice of (−)-chrysene-1,2-dihydrodiol and (+)-chrysene-1,2-diol-3,4-epoxide-2 is greater than that of the corresponding enantiomers [*157, 158*]. These studies clearly indicate that the (1R,2R)-(−)-enantiomer of chrysene-1,2-dihydrodiol and the (+)-chrysene-1,2-diol-3,4-epoxide-2 are proximate and ultimate carcinogenic metabolites.

Among the various monomethylchrysenes, only 5-methylchrysene displays potent carcinogenic activity [*152, 153*]. Hecht *et al.* have provided clear evidence that the bay-region 1,2-diol-3,4-epoxides are the ultimate carcinogenic forms of 5-methylchrysene. The proximate carcinogenic metabolite, 5-methylchrysene-1,2-dihydrodiol, is more tumorigenic in mouse skin than 5-methylchrysene and its other metabolically possible dihydrodiols [*159*]. The tumorigenic activity of various fluorinated derivatives of 5-methylchrysene shows that fluoro substitution at C-1, C-3, and C-12, which interferes with the metabolic formation of the critical bay-region diol-epoxide, leads to diminished or no activity. Substitution instead at C-7 and C-9 has no effect on the carcinogenicity [*160, 161*]. Metabolic studies [*161, 162*] and the identification of a mouse skin DNA adduct arising from a bay-region 1,2-diol-3,4-epoxide of 5-methylchrysene [*163*] provide additional evidence that this hydrocarbon is activated by the proposed diol-epoxide pathway.

8. Benzo[c]phenanthrene

Benzo[c]phenanthrene and its monomethyl derivatives are weak carcinogens by repeated application on mouse skin [*127*]. Benzo[c]phenanthrene is also a weak tumor initiator [*164*]. This symmetric molecule has the possibility of forming two bay-region diol-epoxides. Benzo[c]phenanthrene-3,4-dihydrodiol, the proximate carcinogenic metabolite, is more active than the parent compound, whereas the other two possible dihydrodiols are inactive [*164*]. In these studies the tumorigenicity of the diol-epoxide-1 and diol-epoxide-2 is greater than that of the proximate dihydrodiol and, in fact, greater than any other PAH diol-epoxide tested in mouse skin.

9. Dibenzo[a,h]pyrene and Dibenzo[a,i]pyrene

These two compounds are very potent carcinogens [114, 115]. Both are symmetric and have two identical bay regions. Tumor-initiating activity in mouse skin of the proximate carcinogenic metabolites, DB[a,h]P-1,2-dihydrodiol and DB[a,i]P-3,4-dihydrodiol, is similar to that of their respective parent compounds, whereas the diol-epoxides-2 of DB[ah]P and DB[ai]P, in which the benzylic hydroxyl group and the orirane ring are trans, are less active than the parent compound and the proximate dihydrodiols [165]. In newborn mice, more lung and liver tumors are induced by the proximate dihydrodiols than by the parent compounds or diol-epoxides-2 [165]. The 2,10-difluoro-DB[a,i]P, in which both bay regions are blocked, shows little or no activity in mouse skin and newborn mice [165–167]. Once again it is necessary to invoke multiple mechanisms of activation to explain the potent carcinogenicity of these compounds.

10. 15,16-Dihydro-11-methylcyclopenta[a]phenanthren-17-one

Metabolism and binding studies provide evidence that the bay-region *trans*-3,4-diol-1,2-epoxide of 15,16-dihydro-11-methylcyclopenta[a]phenanthren-17-one is its ultimate carcinogenic metabolite [168, 169]. In addition, the proximate carcinogenic metabolite, *trans*-3,4-dihydrodiol, is more active as a tumor initiator on mouse skin than the parent compound [170]. The 1,11-methano-17-one deriva-

tive, in which the bay region is blocked by a methylene bridge, shows moderately strong tumorigenic activity by initiation–promotion [169]. This result suggests that a mechanism of activation other than vicinal diol-epoxide formation is operative for the bridged derivative.

B. Benzylic Esters of Polycyclic Aromatic Hydrocarbons

Hydroxylation of the benzylic carbon atom of methyl-substituted PAH represents a major metabolic pathway for these compounds [79, 128–130, 148, 171–174]. Conjugation of hydroxymethyl derivatives to form glucuronides, sulfates, phosphates, or acetates can occur, and these polar esters can readily be excreted. This generally represents a major detoxification pathway. However, an ester that bears a good leaving group may be a potential ultimate carcinogen when it can generate a relatively stable carbonium ion capable of binding to cellular nucleophiles [12, 175, 176]. Benzylic carbonium ions of meso-methylated PAH can fulfill this requirement.

Determination of the adducts of 1'-hydroxyestragole [177] and 1'-hydroxysafrole [178] with liver DNA generated by the treatment of mice with the two compounds indicates that reactive allylic–benzylic esters are formed during in vivo metabolism of these compounds.

Carcinogenicity and mutagenicity data as well as binding studies of benzylic esters lead to the conclusion that certain of these compounds, if formed in target tissues, are possible ultimate carcinogens of their parent compounds.

Hydroxymethyl derivatives of PAH are generally carcinogenic, with potencies ranging from weaker to stronger than that of the parent PAH. The potency of these hydroxymethyl derivatives depends on the compound and the tumor model tested. 6-Hydroxymethyl-B[a]P is a stronger inducer of subcutaneous sarcomas in rats than 6-methyl-B[a]P [14]. However, it is a weaker carcinogen than the parent compound when tested by repeated application on mouse skin [173]. The sodium sulfate ester of 6-hydroxymethyl-B[a]P is a more potent carcinogen in mouse skin than 6-hydroxymethyl-B[a]P, whereas the effect of 6-acetoxymethyl-B[a]P does not significantly differ from that of 6-hydroxymethyl-B[a]P [13]. In the 7-methyl-B[a]A series, 7-acetoxymethyl-B[a]A is a stronger carcinogen than 7-hydroxymethyl-B[a]A, and this in turn is more potent than the parent compound, 7-methyl-B[a]A [13]. These results suggest that certain benzylic esters of meso-methyl-substituted PAH may be ultimate carcinogens for the parent compound.

High levels of mutagenicity in Salmonella typhimurium TA98 are observed for the sodium sulfate esters of 6-hydroxymethyl-B[a]P and 7-hydroxymethyl-B[a]A at concentrations of 0.1 to 3 nmol per plate [179]. No mutagenic activity is observed with the nonbenzylic sodium sulfate ester of 7-hydroxyethyl-B[a]A at concentrations of up to 100 nmol per plate [179]. Mutagenic activities are also obtained in the same strain of bacteria with the sulfate esters of 7-hydroxymethyl-12-methyl-B[a]A and 1-hydroxymethylpyrene [180]. In the case of 7-hydroxymethyl-12-methyl-B[a]A mutagenicity is also obtained when the hydroxymethyl derivative is incubated with a 3'-phosphoadenosine 5'-phosphosulfate-generating system and rat liver cytosol, indicating that the effect is mediated by the intermediate reactive sulfate ester. This mutagenesis is inhibited by the presence of glutathione or sulfatase.

Covalent binding of hydroxymethyl derivatives of PAH to DNA is observed when ATP is used to mediate the nonenzymatic reaction [181–183]. Although the mechanism has not been proved, the reaction appears to proceed through the formation of a reactive intermediate phosphate ester [*183*]. Because about 10 times as much 6-hydroxymethyl-B[a]P or 7-hydroxymethyl-12-methyl-B[a]A is bound to DNA in this reaction as 7-hydroxymethyl-B[a]A, the extent of binding appears to depend on the relative stability of the PAH carbonium ion [*181–183*]. The fluorescence spectrum of 6-hydroxymethyl-B[a]P bound to DNA in the reaction is virtually identical to the spectrum of 6-methyl-B[a]P, indicating that the ATP-mediated binding of 6-hydroxymethyl-B[a]P occurs at the 6-methyl group, as expected if it proceeds through formation of a reactive phosphate ester intermediate [*183*].

Although hydroxylated derivatives have been observed to be major metabolites of methylated PAH, in particular 7-methyl-B[a]A and DMBA, clear evidence has not yet been obtained for subsequent binding of the methyl group to cellular macromolecules. However, when mouse skin is treated with 6-methyl-B[a]P and the DNA is isolated, hydrolyzed to deoxyribonucleosides, and analyzed by HPLC, at least one adduct is observed in which 6-methyl-B[a]P is bound through the methyl group to DNA [*74*]. This binding could proceed through hydroxylation and subsequent esterification to form an intermediate that reacts with DNA, or the 6-methyl-B[a]P could be activated by one-electron oxidation to form a radical-cation, which binds to DNA at the methyl group (see Section III,B, p. 337). The significance of each of these two possible mechanisms depends on the enzymes and cofactors present in a particular organ or cell.

V. DISCUSSION

A. Assessment of Enzymology Involved in One-Electron Transfers

Chemical carcinogens, other xenobiotics, and certain endobiotics undergo metabolic activation by two major general mechanisms: two-electron oxidation, or oxygenation, and one-electron oxidation. The major enzyme involved in this process is cytochrome P-450, which, acting as a monooxygenase, is well recognized to catalyze the oxygenation of various compounds. The monooxygenase can also catalyze one-electron oxidation [*15–20*], although the process seems to be more efficiently catalyzed by cytochrome P-450 acting as a peroxidase [*26–28*] or by numerous peroxidases themselves [*29–50*]. In general it is very difficult to discern the efficiency of these two metabolic pathways in the various enzyme systems.

PES-Catalyzed oxidation of B[a]P produces exclusively B[a]P-1,6-, B[a]P-3,6-, and B[a]P-6,12-dione [*51, 52*]. The other common metabolites of B[a]P formed by mammalian microsomes, namely, phenols and dihydrodiols, are not present. Evidence is accumulating that PES also catalyzes the one-

electron oxidation of *N*-hydroxy-2-acetylaminofluorene [*32*], tetramethylbenzidine [*47*], and diethylstilbestrol [*48–50*]. The percentage of B[a]P quinones
formed generally varies with the tissue and conditions used for metabolism. For
example, the B[a]P quinones are 28% of the metabolites produced by rat liver
microsomes with NADPH, whereas with H_2O_2 as cofactor the percentage increases to 50, and with cumene hydroperoxide, to 94. These and other data
summarized in the following paragraphs indicate that the percentage of quinones
generally increases under peroxidative conditions.

Although the precursor of B[a]P quinones has not been isolated, it has been
suggested by both Nagata *et al.* [*184–187*] and Ts'o and collaborators [*188–191*]
to be 6-hydroxy-B[a]P. This hypothesis arises from the detection by ESR spectroscopy of 6-oxy-B[a]P radical when B[a]P is incubated with microsomes. The
same signal is observed by autoxidation or microsomal activation of 6-hydroxy-
B[a]P, and both substrates produce the same B[a]P quinones. On the basis of
these experimental results, the mechanism in Fig. 5 has been proposed. In this
pathway of activation it is assumed that transfer of active oxygen in the formation
of the putative 6-hydroxy-B[a]P occurs via the oxene-type intermediate of the
cytochrome *P*-450-monooxygenase enzyme system. Formation of the quinones
occurs in general by autoxidation of 6-hydroxy-B[a]P, as presented in Fig. 5 for
B[a]P-3,6-dione.

For several reasons we do not think that the insertion of active oxygen in the
first step of B[a]P metabolism to quinones involves the typical oxene intermediate:

1. The formation of B[a]P quinones is almost exclusive when metabolism
   occurs under peroxidatic conditions, namely, by PES [*51, 52*] or cytochrome *P*-450 with cumene hydroperoxide [*26, 27*]. Under these condi-

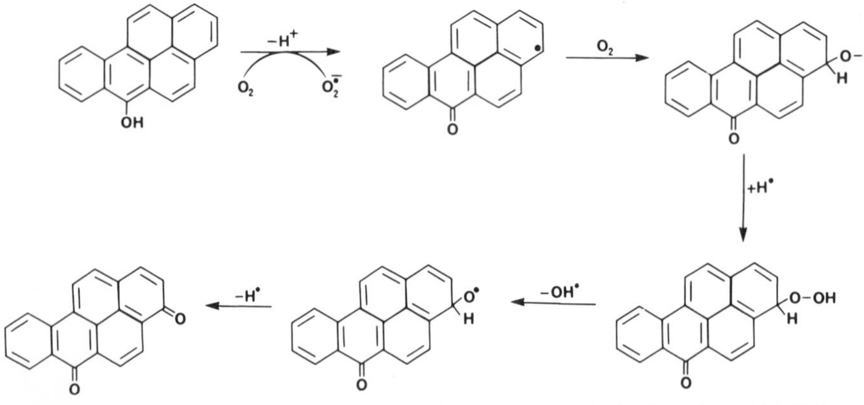

Fig. 5. Autoxidation of 6-hydroxy-B[a]P to B[a]P-diones. Only the formation of B[a]P-3,6-
dione is shown [*188*].

TABLE VIII   Ionization Potentials of PAH and ESR Signal of the
Oxy Radicals

Compound	Ionization potential (eV)[a]	ESR of oxy radical[b]
Benzo[e]pyrene	7.62	−
Dibenz[a,h]anthracene	7.57	−
Pyrene	7.50	−
Anthracene	7.43	−
Benzo[a]pyrene	7.23	+
Dibenzo[a,h]pyrene	6.97	+
Anthanthrene	6.96	+

[a]References [65, 66].
[b]References [187, 189]; −, not detected; +, detected.

tions one-electron transfer is an important mechanism of activation [28, 47–50].

2. The typical oxy radical precursor of quinones has also been observed for DB[ah]P [189] and anthanthrene [187], whereas no oxy radical has been observed for benzo[e]pyrene [189], DB[ah]A [189], pyrene [187], or anthracene [187]. From Table VIII it is clear that the phenoxyl radical is formed only for PAH with relatively low IP, indicating that this metabolic step is related to the capacity of the PAH to donate an electron. A similar cutoff has also been observed for HRP–H_2O_2-catalyzed binding of PAH to DNA (Table III) and carcinogenicity in rat mammary gland (Tables V and VI).

3. Cytochrome P-450-catalyzed oxygenation of sulfide to sulfoxide and sulfoxide to sulfone proceeds via initial formation of a radical-cation by one-electron transfer (p. 325) [18, 19].

All of these results suggest that the first step in the formation of B[a]P quinones involves the transfer of one electron from B[a]P to the cytochrome P-450–(FeO)$^{3+}$ complex (Fig. 6), as in the mechanism proposed by Watanabe et al. [18, 19]. The B[a]P radical-cation would then react with the generated nucleophilic oxygen of the cytochrome P-450–(FeO)$^{2+}$ complex to form the 6-oxy-B[a]P radical, in which the spin density is localized mainly on the oxygen and at positions 1, 3, and 12. Autoxidation of this intermediate, as illustrated in Fig. 5 would generate the three B[a]P quinones.

The formation of B[a]P quinones via an initial one-electron transfer pathway implies that the relative amount of quinones as compared with other B[a]P metabolites represents an index of one-electron oxidation versus oxygenation. Metabolism of B[a]P by numerous animal tissue explants, cultured cells, and cell

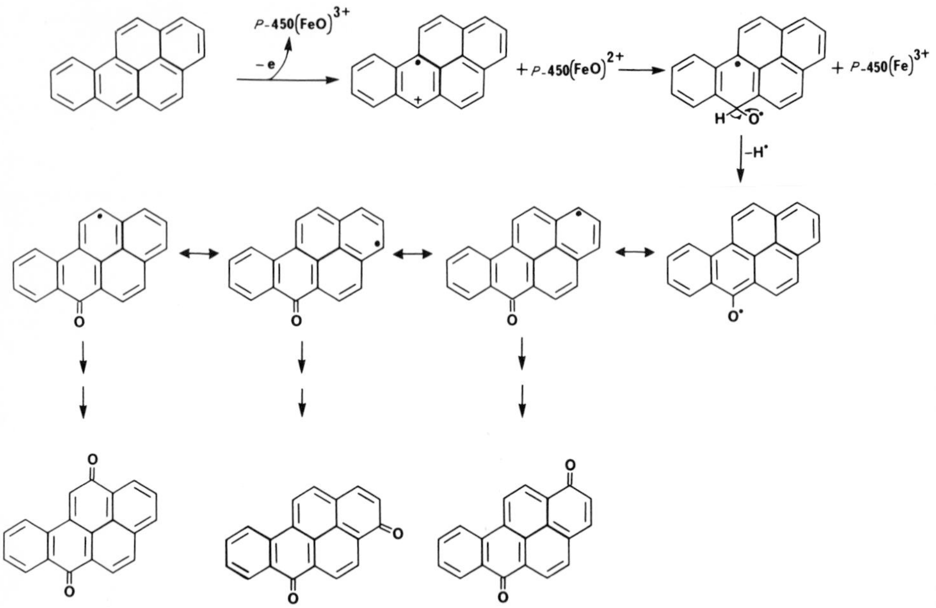

Fig. 6.   Proposed metabolism of B[a]P to B[a]P-diones by one-electron oxidation.

fractions produces B[a]P quinones in variable amounts (Table IX). It is particularly interesting that, in the presence of NADPH, liver nuclei and nuclear membranes from untreated rats produce about 50% quinones, whereas the liver microsomes yield only 26% quinones. Microsomes from hamster placentas, fetal skins, and fetal livers all produce approximately 70% quinones. Pretreatment of rats, mice, or hamsters with B[a]P or MC reduces the proportion of quinone metabolites or has no effect. Finally, metabolism under peroxidative conditions, that is, rat liver microsomes in the presence of H_2O_2 or cumene hydroperoxide and ram seminal vesicle PES with arachidonic acid, results in increased or exclusive formation of B[a]P quinones. A similar compilation of results from metabolism of B[a]P by human explants, cultured cells, and microsomal fractions (Table X) shows that B[a]P quinones are formed in variable amounts. When B[a]P coated on ferric oxide particles is metabolized by cultured human bronchus, approximately half of the metabolites are B[a]P quinones. Similarly high levels of B[a]P quinones are observed in liver and kidney microsomal incubations and with cultured lymphocytes. The results in Tables IX and X indicate that the formation of B[a]P quinones by one-electron oxidation occurs in a wide variety of mammalian cells, and this metabolic pathway is predominant under certain conditions. The relative amount of B[a]P quinones versus other metabolites reflects the tendency of enzyme systems to activate appropriate substrates by one-electron oxidation.

TABLE IX Percentage of B[a]P Quinones after Metabolism by Animal Explants, Cells, or Cell Fractions

System[b]	B[a]P quinones (%)[a] Untreated	Aroclor[c]	β-naphtho-flavone	B[a]P	MC	Phenobarbital
Mouse						
Cultured tracheobronchial epithelium, C57Bl/6N	7 [192]					
Cultured tracheobronchial epithelium, DBA/2N	28 [192]					
Skin microsomes, C57Bl/6N	25 [193]	9 [193]			14 [193]	
Skin microsomes, DBA/2N	11 [193]	20 [193]			15 [193]	
Liver microsomes, C57Bl/6J	46 [194]				32 [194]	
Liver microsomes, DBA/2J	45 [194]				47 [194]	
Rat						
Cultured tracheobronchial epithelium	30 [192]					
Cultured bladder epithelial cells	3 [195]					
Skin microsomes	12 [193]	8 [193]			15 [193]	
Lung microsomes	31 [196]				14 [196]	29 [196]
Liver microsomes	26 [191, 196, 197]				18 [196–198]	28 [27, 196]
Liver microsomes + H$_2$O$_2$					50 [27]	
Liver microsomes + cumene hydroperoxide					94 [27]	
Liver nuclei	47 [199]				21 [199]	27 [94]
Liver nuclear membranes	55 [191]					
Hamster						
Cultured tracheobronchial epithelium	4 [192]					
Liver microsomes	30 [196]				33 [196]	
Pregnant female liver microsomes	30 [200]			21 [200]		
Lung microsomes	44 [196]				37 [196]	
Pregnant female liver microsomes	58 [200]			17 [200]		
Placental microsomes	65 [200]			55 [200]		

(continued)

TABLE IX   (*Continued*)

	B[a]P quinones (%)[a]					
System[b]	Untreated	Aroclor[c]	β-naphtho-flavone	B[a]P	MC	Phenobarbital
Fetal skin microsomes	72 [*200*]			52 [*200*]		
Fetal liver microsomes	66 [*200*]			52 [*200*]		
Rabbit						
Liver microsomes						
Cytochrome P-450$_{LM1}$						9 [*201*]
Cytochrome P-450$_{LM2}$						12 [*201*]
Cytochrome P-450$_{LM3b}$						26 [*201*]
Cytochrome P-450$_{LM4}$			50 [*201*]			38 [*201*]
Cytochrome P-450$_{LM7}$	14 [*201*]					
Ram						
Seminal vesicle PES + arachidonic acid	100 [*52*]					

[a]Numbers in brackets indicate the reference from which value is taken.
[b]Cultured cells were exposed to B[a]P for 24 h, whereas microsomal and nuclear preparations were incubated with B[a]P for 30 min and included NADPH as cofactor unless otherwise indicated.
[c]Induction regimens are presented in the original reference.

B. Multiple Mechanisms of Polycyclic Aromatic Hydrocarbon Carcinogenesis

Many chemical carcinogens, including PAH, display organ specificity in their action. One of the primary reasons for this specificity is derived from the capacity of an organ to activate a procarcinogen to reactive electrophiles capable of triggering the tumor initiation process.

Numerous experimental results concerning activation of PAH to bay-region diol-epoxides has led the scientific community to think that this mechanism is almost exclusive in PAH carcinogenesis. There are, however, a number of carcinogenic PAH that for various reasons cannot be activated by the formation of bay-region diol-epoxides. Anthanthrene, which lacks a bay region, induces skin carcinomas in Swiss mice by repeated application [*211*] and lung carcinomas in rats by implantation [*212*]. Similarly, 6-methyl- and 6,12-dimethylanthanthrenes induce tumors in rats by subcutaneous injection [*213*] and in mouse skin by both repeated application and initiation–promotion [*76*].

The carcinogenicity of 10-aza-B[a]P [*214*], in which nitrogen replaces the carbon atom at the 10-position, cannot be explained by formation of the bay-region diol-epoxide. This mechanism also cannot play a role in the activity of 2-fluoro-7-methyl-B[a]A [*133*], 3-fluoro-7-methyl-B[a]A [*215*], 4-fluoro-7,12-di-

TABLE X   Percentage of B[a]P Quinones after Metabolism by Human Explants, Cells, or Microsomes

System[a]	B[a]P quinones (%)[b]
Cultured esophagus	6 [202], 31 [203]
Cultured trachea	26 [203]
Cultured normal bronchus	3 [202], 12 [203], 17 [204]
Cultured normal bronchus + [³H]B[a]P on ferric oxide particulates for 1 to 6 h	39–61 [205]
Cultured bronchus from lung cancer patient	6 [204]
Cultured peripheral lung	4 [206]
Cultured tracheobronchial epithelium	28 [192]
Cultured pulmonary alveolar macrophages	9 [207]
Cultured lymphocytes	49, 35 [198], 20 [208]
Cultured monocytes	6 [208]
Cultured colon	3 [209], 6 [202], 13 [203], 13 [210]
Cultured duodenum	6 [202]
Cultured bladder epithelial cell lines	0, 5, 14 [195]
Bronchial microsomes[c]	31 [204]
Lung microsomes	34 [196]
Kidney microsomes	57 [196]
Liver microsomes	27 [198], 55 [196]

[a]Cultured cells were exposed to B[a]P in medium for 24 h unless otherwise indicated.
[b]Numbers in brackets indicate the reference from which value is taken.
[c]Microsomes were incubated with B[a]P in the presence of added NADPH for 30 min.

methyl-B[a]A [133, 140], 8-fluoro-MC [77], or 10-fluoro-MC [77], in which the formation of diol-epoxide is blocked by fluoro substitution. Similarly, 1,2,3,4-tetrahydro-DMBA, which displays potent tumor-initiating activity [140], must be activated by another mechanism.

The carcinogenicity of PAH, as well as other compounds, must be considered in terms of two basic mechanisms of enzymatic activation: two-electron oxidation, or oxygenation, and one-election oxidation. From present knowledge it is logical to assume that vicinal bay-region diol-epoxides are the predominant ultimate carcinogens in the latter mechanism. However, other non-bay-region vicinal diol-epoxides, as reported in the activation of B[a]A [216], could play a minor role, and some small contribution may arise from the formation of simple arene oxides. In this connection the carcinogenicity of cyclopenta[cd]pyrene is attributed to the formation of its ultimate 3,4-oxide [8].

Hydroxylation of meso-methylated PAH is a major metabolic reaction. Further conjugation to form an ester that bears a good leaving group may generate a major potential carcinogenic intermediate. Present carcinogenicity and mutagenicity data (see Section III,B, p. 353) indicate that, if these benzylic esters are formed in target tissues, they could be the ultimate carcinogens of their parent compounds.

One-electron oxidation of PAH generates radical-cations, which can bind directly to cellular nucleophiles. Binding studies catalyzed by $HRP-H_2O_2$ and carcinogenicity studies in rat mammary gland, in which one-electron oxidation of PAH seems to play a predominant role, indicate that there exists an IP barrier above which this mechanism of activation cannot occur. Therefore, formation of bay-region diol-epoxides can explain the carcinogenicity of PAH with relatively high IP [65], such as benzo[c]phenanthrene, chrysene, 5-methylchrysene, B[a]A, DB[ah]A, and 15,16-dihydro-11-methylcyclopenta[a]phenanthren-17-one. Most of the potent carcinogenic PAH have IP under 7.35 eV, which has been set as a cutoff value [65, 78]. Thus, compounds such as B[a]P, DMBA, MC, DB[ah]P, and DB[ai]P can be activated both by one-electron oxidation and oxygenation with formation of bay-region diol-epoxides, depending on the enzymes present in the target organ. We have seen that the presence of enzyme systems with peroxidatic activities tend to favor one-electron oxidation. We think that measurement of the relative amount of B[a]P quinones versus other metabolites formed in target cells may provide a guideline for predicting which pathway of carcinogenic activation occurs therein. Once these data are available, binding, carcinogenicity, and metabolism studies can be conducted to delineate the mechanism(s) of activation involved.

ACKNOWLEDGMENTS

We thank the National Institutes of Health for supporting this research through the National Institute of Environmental Health Sciences, Grant R01-ES02145, and the National Cancer Institute, Contract N01 CP05620 and Grants R01 CA25176 and R01 CA32376. We also appreciate the editorial assistance of Ms. Mardelle Susman.

REFERENCES

1. J. A. Miller, *Cancer Res.* **30**, 559–576 (1970).
2. E. C. Miller and J. A. Miller, *Cancer* **47**, 2327–2345 (1981).
3. A. Hamilton, in "Molecular Mechanisms of Oxygen Activation" (O. Hayaishi, ed.), pp. 405–451. Academic Press, New York, 1974.
4. D. Y. Cooper, S. Levin, S. Narasimhulu, O. Rosenthal, and R. W. Estabrook, *Science* **147**, 400–402 (1965).
5. A. Y. H. Lu and M. J. Coon, *J. Biol. Chem.* **243**, 1331–1332 (1968).
6. E. Eisenstadt and A. Gold, *Proc. Natl. Acad. Sci. U.S.A.* **75**, 1667–1669 (1978).
7. A. Gold and E. Eisenstadt, *Cancer Res.* **40**, 3940–3944 (1980).
8. E. Cavalieri, E. Rogan, B. Toth, and A. Munhall, *Carcinogenesis (London)* **2**, 277–281 (1981).
9. M. Nordqvist, D. R. Thakker, H. Yagi, R. E. Lehr, A. W. Wood, W. Levin, A. H. Conney, and D. M. Jerina, in "Molecular Basis of Environmental Toxicity" (R. S. Bhatnager, ed.), pp. 329–357. Ann Arbor Sci. Publ., Ann Arbor, Michigan, 1979.

10. A. W. Wood, W. Levin, R. L. Chang, H. Yagi, D. R. Thakker, R. E. Lehr, D. M. Jerina, and A. H. Conney, in "Polynuclear Aromatic Hydrocarbons" (P. W. Jones and P. Leber, eds.), pp. 531–551. Ann Arbor Sci. Publ., Ann Arbor, Michigan, 1979.
11. A. H. Conney, *Cancer Res.* **42**, 4875–4917 (1982).
12. J. W. Flesher and K. L. Sydnor, *Cancer Res.* **31**, 1951–1954 (1971).
13. E. Cavalieri, R. Roth, and E. Rogan, in "Polynuclear Aromatic Hydrocarbons" (P. W. Jones and P. Leber, eds.), pp. 517–529. Ann Arbor Sci. Publ., Ann Arbor, Michigan, 1979.
14. K. L. Sydnor, C. H. Bergo, and J. W. Flesher, *Chem.-Biol. Interact.* **29**, 159–167 (1980).
15. R. P. Hanzlik and R. H. Tullman, *J. Am. Chem. Soc.* **104**, 2048–2050 (1982).
16. T. L. Macdonald, K. Zirvi, L. T. Burka, P. Peyman, and F. P. Guengerich, *J. Am. Chem. Soc.* **104**, 2050–2052 (1982).
17. O. Augusto, H. S. Beilan, and P. R. Ortiz de Montellano, *J. Biol. Chem.* **257**, 11288–11295 (1982).
18. Y. Watanabe, T. Iyanagi, and S. Oae, *Tetrahedron Lett.* **21**, 3685–3688 (1980).
19. Y. Watanabe, T. Iyanagi, and S. Oae, *Tetrahedron Lett.* **23**, 533–536 (1982).
20. E. J. Rauckman, G. M. Rosen, and J. Cavagnaro, *Mol. Pharmacol.* **21**, 458–463 (1982).
21. E. G. Hrcay and P. J. O'Brien, *Arch. Biochem. Biophys.* **147**, 14–27 (1971).
22. F. F. Kadlubar, K. C. Morton, and D. M. Ziegler, *Biochem. Biophys. Res. Commun.* **54**, 1255–1261 (1973).
23. A. D. Rahimtula and P. J. O'Brien, *Biochem. Biophys. Res. Commun.* **60**, 440–447 (1974).
24. M. J. Coon, *Drug Metab. Dispos.* **9**, 1–4 (1980).
25. R. C. Blake and M. J. Coon, *J. Biol. Chem.* **255**, 4100–4111 (1980).
26. J. Capdevila, R. W. Estabrook, and R. A. Prough, *Arch. Biochem. Biophys.* **200**, 186–195 (1980).
27. R. Renneberg, J. Capdevila, N. Chacos, R. W. Estabrook, and R. A. Prough, *Biochem. Pharmacol.* **30**, 843–848 (1981).
28. B. W. Griffin, C. Marth, Y. Yasukochi, and B. S. S. Masters, *Arch. Biochem. Biophys.* **205**, 543–553 (1980).
29. H. Bartsch and E. Hecker, *Biochim. Biophys. Acta* **237**, 567–578 (1971).
30. H. Bartsch, J. A. Miller, and E. C. Miller, *Biochim. Biophys. Acta* **273**, 40–51 (1972).
31. R. A. Floyd, L. M. Soong, and P. L. Culver, *Cancer Res.* **36**, 1510–1519 (1976).
32. P. K. Wong, M. J. Hampton, and R. A. Floyd, in "Prostaglandins and Cancer" (T. J. Powles, R. S. Bockman, K. V. Honn, and P. Ramwell, eds.), pp. 167–179. Alan R. Liss, Inc., New York, 1982.
33. M. Metzler and J. A. McLachlan, *Biochem. Biophys. Res. Commun.* **85**, 874–884 (1978).
34. T. Sawahata and R. A. Neal, *Biochem. Biophys. Res. Commun.* **109**, 988–994 (1982).
35. B. W. Griffin and P. L. Ting, *Biochemistry* **17**, 2206–2211 (1978).
36. G. Galliani and B. Rindone, *J. Chem. Soc., Perkin Trans. 1* pp. 456–460 (1978).
37. G. Galliani and B. Rindone, *J. Chem. Soc., Perkin Trans. 2* pp. 1–3 (1980).
38. G. Galliani and B. Rindone, *Bioorg. Chem.* **10**, 283–289 (1981).
39. P. D. Josephy, T. Eling, and R. P. Mason, *J. Biol. Chem.* **257**, 3669–3675 (1982).
40. P. D. Josephy, R. P. Mason, and T. Eling, *Carcinogenesis (London)* **3**, 1227–1230 (1982).
41. B. Kalyanaraman and R. P. Mason, *Biochem. Biophys. Res. Commun.* **105**, 217–224 (1982).
42. J. S. Wiseman, J. S. Nichols, and M. X. Kolpak, *J. Biol. Chem.* **257**, 6328–6332 (1982).
43. D. Reigh, M. Stuart, and R. Floyd, *Experientia* **34**, 337–339 (1978).
44. T. V. Zenser, M. B. Mattammal, H. J. Armbrecht, and B. B. Davis, *Cancer Res.* **40**, 2839–2845 (1980).
45. M. B. Mattammal, T. V. Zenser, and B. B. Davis, *Cancer Res.* **41**, 4961–4966 (1981).
46. B. Kalyanaraman, K. Sivarajah, T. E. Fling, and R. P. Mason, *Carcinogenesis (London)* **4**, 1341–1343 (1983).

47. P. D. Josephy, T. E. Eling, and R. P. Mason, *J. Biol. Chem.* **258**, 5561–5569 (1983).
48. G. H. Degen, T. E. Eling, and J. A. McLachlan, *Cancer Res.* **42**, 919–923 (1982).
49. J. A. McLachlan, A. Wong, G. H. Degen, and J. C. Barrett, *Cancer Res.* **42**, 3040–3045 (1982).
50. G. H. Degen, A. Wong, T. E. Eling, J. C. Barrett, and J. A. McLachlan, *Cancer Res.* **43**, 992–996 (1983).
51. L. J. Marnett, G. A. Reed, and J. T. Johnson, *Biochem. Biophys. Res. Commun.* **79**, 569–576 (1977).
52. L. J. Marnett and G. A. Reed, *Biochemistry* **18**, 2923–2929 (1979).
53. M. Wilk, W. Bez, and J. Rochlitz, *Tetrahedron* **22**, 2599–2608 (1966).
54. J. Fried, in "Chemical Carcinogenesis" (P. O. P. Ts'o and J. DiPaolo, eds.), Part A, pp. 197–215. Dekker, New York, 1974.
55. E. Cavalieri and R. Auerbach, *JNCI, J. Natl. Cancer Inst.* **53**, 393–397 (1974).
56. E. M. Menger, R. B. Spokane, and P. D. Sullivan, *Biochem. Biophys. Res. Commun.* **71**, 610–616 (1976).
57. J. Rochlitz, *Tetrahedron* **23**, 3043–3048 (1967).
58. M. Wilk and W. Girke, *J. Natl. Cancer Inst.* **49**, 1585–1597 (1972).
59. W. Caspary, B. Cohen, S. Lesko, and P. O. P. Ts'o, *Biochemistry* **12**, 2649–2656 (1973).
60. S. A. Lesko, P. O. P. Ts'o, and R. S. Umans, *Biochemistry* **8**, 2291–2298 (1969).
61. H. D. Hoffman, S. A. Lesko, Jr., and P. O. P. Ts'o, *Biochemistry* **9**, 2594–2604 (1970).
62. E. Cavalieri and R. Roth, *J. Org. Chem.* **41**, 2679–2684 (1976).
63. E. Cavalieri, R. Roth, and E. G. Rogan, in "Polynuclear Aromatic Hydrocarbons: Chemistry, Metabolism and Carcinogenesis" (R. I. Fruedenthal and P. W. Jones, eds.), Vol. 1, pp. 181–190. Raven Press, New York, 1976.
64. E. G. Rogan, R. Roth, and E. Cavalieri, in "Polynuclear Aromatic Hydrocarbons: Chemistry and Biological Effects" (A. Bjørseth and A. J. Dennis, eds.), pp. 259–265. Battelle Press, Columbus, Ohio, 1980.
65. E. L. Cavalieri, E. G. Rogan, R. W. Roth, R. K. Saugier, and A. Hakam, *Chem.-Biol. Interact.* **47**, 87–109 (1983).
66. E. S. Pysh and N. C. Yang, *J. Am. Chem. Soc.* **85**, 2124–2130 (1963).
67. L. Jeftic and R. N. Adams, *J. Am. Chem. Soc.* **92**, 1332–1337 (1970).
68. G. M. Blackburn, P. E. Taussing, and J. P. Will, *J. Chem. Soc., Chem. Commun.* pp. 907–908 (1974).
69. E. Rogan, R. Roth, P. Katomski, J. Benderson, and E. Cavalieri, *Chem.-Biol. Interact.* **22**, 35–51 (1978).
70. E. G. Rogan, P. A. Katomski, R. W. Roth, and E. L. Cavalieri, *J. Biol. Chem.* **254**, 7055–7059 (1979).
71. M. Koreeda, P. D. Moore, P. G. Wislocki, W. Levin, A. H. Conney, H. Yagi, and D. M. Jerina, *Science* **199**, 778–781 (1978).
72. S. W. Ashurst and G. M. Cohen, *Int. J. Cancer* **27**, 357–364 (1981).
73. W. Baer-Dubrowska and W. Alexander, *Cancer Lett.* **13**, 47–52 (1981).
74. E. G. Rogan, A. Hakam, and E. L. Cavalieri, *Chem.-Biol. Interact.* **47**, 111–122 (1983).
75. E. Cavalieri, E. Rogan, and R. Roth, in "Free Radicals and Cancer" (R. A. Floyd, ed.), pp. 117–158. Dekker, New York, 1982.
76. E. Cavalieri, D. Sinha, and E. Rogan, in "Polynuclear Aromatic Hydrocarbons: Chemistry and Biological Effects" (A. Bjørseth and A. J. Dennis, eds.), pp. 215–231. Battelle Press, Columbus, Ohio, 1980.
77. E. Cavalieri and E. Rogan, in "Polynuclear Aromatic Hydrocarbons: Physical and Biological Chemistry" (M. Cooke, A. J. Dennis, and G. L. Fisher, eds.), pp. 145–155. Battelle Press, Columbus, Ohio, 1982.
78. E. Cavalieri and E. Rogan, in "Polynuclear Aromatic Hydrocarbons: Formation, Metabolism,

and Measurement" (M. Cooke and A. J. Dennis, eds.) pp. 1–26. Battelle Press, Columbus, Ohio 1983.

79. S. S. Hecht, R. Mazzarese, S. Amin, E. LaVoie, and D. Hoffmann, in "Polynuclear Aromatic Hydrocarbons" (P. W. Jones and P. Leber, eds.), pp. 733–752. Ann Arbor Sci. Publ., Ann Arbor, Michigan, 1979.

80. A. Borgen, H. Darvey, N. Castagnoli, T. T. Crocker, R. E. Rasmussen, and I. Y. Wang, *J. Med. Chem.* **16**, 502–506 (1973).

81. P. Sims, P. L. Grover, A. Swaisland, K. Pal, and A. Hewer, *Nature (London)* **252**, 326–328 (1974).

82. D. M. Jerina and J. W. Daly, in "Drug Metabolism-From Microbe to Man" (D. W. Parke and R. L. Smith, eds.), pp. 13–32. Taylor & Francis, London, 1976.

83. D. M. Jerina, R. E. Lehr, H. Yagi, O. Hernandez, P. M. Dansette, P. G. Wislocki, A. W. Wood, R. L. Chang, W. Levin, and A. H. Conney, in "In Vitro Metabolic Activation in Mutagenesis Testing" (F. J. deSerres, J. R. Fouts, J. R. Bend, and R. M. Philpot, eds.), pp. 159–177. Elsevier/North-Holland Biomedical Press, Amsterdam, 1976.

84. M. J. S. Dewar, "The Molecular Orbital Theory of Organic Chemistry," pp. 214–217, 304–306. McGraw-Hill, New York, 1969.

85. J. K. Selkirk, R. G. Croy, P. P. Roller, and H. V. Gelboin, *Cancer Res.* **34**, 3474–3480 (1974).

86. G. Holder, H. Yagi, P. Dansette, D. M. Jerina, W. Levin, A. Y. H. Lu, and A. H. Conney, *Proc. Natl. Acad. Sci. U.S.A.* **71**, 4356–4360 (1974).

87. S. K. Yang, P. P. Roller, and H. V. Gelboin, *Biochemistry* **16**, 3680–3687 (1977).

88. S. K. Yang and H. V. Gelboin, *Biochem. Pharmacol.* **25**, 2221–2225 (1976).

89. D. R. Thakker, H. Yagi, H. Akagi, M. Koreeda, A. Y. H. Lu, W. Levin, A. W. Wood, A. H. Conney, and D. M. Jerina, *Chem.-Biol. Interact.* **16**, 281–300 (1977).

90. S. K. Yang, D. W. McCourt, P. P. Roller, and H. V. Gelboin, *Proc. Natl. Acad. Sci. U.S.A.* **73**, 2594–2598 (1976).

91. S. K. Yang, D. W. McCourt, J. C. Leutz, and H. V. Gelboin, *Science* **196**, 1199–1201 (1977).

92. J. Deutsch, J. Leutz, S. K. Yang, H. V. Gelboin, Y. L. Chiang, K. P. Vatsis, and M. J. Coon, *Proc. Natl. Acad. Sci. U.S.A.* **75**, 3123–3127 (1978).

93. J. Deutsch, K. P. Vatsis, M. J. Coon, J. C. Leutz, and H. V. Gelboin, *Mol. Pharmacol.* **16**, 1011–1018 (1979).

94. H. W. S. King, M. R. Osborne, F. A. Beland, R. G. Harvey, and P. Brookes, *Proc. Natl. Acad. Sci. U.S.A.* **73**, 2679–2681 (1976).

95. A. M. Jeffrey, K. W. Jennette, S. H. Blobstein, I. B. Weinstein, F. A. Beland, R. G. Harvey, H. Kasai, I. Muira, and K. Nakanishi, *J. Am. Chem. Soc.* **98**, 5714–5715 (1976).

96. A. M. Jeffrey, I. B. Weinstein, K. W. Jennette, K. Grzeskowiak, and K. Nakanishi, *Nature (London)* **269**, 348–350 (1977).

97. K. Nakanishi, H. Kasai, H. Cho, R. G. Harvey, A. M. Jeffrey, K. W. Jennette, and I. B. Weinstein, *J. Am. Chem. Soc.* **99**, 258–260 (1977).

98. W. M. Baird and L. Diamond, *Biochem. Biophys. Res. Commun.* **77**, 162–167 (1977).

99. H. Autrup, C. C. Harris, B. F. Trump, and A. M. Jeffrey, *Cancer Res.* **38**, 3689–3696 (1978).

100. H. S. Brown, A. M. Jeffrey, and I. B. Weinstein, *Cancer Res.* **39**, 1673–1677 (1979).

101. W. Levin, A. W. Wood, H. Yagi, P. M. Dansette, D. M. Jerina, and A. H. Conney, *Proc. Natl. Acad. Sci. U.S.A.* **73**, 243–247 (1976).

102. W. Levin, A. W. Wood, H. Yagi, D. M. Jerina, and A. H. Conney, *Proc. Natl. Acad. Sci. U.S.A.* **73**, 3867–3871 (1976).

103. W. Levin, A. W. Wood, P. G. Wislocki, J. Kapitulnik, H. Yagi, D. M. Jerina, and A. H. Conney, *Cancer Res.* **37**, 3356–3361 (1977).

104. I. Chouroulinkov, A. Gentil, P. L. Grover, and P. Sims, *Br. J. Cancer* **34**, 523–532 (1976).

105. T. J. Slaga, A. Viaje, D. L. Berry, and W. Bracken, *Cancer Lett.* **2,** 115–121 (1976).
106. W. Levin, A. W. Wood, R. L. Chang, T. J. Slaga, H. Yagi, D. M. Jerina, and A. H. Conney, *Cancer Res.* **37,** 2721–2725 (1977).
107. T. J. Slaga, W. M. Bracken, A. Viaje, W. Levin, H. Yagi, D. M. Jerina, and A. H. Conney, *Cancer Res.* **37,** 4130–4133 (1977).
108. T. J. Slaga, A. Viaje, W. M. Bracken, D. L. Berry, S. M. Fischer, D. R. Miller, and S. M. Leclerc, *Cancer Lett.* **3,** 23–30 (1977).
109. T. J. Slaga, W. M. Bracken, G. Gleason, W. Levin, H. Yagi, D. M. Jerina, and A. H. Conney, *Cancer Res.* **39,** 67–71 (1979).
110. J. Kapitulnik, W. Levin, A. H. Conney, H. Yagi, and D. M. Jerina, *Nature (London)* **266,** 378–380 (1977).
111. J. Kapitulnik, P. G. Wislocki, W. Levin, H. Yagi, D. M. Jerina, and A. H. Conney, *Cancer Res.* **38,** 354–358 (1978).
112. J. Kapitulnik, P. G. Wislocki, W. Levin, H. Yagi, D. R. Thakker, H. Akagi, M. Koreeda, D. M. Jerina, and A. H. Conney, *Cancer Res.* **38,** 2661–2665 (1978).
113. M. K. Buening, P. G. Wislocki, W. Levin, H. Yagi, D. R. Thakker, H. Akagi, M. Koreeda, D. M. Jerina, and A. H. Conney, *Proc. Natl. Acad. Sci. U.S.A.* **75,** 5358–5361 (1978).
114. "IARC Monographs on the Evaluation of Carcinogenic Risk of Chemical to Man," Vol. 3. International Agency for Research on Cancer, Lyon, 1973.
115. A. Dipple, *in* "Chemical Carcinogens" (C. E. Searle, ed.), pp. 245–314. Am. Chem. Soc., Washington, D.C., 1976.
116. T. J. Slaga, G. T. Bowden, J. D. Scribner, and R. K. Boutwell, *J. Natl. Cancer Inst.* **53,** 1337–1340 (1974).
117. B. Tierney, A. Hewer, A. D. MacNicoll, P. G. Gervasi, H. Rattle, C. Walsh, P. L. Grover, and P. Sims, *Chem.-Biol. Interact.* **23,** 243–257 (1978).
118. D. R. Thakker, W. Levin, H. Yagi, D. Ryan, P. E. Thomas, J. M. Karle, R. E. Lehr, D. M. Jerina, and A. H. Conney, *Mol. Pharmacol.* **15,** 138–153 (1979).
119. D. R. Thakker, W. Levin, H. Yagi, S. Turujiman, D. Kapadia, A. H. Conney, and D. M. Jerina, *Chem.-Biol. Interact.* **27,** 145–161 (1979).
120. D. R. Thakker, W. Levin, H. Yagi, A. H. Conney, and D. M. Jerina, *in* "Biological Reactive Intermediates II" (R. Snyder, D. Parke, J. J. Kocsis, D. J. Jollow, G. G. Gibson, and C. Witmer, eds.), Part A, pp. 525–539. Plenum, New York, 1982.
121. A. W. Wood, W. Levin, R. L. Chang, R. E. Lehr, M. Schaefer-Ridder, J. M. Karle, D. M. Jerina, and A. H. Conney, *Proc. Natl. Acad. Sci. U.S.A.* **74,** 3176–3179 (1977).
122. T. J. Slaga, E. Huberman, J. K. Selkirk, R. G. Harvey, and W. M. Bracken, *Cancer Res.* **38,** 1699–1704 (1978).
123. W. Levin, D. R. Thakker, A. W. Wood, R. L. Chang, R. E. Lehr, D. M. Jerina, and A. H. Conney, *Cancer Res.* **38,** 1705–1710 (1978).
124. P. G. Wislocki, J. Kapitulnik, W. Levin, R. Lehr, M. Schaefer-Ridder, J. M. Karle, D. M. Jerina, and A. H. Conney, *Cancer Res.* **38,** 693–696 (1978).
125. P. G. Wislocki, M. K. Buening, W. Levin, R. E. Lehr, D. R. Thakker, D. M. Jerina, and A. H. Conney, *J. Natl. Cancer Inst.* **63,** 201–204 (1979).
126. W. F. Dunning and M. P. Curtis, *J. Natl. Cancer Inst.* **25,** 387–391 (1960).
127. J. L. Stevenson and E. VanHaam, *Am. Inst. Hyg. Assoc. J.* **26,** 475–578 (1965).
128. B. Tierney, A. Hewer, C. Walsh, P. L. Grover, and P. Sims, *Chem.-Biol. Interact.* **18,** 179–193 (1977).
129. S. K. Yang, M. W. Chou, and P. P. Fu, *in* "Polynuclear Aromatic Hydrocarbons: Chemistry and Biological Effects" (A. Bjørseth and A. J. Dennis, eds.), pp. 645–662. Battelle Press, Columbus, Ohio, 1980.
130. S. K. Yang, M. W. Chou, and P. P. Fu, *in* "Chemical Analysis and Biological Fate: Poly-

nuclear Aromatic Hydrocarbons'' (M. Cooke and A. J. Dennis, eds.), pp. 253–264. Battelle Press, Columbus, Ohio, 1981.

131. P. Vigny, M. Duquesne, H. Coulomb, C. Lacombe, B. Tierney, P. L. Grover, and P. Sims, *FEBS Lett.* **75**, 9–12 (1977).

132. I. Chouroulinkov, A. Gentil, B. Tierney, P. Grover, and P. Sims, *Cancer Lett.* **3**, 247–253 (1977).

133. R. G. Harvey and F. B. Dunne, *Nature (London)* **273**, 566–568 (1978).

134. S. K. Yang, M. W. Chou, and P. P. Roller, *J. Am. Chem. Soc.* **101**, 237–239 (1979).

135. P. Vigny, M. Duquesne, H. Coulomb, B. Tierney, P. L. Grover, and P. Sims, *FEBS Lett.* **82**, 278–282 (1977).

136. C. A. H. Bigger, J. E. Tomaszewski, and A. Dipple, *Biochem. Biophys. Res. Commun.* **80**, 229–235 (1978).

137. T. J. Slaga, G. L. Gleason, J. DiGiovanni, K. B. Sukumaran, and R. G. Harvey, *Cancer Res.* **39**, 1934–1936 (1979).

138. P. G. Wislocki, K. M. Gadek, M. W. Chou, S. K. Yang, and A. Y. H. Lu, *Cancer Res.* **40**, 3661–3664 (1980).

139. E. Huberman and T. J. Slaga, *Cancer Res.* **39**, 411–414 (1979).

140. J. DiGiovanni, L. Diamond, J. M. Singer, F. B. Daniel, D. T. Witiak, and T. J. Slaga, *Carcinogenesis* **3**, 651–655 (1982).

141. D. R. Thakker, W. Levin, A. W. Wood, A. H. Conney, T. A. Stoming, and D. M. Jerina, *J. Am. Chem. Soc.* **100**, 645–647 (1978).

142. D. R. Thakker, W. Levin, T. A. Stoming, A. H. Conney, and D. M. Jerina, *Carcinog.—Compr. Surv.* **3**, 253–264 (1978).

143. W. Levin, M. K. Buening, A. W. Wood, R. L. Chang, D. R. Thakker, D. M. Jerina, and A. H. Conney, *Cancer Res.* **39**, 3549–3553 (1979).

144. P. Sims, *Int. J. Cancer* **2**, 505–508 (1967).

145. E. Cavalieri, R. Roth, J. Althoff, C. Grandjean, K. Patil, S. Marsh, and D. McLaughlin, *Chem.-Biol. Interact.* **22**, 69–81 (1978).

146. I. Chouroulinkov, A. Gentil, B. Tierney, P. L. Grover, and P. Sims, *Int. J. Cancer* **24**, 455–460 (1979).

147. E. L. Kennaway and I. Hieger, *Br. Med. J.* **1**, 1044–1046 (1930).

148. P. Sims, *Biochem. Pharmacol.* **19**, 795–818 (1970).

149. M. Nordqvist, D. R. Thakker, W. Levin, H. Yagi, D. E. Ryan, P. E. Thomas, A. H. Conney, and D. M. Jerina, *Mol. Pharmacol.* **16**, 643–655 (1979).

150. M. K. Buening, W. Levin, A. W. Wood, R. L. Chang, H. Yagi, J. M. Karle, D. M. Jerina, and A. H. Conney, *Cancer Res.* **39**, 1310–1314 (1979).

151. T. J. Slaga, G. L. Gleason, G. Mills, L. Ewald, P. P. Fu, H. M. Lee, and R. G. Harvey, *Cancer Res.* **40**, 1981–1984 (1980).

152. D. Hoffmann, W. E. Bondinell, and E. L. Wynder, *Science* **183**, 215–216 (1974).

153. S. S. Hecht, W. E. Bondinell, and D. Hoffmann, *J. Natl. Cancer Inst.* **53**, 1121–1133 (1974).

154. A. D. MacNicoll, P. L. Grover, and P. Sims, *Chem.-Biol. Interact.* **29**, 169–188 (1980).

155. M. Nordqvist, D. R. Thakker, K. P. Vyas, H. Yagi, W. Levin, D. E. Ryan, P. E. Thomas, A. H. Conney, and D. M. Jerina, *Mol. Pharmacol.* **19**, 168–178 (1981).

156. W. Levin, A. W. Wood, R. L. Chang, H. Yagi, H. D. Mah, D. M. Jerina, and A. H. Conney, *Cancer Res.* **38**, 1831–1834 (1978).

157. M. K. Buening, W. Levin, J. M. Karle, H. Yagi, D. M. Jerina, and A. H. Conney, *Cancer Res.* **39**, 5063–5068 (1979).

158. R. L. Chang, W. Levin, A. W. Wood, H. Yagi, M. Tada, K. P. Vyas, D. M. Jerina, and A. H. Conney, *Cancer Res.* **43**, 192–196 (1983).

159. S. S. Hecht, A. Rivenson, and D. Hoffmann, *Cancer Res.* **40**, 1396–1399 (1980).

368   **Ercole L. Cavalieri and Eleanor G. Rogan**

160. S. S. Hecht, N. Hirota, M. Loy, and D. Hoffmann, *Cancer Res.* **38**, 1694–1698 (1978).
161. S. S. Hecht, E. LaVoie, R. Mazzarese, N. Hirota, T. Ohmori, and D. Hoffmann, *J. Natl. Cancer Inst.* **63**, 855–861 (1979).
162. S. S. Hecht, E. LaVoie, R. Mazzarese, S. Amin, V. Bedenko, and D. Hoffmann, *Cancer Res.* **38**, 2191–2194 (1978).
163. A. A. Melikian, E. J. LaVoie, S. S. Hecht, and D. Hoffmann, *Cancer Res.* **42**, 1239–1242 (1982).
164. W. Levin, A. W. Wood, R. L. Chang, Y. Ittah, M. Croisy-Delcey, H. Yagi, D. M. Jerina, and A. H. Conney, *Cancer Res.* **40**, 3910–3914 (1980).
165. R. L. Chang, W. Levin, A. W. Wood, R. E. Lehr, S. Kumar, H. Yagi, D. M. Jerina, and A. H. Conney, *Cancer Res.* **42**, 25–29 (1982).
166. E. Boger, R. F. O'Malley, and D. J. Sardella, *J. Fluorine Chem.* **8**, 513–525 (1976).
167. S. S. Hecht, E. J. LaVoie, V. Bendenko, L. Pingaro, S. Katayama, D. Hoffmann, D. J. Sardella, E. Boger, and R. E. Lehr, *Cancer Res.* **41**, 4341–4345 (1981).
168. M. M. Coombs, A.-M. Kissonerghis, J. A. Allen, and C. W. Vose, *Cancer Res.* **39**, 4160–4165 (1979).
169. M. M. Coombs, T. S. Bhatt, D. C. Livingston, S. W. Fisher, and P. J. Abbott, *in* "Chemical Analysis and Biological Fate: Polynuclear Aromatic Hydrocarbons" (M. Cooke and A. J. Dennis, eds.), pp. 63–73. Battelle Press, Columbus, Ohio, 1981.
170. M. M. Coombs and T. S. Bhatt, *Carcinogenesis (London)* **3**, 449–451 (1982).
171. J. W. Flesher and K. L. Sydnor, *Int. J. Cancer* **11**, 433–437 (1973).
172. J. DiGiovanni, T. J. Slaga, D. L. Berry, and M. R. Juchau, *Drug Metab. Dispos.* **5**, 295–301 (1977).
173. E. Cavalieri, R. Roth, C. Grandjean, J. Althoff, K. Patil, S. Liakus, and S. March, *Chem.-Biol. Interact.* **22**, 53–67 (1978).
174. S. K. Yang, M. W. Chou, P. G. Wislocki, and A. Y. H. Lu, *in* "Polycyclic Aromatic Hydrocarbons: Chemistry and Biological Effects" (A. Bjørseth and A. J. Dennis, eds.), pp. 733–752. Battelle Press, Columbus, Ohio, 1980.
175. A. Dipple, P. D. Lawley, and P. Brookes, *Eur. J. Cancer* **4**, 493–506 (1968).
176. R. K. Natarajan and J. W. Flesher, *J. Med. Chem.* **16**, 714–715 (1973).
177. D. H. Phillips, J. A. Miller, E. C. Miller, and B. Adams, *Cancer Res.* **41**, 176–186 (1981).
178. D. H. Phillips, J. A. Miller, E. C. Miller, and B. Adams, *Cancer Res.* **41**, 2664–2671 (1981).
179. P. G. Wislocki, R. W. Roth, R. L. Saugier, S. L. Johnson, E. G. Rogan, and E. L. Cavalieri, *Carcinogenesis (London)* (in press).
180. T. Watabe, T. Ishizuka, M. Isobe, and N. Ozawa, *Science* **215**, 403–405 (1982).
181. J. W. Flesher and L. K. Tay, *Res. Commun. Chem. Pathol. Pharmacol.* **22**, 345–355 (1978).
182. L. K. Tay, K. L. Sydnor, and J. W. Flesher, *Chem.-Biol. Interact.* **25**, 35–44 (1979).
183. E. G. Rogan, R. W. Roth, P. A. Katomski-Beck, J. R. Laubscher, and E. L. Cavalieri, *Chem.-Biol. Interact.* **31**, 51–63 (1980).
184. C. Nagata, M. Inomata, M. Kodama, and Y. Tagashira, *Gann* **59**, 289–298 (1968).
185. C. Nagata, Y. Tagashira, and M. Kodama, *in* "Chemical Carcinogenesis" (P.O.P. Ts'o and J. A. DiPaolo, eds.), Part A, pp. 87–111. Dekker, New York, 1974.
186. T. Kimura, M. Kodama, and C. Nagata, *Biochem. Pharmacol.* **26**, 671–674 (1977).
187. C. Nagata, M. Kodama, Y. Ioki, and T. Kimura, *in* "Free Radicals and Cancer" (R. A. Floyd, ed.), pp. 1–62. Dekker, New York, 1982.
188. R. J. Lorentzen, W. J. Caspary, S. A. Lesko, and P. O. P. Ts'o, *Biochemistry* **14**, 3970–3977 (1975).
189. S. Lesko, W. Caspary, R. Lorentzen, and P. O. P. Ts'o, *Biochemistry* **14**, 3978–3984 (1975).
190. P. O. P. Ts'o, W. J. Caspary, and R. J. Lorentzen, *in* "Free Radicals in Biology" (W. A. Pryor, ed.), Vol. 3, pp. 251–303. Academic Press, New York, 1977.
191. S. A. Lesko, R. J. Lorentzen, and P. O. P. Ts'o, *in* "Polycyclic Hycrocarbons and Cancer"

(H. V. Gelboin and P. O. P. Ts'o, eds.), Vol. 1, pp. 261–269. Academic Press, New York, 1978.

192. H. Autrup, F. C. Wefald, A. M. Jeffrey, H. Tate, R. D. Schwartz, B. F. Trump, and C. C. Harris, *Int. J. Cancer* **25**, 293–300 (1980).

193. R. D. Bickers, H. Mukhtar, and S. K. Yang, *in* "Polynuclear Aromatic Hyrocarbons: Physical and Biological Chemistry" (M. Cooke, A. J. Dennis, and G. L. Fisher, eds.), pp. 121–131. Battelle Press, Columbus, Ohio, 1982.

194. G. M. Holder, H. Yagi, D. M. Jerina, W. Levin, A. Y. H. Lu, and A. H. Conney, *Arch. Biochem. Biophys.* **170**, 557–566 (1975).

195. H. Autrup, R. C. Grafstrom, B. Christensen, and J. Kieler, *Carcinogenesis* **2**, 763–768 (1981).

196. R. A. Prough, V. W. Patrizi, R. T. Okita, B. S. S. Masters, and S. W. Jakobsson, *Cancer Res.* **39**, 1199–1206 (1979).

197. S. K. Yang, J. K. Selkirk, E. V. Plotkin, and H. V. Gelboin, *Cancer Res.* **35**, 3642–3650 (1975).

198. J. K. Selkirk, R. G. Croy, J. P. Whitlock, Jr., and H. V. Gelboin, *Cancer Res.* **35**, 3651–3655 (1975).

199. E. Bresnick, T. A. Stoming, J. B. Vaught, D. R. Thakker, and D. M. Jerina, Arch. Biochem. Biophys. **183**, 31–37 (1977).

200. I. Y. Wang, R. E. Rasmussen, and T. T. Crocker, *Life Sci.* **15**, 1291–1300 (1974).

201. J. Deutsch, J. C. Leutz, S. K. Yang, H. V. Gelboin, Y. L. Chiang, K. P. Vatsis, and M. J. Coon, *Proc. Natl. Acad. Sci. U.S.A.* **75**, 3123–3127 (1978).

202. H. Autrup, R. C. Grafstrom, M. Brush, J. F. Lechner, A. Hansen, B. F. Trump, and C. C. Harris, *Cancer Res.* **42**, 934–938 (1982).

203. H. Autrup, A. M. Jeffrey, and C. C. Harris, *in* "Polynuclear Aromatic Hydrocarbons: Chemistry and Biological Effect" (A. Bjørseth and A. J. Dennis, eds.), pp. 89–105. Battelle Press, Columbus, Ohio, 1980.

204. C. C. Harris, H. Autrup, G. Stoner, S. K. Yang, J. C. Leutz, H. V. Gelboin, J. K. Selkirk, R. J. Connor, L. A. Barrett, R. T. Jones, E. McDowell, and B. F. Trump, *Cancer Res.* **37**, 3349–3355 (1977).

205. H. Autrup, C. C. Harris, P. W. Schafer, B. F. Trump, G. D. Stoner, and I. C. Hsu, *Proc. Soc. Exp. Biol. Med.* **161**, 280–284 (1979).

206. G. D. Stoner, C. C. Harris, H. Autrup, B. F. Trump, E. W. Kingsbury, and G. A. Myers, *Lab. Invest.* **38**, 685–692 (1978).

207. H. Autrup, C. C. Harris, G. D. Stoner, J. K. Selkirk, P. W. Schafer, and B. F. Trump, *Lab. Invest.* **38**, 217–224 (1978).

208. P. Okano, H. N. Miller, R. C. Robinson, and H. V. Gelboin, *Cancer Res.* **39**, 3184–3193 (1979).

209. G. M. Cohen, R. C. Grafstrom, E. M. Gibby, L. Smith, H. Autrup, and C. C. Harris, *Cancer Res.* **43**, 1312–1315 (1983).

210. H. Autrup, C. C. Harris, B. F. Trump, and A. M. Jeffrey, *Cancer Res.* **38**, 3689–3696 (1978).

211. E. Cavalieri, P. Mailander, and A. Pelfrene, *Z. Krebsforsch.* **89**, 113–118 (1977).

212. R. P. Deutsch-Wenzel, H. Brune, G. Grimmer, G. Dettbarn, and J. Misfeld, *J. Natl. Cancer Inst.* **71**, 539–543 (1983).

213. A. Lacassagne, N. P. Buu-Hoi, and F. Zajdela, *C. R. Hebd. Seances Acad. Sci.* **246**, 1477–1480 (1958).

214. A. Lacassagne, N. P. Buu-Hoi, F. Zaydela, and P. Marille, *C. R. Hebd. Seances Acad. Sci.* **258**, 3387–3389 (1964).

215. J. A. Miller and E. C. Miller, *Cancer Res.* **23**, 229–239 (1963).

216. C. S. Cooper, O. Ribeiro, A. Hewer, C. Walsh, K. Pal, P. L. Grover, and P. Sims, *Carcinogenesis (London)* **1**, 233–243 (1980).

CHAPTER **11**

Antioxidants, Aging, and Longevity

Richard G. Cutler

Gerontology Research Center
National Institute on Aging
Baltimore City Hospitals
Baltimore, Maryland

I. INTRODUCTION

Several hundred years ago few people lived long enough to suffer from the general disabling effects of aging. Death was usually related to infectious diseases, predators, improper nutrition, or accidents. Today in the developed countries of the world, the underlying effects of aging now account for most disease

FREE RADICALS IN BIOLOGY, VOL. VI

ISBN 0-12-566506-7

and death. It is therefore becoming apparent that any effective treatment of the health problems that affect most people must deal with the aging process itself. Yet very little research is being carried out to establish the scientific basis necessary to develop intervention therapies for countering the disabling effects of aging. Part of this problem is the ubiquity and complexity of the aging process.

Studies presented in this chapter, however, suggest that the effects of aging might be subject to intervention without the necessity of understanding all the complexities of the aging process. These studies are based on investigations of the biological basis of human longevity, where research is focused on learning what it is in the biological makeup of humans that gives this species a life span almost twice that of other mammalian species.

The experimental work presented here has been guided by a working hypothesis of aging and longevity for the mammalian species. The two major postulates of this hypothesis are that (a) aging is passive in nature, being a result of pleiotropic effects of normal metabolism and development and (b) longevity is active in nature, being a result of processes (some intrinsically involved in metabolism and development, but others completely separate) that reduce the aging effects of metabolism and development. Thus, a unique aspect of this hypothesis is the prediction of specific longevity determinant processes that are often separate from the biological makeup of the organism or the processes causing aging. The processes determining aging and longevity are postulated to be common in all species independent of a species' life span potential (LSP). A species' LSP is therefore the result of the relative intensity of expression of the aging processes in relation to the intensity of expression of the longevity determinant processes that evolved to counter these causes of aging.

In this chapter I review evidence for the effects of oxygen metabolism producing toxic free radicals and related active oxygen species as a possible contributor to the aging process and endogenous antioxidants as possible longevity determinant processes. Reviews of other aspects of this work can be found in other publications [1–7].

II. WORKING HYPOTHESIS OF AGING AND LONGEVITY FOR MAMMALIAN SPECIES

A. Human Aging and Longevity

Typical human survival curves at different historical times are shown in Fig. 1. Survival data such as these have established that the mean life span of humans for most of human recorded history has been ~20–30 years, few individuals having lived much beyond 40 years of age (Table I). The relatively recent increase in mean life span over the past 400 years or so has been largely the result

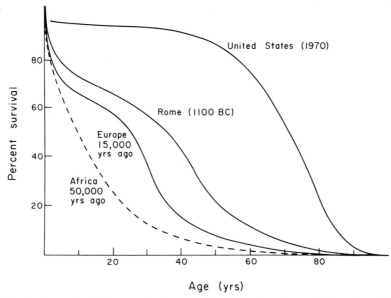

Age (yrs)

Fig. 1. Percent survival curves for humans under different environmental hazard conditions. Note different 50% mean values but constant LSP values of about 100 years. Taken in part from Acsàdi and Nemeskéri [8], Comfort [9], and Strehler [10].

of decreasing environmental hazards. An example is the reduced incidence of infectious diseases by the implementation of sanitary and hygienic conditions. The decrease in these environmental hazards, however, has not affected the natural intrinsic aging rate of humans, as evidenced by the fact that the maximum LSP of human populations is remarkably constant over a wide range of mean life spans (see Fig. 1). The increase in mean life span resulted in a progression from a population of essentially nonaged individuals to a population in which more than 10% of the individuals are now over 65 years of age. Thus, improved living conditions and advances in the biomedical sciences have given us additional years to live—but not additional healthier years [6, 7].

Some of the evidence leading to this conclusion is shown in Fig. 2, indicating the age-dependent loss of a number of physiological functions in humans. From about the age of 20 to 30 years there is an age-dependent, steady, linear decline from maximum capacity in the performance of most physiological functions. The slope of these curves collectively defines the aging rate of the species, the maximum LSP of a species being inversely proportional to the physiological aging rate. In Fig. 3, the age-specific death rate of humans from several different neoplasms is shown to illustrate how sharply the onset frequency of age-related diseases increases with age. As with many other diseases, up to the age of about

TABLE I   Average Mortality of Man in the Past and Present Environment[a]

Time period	Average chronoage at 50% survival (years)	Maximum life span potential (years)
Würm (~70,000–30,000 years ago)	29.4	69–77
Upper Paleolithic (~30,000–12,000 years ago)	32.4	95
Mesolithic (~12,000–10,000 years ago)	31.5	95
Neolithic Anatolian (~10,000–8,000 years ago)	38.2	95
Classic Greece (1100 B.C.–1 A.D.)	35	95
Classic Rome (753 B.C.–476 A.D.)	32	95
England (1276 A.D.)	48	95
England (1376–1400)	38	95
United States (1900–1902)	61.5	95
United States (1950)	70.0	95
United States (1970)	72.5	95

[a]Data taken in part from Deevey [11].

30 years the onset frequency of cancer is relatively low, but at later ages it increases dramatically.

These survival data and the age-dependent loss of physiological function and onset frequency of diseases indicate that LSP or aging rate is a species or genetically determined characteristic not significantly influenced by environmental hazards or life styles such as nutrition or exercise. Humans appear to be designed to live for about 30 years at maximum physiological vigor and free of most age-related diseases. These data also demonstrate the vast complexity of the aging process, affecting essentially all functions of an organism.

The complex nature and ubiquity of aging have a significant impact on the likely consequences of conventional biomedical therapies on diseases associated with aging. This is illustrated in Table II, where it is predicted what the gain in average life span of people in the United States would be if some of the major causes of death today were completely eliminated. It is seen that, with the complete elimination of each of these diseases, people of age 65 would gain about the same increase in mean life span as those that had just been born. This is

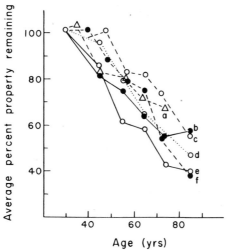

Fig. 2. Decline of human physiological processes as a function of age. Maximum capacity to function is assigned 100% at age 30 years. a, Cardiac index; b, vital capacity; c, standard glomerular filtration rate; d, standard renal plasma flow; e, maximum breathing capacity; f, standard renal function. Taken in part from Kohn [12].

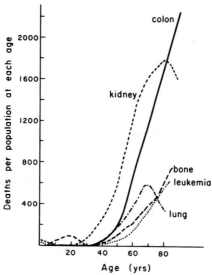

Fig. 3. Age-specific death rates from several neoplasms: colon, lung, and leukemia per 10^6 people; bone and kidney per 10^7 people. Taken in part from Kohn [12].

TABLE II   Gain in Expectancy of Life at Birth and at Age 65 Due to Elimination of Various Causes of Death[a]

Cause of death	Gain in expectancy of life if cause were eliminated (years)	
	At birth	At age 65
Major cardiovascular–renal diseases	10.9	10.0
Heart disease	5.9	4.9
Vascular diseases affecting central nervous system	1.3	1.2
Malignant neoplasms	2.3	1.2
Accidents other than by motor vehicles	0.6	0.1
Motor vehicle accidents	0.6	0.1
Influenza and pneumonia	0.5	0.2
Infectious diseases (excluding tuberculosis)	0.2	0.1
Diabetes mellitus	0.2	0.2
Tuberculosis	0.1	0.0

[a]From life tables published by the National Center of Health Statistics, USPHS and U.S. Bureau of the Census [13].

because the people suffering from these diseases are usually 65 years of age or older. However, of equal interest is the relatively small increase in mean life span (about 1–3 years) predicted if death from all types of malignant neoplasms were completely eliminated. The explanation for such a small increase is related to the ubiquity of aging, affecting all biological functions of an organism. If one type of disease such as cancer were eliminated, another type of disease would simply replace it in a matter of years as the major killer.

These data make the point that a piecemeal approach to the elimination of specific age-related diseases would do little to increase the average life span of people living today. Moreover, the years that would be added would most likely be merely additional years rather than healthy, more enjoyable years of life. The only means of truly adding years of good health would appear to be uniformly postponing most of the disabling effects of aging of the entire organism. This conclusion is not new but its achievement has seemed impossible in the past in view of the vast complexity of the aging processes. There is, however, an important fact bearing on this problem that has not been adequately explored. In spite of the complexity of aging, different mammalian species have different LPSs. What then might be the biological basis for their differences in longevity?

B. Life Span Potential of Mammalian Species

When one follows the survival of a given population of individuals, the age at which the last member of the population died is eventually reached. If a sufficient number of people in the original population were being studied and the living conditions were adequate, then the age when this last member died is defined as the maximum LSP of that species. A typical survival curve indicating the maximum LSP of about 100 years for humans is shown in Fig. 1. An important point in this figure is that maximum LSP is relatively unaffected by different environmental hazard conditions as compared with mean life span. This makes the determination of maximum LSP more reliable and indicates that it is a species' characteristic.

LSP is defined in this chapter as maximum LSP. The innate ability of a species to resist the general effects of aging has been found to be proportional to a species' LSP. It is also found that the average slope of the loss of physiological functions (Fig. 2) is inversely related to a species LSP. Thus, LSP appears to be inversely proportional to a species' aging rate.

Values of LSP are estimated from mortality data of different animals kept in captivity, where most of the normal environmental hazards to which the animals would have been exposed if living in the wild are removed. The living conditions of animals that are kept in zoos and other captivity facilities have been remarkably improved in the past 50 years, and today we have reasonable estimates of LSPs of a wide variety of mammalian species. Of course, there is room for improvement, so LSP estimates are still probably no more accurate than about ±25% on the average. LSP estimates are considered more reliable on a comparative rather than an absolute basis.

Comparative studies of mammalian species have shown that LSPs of mammalian species range over 30-fold (Tables III and IV). These estimates were compiled from the data of over 100 zoos throughout the world [7]. It also appears that the biological basis of these different LSPs is different physiological aging rates uniformly affecting all biological functions of the organism. Species differences in LSP cannot be explained by differences in specific physiological systems, such as a weak cardiovascular system or poor kidney function. Thus, mammalian species appear to have characteristic LSPs or aging rates but a qualitatively common aging process affecting all biological systems.

This result is expected on considering the remarkable biological similarities of mammalian species, particularly in the primate species [14, 15]. Indeed, it has been proposed that all mammalian species share a common spectrum of structural genes and that speciation is largely a result of a difference in timing and degree of expression of these structural genes [16]. Thus, different species arise largely because of differences in regulatory and not structural genes. If this

TABLE III   Primate Longevity: Rank Order According to Life Span Potential

Genus	Common name	LSP (years)	LEP (kcal/g)
Homo	Human	90	815
Pongo	Orangutan	50	447
Pan	Chimpanzee	48	469
Gorilla	Gorilla	43	309
Cebus	Capuchin	42	804
Hylobates	Gibbon	35	569
Mandrilla	Mandrill	35	421
Papio	Baboon	35	394
Lemur	Ring-tailed lemur	35	743
Macaca	Rhesus monkey	34	517
Cercocebus	Mangabey	33	501
Presbytis	Langur	30	388
Ateles	Spider monkey	30	524
Lagothrix	Woolly monkey	30	515
Symphalangus	Siamang	25	354
Cercopithecus	Guenon	25	394
Colobus	Guereza	25	391
Daubentonia	Aye-aye	23	642
Erythrocebus	Patas	22	333
Perodicticus	Potto	22	537
Alouatta	Howler monkey	20	371
Saguinus	Tamarin	20	643
Aotus	Night monkey	20	530
Propithecus	Sifaka	20	360
Saimiri	Squirrel monkey	18	485
Galago	Galago	17	419
Callithrix	Common marmoset	15	535
Pithecia	Saki	15	373
Hapalemur	Gentle lemur	15	300
Loris	Slender loris	15	500
Nycticebus	Slow loris	15	355
Tarsius	Tarsier	15	635
Tupaia	Tree shrew	13	517

indeed is true, it follows then that LSP, which is also a species characteristic, might also be determined by a common set of structural genes found in all mammalian species, and the extent of LSP determined by their degree of expression, as governed by regulatory genes. This type of evidence and reasoning has led me to the idea that specific longevity determinant genes might exist that can control a species' LSP over at lease a 30-fold range in spite of the vast complexity of the aging process [1, 2, 7, 17, 18].

An estimate of the number of regulatory genes that might be involved in determining the LSPs of primates species was made by determining the rate at which LSP evolved during primate evolution leading to the human [2, 18]. These studies led to the conclusion that less than 0.5% of the informational content of the genome was involved in the recent increase of hominid LSP leading to the human. This conclusion supports the concept that specific longevity determinant genes do exist and further indicates that they are relatively few in number.

TABLE IV   Primate Longevity: Rank Order According to Life Span Energy

Genus	Common name	LSP (years)	LEP (kcal/g)
Homo	Human	90	815
Cebus	Capuchin	42	804
Lemur	Ring-tailed lemur	35	743
Saguinus	Tamarin	20	643
Daubentonia	Aye-aye	23	642
Tarsius	Tarsier	15	635
Hylobates	Gibbon	35	569
Perodicticus	Potto	22	537
Callithrix	Common marmoset	15	535
Aotus	Night monkey	20	530
Ateles	Spider monkey	30	524
Macaca	Rhesus monkey	34	517
Lagothrix	Woolly monkey	30	515
Tupaia	Tree shrew	13	512
Cercocebus	Mangabey	33	501
Loris	Slender loris	15	500
Saimiri	Squirrel monkey	18	485
Pan	Chimpanzee	48	469
Pongo	Orangutan	50	447
Mandrilla	Mandrill	35	421
Galago	Galago	17	419
Cercopithecus	Guenon	25	394
Papio	Baboon	35	394
Colobus	Guereza	25	391
Presbytis	Langur	30	388
Pithecia	Saki	15	373
Alouatta	Howler monkey	20	371
Propithecus	Sifaka	20	360
Nycticebus	Slow loris	15	355
Symphalangus	Siamang	25	354
Erythrocebus	Patas	22	333
Gorilla	Gorilla	43	309
Hapalemur	Gentle lemur	15	300

C. Life Span Energy Potential of Mammalian Species

A common characteristic of animals living in their natural ecological niche is that they are always killed on the average before aging sets in to reduce significantly their biological performance. This was shown to be true for humans, whose mean life span was about 30 years only a few hundred years ago and for a number of other mammalian species still living in the natural ecological niche [2, 7]. Studies on the survival of these animals in the wild have shown that LSP appears to have evolved just to the point where aging plays no role in determining mean life span. This observation is important in that it rules out a popular class of aging hypotheses called genetic-programmed aging. These hypotheses postulate the existence of aging genes or death genes that have evolved for the purpose of limiting life span for the good of the species or individual. However, we find that normal environmental hazards killed most individuals before they ever reached a biological old age and that survival, not excess longevity, was clearly the problem at hand.

If aging is not genetically programmed in an active manner, the only alternative is that aging must be the result of normal biological processes of the organism: that is, aging is the result of passive, pleiotropic processes. Two classes of normal biological processes that might cause aging have been postulated [2, 6, 7]. One is related to developmental processes, and the other to metabolic processes not related to development. In this chapter I review some of our work concerning this latter class where oxygen-related metabolism as a cause of aging is considered and the longevity determinant processes that have evolved to counter its aging effects.

The earliest data suggesting that oxygen-related metabolism may be a cause of aging come from the relationship of a species' LSP to the species' basal specific metabolic rate (SMR) [1, 17, 19]. Basal SMR has been shown to be related to a species' average daily metabolism and is determined by measuring the rate of oxygen consumption per unit body weight [20]. In Fig. 4 we see that for most mammalian species, the product of LSP × SMR is remarkably similar. This product has been called life span energy potential (LEP) and is about 220 kcal/g. However, there are a few species, such as the cat and nonhuman primates, that have larger LEP values of about 458 kcal/g. A third class of species, consisting of humans, capuchins, and lemurs, have the highest LEP values of about 781 kcal/g.

These data show a strong correlation between rate of oxygen consumption per body weight and aging rate, suggesting that oxygen utilization is somehow related to the processes of aging. If this is true, then species having higher than normal LEP values, such as primate species and particularly humans, would be expected to be more resistant to the aging effects of oxygen utilization.

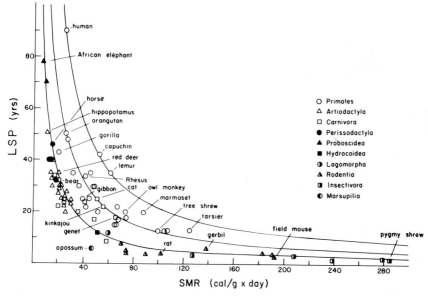

Fig. 4. The LEPs of mammalian species. Data based on SMR (calories per gram × day) and LSP (years) where 90% mortality occurs. The LSP values were taken from previously published data of Spector [21], Jones [22, 23], Holliday et al. [24], Napier and Napier [25], Altman and Dittmer [26, 27], Bauchot and Stephan [28], Jerison [29], Walker [30], Cutler [1, 4], Bowden and Jones [31], and Tolmasoff et al. [20]. Data base represents 77 mammalian species chosen on the basis of reliable SMR and LSP values. LEP = life span × SMR; LEP = 220 kcal/g (most nonprimate species; 458 kcal/g (most primate species); 781 kcal/g (human, capuchin, lemur).

These data also indicate that humans have the highest LEP value of all mammalian species (about 813 kcal/g) as well as the longest LSP (about 90 years). Considering both the LSP and LEP values, it is clear that humans are extraordinarily unique in longevity qualities. Thus not only is LSP a measure of longevity, but it now appears that a species' LEP value is an equally important measure of longevity. Estimates of LSP and LEP for primate species according to their phylogenetic relationships are shown in Fig. 5 for comparative purposes of their innate longevity potentials.

III. ANTIOXIDANTS AS LONGEVITY DETERMINANTS

A. Pleiotropic Nature of Oxygen Metabolism

How is the rate of oxygen metabolism related to aging rate? A possible answer is provided by the substantial amount of literature describing the toxicity of various by-products of oxygen metabolism [33–36]. For example, it is well known that oxygen metabolism produces free radicals, aldehydes, and a wide

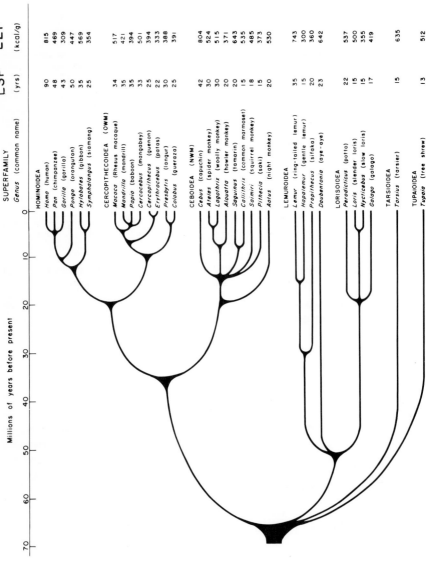

Fig. 5. Phylogenetic relationship of LSP and LEP estimates for the primate species. Data represent estimates taken from literature

range of peroxides that are highly toxic and could therefore be involved in aging [*37, 38*].

Oxygen metabolism is another good example of the pleiotropic nature of aging, being essential for life, on the one hand, but also having long-term toxic effects, on the other. For life to be possible in aerobic organisms, the toxic effects of oxygen had to be reduced [*2, 6, 17*]. Thus, a number of protective measures evolved that are most likely reflected in much of the morphology and biochemistry of an organism. For example, considering the morphological design of a cell, the compartmentalization of some of the most toxic reactions into a single organelle is a protective measure. The mitochondria and the nuclear membrane are good examples; reactions in the mitochondria involving toxic intermediates and by-products of oxygen metabolism are kept away from other, more sensitive regions of the cell such as the nucleus. Another protective measure was the evolution of antioxidants, which remove free radicals before they can destroy important structures in the cell.

There are a number of defense and protective mechanisms in a cell that are essential for defending the organism against the toxic side effects of normal oxygen metabolism, and many others are likely to be discovered [*6, 34, 35*]. All of these defense processes are therefore potential longevity determinants. It is also important to note that similar reactions utilizing oxygen appear to be present in all mammalian species, producing the same spectra of free radicals and thus resulting in the same spectra of aging processes. In addition, all mammalian species appear to have the same types of defense processes, independent of their LSP.

Thus, oxygen metabolism fits the prediction of the properties that causes of aging would have, and antioxidants fit the prediction of the properties to be expected for longevity determinants. If antioxidants are indeed longevity determinants, then the levels of antioxidants in longer lived species should be higher than the levels of the same antioxidants in shorter lived species. The higher tissue levels would be predicted to have evolved by changes in the regulatory genes controlling the levels of the antioxidant defense processes.

B. Dysdifferentiative Nature of Aging

Insight into potential longevity determinants could be gained if we knew more about the basic nature of the aging process itself. According to our working hypothesis, aging is pleiotropic in nature, but this is the cause of aging, not the aging process itself. Most hypotheses of aging postulate an accumulation of damage or wear-and-tear process. In these hypotheses it is implied that aging results from the accumulation of damage such that cells and other body processes cannot properly function. How the accumulation of damage actually causes aging is rarely mentioned, but one popular idea is that the accumulation of altered proteins causes aging [5]. However, the common aging mechanism in all of these hypotheses is that, if a cell cannot operate properly because of free-radical damage, then aging results. Most evidence of which the author is aware does not support this accumulative-damage model of aging. No cellular damage of any type that can account for aging has been found to accumulate with age [5, 6].

Although the accumulative-damage concept of aging may yet prove to be at least partially correct, the phenomenon that most satisfactorily accounts for aging is the possibility that, with time, many cells within an organism gradually drift away from their proper state of differentiation [5, 6, 39]. This change in differentiation could be a result of an accumulation of genetic damage, but not necessarily. For example, the changes occurring in the differentiated state of cells during normal development are not likely to be the result of an accumulation of genetic damage, but the result of changes occurring in gene regulation, as influenced by a large number of intracellular and extracellular factors. Thus, the same type of processes that cause changes in the differentiated state during normal developmental processes could also cause (but at a much slower pace) improper states of differentiation.

In this model of aging, then, the properties of cells do change with age, but the change may not be due to an accumulation of damage resulting in an inability to perform properly. Instead, aging could be the result of the spontaneous nature of cells to acquire different states of differentiation after normal developmental processes have stopped. Specific examples of this process, called *dysdifferentiation,* are the synthesis of hemoglobin in neuron cells, the appearance of endogenous viral gene products in nontarget cells, and the change in the concentration ratios of specific isoenzymes as a function of increasing age. Changes with age in cell morphology, hormone receptor densities, and hair are other examples [5, 6]. Experimental evidence from our laboratory supporting this hypothesis is the age-dependent increase of RNA species similar in sequence to α- and β-globin and to mouse leukemia and mammary tumor viruses in brain of mice [39, 40]. These data indicate an increase in improper gene expression (a derepression) with increasing age.

Basic to the concept of dysdifferentiation as a primary aging process is that it may result from the normal dynamic nature of DNA, involving exchanges of DNA sequences and rearrangements of DNA segments (transposon migration) rather than the accumulation of point mutations in the DNA. These types of DNA changes are known to affect the proper regulation of genes and can be produced by extremely low levels of mutagens such as is produced by normal levels of oxygen metabolism [41–43]. The mutagens used in these studies have been shown to affect the proper differentiated state of cells at levels far below those necessary to cause significant cell damage. Thus, the most important long-term effect of low levels of free radicals commonly found in normal cells may be their effect on the normal differentiated state of cells by epigenetic-like mechanisms rather than by an accumulation of point mutations or other types of cellular damage.

It follows that, if aging is dysdifferentiation, then longevity of a species must be related to the innate stability of the cells of the organism against dysdifferentiation; that is, cells from longer-lived species would be expected to be intrinsically more stable against the effects of mutagenic-like agents arising from an endogenous origin that are known to dysdifferentiate cells. This is what is found experimentally; human cells have yet to be transformed, for example, by a chemical mutagen, but cells from shorter lived species such as mice are relatively easy to transform by the same mutagens [5, 6].

What, then, is the mechanism(s) acting to stabilize the differentiated state of cells? Essentially nothing is known in this area. The present research front of molecular biology involves learning how genes are turned off and on during development and not how gene expression is stabilized after development is complete. However, if free radicals and other mutagenic agents cause dysdifferentiation of cells, then protective agents such as antioxidants that act against these mutagens would form one class of stabilizers of the differentiated state of cells. Thus, the most important role of antioxidants may not be to prevent the gross accumulation of damage throughout the cell but instead to protect or stabilize the differentiated state of cells so that proper cell function can be maintained for longer periods of time. This is possible if the differentiated state of cells is the more sensitive target of the effects of free radicals and other mutagens, and this appears to be the case.

Thus, aging may largely be a dysdifferentiation process and longevity a result determined by those processes stabilizing the proper differentiated state of cells. The problem of what determines human longevity has now been reduced to the problem of what enables human cells to maintain their proper state of differentiation for as long as they do.

C. Evaluation of Antioxidants as Longevity Determinants

The antioxidants that are found naturally in tissues represent a potential class of longevity determinants that could be associated with a species' LSP or LEP. According to the working hypothesis, it would be expected that all mammalian species would have common antioxidants and that their LSPs would be positively correlated with the tissue concentrations of these antioxidants. Thus, antioxidants are evaluated as potential longevity determinants by comparing their concentration in a given tissue as a function of a species' LSP. This comparative analysis is done primarily with primate species, but I also include the rodent species *Mus* (field mouse) with an LSP of 3.5 years and LEP of 232 kcal/g and

Peromyscus (deer mouse) with an LSP of 8 years and LEP of 440 kcal/g to provide examples of species with unusually short LSPs and low LEPs.

1. Controls

Identification of a potential longevity determinant such as an antioxidant is established by a positive correlation between the levels of the potential longevity determinant and LSP or LEP. However, for such an identification, one must be reasonably sure that all enzymes or other compounds that are clearly not longevity determinants do not also correlate positively with LSP and LEP of mammalian species. It is well known that body size, SMR, and rate of development are related to LSP in mammals and other species [*44–47*]. Thus, the enzymes and other compounds that are involved in these processes would be expected to correlate with LSP in some manner. Most of these compounds, however, are not likely to be important as longevity determinants, so a positive correlation with LSP is clearly not proof that a given compound is a longevity determinant. This is why we call such compounds for which a positive correlation is found potential longevity determinants. Further testing for cause and effect associations is necessary to prove the existence of a longevity determinant.

In evaluating potential longevity determinants, it is helpful if one knows a mechanism by which they might act to govern life span. Thus, finding a positive correlation with an antioxidant (e.g., superoxide dismutase) is more meaningful than a correlation with albumin.

To evaluate the significance of a correlation with LSP or LEP, a number of enzymes and other factors that were not thought to be longevity determinants were evaluated using values found in the literature for mammalian species [*26, 48–55*]. No significant correlation in the concentration of the following substances was found with LSP or LEP:

1. Tissue enzymes: random assortment of 57 different enzymes analyzed
2. Vitamins (whole blood): thiamine, riboflavin, pyridoxal, cyanocobalamine, nicotinic acid, and pantothenic acid

An example of these studies is given in Fig. 6, which shows levels of common enzymes of metabolism as a function of species LSP and LEP. For these and similar data, it is clear that all enzyme activities or other compounds do not correlate positively with LSP or LEP and that such positive correlations are likely to represent exceptional cases.

2. Superoxide Dismutase

Superoxide dismutase (SOD) is one of the most important defense enzymes against the toxic effects of oxygen metabolism [*35*]. Tissue concentrations of

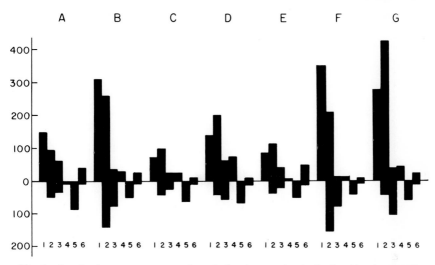

Fig. 6. Levels of common enzymes of metabolism in cytosol and mitochondria of mammalian species having different LSPs. Numbers 1–6 correspond to the enzymes lactate dehydrogenase, malate dehydrogenase, glutamate-oxaloacetate transaminase, glucose-1-phosphate thymidyltransferase, glutamate dehydrogenase, and isocitrate dehydrogenase, respectively. Bars above the line are for cytosol, and those below line are for mitochondrial. Letters A–G correspond to the species listed in the following tabulation:

	Species	LSP (years)	LEP (kcal/g)
A	Human	90	815
B	Marmoset	20	643
C	Horse	46	235
D	Dog	20	255
E	Guinea pig	8	204
F	Rat	4	115
G	Mouse	3.5	232

Non-life-span data from Mattenheimer [53].

SOD were measured in brain, liver, and heart tissues of 12 primate and 2 rodent species [20]. An excellent correlation was found between the ratio of SOD per SMR and LSP, as shown in Fig. 7 for liver tissue. This correlation implies that

$$SOD/SMR = kLSP$$

or

$$SOD = kLSP \times SMR$$

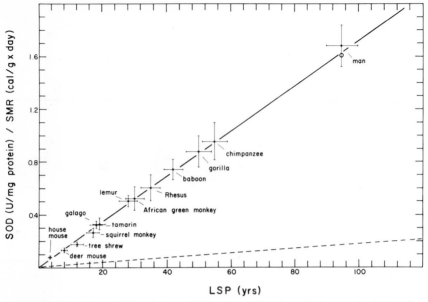

Fig. 7.   Correlation of SOD activity per SMR against LSP in liver of primate and rodent species. Correlation coefficient, $r = .998$; $p \leq .001$. Data from Tolmasoff *et al.* [20].

which gives

$$SOD = kLEP$$

Thus, we find that a species' LEP is proportional to the level of SOD in its tissue. This is shown directly in Fig. 8, where SOD levels in liver are shown to be directly proportional to a species' LEP. This makes sense if the ratio of production of superoxide free radicals per amount of oxygen consumed is a constant for the species investigated. Because of the remarkable linearity of the correlation found for SOD per SMR with LSP, this ratio may indeed be a constant, indicating that the mechanisms of oxygen metabolism are remarkably similar in mammalian species. It is concluded that the total amount of oxygen that a tissue uses over a life span is directly proportional to the amount of SOD protection that tissue has against the toxic by-products of oxygen metabolism.

The discovery that SOD tissues levels are proportional to species' LEPs suggests a biochemical basis for the large LEP value of humans. Moreover, it has been found that the SOD enzyme has a similar structure in a number of mammalian species, so the high SOD levels found in human tissue are likely to be the result of more SOD enzyme, not the result of a more efficient enzyme [20]. Thus, the mechanism for increased LEP and consequently increased LSP appears

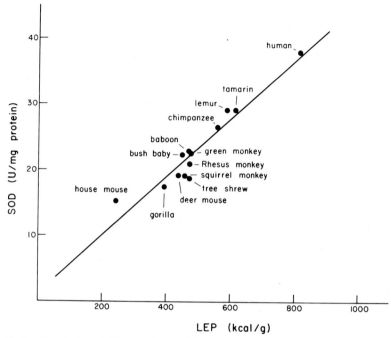

Fig. 8. Correlation of SOD activity per LEP in liver. Correlation coefficient, $r = .952$; $p \leq .001$. Data from Cutler [56].

to involve a regulatory gene change, resulting in higher cellular concentrations of SOD. This result was predicted on the basis of comparative biology and evolutionary considerations previously discussed.

Not only is SOD an antioxidant, but also its only biological function is to act as an antioxidant. However, tissues contain many other compounds that may have antioxidant properties but that may not be acting biologically as antioxidants. Thus, a useful means of evaluating whether these compounds are biologically important antioxidants that contribute to a species' LEP value is to determine whether a correlation exists between their concentration per SMR and the species' LSP. Such a positive correlation is possible if either the antioxidant increases and/or the SMR decreases with increasing LSP.

More recent measurements of the ratio of Mn-SOD to total SOD (Cu,Zn-SOD + Mn-SOD) as a function of a species' LSP have been made. Typical data are shown in Fig. 9, where it is found that (*a*) primate species appear to have considerably higher levels of Mn-SOD compared with Cu,Zn-SOD than other species such as rodents and (*b*) the ratio of Mn-SOD to total SOD increases with LSP in brain but not liver in primate species.

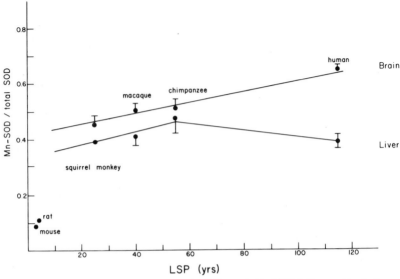

Fig. 9. Ratio of Mn-SOD to total SOD (Mn-SOD + Cu,Zn-SOD) in liver and brain tissues of primate species as a function of LSP. Note comparison of ratio with rat and mouse [60]. Measurements made as previously described [20].

3. Uric Acid

Uric acid is a by-product of purine metabolism and has usually been thought to be a waste product with no biological function [57, 58]. More recently, it has been found that uric acid (or its sodium salt urate) is an excellent antioxidant that is capable of protecting membranes from lipid peroxidation [59]. An evaluation of urate as a potential longevity determinant was made by correlating plasma and tissue levels of uric acid with species LSP and LEP values. Figure 10 shows the correlation between plasma urate levels and LSP in primates, and Fig. 11 shows that in nonprimate species. These observations indicate that urate may be an important antioxidant biologically [59] and a longevity determinant [6, 7].

All of these plasma urate values were taken from the literature, and so we determined urate levels in our laboratory from plasma and other tissues. In Fig. 12 plasma urate concentration per SMR is shown for selected primate species, and again a good positive correlation is found. Urate levels were then determined in brain tissues and, as shown in Fig. 13, an excellent correlation was found between urate and LEP for primate species. These data imply that because of the unusually high levels of urate in tissues and plasma in humans, urate may play an important role as an antioxidant by contributing to the human LEP.

In addition to being a potentially important antioxidant, uric acid may also be a

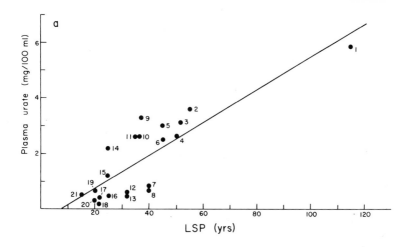

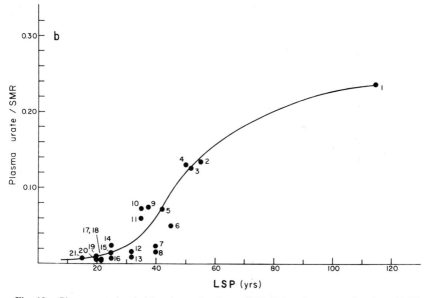

Fig. 10. Plasma urate levels (a) and urate levels per SMR (b) in primates as a function of LSP. Plasma urate urate and SMR values were taken from the literature. Specific identifications: 1, human, 2, chimpanzee; 3, orangutan; 4, gorilla; 5, gibbon; 6, capuchin; 7, macaque; 8, baboon; 9, spider monkey; 10, Siamang gibbon; 11, wooly monkey; 12, langur; 13, grivet; 14, tamarin; 15, squirrel monkey; 16, night monkey; 17, potto; 18, patas; 19, galago; 20, howler monkey; 21, tree shrew. Plasma urate levels as a function of LSP are shown in panel (a), where the correlation coefficient r = .82, $p \leq$ .001. Plasma urate levels per SMR are shown in panel (b), where a positive correlation coefficient is found [60].

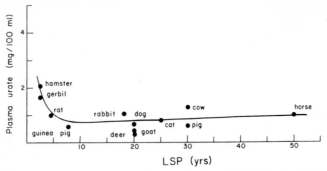

Fig. 11. Plasma urate in nonprimate species as a function of LSP. Note that, for these species, LEP values are approximately equal.

nervous system stimulant [6, 7, 59]. The structure of uric acid is similar to that of caffeine and other neural stimulants. Indeed, it has long been known that men who have suffered from gout are frequently remarkably successful. It is also known that the probability of suffering from gout is related to the serum uric acid level: The higher the uric acid level, the higher the probability of suffering from the symptoms of gout. Experiments to test the possible relationship between serum uric acid levels and intelligence, achievement, and the need for achievement were performed in individuals 12 to 18 years of age. A highly significant correlation was found with each behavioral category [61]. Also, longer lived species in general appear to be more intelligent and more dependent on learned

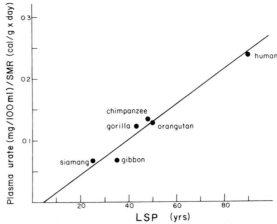

Fig. 12. Plasma urate levels per SMR as a function of LSP for the Hominoidae primate species. Correlation coefficient $r = .98$; $p \leq .001$ [60].

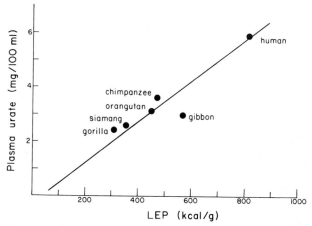

Fig. 13. Plasma urate levels as a function of LEP in Hominoidae primate species. Correlation coefficient $r = .93$; $p \leq .01$ [*60*].

behavior than instinctive behavior. Thus, uric acid may have two important effects in primates, one as an antioxidant and the other as a neural stimulant.

Plasma and tissue levels of uric acid are determined in part by the tissue levels of uricase, an enzyme that degrades uric acid, and the capacity of the kidney to excrete uric acid from the blood. A study of the levels of these two regulatory processes in primate species as a function of their tissue and plasma uric acid concentrations and phylogenetic relationship indicates that only a few genetic regulatory alterations are required to account for the evolutionary increase of urate in primate species with increasing LSP. The important finding in this work was that the increase in uric acid levels with increased LSP was related to a loss of uricase enzyme activity. For example, humans have the highest tissue levels of urate and have no uricase.

Thus, we find that all mammalian species have urate and that the mechanism of increase in urate may have been a regulatory gene alteration, as may also be the case for SOD. In this case, however, the increase in urate was the result of a decrease and eventual loss of an enzyme activity. This is a process that has a high probability of occurrence and therefore could evolve at a high rate.

4. Carotenoids

Carotenoids are synthesized only in plants and are used to protect plants from the free radicals generated during photosynthesis [*62*]. β-Carotene (one of the carotenoids) has been thought to be of value to humans and other species only as a precursor of vitamin A. More recently, however, β-carotene has been found to have excellent antioxidant properties [*63, 64*]. Also, people with unusually low tissue levels of β-carotene are unusually prone to developing certain types of cancers [65]. For this reason there is considerable interest in the pharmacological prevention of cancer by retinoids [66–68].

We measured carotenoid and vitamin A concentrations in serum as a function of LSP. These data are shown in Figs. 14 and 15, respectively, and in Table V. Although a good correlation was found for the carotenoids, the correlation was not significant for vitamin A in species with LSPs above ~30 years.

Carotenoids are also found in all tissues of the body, and the levels in brain as a function of LEP are shown in Fig. 16 for human, *Macaca* (rhesus monkey), and baboon. Vitamin A levels in the brain did not show as significant a correlation with species LSP or LEP values (data not shown). These data suggest that the carotenoids may be more important in our diet than vitamin A, acting both as an antioxidant and as a precursor of vitamin A. Thus, carotenoids may be a better supplement to our diet than vitamin A.

The mechanism of tissue regulation of the carotenoids is not fully understood, but the level of the enzyme carotenase in the intestine plays an important role in determining tissue levels of β-carotene [69, 70]. This enzyme cleaves the β-carotene molecule in half, forming vitamin A. Although we have not yet measured the level of carotenase in different species, it is predicted that human levels of carotenase are unusually low and that in general an inverse correlation between β-carotene tissue levels and carotenase activity exists. If so, then tissue levels of β-carotene (like those of SOD and urate) would be determined by regulatory genes, and an increase in carotene would be a result of a loss of carotenase activity.

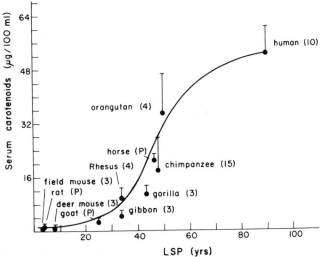

Fig. 14.   Serum level of carotenoids as a function of LSP. Numbers in parentheses represent the number of individuals used in each determination [60]. (P), Plasma.

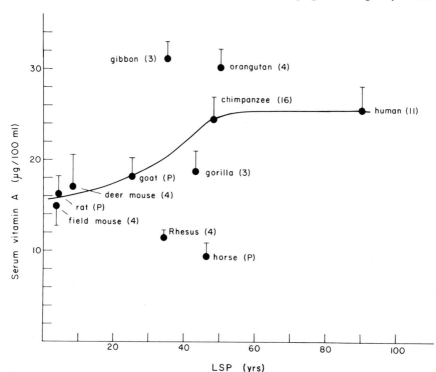

Fig. 15.   Serum level of vitamin A as a function of LSP. Numbers in parentheses represent the number of individuals used in each determination [*60*]. (P), Plasma.

TABLE V   Serum Levels of β-Carotene and Vitamin A in Mammalian Species as a Function of Life Span Potential and Life Span Energy Potential[a]

Species	LSP (years)	LEP (kcal/g)	C (mg/100 ml)	VA (mg/100 ml)	C/VA
Human	90	815	52.0	25.5	2.0
Orangutan	50	447	35.5	30.0	1.18
Horse	46	235	22.0	9.4	2.34
Chimpanzee	48	469	18.0	24.5	0.734
Gorilla	43	309	11.0	18.8	0.585
Rhesus	34	517	10.0	11.5	0.869
Gibbon	35	569	4.5	31.0	0.145
Goat	25	277	3.0	18.2	0.164
Deer mouse	8	440	1.0	17.0	0.0588
Rat	4	115	2.5	16.2	0.154
Field mouse	3.5	232	0.7	15.0	0.0466

[a]β-Carotene (C) and vitamin A (VA) levels taken from the literature [*49–51*]. C/VA versus LSP, $r = .741$, $n = 11$, $p \leq .01$; C/VA versus LEP, $r = .342$, $n = 11$, $p \leq$ not significant.

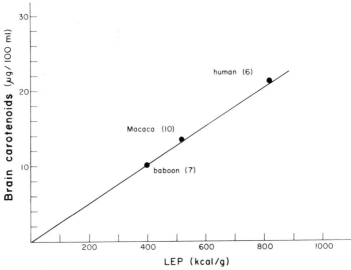

Fig. 16.   Brain levels of carotenoids as a function of LEP in human, *Macaca*, and baboon. Correlation coefficient $r = .605; p \leq .010$. Numbers in parentheses represent the number of animals used for each determination [*60*].

5. α-Tocopherol

α-Tocopherol, or vitamin E, is a well-known tissue antioxidant [*71*]. The importance of vitamin E to human health maintenance, however, is highly controversial [*72*]. To ascertain the possible role of vitamin E in determining human LSP, plasma levels of vitamin E were evaluated as a function of LSP (Fig. 17) and per SMR as a function of LSP (Fig. 18). It was found that human serum has higher plasma levels of vitamin E than serum of shorter lived species and that vitamin E per SMR versus LSP shows an improved positive correlation as compared with vitamin E versus LSP. These data therefore support α-tocopherol as a potential longevity determinant that has an important function as an antioxidant in determining LEP values. Little is known about the genetic or biochemical basis of α-tocopherol regulation, so it remains to be seen if the high plasma levels in humans are a result of changes in regulatory gene expression.

6. Ascorbic Acid

Ascorbic acid, or vitamin C, has long been advocated to be essential for human health and longevity. Indeed, the inability of humans to synthesize ascorbic acid, in contrast to many other species, has been considered a serious human

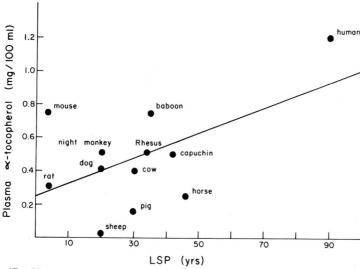

Fig. 17.   Plasma levels of vitamin E as a function of LSP in mammalian species. Correlation coefficient $r = .554$; $p \leq .050$. Non-life-span data from Altman and Dittmer [49] and Bernischke et al. [54].

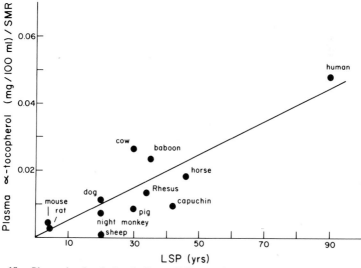

Fig. 18.   Plasma levels of vitamin E per SMR as a function of LSP in mammalian species. Correlation coefficient $r = .864$; $p \leq .001$. Non-life-span data from Altman and Dittmer [49] and Bernischke et al. [54].

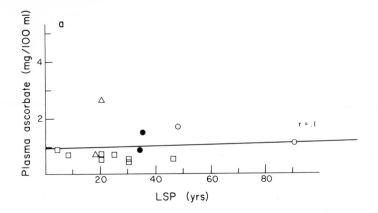

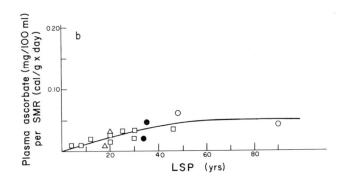

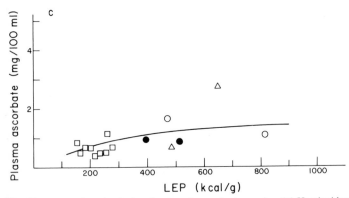

Fig. 19. Plasma ascorbate levels in primate and nonprimate species. (○) Hominoidae, (●) Old World monkeys, (△) New World monkeys, (□) nonprimate mammals. Ascorbate data taken from the literature [60].

TABLE VI   Ascorbate Levels in Cerebrospinal Fluid of Mammalian Species[a]

Species	LSP (years)	LEP (kcal/g)	Ascorbate[b] (mg/100 ml)	LGO[b]
Human	90	815	2.47	No
Horse	49	233	1.70	Yes
Macaca	40	512	2.30	No
Cat	30	457	3.80	Yes
Dog	20	255	6.60	Yes

[a]Ascorbate data from Altman and Dittmer [49–51].
[b]LGO, Presence of the enzyme L-gulonolactone oxidase.

hereditary defect [73–75]. Ascorbic acid is used in a number of metabolic processes including collagen synthesis, but it may also act as an antioxidant. On evaluating ascorbic acid as a potential longevity determinant by searching the literature for data on ascorbic acid levels in different mammalian tissues, the author found no significant positive correlation with species LSP or LEP values.

Figure 19 shows plasma ascorbate concentration as a function of LSP and LEP and plasma ascorbate concentration per SMR as a function of LSP for primates and nonprimate species. These data clearly show that human plasma is not unusually high in ascorbate as compared with that of shorter lived species. In fact it appears that longer lived species actually have less ascorbate. These data are shown in Table VI for cerebrospinal fluid, Table VII for the adrenal gland, and Table VIII for liver. For eye lens (Table IX) and whole brain tissue (Table X), species levels of ascorbate appear to be about the same, regardless of LSP or LEP. It is important to note, however, that ascorbate levels do decrease significantly with age in most tissues (see Table XI for nervous system in humans), in this regard ascorbate may play a role in aging rather than longevity.

TABLE VII   Ascorbate Levels in Adrenal Gland of Mammalian Species

Species	LSP (years)	LEP (kcal/g)	Ascorbate[a] (mg/100 g)	LGO[b]
Human	90	815	44.8	No
Macaca	40	512	112.4	No
Guinea pig	8	204	90.8	No
Rat	4	151	39.20	Yes

[a]Ascorbate data from Altman and Dittmer [49–51].
[b]LGO, Presence of the enzyme L-gulonolactone oxidase.

TABLE VIII Ascorbate Levels in Liver of Mammalian Species

Species	LSP (years)	LEP (kcal/g)	Ascorbate (mg/100 g)[a]	LGO[b]
Human	90	815	13.5	No
Macaca	35	512	14.9	No
Cow	30	164	26.1	Yes
Pig	30	219	11.2	Yes
Chicken	30	—	36.8	Yes
Rabbit	18	—	27.0	Yes
Guinea Pig	8	204	16.6	No
Rat	4	151	32.9	Yes

[a] Ascorbate data from Altman and Dittmer [49–51].
[b] LGO, Presence of the enzyme L-gulonolactone oxidase.

TABLE IX Ascorbate Levels in Eye Lens of Mammalian Species

Species	LSP (years)	LEP (kcal/g)	Ascorbate (mg/100 ml)[a]		LGO[b]
			Young adult	Old adult	
Human	90	815	30	20	No
Cow	30	164	35	30	Yes
Rabbit	12	278	20	9	Yes
Rat	4	151	22	7	Yes

[a] Ascorbate data from Altman and Dittmer [49–51].
[b] LGO, Presence of the enzyme L-gulonolactone oxidase.

TABLE X Ascorbate Concentration in Whole Brain Tissues of Mammalian Species as a Function of Age

Species	LSP (years)	Ascorbate (mg % wet wt)[a]	
		Young	Adult
Mouse	3	41	27
Rat	4	12	8
Guinea pig	8	15	6
Rabbit	12	39	20
Cow	30	40	20

[a] Ascorbate data taken from Kirk [76].

TABLE XI   Ascorbate Concentration in the Human Central Nervous System as a Function of Age

Age	N^a	Cerebral cortex[b]	Cerebellar cortex[b]	Spinal cord[b]
3–5 months (fetus)	7	65	37	16
1.5–9 months	6	23	31	17
2–8 years	4	18	23	16
25–56 years	16	10	13	13
65–82 years	12	7.5	9	11

[a]N, Number of individuals.
[b]Values are expressed as milligrams per 100 g wet weight tissue. Data from Kirk [76].

These data suggest that ascorbate may not have played an important role in determining the unusually high LSP or LEP values for humans. Indeed, it appears that ascrobate may even be harmful, as indicated by the lower ascorbate levels in some tissues of longer lived species and of the generation of active oxygen species in the presence of Fe^{2+} and ascorbate [6, 7]. Clearly, there is no indication that the inability of humans to synthesize ascorbate is a disadvantage, and perhaps the loss of this synthesis capacity was an advantage or at least has a neutral effect.

An interesting finding related to this possibility is that urate is very effective in removing Fe^{2+} from tissues (personal communication from P. Hochstein). The Fe^{2+} ion may be involved in the generation of a number of toxic compounds, such as by the iron-catalyzed Haber–Weiss mechanism [35, 77, 78]. Also, in the presence of oxygen and Fe^{2+}, ascorbate is transformed to the ascorbate free radical, which is highly toxic. Urate, therefore, may protect against the toxic effects of ascorbate and, by preventing ascorbate radical formation, may act to preserve the use of ascorbate for nonantioxidant-related functions such as in the synthesis of collagen. Thus, with higher urate levels, less ascorbate is needed.

7. Glutathione

Glutathione is thought to be one of the most important tissue antioxidants [79, 80]. However, when tissue levels of glutathione were compared with species LSP or LEP values, a negative correlation was found. These data are shown in Table XII for whole blood and Table XIII for lens. A similar negative correlation was found in other tissues, as shown in Table XIV.

TABLE XII   **Whole-Blood Glutathione Levels in Mammalian Species**[a]

Species	LSP (years)	LEP (kcal/g)	Total (mg/100 ml)	Oxidized (mg/100 ml)	Reduced (mg/100 ml)
Human	90	815	36.8	31	4
Horse	49	152	60	50	—
Baboon	35	512	49	—	—
Cow	30	153	46	40	6
Pig	30	219	36	—	—
Sheep	25	186	26	—	—
Dog	20	268	31	29	—
Rabbit	12	257	45	35	—
Guinea pig	8	204	127	—	—
Rat	4	152	120	—	—
Mouse	3	232	102	—	—

[a]Non-life-span data from Altman and Dittmer [49–51].

These data indicate that glutathione is not a longevity determinant. One reason for this may be that metabolic reactions involving glutathione may be toxic, such as in the production of the thio free radical. This prediction has found support from studies showing that glutathione in the presence of tissue extracts is positive in the Ames mutagenicity test [82].

8. Glutathione Peroxidase

Glutathione peroxidase has also been considered to be a major protective enzyme against the accumulation of peroxides [83]. However, tissue levels of glutathione peroxidase also showed a decrease with increased LSP. These results are shown in Table XV for liver and brain, Table XVI for blood (red blood cells), and Table XVII for liver (selenium and nonselenium activity). This is true both

TABLE XIII   **Concentration of Free and Protein-Bound Glutathione in the Lens of Mammalian Species**

Species	LSP (years)	Glutathione[a] (μmol/g wet wt)	Mixed disulfide[a] (μmol/g wet wt)
Human	90	2.20	1.20
Cow	30	14.0	0.20
Rabbit	12	12.0	1.30
Rat	4	7.9	0.96

[a]Data from Reddy et al. [81].

TABLE XIV   Glutathione Concentration in Tissues of Mammalian Species[a]

Species	LSP (years)	LEP (kcal/g)	Relative glutathione concentration				
			Brain	Heart	Liver	Kidney	Spleen
Human	90	815	12.8 ± 3.45 $n = 5$	11.3 ± 1.95 $n = 4$	13.0 ± 9.41 $n = 4$	7.42 ±2.89 $n = 4$	6.2 $n = 1$
Baboon	35	394	17.4 ± 2.68 $n = 4$	15.7 ± 4.59 $n = 2$	36.8 ±13.0 $n = 4$	8.66 ±6.35 $n = 3$	7.66 ± 6.78 $n = 3$
Pig-tailed macaque	34	517	17.5 ± 1.41 $n = 2$	13.0 ± 0.981 $n = 3$	42.6 ± 2.46 $n = 3$	4.30 ±1.13 $n = 2$	13.1 ± 4.38 $n = 2$
Deer mouse	8	440	54.5 ± 2.1 $n = 2$	—	—	—	—
Field mouse	3	232	49.0 ± 1.4 $n = 2$	—	—	—	—

[a]LSP versus glutathione (GSH), $r = -.803$ (brain); LEP versus GSH, $r = -.636$. Taken from Cutler [60].

TABLE XV   Glutathione Peroxidase Activity in Liver and Brain of Mammalian Species[a]

Species	N[b]	LSP (years)	LEP (kcal/g)	Enzyme activity[c]	
				Liver	Brain
Mouse	5	3	232	1140	23 + 3
Rat	5	4	152	153 ± 23	5
Guinea pig	5	6	206	57	14 ± 4
Rabbit	5	12	257	381	20 ± 5
Dog	2	20	268	ND[d]	3
Cow	5	30	153	ND[d]	5 + 1

[a]Non-life-span data from De Marchena et al. [84].
[b]N, Number of animals used for determination.
[c]Activity of cytosol fraction: mean + SD. Nanomoles glutathione oxidized per minute × milligram protein. LSP versus glutathione peroxidase (GPX), $r = -.315$ (liver), $r = -.551$ (brain), LEP versus GPX, $r = .476$ (liver), $r = .412$ (brain).

TABLE XVI   Blood Selenium and Glutathione Peroxidase Activity[a]

Species	LSP (years)	LEP (kcal/g)	SE[b] (ppm)	GPX[b,c]	GPX/Se
Rat (6 male, 3 female adults)	4	152	0.0775 ± 0.060 $n = 4$	120 ± 100 $n = 4$	1529 ± 161
Sheep (all adult females)	20	186	0.115 ± 0.102 $n = 6$	152 ± 143 $n = 6$	1281 ± 73.6
Rhesus (all adult females)	34	512	0.72 ± 0.054 $n = 5$	19.8 ± 18 $m = 5$	257 ± 71.9
Human (2 male, 8 female adults)	90	815	0.10 $n = 10$	19.0 $n = 10$	190

[a]Selenium (Se) versus LSP, $r = .254$; Se versus LEP, $r = -.010$; glutathione peroxidase (GPX) versus LSP, $r = -.720$; GPX versus LEP, $r = -.893$; GPX/Se versus LSP, $r = -.812$; GPX/Se versus LEP, $r = -.928$.

[b]Data from Butler *et al.* [*85*].

[c]Glutathione peroxidase activity expressed as nanomoles NADPH oxidized per minute × milligram hemoglobin.

TABLE XVII   Glutathione Peroxidase Activity in Liver[a,b]

Species (adult)	LSP (years)	Se GPX	Non-Se GPX	Total GPX activity	Non-Se GPX (% of total)
Hamster	3	26.0 + 2.8	19.1 ± 1.3 $n = 3$	45.1	43
Rat	4	19.6 + 4.5	8.4 + 1.4 $n = 3$	28.0	35
Guinea pig	8	0	7.3 ± 0.7 $n = 2$	7.3	100
Sheep	20	3.8 + 0.9	17.6 ± 2.7 $n = 3$	21.4	81
Pig	30	2.5 + 5	6.0 ± 2.4 $n = 3$	8.5	67
Chicken	30	1.5 + 0.02	3.4 ± 0.2 $n = 2$	4.9	70
Human	90	1.3 + 0.32	8.4 ± 1.2 $n = 6$	9.7	84

[a]Non-life-span data from Lawrence and Burk [*86*].

[b]Selenium-dependent glutathione peroxidase (Se GPX) versus LSP, $r = -.489$; glutathione *S*-transferase (non-Se GPX) versus LSP, $r = -.269$; total GPX versus LSP, $r = -.469$. One unit of activity is 1 μmol NADPH oxidized per minute.

for the selenium and non-selenium-dependent types of glutathione peroxidase. Thus, according to our criteria, glutathione peroxidase is not an important longevity determinant in mammalian species.

9. Glutathione *S*-transferase

Glutathione *S*-transferases are important enzymes for detoxification reactions [*87*], and one might expect that longevity would be related to higher levels of these enzymes. These enzymes are also known as the nonselenium glutathione peroxidases. Figure 20 shows that, in the liver of humans, chimpanzees, and rhesus monkeys, glutathione *S*-transferase concentrations decrease with increasing LSP. Similar results are shown in Table XVIII for other species.

The negative correlations found for glutathione, glutathione peroxidase, and glutathione *S*-transferase with LSP are similar to the results of Schwartz and coworkers [*89, 90*], showing that the activation of mutagenic chemicals decreases with increasing LSP. Consistent with this finding is that the levels of *P*-450 and *P*-448 enzymes also decrease with LSP [*89, 90*]. These results are shown in Table XIX.

There is evidence that mixed-function oxidation systems produce active oxygen products capable of inactivating proteins [*91, 92*]. Thus, the decrease in activity of these detoxification systems in longer lived species may be essential in

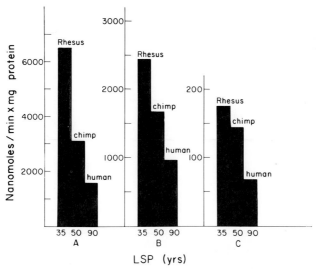

Fig. 20. Glutathione *S*-transferase activity in liver as a function of LSP in primates. (A) 100,000 *g* supernatant; (B) 9000 *g* supernatant, (C) microsomes. Non-life-span data taken from Summer and Greim [*88*].

TABLE XVIII   Percent Glutathione *S*-Transferase to Total Glutathione Peroxidase Activity in Liver[a,b]

Species	LSP (years)	SMR (cal/g × day)	LEP (kcal/g)	Non-SE GPX (% of total GPX)	% Non-SE GPX/SMR
Hamster	3	118	129	43	0.364
Rat	4	104	152	35	0.336
Guinea pig	8	69.8	204	100	1.43
Sheep	20	25.6	186	81	3.16
Pig	30	20	219	67	3.36
Human	90	24.8	815	84	3.38

[a]Non-life-span data from Lawrence and Burk [86].

[b]LSP versus % glutathione *S*-transferase (non-Se GPX), $r = .131$; LEP versus % non-Se GPX, $r = .393$; LSP versus % non-Se GPX/SMR, $r = .701$.

the evolution of longer LSPs. What has replaced these detoxification systems, if any, remains to be answered.

10. Ceruloplasmin

Ceruloplasmin has been suggested to be the major antioxidant in blood plasma [77, 78, 93–96]. The relation of ceruloplasmin to LSP and LEP in primate species is positive but not highly significant (Table XX). Looking at nonprimate species in Table XXI, where humans are included, we see that ceruloplasmin is not supported as an important longevity determinant.

TABLE XIX   Cytochrome *P*-450 and *P*-448 Levels in Cultured Fibroblast Cells from Selected Species[a,b]

Species	LSP (years)	LEP (kcal/g)	*P*-448[c]	*P*-450[c]	*P*-448/*P*-450
Human	90	815	Not detectable	3.0	0
Elephant	78	194	1.5 ± 0.5	3.5 ± 0.5	0.428
Cow	30	164	8.6 ± 1.6	12.7 ± 4.2	0.677
Rabbit	12	257	11.8 ± 1.1	10.5 ± 0.9	1.12
Rat	4	115	19.7 ± 1.7	9.9 ± 4.3	2.01
Mouse	3.5	232	22.2 ± 2.0	10.0 ± 1.0	2.22

[a]Non-life-span data from Pashko and Schwartz [90].

[b]LSP versus *P*-448, $r = -.934$, $n = 6$, $p \leq .010$; LSP versus *P*-450, $r = -.870$, $n = 6$, $p \leq .050$; LSP versus *P*-448/*P*-450, $r = -.888$, $n = 6$, $p \leq .020$.

[c]Cytochrome *P*-448 and *P*-450 levels are expressed as picomoles per milligram of microsomal protein. Tissue culture cells were all taken from early passages (3–6).

TABLE XX   Plasma Ceruloplasmin Levels in Primates as a Function of Life Span Potential[a,b]

Species	LSP (years)	LEP (kcal/g)	CP (A_{530}, mU)	CP/mg protein
Human	90	815	0.550	0.0785
Capuchin	42	804	0.490	0.0671
Spider monkey	30	524	0.600	0.075
Night monkey	20	530	0.630	0.0875
Howler monkey	20	371	0.215	0.0405
Squirrel monkey	18	485	0.125	0.0189

[a]Non-life-span data from Seal [97].
[b]LSP vs. ceruloplasmin (CP), $r = 0.445$; LEP versus CP, $r = 0.511$.
[c]Measured as p-phenylenediamine oxidase activity.

IV. AUTOXIDATION OF TISSUES AS A FUNCTION OF SPECIES LIFE SPAN POTENTIAL

In addition to the many known tissue antioxidants, there are probably many other important ones of which we are not yet aware. Thus, to test fully the hypothesis that human LSP is determined in part by tissue levels of antioxidants, it is necessary to determine if indeed human tissues are unusually resistant to the toxic effects of free radicals and other active oxygen species. That is, simply because we have shown a correlation between a few specific antioxidants and LSP or LEP, we have not proved that the net antioxidant capacity of longer lived species is greater than that of shorter lived species. This is particularly true in view of the fact that tissue levels of two antioxidants (ascorbate and glutathione) appear to decrease with LSP.

An estimate of the net antioxidant protection of tissues was made by measur-

TABLE XXI   Plasma Ceruloplasmin Concentration in Mammalian Species

Species	LSP (years)	Ceruloplasmin (mg/100 ml)[a]	Ceruloplasmin-bound copper (μg/100 ml)[a]
Human	90	35.6	121
Cow	30	20.3	69
Sheep	25	26.5	90
Pig	25	35.3	120
Dog	20	17.4	59
Rat	4	35.9	122

[a]Non-life-span data from Evans and Wiederanders [98].

ing their autoxidation rate. This was done by homogenizing a tissue in a neutral aqueous solution and then placing the homogenate in a shaking water bath at the normal body temperature of 37°C. The extent of the autoxidation process was measured by determining the amount of malonaldehyde or thiobarbituric acid-reacting material (TBARM) produced using the thiobarbituric acid (TBA) assay [99].

The rate of autoxidation of a tissue has been found to depend on how much peroxidizable material, such as unsaturated fatty acids and the net level of antioxidants, is present [100, 101]. Thus, the rate at which a tissue homogenate undergoes autoxidation, as determined by the TBA assay, is a reflection of the intrinsic susceptibility of that tissue homogenate to reaction with active oxygen species. If human LSP is a result in part of the innate capacity of tissues to protect themselves from the toxic effects of oxygen, then we would expect to find human tissues to be the most resistant to autoxidation.

Figure 21 shows typical data of the autoxidation of whole-brain homogenates in air of different species. It was found that, indeed, the human brain homogenate was extraordinarily resistant to autoxidation and that shorter lived species were, in proper rank order, more susceptibel to autoxidation. The TBARM was found to increase linearly over the time period of 0 to 240 min (see Fig. 21), so that the rate of autoxidation was calculated as the slope of this linear function.

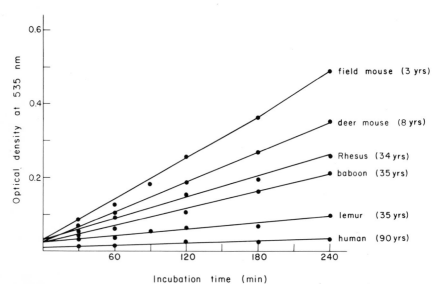

Fig. 21. Autoxidation of whole-brain homogenate in air from mammalian species as a function of LSP. Optical density is proportional to malonaldehyde concentration per gram wet weight of tissue [60].

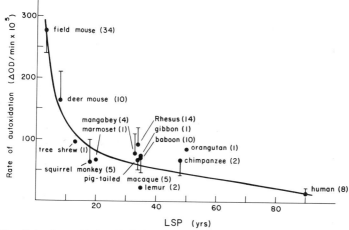

Fig. 22.   Rate of autoxidation of whole-brain homogenate in air from mammalian species as a function of LSP. Numbers in parentheses represent the number of individuals used in each determination. All-species correlation coefficient, $r = -.652, p \leq .02$. Primates, $r = -.609: p \leq 0.05$, OD, optical density [60].

Figure 22 shows the rate of brain autoxidation in different species as a function of LSP; a fairly good correlation is found. Figure 23 shows the rate of autoxidation against LEP; here, the correlation is better. An examination of the rate of autoxidation of brain homogenate as a function of the ln(LEP) produces an excellent linear correlation, as shown in Fig. 24. Similar results were found for kidney tissues, but other tissues have not been tested yet.

An important control in these studies is the similarity of plateau levels of autoxidation of the tissues (the maximum amount of autoxidation obtainable) for all species. Plateau levels for most species were found on incubating the homogenates up to 20 hours. For humans, no plateau level was found after 20 hs, but an extrapolation indicates approximately an equal level to other primates. *Peromyscus* and *Mus* had slightly higher plateau levels, with those of *Mus* being higher than those of *Peromyscus*. Thus, the composition of the tissues, in terms of the amount of autoxidizable material, appears to be remarkably similar in the different species. Therefore, because it is known that the rate of autoxidation is dependent on both the amount of peroxidizable material and on levels of antioxidants in the tissues, these control data indicate that the differences in rate of autoxidation of the tissues is due largely to differences in net levels of antioxidant protection.

These data establish the most convincing demonstration yet that human LSP and the LSPs of other mammalian species are determined, at least in part, by the innate resistance of the tissues to autoxidation. The resistance of the tissues to

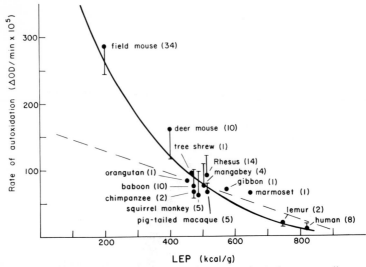

Fig. 23. Rate of autoxidation of whole-brain homogenate in air from mammalian species as a function of LEP. Numbers in parentheses represent the number of individuals used in each determination. All-species correlation coefficient, $r = -.851; p \leq .001$. Primates, $r = -.853, p \leq .001$, OD, optical density [60].

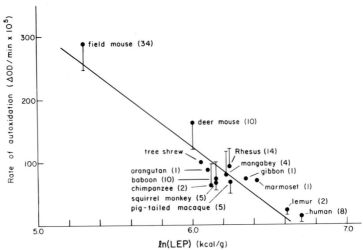

Fig. 24. Rate of autoxidation of brain homogenate as a function of ln(LEP). Correlation coefficient $r = -.934; p \leq .001$, OD, optical density [60].

autoxidation is likely to be a result of higher levels of the same types of antioxidants found in all mammalian species, which would be proportional to a species' LEP.

V. PEROXIDES IN TISSUES AS A FUNCTION OF LIFE SPAN POTENTIAL

Some of the toxic products related to oxygen metabolism are peroxides [33–36]. High levels of lipid peroxides have been related to a number of degenerative diseases, such as diabetes, atherosclerosis, and retinal degeneration [102]. Many peroxides are produced when unsaturated fatty acids undergo oxidation, producing, for example, malonaldehyde among other peroxides. The amount of peroxide in a tissue might therefore be expected to be related to the rate of oxygen utilization, the amount of peroxidizable material, and the level of antioxidant defense in a tissue. Thus, it might be expected that longer lived species would have lower serum and tissue levels of maldonaldehyde and other peroxides throughout their life span. Levels of maldonaldehyde have been measured using the TBA assay in serum samples taken from a number of mouse strains. When these data are examined as a function of LSP for these mouse strains, as shown in Fig. 25, an excellent inverse correlation between serum TBARM levels and mean life span of these mouse strains is found.

It is important to note that prostaglandin endoperoxides as well as malonaldehyde are TBA positive [105–107]. Thus, it may be that a complex array of peroxides are being measured by the TBA assay here rather than malonaldehyde alone, which is more appropriately called TBARM, as previously noted.

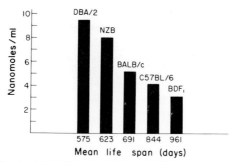

Fig. 25. Plasma levels of TBARM as a function of LSP in mouse strains. All animals were 5 weeks of age. Non-life-span taken from Myers [103] and Nakakimura et al. [104].

Because these mouse strains were chosen randomly and were not selected for a correlation of LSP with serum peroxide levels, it appears that mutations affecting antioxidant levels (or at least serum peroxide levels) may be a highly probable event and a cause of life span differences among these strains. This is of considerable interest in view of the many other possible ways in which mutations can shorten life span. This raises the possibility that many human inherited diseases in which LSP is significantly lower may have a similar basis. Thus, defects in the maintenance of proper levels of longevity determinants such as antioxidants may be important not only to dysfunctions clearly related to aging but also to many other types of dysfunctions not normally being related to the aging process.

Plasma levels of TBARM in mammalian species, as found in the literature, are shown in Fig. 26, which indicates that human levels were not exceptionally low as compared with those of shorter lived species. However, it is of interest that human diabetics have unusually high TBARM levels [*108, 109*]. The diabetic patient has many symptoms similar to those associates with an accelerated aging phenomenon. In addition, human serum TBARM levels do appear to increase with age, suggesting that antioxidant defense systems may decrease with age [*108, 109*].

Measurements of serum TBARM levels using the TBA assay [*110, 111*] were made using primate species plus *Mus* and *Peromyscus* rodent species. Typical data are shown in Fig. 27, indicating that there is a fairly good inverse correlation between serum TBARM levels and increasing LSP, but again human levels are not exceptionally low. The correlation is slightly better when TBARM values are plotted as a function of species LEP values, as shown in Fig. 28. Similar results were found for brain tissues. Thus, TBARM, and perhaps serum peroxide levels, may be a fair indicator of longevity determinant levels in different species.

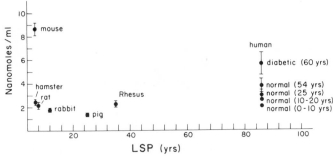

Fig. 26.   Plasma levels of TBARM in mammalian species. Peroxide levels determined by the TBA assay. Non-life-span data from Nakakimura *et al.* [*104*].

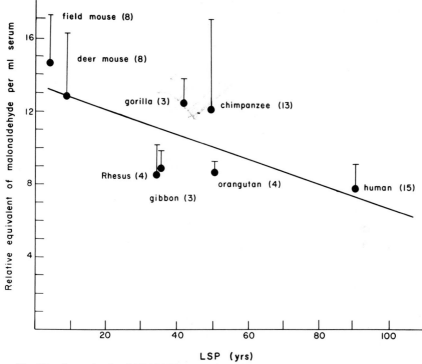

Fig. 27.   Serum levels of TBARM in mammalian species as a function of LSP. Numbers in parentheses represent the number of individuals used for each test. Correlation coefficient, $r = -.692$; $p \leq .050$ [60].

VI. COMPENSATIONAL NATURE OF ENDOGENOUS ANTIOXIDANT CONCENTRATIONS IN TISSUES

One serious difficulty in the hypothesis that antioxidants are important longevity determinants consists of the large concentration differences in the tissues of different individuals who appear to have similar LSPs. For example, we have found twofold differences among human individuals for SOD [20]. Similar ranges of values were found for serum urate and α-tocopherol levels in humans, whereas the intrinsic aging rate of different human individuals appears to be extraordinarily similar, with a variation approximately equal to that found in height differences in a given race [7]. Also, it could be argued that such an important characteristic as longevity would not be expected to be influenced by an individual's diet. Thus, because tissue levels of urate, carotenoids, and vi-

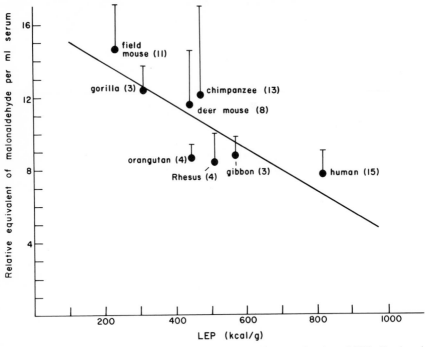

Fig. 28. Serum levels of TBARM in mammalian species as a function of LEP. Numbers in parentheses represent the number of individuals used in each test. Correlation coefficient, $r = -.804$; $p \leq 0.020$ [60].

tamin E all depend on diet composition to some extent, how could they possibly be longevity determinants?

Finally, the most convincing evidence that antioxidants do not play an important role in determining longevity has been the experimental results showing that antioxidant supplementation of the diet does not significantly increase the LSP of normal long-lived wild-type laboratory animals, although mean life span has been increased to a small extent [37, 38]. Data demonstrating this argument are shown in Fig. 29, where it can be seen that vitamin E supplementation of the diet failed to increase either the mean life span or LSP of mice even though the rate of accumulation of lipofuscin pigment in tissues was decreased twofold [112]. If antioxidants are important in determining longevity, then at least a doubling of life span should be evident if the LEP of a mouse, for example, were increased from 232 to 815 kcal/g.

A possible answer to these arguments against the biological significance of free radicals, which may prove important to all of our future studies concerning the genetic and biochemical nature of longevity determinants, is now proposed.

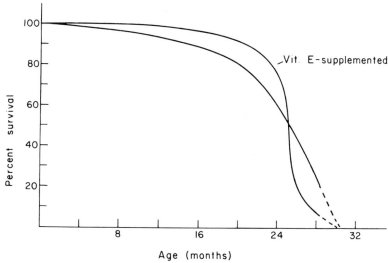

Fig. 29. Effects of vitamin E supplementation of the diet of rats on life span. Data adapted from Blackett and Hall [*112*].

The proposal is that the net antioxidant levels of tissues are under compensatory or homeostatic control by the regulatory adjustment of individual antioxidants. Because it is the net protective antioxidant level of a tissue that would be expected to determine aging rate, then it is this net level and not the level of individual antioxidants that would be expected to determine a species' LSP.

This compensatory model of maintenance of tissue levels of antioxidants is based on the concept that antioxidants have overlapping capabilities by which one antioxidant can partially replace another [6, 7]. Typical data supporting this prediction are shown in Table XXII, where it is found that, on maintaining a rat on a long-term vitamin E-deficient diet, glutathione peroxidase, glutathione reductase, and SOD levels dramatically increase, apparently in order to compensate for the vitamin E-deficiency and to minimize its effects. Other data showing that antioxidant levels frequently increase in tissues placed under oxidative stress have been reviewed [6, 7].

If this model is correct, we would expect that supplementation of vitamin E in the diet at unusually high levels would similarly depress the levels of other antioxidants, thereby diminishing the net gain in antioxidant protection that vitamin E could potentially offer. Thus, over a wide range of dietary uptakes of antioxidants or oxidative stresses, the net tissue antioxidant protection level would be expected to remain fairly constant.

The compensatory model would explain the remarkable uniformity of aging rate found among different human individuals in spite of large variations of

TABLE XXII   **Effect of Long-Term α-Tocopherol Deficiency in the Rat on Other Antioxidants and Related Enzymes**[a,b]

Tissue	α-Tocopherol	Glutathione peroxidase	Glutathione reductase	G-6-PD	SOD	Catalase	Ascorbate
			Fraction of control				
Red blood cells	0.72	0.80	0.82	1.38	1.40	1.10	—
Plasma	0.045	0.70	NS	—	—	—	0.61
Heart	—	NS	1.26	1.78	1.42	1.37	—
Liver	0.77	0.69	NS	1.63	1.41	NS	0.77
Kidney	—	0.89	NS	NS	NS	NS	—
Testis	—	1.35	1.33	1.26	—	1.52	—
Muscle	—	1.21	1.31	3.64	1.75	NS	—

[a]Weanling Sprague–Dawley rats on 100 IU vitamin E per kilogram food for 12 months. Adapted from data of Chen et al. [113].
[b]G-6-PD, Glucose-6-phosphate dehydrogerase; NS, not significant.

dietary uptake of specific antioxidants or endogenous synthesis levels of specific enzymatic antioxidants. This model also explains the failure of antioxidant-supplemented diets to increase LSP significantly; the reason is that little effective net gain of antioxidant protection in the tissue was ever achieved in normal wild-type animals. Short-lived strains of mice do appear, however, to benefit from antioxidant supplementation, and this may be because they are abnormally deficient in antioxidant synthesis, antioxidant absorption, or compensatory control of antioxidants. Thus, large antioxidant supplementation of the diet does not appear to be the method of choice to enhance net tissue antioxidant levels for normal healthy individuals, although supplementation of antioxidants in the diet may prove to be of value for a number of diseases where a specific deficiency exists and to gain additional protection of the digestive tract from environmental toxins [36]. However, to increase antioxidant protection to a net higher level in the tissues of normal individuals, requires some means of intervening in the antioxidant compensatory regulatory mechanism. To accomplish this, we must learn how tissue levels of antioxidants are regulated.

VII. REGULATION OF THE OXIDATIVE ENVIRONMENT OF A CELL

There is now some evidence, as shown in this chapter, that some antioxidants may be longevity determinants. Many of the antioxidants (including those that are not longevity determinants) are likely to be involved in a compensatory–

homeostatic-like regulatory system designed to reduce the concentration of reactive oxygen species in the cell to a species' characteristic level over a wide range of life styles. It is this mean tissue level of antioxidants, which is probably unique for each differentiated cell type, which is proposed to govern the rate of dysdifferentiation of the cell and thus the aging rate or LSP of the organism.

Inadequacies of endogenous synthesis of antioxidants or other components of the antioxidant compensatory system could be involved in producing specific types of disease processes, some of which might result in accelerated aging syndromes. Such may be the case for a number of inbred mice strains (see Fig. 25) as well as a number of human inherited diseases such as ataxia telangiectasia, Bloom's syndrome, and collagen diseases [*114*, *115*].

Species having different LSPs would be predicted to have unique set points to maintain a given mean level of active oxygen species within their cells about which the compensational regulation of antioxidant levels would operate. The level of this set point (level of active oxygen species) would be inversely related to the species' LSP. The lower the set point, the lower would be the mean value of active oxygen species in a cell and the greater the LSP. This set point would be maintained by governing the species' net antioxidant protection per SMR, which has been found experimentally to be proportional to LSP.

What are the biochemical mechanisms for maintaining this set point or, more specifically, what are the regulatory mechanisms for maintaining tissue levels of antioxidants so high in humans as compared with those of shorter-lived species? An understanding of this regulatory process may lead to the discovery of a means of intervention and the development of therapeutic methods to correct a number of human hereditary diseases as well as to prolong the healthy years of the human life span.

A model is presented in this section for the regulation of tissue levels of antioxidants. It is based on the properties of three substances that are unusually sensitive to the effects of active oxygen species. These substances are (*a*) arachidonic acid, an essential unsaturated fatty acid (20:4) that is extremely sensitive to lipid peroxidation and is the only precursor of prostaglandin, prostacyclin, thromboxane, and the leukotrienes; (*b*) prostaglandin cyclooxygenase, the first enzyme in the conversion of arachidonic acid to prostaglandin endoperoxide (PGG_2) (cyclooxygenase is unusual because it is sensitive to activation by low levels of hydroperoxides and inactivation by high levels of hydroperoxides and number of active oxygen species); and (*c*) guanylate cyclase, which is the enzyme catalyzing the synthesis of cGMP and is very sensitive to activation by a large number of active oxygen species [*116–123*].

The guanylate cyclase–arachidonate–cyclooxygenase model (GAC model) is diagrammed in Fig. 30. The basis of the model, as already noted, is the unusual sensitivity of three constituents to their oxidation–reduction environment and the capacity of these constituents when altered by their oxidation–reduction environ-

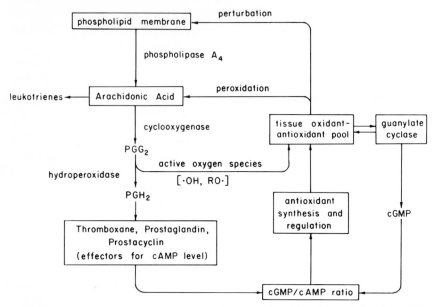

Fig. 30. Guanylate cyclase–arachidonate–cyclooxygenase model for the regulation of the oxidation–reduction potential of a cell. Data taken in part from Cutler [56].

ment to effect changes in the intracellular level of the cGMP/cAMP ratio and thus the synthesis rates of antioxidants.

The amount of arachidonic acid produced and/or the activity of cyclooxygenase determines the rate of prostaglandin H_2 (PGH$_2$) synthesis, which then determines the rate of prostaglandin synthesis and in turn the cellular levels of cAMP. Active oxygen species will peroxidize arachidonic acid, resulting in a decrease in the arachidonic acid available for synthesis of PGH$_2$ and lower cellular levels of cAMP. Active oxygen species also inactivate cyclooxygenase and lower cAMP levels. In contrast, antioxidants tend to lower the levels of the active oxygen species and thus increase the overall arachidonic acid availability and/or increase the activity of cyclooxygenase, resulting in a higher rate of PGH$_2$ synthesis and in turn a higher cellular level of cAMP. Important to this scheme is that, in the synthesis of PGG$_2$ from arachidonic acid, an active oxygen species is produced and this adds to the tissue oxidation–reduction load of the cell. Thus, a self-producing negative feedback process stabilizes the overall prostaglandin synthesis pathway by being sensitive to the intercellular levels of active oxygen species [124].

In contrast to cyclooxygenase, guanylate cyclase is activated by active oxygen species. An increase in guanylate cyclase activity results in an increase in cellular levels of cGMP. Thus, the overall result is a "push–pull effect"; if the intercellular oxidative environment increases above the correct set point of the cells, then the cGMP level will increase and the cAMP level will decrease, resulting in a large net increase in the cGMP/cAMP ratio. Conversely, if antioxidant levels are in excess (a lower than normal level of active oxygen species), this push–pull effect will result in a net decrease in the cGMP/cAMP ratio. The rate of antioxidant synthesis is proposed to be determined by the ratio of intracellular levels of cGMP to those of cAMP.

All of the details of this hypothesis may not be correct, but the essential feature is that certain substances or enzymes may detect endogenous levels of active oxygen species, which can in turn govern gene activity. In addition, the model suggests specific experiments that can easily test its validity. One of the most interesting features of the GAC model is that it suggests methods of tricking the regulatory system to react as if levels of active oxygen species were too low or too high. This feature could lead to methods of enhancing or inhibiting endogenous antioxidant synthesis by nontoxic means. For example, the inhibition of cyclooxygenase by indomethacin or aspirin or many of the other nonsteroid antiinflammatory agents mimics the condition of high intercellular levels of active oxygen species. Thus, the resultant decrease in prostaglandin and cAMP synthesis would increase the cGMP/cAMP ratio and be predicted to stimulate increased antioxidant synthesis. This leads to one of the key predictions of the model, which is that many antinflammatory agents may act not only by reducing the level of prostaglandin endoperoxides and related active oxygen species, but also by enhancing levels of tissue antioxidants.

Another important feature of the model is the prediction of the consequences of adding large amounts of antioxidants. In this case the effect would be a decrease in the cGMP/cAMP ratio, resulting in a decrease in endogenously synthesized antioxidants. Thus, the model accounts for the fact that feeding high levels of antioxidants in the diet would depress endogenous synthesis of antioxidants and therefore not result in a net gain of antioxidant protection. Some of the experimental data supporting the GAC model follow.

The activity of guanylate cyclase is increased by its interactions with free radicals and a number of mutagenic agents [117, 118]. Cigarette smoke activates guanylate cyclase and causes the level of guanosine 3′,5′-monophosphate in tissue to increase, the active component of the smoke being nitric oxide [125]. Estrogens and progesterone increase guanylate cyclase activity, and this may be related to their possible promoter effect in cancer [126]. In fact, there is a substantial amount of data linking an increase in cGMP to cell injury and to the increase in detoxification of chemical carcinogens as well as natural toxins [127].

Adenylate cyclase is activated in rat brain tissue by lipid peroxidation reactions [128], so perhaps both adenylate and guanylate play a role in detecting cell injury. Also of interest is the observation that plasma levels of insulin and glucagon are correlated with hepatocyte levels of guanylate cyclase, suggesting some type of regulatory linkage [*129*]. This may be important in view of the fact that the β-cells of the islets of Langerhans appear to be unusually sensitive to the toxic effects of free radicals [*130*].

The concept that changes in the oxidation–reduction state of a cell may be linked to guanylate cyclase activity in the cell, and thus cGMP levels, is supported by the reversible inactivation of guanylate cyclase by disulfides [*131*]. In addition, when the nematode *Caenorhabditis elegans* was exposed to cGMP in the growth medium, a significant increase in longevity was observed (33–42 days) [*132*]. Possibly related to the effects of altering the intercellular cGMP/cAMP ratio is the evidence that the addition of cAMP to cell cultures or the injection of cAMP into mice enhances their resistance to radiation by a yet undiscovered mechanism [*133–137*]. Guanylate cyclase activity has also been found to increase with age in the skeletal muscle of the rhesus monkey [*138*].

To determine whether guanylate cyclase plays an important role in the detection and genetic regulation of longevity determinants in the cell, the activity of soluble guanylate cyclase was measured in the brain tissue of primate species. The results showed an inverse correlation of guanylate cyclase activity with LSP [*7, 20*]. This would be the expected result if the GAC model were correct, and the cGMP/cAMP ratio would be proportional to the active oxygen set point.

Catecholamines have antioxidant properties and stimulate prostaglandin synthesis [*139, 140*]. This stimulation can be prevented by monamine oxidase inhibitors [*141*]. Catecholamines are degraded by monamine oxidation, producing H_2O_2, which in turn activates cyclooxygenase. These reactions are known to occur, for example, in brain tissue mitochondria. Some evidence suggests that a key factor in determining intercellular levels of antioxidants may be H_2O_2, and so the capacity of H_2O_2 to modulate the activity of cycloxygenase may be important in this respect [*142*].

Uric acid, as well as other antioxidants such as epinephrine, phenol, tryptophan, and hydroquinone stimulate prostaglandin synthesis [*143–145*]. Because uric acid is an antioxidant [*59*], this result is consistent with urate protecting cyclooxygenase from being inactivated. The result is that urate increases prostaglandin synthesis, resulting in a decreased cGMP/cAMP ratio and less antioxidant synthesis. Unusually high levels of uric acid, however, may lower the synthesis rate of the other antioxidants to the point that free-radical pathology appears. This would be consistent with the enhancement of arterial thrombo formation by uric acid [*145*] and some of the symptoms associated with gout.

Prostaglandin endoperoxides (PGG_2 and PGH_2) and hydroperoxides and fatty acid hydroperoxides have been found to enhance the activity of guanylate cyclase [*146*]. Thus, the active oxygen species can decrease or increase the activity of cyclooxygenase, depending on their concentration, and increase the activity of guanylate cyclase.

Levels of prostaglandins, particularly the E series as PGE_2, are elevated in cancerous cells [*120, 147*]. In addition, cancer cells have a high innate rate of dysdifferentiation (metastases), reflecting an inability to maintain a stable differentiated state. This characteristic is consistent with the fact that the levels of antioxidants such as catalase and SOD are normally low in cancer cells.

An unusually high level of prostaglandin synthesis would be characteristic of normal cells having higher than normal levels of antioxidants, but in fact cancerous cells have a deficiency in antioxidant protection. This implies that these cells are trying to overcome the antioxidant deficiency but that there is a defect in the feedback control system such that the high levels of prostaglandin synthesis fail to stimulate sufficient antioxidant synthesis. Thus, a common defect in cancer cells might involve the antioxidant regulatory system of the cell. Similar regulatory defects might also be involved in Alzheimer's, Parkinson's, and Hodgkin's diseases, which are also characterized by free-radical-associated pathology and high levels of prostaglandin synthesis [*7, 148–151*].

Tumor promoters such as 12-O-tetradecanoylphorbol 13-acetate (TPA) enhance prostaglandin synthesis through phospholipase A_2 activation. Consequently, inhibition of phospholipase A_2 protects cells from TPA promotion. One way in which tumor promoters function is by increasing intercellular levels of active oxygen species. Thus, scavengers against active oxygen species such as HO· and $O_2^{\overline{\cdot}}$ inhibit the effects of tumor promoters [*115, 152, 153*].

One of the interesting effects of the tumor promoter TPA is that it enhances a clastogenic factor similar to that found in the chromosomal breakage disorder of ataxia telangiectasia and Bloom's syndrome. In these disorders, SOD can decrease the levels of this clastogenic factor and the severity of the disease [*114, 115*].

These results suggest that tumor promoters interfere with the normal operation of the antioxidant regulatory system of a cell. The promoter TPA also appears to interfere with the antioxidant regulatory system such that the cell reacts as if there were excess antioxidant protection. This is achieved by activating the phospholipase A_2 enzyme, as already described, which is one effect of TPA, resulting in an increase in prostaglandin synthesis and thus a decrease in antioxidant synthesis. The end result is that the cell responds by lowering its normal level of antioxidant protection, thus increasing the lifetime of the toxic clastogenic factor. This clastogenic factor would be normally produced in cells but rendered inactive by SOD and other antioxidants before it had a chance to cause cell damage.

Tumor promoters also have a wide spectrum of other effects on cells, such as changes in membrane and gene expression. These results suggest that tumor promoters enhance the alteration of the proper differentiated state of cells. This property would also be consistent with the fact that tumor promoters lower normal antioxidant defense levels within a cell and would support the general hypothesis that the stability of the differentiated state is in part determined by the antioxidant levels. Consistent with this view is the finding that α-tocopherol also reduces the effects of the tumor promoter TPA on dysdifferentiation of cells [153, 154].

These effects of tumor promoters are consistent and support the hypothesis that aging is a result of dysdifferentiation, that cancer is a special case of dysdifferentiation, and that aging rate is governed in part by antioxidant levels in a cell.

Finally, these data support the concept that ataxia telangiectasia and Bloom's syndrome represent human inherited diseases involving improper regulation of antioxidant levels in cells. From this, it would be predicted that deficiency in antioxidant levels and/or the capacity of cells to respond in a compensatory manner to oxidative stress is likely to be important in a number of other human inherited diseases.

Feeding vitamin E to animals affects prostaglandin levels (PGE_2 and PGF_2) in an inverse manner [71, 155, 156]; that is, α-tocopherol inhibits prostaglandin synthesis and thus lowers platelet aggregation. However, prostaglandin inhibitors such as indomethacin and aspirin also reduce platelet aggregation by inhibiting cyclooxygenase activity and thus prostaglandin synthesis. These results suggest that such inhibition of prostaglandin synthesis has properties similar to those of α-tocopherol. This idea was supported by the observation that aspirin prevented anemia and thrombocythemia in vitamin E-deficient animals [71]. However, other physiological effects of vitamin E deficiency were not reduced by aspirin. These data support the concept of partial compensation among antioxidants and that inhibition of prostaglandin synthesis by drugs results in a further enhancement of other antioxidants in vitamin E-deficient animals. It is important to note that higher than normal levels of vitamin E exert a suppressive effect on prostaglandin synthesis, and lower than normal levels enhance protaglandin synthesis. The first effect is likely to occur by way of protection of the cyclooxygenase from inactivation, and the second effect by enhancement of the lipoxygenation of arachidonic acid [157, 158].

A key prediction of the GAC model is that inhibition of prostaglandin synthesis should increase tissue levels of antioxidants. Antiinflammatory agents such as indomethacin or aspirin would be predicted to decrease the susceptibility of a tissue to lipid peroxidation. Evidence that this is indeed the case was found when lipid peroxidation levels were measured in the serum of rabbits 6 hr after the intravenous injection of aspirin (50 mg/kg) [107]. The result was a twofold decrease in serum lipid peroxidation level, as measured by the TBA reaction. However, this result might be related more to a decrease in prostaglandin endoperoxide synthesis than to an increase in antioxidant levels.

The effects of aspirin on the life span of *Drosophila melanogaster* have been studied, and an increase in life span (both mean and maximum life span) of over 20% was found [159]. It is interesting that a similar extension of life span was observed upon the addition of corticosterol to the diet of flies [*159*]. Aspirin given to mice, however, did not significantly increase life span and at one concentration significantly shortened life span [*160*]. This result may indicate the critical dosage effect of aspirin, a phenomenon similar to its anticoagulation effects. Both of these agents, aspirin and corticosteroids, are known to be both membrane stabilizers and antiinflammatory agents. Thus, the life-span-increasing effects could have been related to an increased level of tissue antioxidants.

In conclusion, I have presented evidence that, antioxidants have sufficient overlap in protective properties to provide effective compensation if a deficiency of a specific antioxidant exists. Such a compensatory action of antioxidants could account for the ability of an organism to maintain its characteristic LSP over a wide range of levels of antioxidants in the diet, endogenous synthesis levels of antioxidants, and exposure to oxidative stress. A model to account for the maintenance of the general level of antioxidant protection based on the sensitivity of arachidonic acid to lipid peroxidation, cyclooxygenase to inactivation, and guanylate cyclase to activation by active oxygen species, has been presented. This model may prove useful in evaluating a possible new class of genetic diseases, of which Alzheimer's disease, Parkinson's disease, Hodgkin's disease, Bloom's syndrome, and ataxia telangiectasia may be examples, as well as some of the shorter lived mouse strains. In addition, certain types of cancer may be associated with a defect in the postulated antioxidant regulatory system. In all these dysfunctions it is suggested that an impairment of the proper regulation of a specific class of longevity determinants, the antioxidants, is involved. This model may be helpful in the achievement of a rational basis for treating these dysfunctions and points to some of the genetic and biochemical processes that could have been affected during the evolution of human longevity.

ACKNOWLEDGMENT

Thanks are expressed to Edith Cutler for technical and clerical assistance. Her help was made possible by the Glenn Foundation for Medical Research.

REFERENCES

1. R. G. Cutler, *J. Hum. Evol.* **5**, 169 (1976).
2. R. G. Cutler, *Interdiscip. Top. Gerontol.* **9**, 83 (1976).
3. R. G. Cutler, *in* "The Biology of Aging" (J. A. Behnke, C. E. Finch, and G. B. Moment, eds.), p. 311. Plenum, New York, 1978.
4. R. G. Cutler, *Gerontology* **25**, 69 (1979).
5. R. G. Cutler *in* "The Aging Brain" (E. Giacobini, G. Giacobini, G. Filogamo, and A. Vernadakis, eds.), Vol. 20, p. 1. Raven Press, New York (1982).
6. R. G. Cutler, *in* "Testing the Theories of Aging" (R. Adelman and G. Roth, eds.), p. 25. C. R. C. Press, Boca Raton, Florida 1982.
7. R. G. Cutler, *in* "Aging and Cell Structure, Vol. 2" (J. E. Johnson, ed.). Plenum, New York (in press).

8. G. Acsàdi and J. Nemeskéri, "History of Human Lifespan and Mortality." Akadémiai Kaidó, Budapest, 1970.
9. A. Comfort, "The Biology of Senescence." Am. Elsevier, New York, 1978.
10. B. L. Strehler, "Time, Cells and Aging." Academic Press, New York, 1978.
11. E. S. Deevey, *Q. Rev. Biol.* **22**, 283 (1947).
12. R. R. Kohn, "Principles of Mammalian Aging." Prentice-Hall, Englewood Cliffs, New Jersey, 1978.
13. National Center of Health Statistics, "Some Demographic Aspects of Aging in the United States." USPHS and U.S. Bureau of the Census, Washington, D.C., 1973.
14. A. H. Schultz, *in* "Perspectives on Human Evolution" (S. L. Washburn and P. C. Jay, eds.), p. 122. Holt, Rinehart & Winston, New York, 1968.
15. M. C. King and A. C. Wilson, *Science* **188**, 107 (1976).
16. A. C. Wilson, S. S. Carlson, and T. J. White, *Annu. Rev. Biochem.* **46**, 573 (1977).
17. R. G. Cutler, *Adv. Gerontol. Res.* **4**, 219 (1972).
18. R. G. Cutler, *Proc. Natl. Acad. Sci. U.S.A.* **72**, 4664 (1975).
19. G. A. Sacher, *Ciba Found. Colloq. Ageing* **5**, 115 (1959).
20. J. M. Tolmasoff, T. Ono, and R. G. Cutler, *Proc. Natl. Acad. Sci. U.S.A.* **77**, 2777 (1980).
21. W. S. Spector, ed., "Handbook of Biological Data," Div. Biol. Agric., Natl. Acad. Sci.–Natl. Res. Counc., Washington, D.C., 1956.
22. M. L. Jones, *Lab. Primate Newsl.* **1**, 3 (1962).
23. M. L. Jones, *Int. Zoo Yearb.* **8**, 183 (1968).
24. M. A. Holliday, D. Potter, A. Jarrah, and S. Bearg, *Pediatr. Res.* **1**, 185 (1967).
25. J. R. Napier and P. H. Napier, "A Handbook of Living Primates," Academic Press, New York, 1967.
26. P. Altman and D. Dittmer, eds., "Biological Handbook." Fed. Am. Soc. Exp. Biol., Bethesda, Maryland, 1968.
27. P. Altman and D. Dittmer, eds., "Biology Data Book." Fed. Am. Soc. Exp. Biol., Bethesda, Maryland, 1972.
28. R. Bauchot and H. Stephan, *Mammalia* **33**, 225 (1969).
29. H. J. Jerison, "Evolution of the Brain and Intelligence." Academic Press, New York, 1973.
30. E. P. Walker, "Mammals of the World." Johns Hopkins Univ. Press, Baltimore, Maryland, 1975.
31. D. Bowden and M. L. Jones, *in* "Aging in Nonhuman Primates" (D. M. Bowden, ed.), p. 1. Van Nostrand-Reinhold, Princeton, New Jersey, 1979.
32. S. S. Flower, *Proc. Zool. Soc. London* **10**, 145 (1931).
33. W. A. Pryor, *in* "Free Radicals in Biology" (W. A. Pryor, ed.), Vol. 1, p. 1. Academic Press, New York, 1976.
34. J. F. Mead, *in* "Free Radicals in Biology" (W. A. Pryor, ed.), Vol. 1, p. 51. Academic Press, New York, 1976.
35. I. Fridovich, *in* "Free Radicals in Biology" (W. A. Pryor, ed.), Vol. 1, p. 239. Academic Press, New York, 1976.
36. S. D. Aust and B. A. Svingen, *in* "Free Radicals in Biology" (W. A. Pryor, ed.), Vol. 5, p. 1. Academic Press, New York, 1982.
37. D. Harman, *Proc. Natl. Acad. Sci. U.S.A.* **78**, 7128 (1981).
38. D. Harman, *in* "Free Radicals in Biology" (W. A. Pryor, ed.), Vol. 5, p. 255. Academic Press, New York, 1982.
39. T. Ono and R. G. Cutler, *Proc. Natl. Acad. Sci. U.S.A.* **75**, 4431 (1978).
40. D. L. Florine, T. Ono, R. G. Cutler, and M. J. Getz, *Cancer Res.* **40**, 519 (1980).
41. M. J. Fahmy and O. G. Fahmy, *Cancer Res.* **40**, 3374 (1980).
42. M. J. Fahmy and O. G. Fahmy, *Cancer Res.* **43**, 801 (1983).

43. M. J. Fahmy and O. G. Fahmy, *Teterogen., Carcinogen. Mutagen.* **3,** 27 (1983).
44. H. O. Kunkel, J. F. Spalding, G. de Francis, and M. F. Futrell, *Am. J. Physiol.* **86,** 203 (1956).
45. W. R. Stahl, *Science* **137,** 205 (1962).
46. B. Emmitt and P. W. Hochachka, *Respir. Physiol.* **45,** 273 (1981).
47. S. L. Lindstedt and W. A. Calder, *Q. Rev. Biol.* **56,** 1 (1981).
48. E. C. Albritten, ed., "Standard Values in Blood." Fed. Am. Soc. Exp. Biol., Bethesda, Maryland, 1952.
49. P. Altman and D. Dittmer, eds., "Blood and Other Body Fluids." Fed. Am. Soc. Exp. Biol., Bethesda, Maryland, 1961.
50. P. Altman and D. Dittmer, eds., "Biological Handbook. Growth." Fed. Am. Soc. Exp. Biol., Bethesda, Maryland, 1962.
51. P. Altman and D. Dittmer, eds., "Biology Data Book," 2nd ed. Fed. Am. Soc. Exp. Biol., Bethesda, Maryland, 1974.
52. M. Dixon and E. C. Webb, eds., "Enzymes." Academic Press, New York, 1964.
53. H. Mattenheimer, "Mattenheimer's Clinical Enzymology: Principles and Applications." Ann Arbor Sci. Publ., Ann Arbor, Michigan, 1971.
54. K. Bernirschke, F. M. Garner, and T. C. Jones, eds., "Pathology of Laboratory Animals," Springer-Verlag, Berlin and New York, 1978.
55. B. M. Mitruka and H. M. Rawnsley, "Clinical, Biochemical and Hematological Reference Values in Normal Experimental Animals and Normal Humans," Masson, New York, 1981.
56. R. G. Cutler, *Gerontology* **29,** 113 (1983).
57. J. B. Wyngaarden and W. N. Kelley, *Contemp. Metab.* **1,** 86 (1979).
58. J. E. Seegmiller, *Contemp. Metab.* **1,** 1 (1979).
59. B. N. Ames, R. Cathcart, E. Schwiers, and P. Hochstein, *Proc. Natl. Acad. Sci. U.S.A.* **78,** 6858 (1981).
60. R. G. Cutler, submitted for publication.
61. J. M. Keehner, *Diss. Abstr.* 1283–A (1979).
62. C. S. Foote, *in* "Free Radicals in Biology" (W. A. Pryor, ed.), Vol. 2, p. 85. Academic Press, New York, 1976.
63. N. I. Krinsky, *in* "The Science of Photomedicine" (J. D. Regan and J. A. Parrish, eds.), p. 397. Plenum, New York, 1982.
64. N. I. Krinsky and S. M. Deneke, *JNCI, J. Natl. Cancer Inst.* **69,** 205 (1982).
65. R. Peto, R. Doll, J. D. Buckley, and M. B. Sporn, *Nature (London)* **290,** 201 (1981).
66. M. B. Sporn, *in* "Carcinogenesis: Mechanics of Tumor Promotion and Cocarcinogenesis" (T. J. Slaga, A. Sivak, and R. K. Boutwell, eds.), Vol. 2, p. 545. Raven Press, New York, 1978.
67. M. B. Sporn and D. L. Newton, *Fed. Proc., Fed. Am. Soc. Exp. Biol.* **38,** 2528 (1979).
68. R. Litan, *Biochim. Biophys. Acta* **605,** 33 (1980).
69. T. W. Goodwin, "Carotenoids: Their Comparative Biochemistry." Chem. Publ. Co., New York, 1954.
70. O. Isler, "Carotenoids." Birkhaueser, Basel, 1971.
71. L. J. Machlin, *in* "Tocopherol, Oxygen and Biomembranes" (C. deDuve and O. Hayaishi, eds.), p. 179. Academic Press, New York, 1978.
72. L. J. Machlin, *in* "Vitamin E: A Comprehensive Treatise" (L. J. Machlin, ed.), p. 637. Dekker, New York, 1980.
73. S. Lewin, "Vitamin C: Its Molecular Biology and Medical Potential." Academic Press, New York, 1976.
74. I. B. Chatterjee, *World Rev. Nutr. Diet.* **30,** 69 (1978).
75. H. K. Naito, *in* "Nutritional Elements and Clinical Biochemistry" (M. A. Brewster and H. K. Naito, eds.), p. 69. Plenum, New York, 1979.

76. J. E. Kirk, *Vitam. Horm. (N.Y.)* **20,** 82 (1962).
77. J. M. C. Gutteridge, *Ann. Clin. Biochem.* **15,** 293 (1978).
78. J. M. C. Gutteridge, R. Richmond, and B. Halliwell, *FEBS Lett.* **112,** 269 (1980).
79. A. Meister, *Biochem. Soc. Trans.* **10,** 78 (1982).
80. A. Meister, *Science* **220,** 471 (1983).
81. V. N. Reddy, F. J. Giblin, and H. Matsuda, *in* "Red Blood Cell and Lens Metabolism" (S. K. Srivastava, ed.), p. 139. Elsevier/North-Holland, Amsterdam, 1980.
82. H. Glatt, M. Protic-Sabjic, and F. Oesch, *Science* **220,** 961 (1983).
83. L. Flohe, *in* "Free Radicals in Biology" (W. A. Pryor, ed.), Vol. 5, p. 223. Academic Press, New York, 1982.
84. O. De Marchena, M. Guarnieri, and G. McKhann, *J. Neurochem.* **22,** 773 (1974).
85. J. A. Butler, P. D. Whanger, and M. J. Tripp, *Am. J. Clin. Nutr.* **36,** 15 (1982).
86. R. A. Lawrence and R. F. Burk, *J. Nutr.* **108,** 211 (1978).
87. W. B. Jakoby, *Adv. Enzymol. Relat. Areas Mol. Biol.* **46,** 383 (1978).
88. K.-H. Summer and H. Greim, *Biochem. Pharmacol.* **30,** 1719 (1981).
89. A. G. Schwartz and A. Perantoni, *Cancer Res.* **35,** 2482 (1975).
90. L. L. Pashko and A. G. Schwartz, *J. Gerontol.* **37,** 38 (1982).
91. R. P. Mason, *in* "Free Radicals in Biology" (W. A. Pryor, ed.), Vol. 5, p. 161 Academic Press, New York, 1982.
92. L. Fucci, C. N. Oliver, M. J. Coon, and E. R. Stadtman, *Proc. Natl. Acad. Sci. U.S.A.* **80,** 1521 (1983).
93. C. W. Denko, *Agents Actions* **9,** 333 (1979).
94. I. M. Goldstein, H. B. Kaplan, H. S. Edelson, and G. Weissmann, *J. Biol. Chem.* **254,** 4040 (1979).
95. A. Plonka and D. Metodiewa, *Biochem. Biophys. Res. Commun.* **95,** 978 (1980).
96. E. Frieden, *Ciba Found. Symp.* [N.S.] **79,** 93 (1980).
97. U.S. Seal, *Comp. Biochem. Physiol.* **13,** 143 (1964).
98. G. W. Evans and R. E. Wiederanders, *Am. J. Physiol.* **213,** 1183 (1967).
99. H. Ohkawa, N. Ohishi, and K. Yagi, *J. Lipid Res.* **19,** 1053 (1978).
100. J. G. Bieri and A. A. Anderson, *Arch. Biochem. Biophys.* **90,** 105 (1960).
101. A. A. Barber and F. Bernheim, *Adv. Gerontol. Res.* **2,** 355 1967.
102. T. Suematsu, T. Kamada, H. Abe, S. Kikuchi, and K. Yagi, *Clin. Chim. Acta* **79,** 267 (1977).
103. D. D. Myers, *in* "Birth Defects on Aging" (D. Bergsma, D. E. Harrison, and N. E. Paul, eds.), p. 41. Alan R. Liss, Inc., New York, 1978.
104. H. Nakakimura, M. Kakimoto, S. Wada, and K. Mizuno, *Chem. Pharm. Bull.* **28,** 2101 (1980).
105. W. A. Pryor and J. P. Stanley, *J. Org. Chem.* **40,** 3615 (1975).
106. W. A. Pryor, J. Stanley, and E. Blair, *Lipids* **11,** 370 (1976).
107. O. Hayaishi and T. Shimizu, *in* "Lipid Peroxides in Biology and Medicine" (K. Yagi, ed.), p. 41. Academic Press, New York, 1982.
108. Y. Sato, N. Hotta, N. Sakamoto, S. Matsuoka, N. Ohishi, and K. Yagi, *Biochem. Med.* **21,** 104 (1979).
109. I. Nishigaki, M. Hagihara, H. Tsunekawa, M. Maseki, and K. Yagi, *Biochem. Med.* **25,** 373, 1981.
110. K. Yagi, *Biochem. Med.* **15,** 212 (1976).
111. M. Uchiyama and M. Mihara, *Anal. Biochem.* **86,** 271 (1978).
112. A. D. Blackett and D. A. Hall, *Gerontology* **27,** 133 (1981).
113. L. H. Chen, R. R. Thacker, and C. K. Chow, *Nutr. Rep. Int.* **22,** 873 (1980).
114. I. Emerit and P. Cerutti, *Proc. Natl. Acad. Sci. U.S.A.* **78,** 1868 (1981).
115. I. Emerit and P. Cerutti, *Proc. Natl. Acad. Sci. U.S.A.* **79,** 7509 (1982).

116. C. K. Mittal and F. Murad, *J. Cyclic Nucleotide Res.* **3,** 381 (1977).
117. F. Murad, C. K, Mittal, W. P. Arnold, S. Katsuki, and H. Kimura, *Adv. Cyclic Nucleotide Res.* **9,** 145 (1978).
118. F. Murad, W. P. Arnold, C. K. Mittal, and J. M. Braughler, *Adv. Cyclic Nucleotide Res.* **11,** 175 (1979).
119. F. Kuehl and R. W. Egan, *Science* **210,** 978 (1980).
120. R. A. Karmali, *Prostaglandins Med.* **5,** 11 (1980).
121. D. Gemsa, H.-G. Leser, M. Seitz, W. Deimann, and E. Barlin, *Mol. Immunol.* **19,** 1287 (1982).
122. H. Iida, A. Imai, Y. Nozawa, and T. Kimura, *Biochem. Med.* **28,** 365 (1982).
123. G. Pekoe, K. Vandyke, D. Peden, H. Mengoli, and D. English, *Agents Actions* **12,** 371 (1982).
124. P. H. Gale and R. W. Egan *in* "Free Radicals in Biology" (W. A. Pryor, ed.), Vol. 6. Academic Press, New York pp. 1–38.
125. W. P. Arnold, R. Aldred, and F. Murad, *Science* **198,** 934 (1977).
126. D. L. Vesely and D. E. Hill, *Endocrinology* **107,** 2104 (1980).
127. F. R. DeRubertis and P. Craven, *Adv. Cyclic Nucleotide Res.* **12,** 97 (1980).
128. A. Baba, E. Lee, A. Ohta, T. Tatsuno, and H. Iwata, *J. Biol. Chem.* **256,** 3679 (1981).
129. H. S. Earp, *J. Biol. Chem.* **255,** 8979 (1980).
130. R. K. Crouch, S. E. Gandy, G. Kimsey, R. A. Galbraith, G. M. P. Galbraith, and M. G. Buse, *Diabetes* **30,** 235 (1981).
131. H. J. Brandwein, J. A. Lewicki, and F. Murad, *J. Biol. Chem.* **256,** 2958 (1981).
132. J. D. Willett, I. Rahim, M. Geist, and B. M. Zuckerman, *Age* **3,** 82 (1980).
133. S. Lehnert, *Radiat. Res.* **62,** 107 (1975).
134. S. Lehnert, *in* "Cell Survival After Low Doses of Radiation: Theoretical and Clinical Implications" (T. Alper, ed.), p. 226. Wiley, New York, 1975.
135. S. Lehnert, *Int. J. radiat. Oncol., Biol., Phys.* **5,** 825 (1979).
136. S. Lehnert, *Radiat. Res.* **78,** 1 (1979).
137. N. B. Dubravsky, N. Hunter, K. Mason, and H. R. Withers, *Radiology* **126,** 799 (1978).
138. C. H. Beatty, P. T. Herrington, M. K. Hoskins, and R. M. Bocek, *Age* **5,** 1 (1982).
139. G. Cohen and R. E. Heikkila, *in* "Superoxide and Superoxide Dismutases" (A. M. Michelson, J. M. McCord, and I. Fridovich, eds.), p. 351. Academic Press, New York.
140. G. Cohen, *in* "Pathology of Oxygen (A. P. Autor, ed.), p. 115. Academic Press, New York, 1982.
141. A. Seregi, P. Serfozo, and Z. Mergl, *J. Neurochem.* **40,** 407 (1983).
142. A. P. Autor, *Biochem. Soc. Trans.* **10,** 75 (1982).
143. N. Ogino, S. Yamamoto, O. Hayaishi, and T. Tokuyama, *Biochem. Biophys. Res. Commun.* **87,** 184 (1979).
144. C. Deby, G. Deby-Dupont, F.-X. Noel, and L. Lavergne, *Biochem. Pharmacol.* **30,** 2243 (1981).
145. R. H. Bourgain, C. Deby, G. Deby-Dupont, and R. Andries, *Biochem. Pharmacol.* **31,** 3011 (1982).
146. G. Graff, J. H. Stephenson, D. B. Glass, M. K. Haddox, and N. D. Goldberg, *J. Biol. Chem.* **253,** 7676 (1978).
147. J. S. Goodwin, G. Husby, G. and R. C. Williams, *Cancer Immunol. Immunother.* **8,** 3 (1980).
148. J. T. Coyle, D. L. Price, and M. R. DeLong, *Science* **219,** 1184 (1983).
149. F. M. Sinex and C. R. Merrill, eds., "Alzheimer's Disease, Down's Syndrome, and Aging," Ann. N.Y. Acad. Sci., Vol. 396. N.Y. Acad. Sci., New York, 1982.
150. L. M. Ambani, M. H. Van Woert, and S. Murphy, *Arch. Neurol. (Chicago)* **32,** 114 (1975).
151. Y. Noda, P. L. McGeer, and E. G. McGeer, *Neurobiol. Aging* **3,** 173 (1982).

152. A. Novogrodsy, A. Ravid, A. L. Rubin, and K. H. Stenzel, *Proc. Natl. Acad. Sci. U.S.A.* **79,** 1171 (1982).
153. C. Borek and W. Troll, *Proc. Natl. Acad. Sci. U.S.A.* **80,** 1304 (1983).
154. K. Ohuchi and L. Levine, *Biochim. Biophys. Acta* **619,** 11 (1980).
155. W. C. Hope, C. Dalton, L. F. Machlin, R. J. Filipski, and F. M. Vane, *Prostaglandins* **10,** 557 (1975).
156. M. K. Horwitt, *Am. J. Clin. Nutr.* **29,** 569 (1976).
157. E. J. Goetzl, *Nature (London)* **288,** 183 (1980).
158. M. A., Valentovic, C. Gairola, and W. C. Lubawy, *Prostaglandins* **24,** 215 (1982).
159. R. Hochschild, *Exp. Gerontol.* **6,** 133 (1971).
160. R. Hochschild, *Gerontologia* **19,** 271 (1973).

Index

)